Conference Proceedings of the Society for Experimental Mechanics Series

Series Editor

Kristin B. Zimmerman, Ph.D.
Society for Experimental Mechanics, Inc.,
Bethel, CT, USA

More information about this series at http://www.springer.com/series/8922

Shamim Pakzad
Editor

Dynamics of Civil Structures, Volume 2

Proceedings of the 36th IMAC, A Conference and Exposition on Structural Dynamics 2018

Springer

Editor
Shamim Pakzad
Department of Civil and Environmental Engineering
Lehigh University
Bethlehem, PA, USA

ISSN 2191-5644 ISSN 2191-5652 (electronic)
Conference Proceedings of the Society for Experimental Mechanics Series
ISBN 978-3-030-08988-7 ISBN 978-3-319-74421-6 (eBook)
https://doi.org/10.1007/978-3-319-74421-6

Printed on acid-free paper

This Springer imprint is published by the registered company Springer International Publishing AG part of Springer Nature.
The registered company address is: Gewerbestrasse 11, 6330 Cham, Switzerland

Preface

Dynamics of Civil Structures represents one of nine volumes of technical papers presented at the 36th IMAC, a Conference and Exposition on Structural Dynamics, organized by the Society for Experimental Mechanics, and held in Orlando, Florida, February 12–15, 2018. The full proceedings also include volumes on Nonlinear Dynamics; Model Validation and Uncertainty Quantification; Dynamics of Coupled Structures; Special Topics in Structural Dynamics; Structural Health Monitoring, Photogrammetry & DIC; Rotating Machinery, Vibro-Acoustics and Laser Vibrometry; Sensors and Instrumentation, Aircraft/Aerospace and Energy Harvesting; and Topics in Modal Analysis & Testing.

Each collection presents early findings from analytical, experimental, and computational investigations on an important area within Structural Dynamics. Dynamics of Civil Structures is one of these areas which cover topics of interest of several disciplines in engineering and science.

The Dynamics of Civil Structures Technical Division serves as a primary focal point within the SEM umbrella for technical activities devoted to civil structures analysis, testing, monitoring, and assessment. This volume covers a variety of topics including structural vibrations, damage identification, human-structure interaction, vibration control, model updating, modal analysis of in-service structures, innovative measurement techniques and mobile sensing, and bridge dynamics, among many others topics.

Papers cover testing and analysis of different kinds of civil engineering structures such as buildings, bridges, stadiums, dams, and others.

The organizers would like to thank the authors, presenters, session organizers, and session chairs for their participation in this track.

Bethlehem, PA, USA — Shamim Pakzad

Contents

Chapter 1
Experimental Modal Analysis Study of a Standing Soldier and Rifle System

Razvan Rusovici, Joshua Drew, Brian Fischer, George Kontis, Terrence F. Rice, Francis J. Battersby, and Michael Pavlisak

Abstract The response of the human body to shock and vibration has been a subject of interest to many researchers in the aerospace and automotive industry. In a new study, an experimental modal analysis of a rifle-armed, standing soldier in a standard firing position was performed. The purpose of the study was to determine the modes of vibration of the soldier-weapon system in order to gain an understanding of its response during the firing event. The weapon firing accuracy, especially during closely-repeated semi-automatic or fully-automatic fire, as well as the energy transmitted to the body of the soldier depend not only on the weapon itself but also on the soldier body's dynamic characteristics. Rapid weapon fire does not allow a soldier to consciously control muscles needed to bring back the weapon to its original position, so the weapon location after each fire is influenced more by the dynamic characteristics of the human-rifle system. The experimental modal analyses were performed using multi-averaged, impact-force and electrodynamic shaker force excitation (mainly sine sweeps) and roving triaxial acceleration response at various locations on the body. It was observed that testing of human subjects poses significant difficulties since an increasing number of measurement averages could lead to muscle fatigue and ensuing tremors that could negatively influence coherence. The soldier stance during the tests could also change due to the unconscious need to adjust to a more comfortable body position. The positioning of the accelerometers was difficult since attachment could be made to skin only. While there was in general large variability in soldier size, mass and body strength, the study allowed the identification of some lower modes of interest which appeared to have the same mode shape, albeit at different frequencies, for all the various individuals tested. The signals were acquired with a National Instruments hardware and processed using ModalView software. For an average size soldier the first mode occurred at approximately 2 Hz. The first mode shape exhibited combined bending (backward) and twisting characteristics which are generally seen in a "up and to the right" motion of a right-shoulder held weapon during firing. More modes in the 0–20 Hz range were identified.

Keywords Human vibration · Experimental modal analysis · LSCF · Weapon · Impact hammer

1.1 Introduction

Humans are exposed to large variety of shock and vibration sources. In many cases, these sources may be significant enough in magnitude and frequency to affect the health or the job performance of the individual exposed to them. The response of the human body to shock and vibration has been a subject of interest for several decades and includes several studies by starting with NASA [1] and ongoing with several others that found the natural frequencies and mode shapes of human in various positions: References [2–6] are examples of such work. While some research addressed part of the human body [2, 5, 6], others investigated the entire body in various postures [1, 3, 4, 7]. The experimental approach to defining the modal characteristics of the human body have been approached using both experimental modal analysis (EMA) and operational modal analysis (OMA).

R. Rusovici (✉)
Florida Institute of Technology, Melbourne, FL, USA
e-mail: rrusovic@fit.edu

J. Drew · B. Fischer · G. Kontis
Knights Armament Company, Titusville, FL, USA

T. F. Rice · F. J. Battersby · M. Pavlisak
US ARMY ARDEC, Picatinny Arsenal, NJ, USA

S. Pakzad (ed.), *Dynamics of Civil Structures, Volume 2*, Conference Proceedings of the Society for Experimental Mechanics Series, https://doi.org/10.1007/978-3-319-74421-6_1

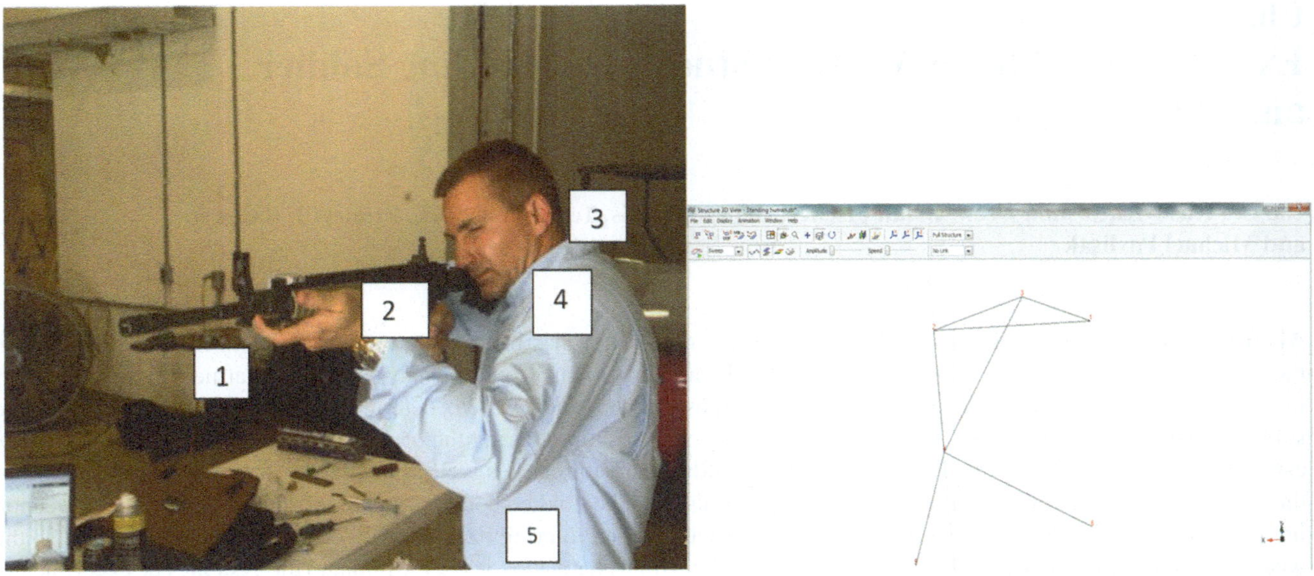

Fig. 1.1 (**a**, **b**) Standing subject and "Stick" model showing measurement points

The goal of the hereby presented work was to investigate the feasibility of finding the modal characteristic of a rifle-armed soldier in the upright firing position, in light of the many variables associated with individual human subjects (height, stance in firing position, grip force, body mass, muscle tone etc.). A thorough understanding of the weapon-soldier system's modal characteristic was necessary to determine, how much energy could the weapon fire transmit into the human body during the firing event, and how would the human-rifle system react during firing

1.2 Methods and Results

This section describes the methods and equipment used to perform the modal testing of the rifle-armed, human subject in standing firing position. The experimental modal analyses (EMA) were conducted using either impact-hammer excitation or electrodynamic shaker excitation (using swept sine signal input). The study used National Instruments ModalView™ software to gather (via a cDAQ 9174 National Instruments 8-channel card) and analyze data.

Four human subjects underwent the studies. Two subjects were 178-cm tall and had 80-kg masses ("average" subjects), one subject was 160-cm tall and had a 50 kg mass (small subject), while the last test subject was 190-cm tall and had 110-kg mass (large subject). The human subjects had different abdominal muscles tone. All human tests subjects were volunteers and were not harmed during the tests. All subjects were required to hold their initial position during the entire testing protocol and maintain the same grip on the weapon or mock weapon.

1.2.1 Impact-Hammer Excitation

A set of EMA's were performed for a standing human subject, armed with a rifle (Fig. 1.1a). A roving response, fixed-impact technique was applied. A corresponding model of the human used in both impact-hammer and shaker-excitation analyses is shown in Fig. 1.1b.

Accelerometer data was acquired at all measurement points, which included gun tip (point 1, where impact hammer excitation was applied in the lateral or negative Y direction), gun trigger (point 2), gun stock (point 3), opposing shoulder (point 4 placed on the left shoulder as the gun was held to the right shoulder), waist (point 5). Additional points (point 6 and 7, feet locations) were considered to be fixed. Once again, the frequency analysis range was in the low frequency range (1 to 20 Hz). All excitation was imparted with a small modally-tuned hammer PCB Piezotronics 086C01 (11.2 mV/N), and all data was collected using a small triaxial accelerometer PCB Piezotronics type 356A17 (500 mV/g sensitivity and 0.5 to 3000 Hz range), as shown in Fig. 1.2. The spatial orientation of the accelerometers was carefully recorded and input into ModalView Software.

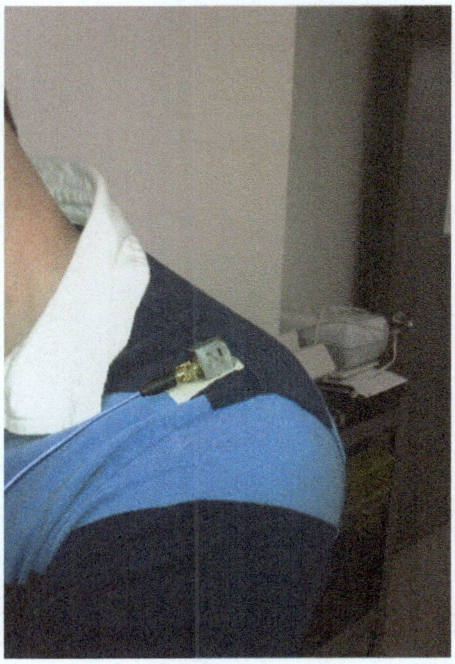

Fig. 1.2 Triaxial accelerometer placement on subject shoulder

The impact excitation was applied at the tip of the gun (Point 1) in the negative Y direction (laterally). Data was acquired at 16384 samples per second with a 0.2 Hz frequency resolution. Seven impacts and the resulting responses were root-mean square (rms) averaged in order to obtain frequency response functions (FRF). Longer testing could not have been conducted due to muscle fatigue considerations that could introduce muscle tremors and worsen the coherence.

The modes were found initially using visual identification on FRF plots followed by a "Quick-Fit" polynomial curve fit. Then, the modes were confirmed using a Least Square Complex Function (LSCF) stabilization chart with model order of 40 and a Complex Mode Indicator Function (CMIF). The LSCF estimator was chosen since it has the advantage of producing clear stabilization diagrams from which possible close, heavily damped modes could be found.

The natural frequencies, the modal assurance criterion 3-D, and data charts in the low frequency range are shown in Fig. 1.3. The coherence was acceptable in the chosen 0.5–20 Hz frequency range.

The first mode shape shown in Figs. 1.4 and 1.5 were found to be two different important mode shapes that could affect shooting accuracy when excited by the weapon fire: one pitching (up and down) mode at 2.2 Hz and one yawing (side-to-side) mode at 7.1 Hz.

1.2.2 EMA-Shaker Excitation

In order to verify the impact hammer-generated modal results, another series of EMA's were performed on standing human shooters. It was thought that the superior displacement provided by a shaker system could be significantly greater than the motion generated by any potential tremors of the subject elicited by fatigue. The swept-sine, force excitation was provided by a TA-206, Unholtz-Dickie 1200-lbf shaker over a low frequency range (0 to 20 Hz). Furthermore, several human subjects were tested (small, medium and large individual).

The force was applied to the gun tip via a stinger connected to the shaker. The stinger formed a 45° angle with the gun's X, Y and Z directions (X was oriented along the axis of the gun, Y was oriented vertically), as shown in Fig. 1.6. The stinger was positioned as such in order to excite all modes in the target frequency range. The accelerometers were positioned at locations depicted in Fig. 1.1b. The force input was acquired using a 208C02 PCB Piezotronics force gauge.

Data was again acquired at 16384 samples per second with a 0.033 Hz frequency resolution. The results of seven sine-sweeps and the resulting accelerometer responses were rms averaged in order to obtain FRF's.

The modes were found using a Least Square Complex Function (LSCF) stabilization chart with model order of 40 and a Complex Mode Indicator Function (CMIF) [8]. The LSCF estimator was chosen since it has the advantage of producing clear stabilization diagrams from which possible close, heavily damped modes could be found.

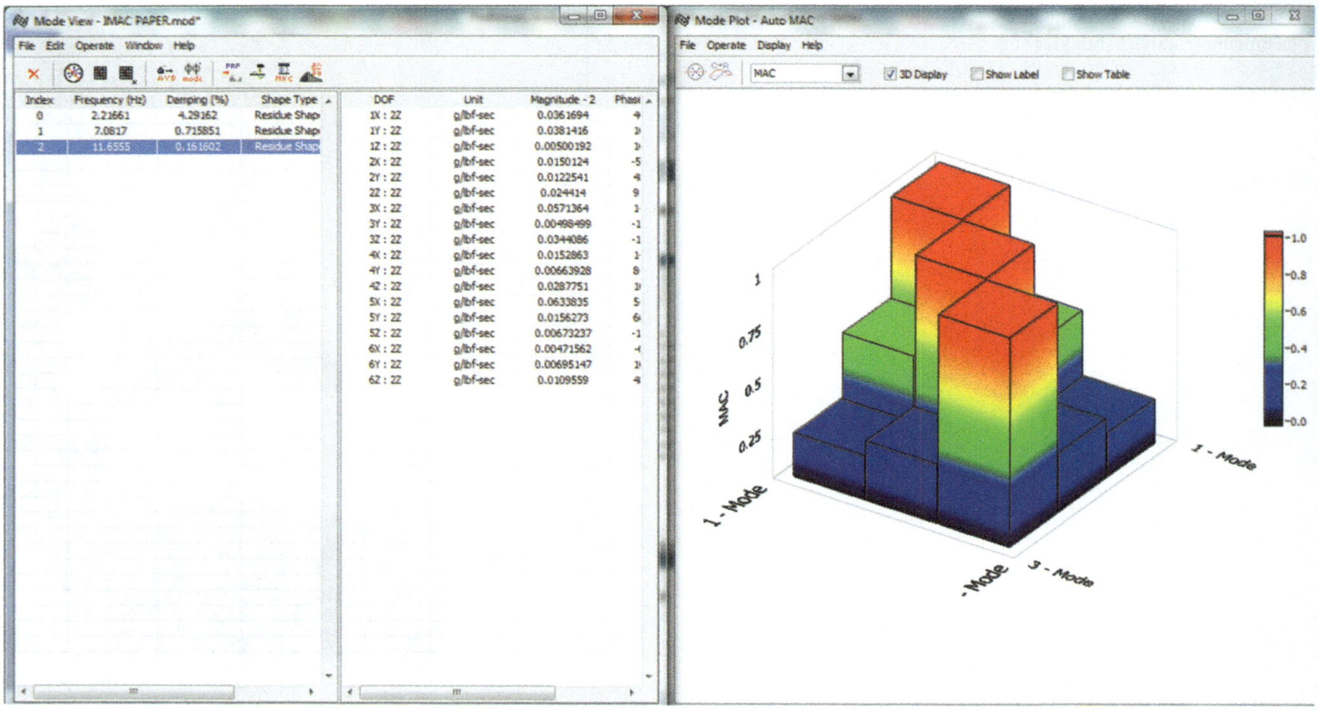

Fig. 1.3 Modes and modal assurance criterion

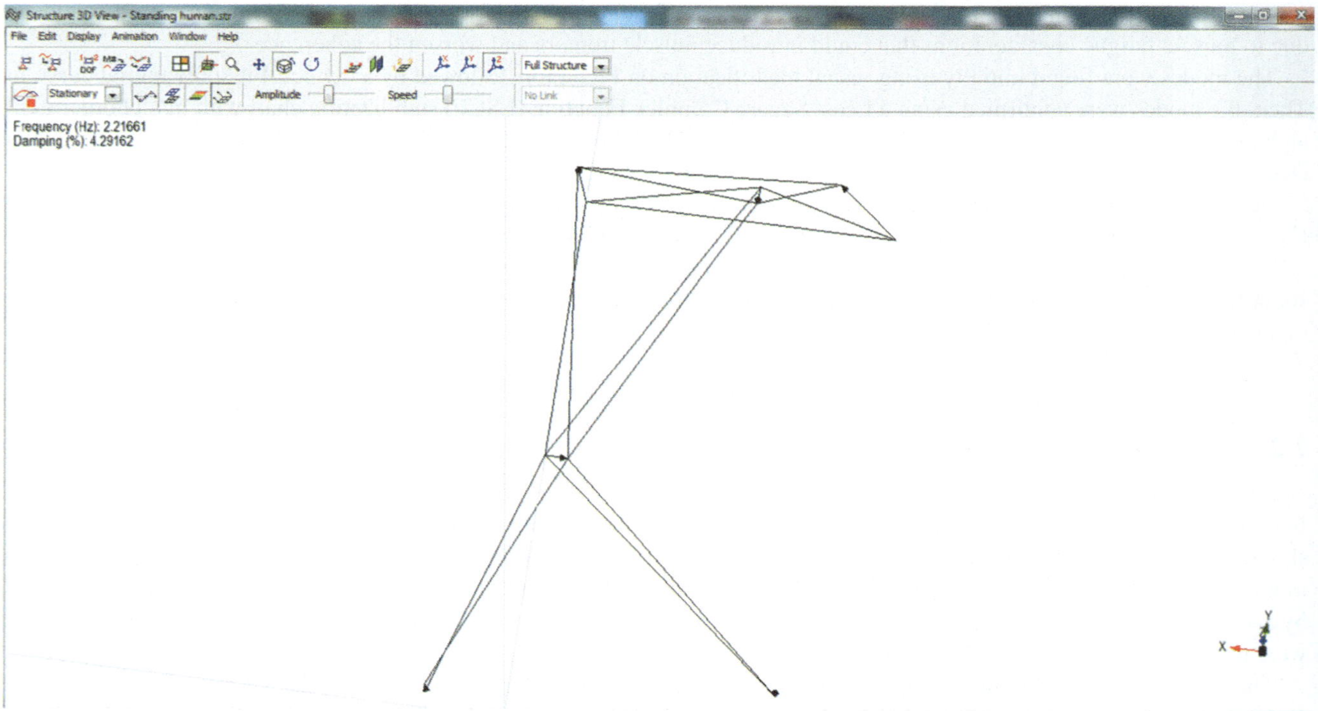

Fig. 1.4 Pitching mode at 2.2 Hz

In a further set of measurements, the variation of natural frequencies and mode shapes with human subject feature was investigated. Three types of subjects were tested: small, average and large. Table 1.1 shows the first natural frequencies of the subjects.

The first mode shape (pitching) appears around 2 Hz for all subjects. However, the next two modes (yawing and shoulder movement) show that the body size and muscle tone (the large subject had the least amount of muscle tone) influence greatly

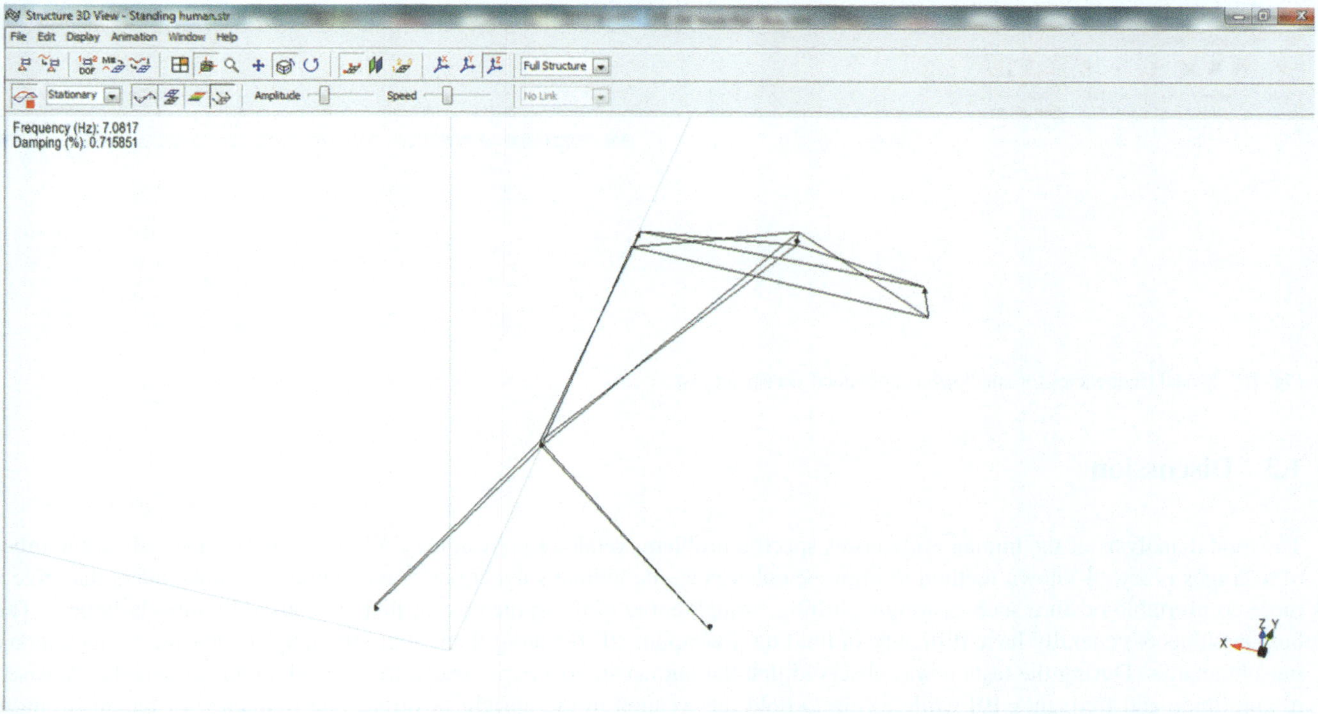

Fig. 1.5 Yawing mode at 7.1 Hz

Fig. 1.6 Mock weapon with stinger and shaker setup

Table 1.1 Variation of modal frequencies with subject size

	Small	Large	Medium	Mode description
1	2.8 Hz	1.87 Hz	2.3 Hz	Pitching (up and down motion
2	6.6 Hz	4.1 Hz	7.27 Hz	Yawing (sideways rocking motion)

the natural frequencies. The large individual has a second mode (yawing) at a frequency significantly lower than the other subjects, who were in much better physical shape.

Modal frequencies are shown in Fig. 1.7 for the small subject.

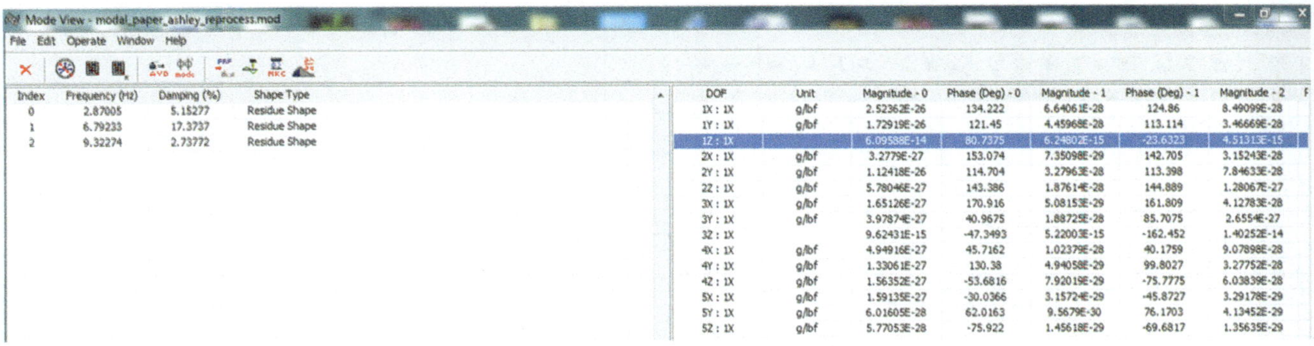

Fig. 1.7 Modal frequencies for small subject obtained via shaker test

1.3 Discussion

The modal analysis of the human body poses specific problems related to physiology. While gathering a significant number of averages is a well-known method to improve coherence, the human subject performs involuntary movements that give a more comfortable posture such as weight shifting, straightening of the spine etc., or that are caused by muscle fatigue. The human subjects generally have difficulty in holding a weapon still for more than, approximately, 1 min due to involuntary muscle tetanus. During the tests it was observed that the human subject experienced tremors which progressively increased in amplitude and frequency [9] while trying to hold the weapon in the upright position. The frequency of tetanus tremors movements begins around 2 Hz and increases to 5 Hz and beyond (complete tetanus) in matter of seconds [9]. The tetanus-induced movements, if occurring during testing, provide a source of significant noise in the input data. All subjects were instructed to hold their positions during testing, but that was not always possible, especially during the shaker excitation tests. The larger shaker displacements could displace the subjects very slightly from their initial stance. There was variability in frequency and damping for the same mode according to body mass, muscle strength, height and posture. Not all subjects could attain the very same posture. The yawing mode's frequency, for example, was intuitively correlated with abdominal muscle strength; the abdominal muscles formed a torsional spring.

The two modes of most interest were the first two, whose mode shapes resembled the motion of a standing, rifle-armed shooter when firing: a combined upward vertical motion with a rotation towards the exterior. It was observed during the tests that various shooters can adjust their stance to minimize this combined motion in order to increase fire accuracy. The conscientious adjustment of stance was another source of error in repeated tests, as natural frequencies vary with a leaning-into the weapon, for example.

The study provided some interesting data regarding the human-weapon system. The EMA's .gave a good insight into what frequencies could be expected and what the mode shapes look like. Future human-weapon research will look at OMA's as another solution method to perform the modal analysis; initial tests have started. These tests, as was noted, suffer from the same issues with repeatability: shooter's position after weapon fire and recoil.

References

1. Piersol, A.G., Paez, T.L. (eds.): Harris' Shock and Vibration Handbook, 6th edn. McGraw Hill Publications (2010)
2. Adewusi, S., Thomas, M., Vu, V., Li, W.: Modal parameters of the human hand-arm using finite element and operational modal analysis. Mech. Ind. **15**, 541–549 (2014)
3. Hostens, I., Ramon, H.: Descriptive analysis of combine cabin vibrations and their effect on the human body. J. Sound Vib. **266**, 453–464 (2003)
4. Kitazaki, S., Griffin, M.J.: Resonance behaviour of the seated human body and effects of posture. J. Biomech. **31**, 143–149 (1998)
5. Hobatho, M.C., Darmana, R., Pastor, P., Barrau, J.J., Laroze, S., Morucci, J.P.: Development of a three-dimensional finite element model of a human tibia using experimental modal analysis. J. Biomech. **24**, 371–383 (1991)
6. Munera, M., Chiementin, X., Murer, S., Bertucci, W.: Model of the risk assessment of hand-arm system vibrations in cycling: case of cobblestone road. Proc. Inst. Mech. Eng. Part P. **229**, 231–238 (2015)
7. Thuong, O., Griffin, M.J.: The vibration discomfort of standing persons: 0.5–16 Hz fore-and-aft, lateral, and vertical vibration. J. Sound Vib. **330**, 816–826 (2011)
8. Phillips, A.W., Allemang, R.J.: The Complex Mode Indicator Function (CMIF) as a parameter estimation method. In: Proceedings, International Modal Analysis Conference, pp. 705–710 (1998)
9. Barrett, K.E., Barman, S.M., Boitano, S., Brooks, H.L.: Ganong's Review of Medical Physiology, 23rd edn. McGraw-Hill Medical (2010)

Chapter 2
Performance Characterization of Modal Identification Algorithms, the Case of Automated Modal Analysis of Palazzo Lombardia

Marcello Vanali, Marta Berardengo, Stefano Manzoni, Giorgio Busca, and Elena Mola

Abstract In this paper the study of the analysis of the data coming from the permanent monitoring system of Palazzo Regione Lombardia, one of the tallest buildings in Milano is presented. The system works acquiring both the vibrations of the building and its static behaviour. The vibration data are used to carry out continuous modal analysis to the aim of detecting possible changes of the structure. Two algorithms for modal parameter analysis are presented and compared through Monte Carlo simulations. The results of this comparison allowed to choose the best data analysis procedure to be used on the real data measured under operating conditions. The first results coming from the modal parameter extraction are presented.

Keywords OMA · Health Monitoring · Vibration · Uncertainty estimation · Continuous monitoring

2.1 Introduction

In recent times the monitoring of civil structures has become more and more interesting and affordable thanks to the continuous technological development. Measurement devices and transducers have become cheaper and cheaper, together with the improvement of their metrological characteristics, while data acquisition and storage system are available as off the shelf solutions. This allows to instrument huge structures with many sensors with different aims: (i) the prompt detection of changes of the structure behaviour, (ii) the monitoring of comfort of people living/working in the monitored building, (iii) the check of the structure behaviour after exceptional events such as earthquakes or severe thuderstorms.

Continuous monitoring of civil structures still requires to face a number of challenges: e.g. efficient management and visualization of the data requires implementing a tailored data pre-processing, the check of sensor stability over long periods of time. Nowadays the redundancy of transducers prevents loss of data on one side, but also requires to find synthetic indexes and features able to provide clear information to the building facility management. In this scenario, this paper presents the analysis of the data coming from the monitoring system of Palazzo Regione Lombardia [1], which is the high-rise building (height of 161 m) seat of the regional government. The monitoring system includes accelerometers, clinometers and transducers for environmental variables (e.g. wind strength and direction, temperature).

According to literature, structural monitoring and consequently damage identification could be performed at five different levels of detail: simply detecting the presence of damage and its localization, evaluating the type of damage, quantifying its entity and finally estimating the residual life [2, 3].

One of the approaches used for monitoring the present building is based on a continuous modal parameter extraction using the accelerometer signals, coupled to the analysis of the data reckoned by means of clinometers. The modal data are able to describe the dynamics of the structure, while the clinometers are employed to describe its static behaviour. In this paper the possibility of using modal parameters as a damage warning feature is investigated [4].

The modal parameter extraction is carried out by using Operational Modal Analysis (OMA), because the environmental excitation (e.g. wind and traffic), which cannot be measured, is used as source to excite the system [5–7].

M. Vanali (✉) · M. Berardengo
Dipartimento di Ingegneria e Architettura, Università degli studi di Parma, Parma, Italy
e-mail: marcello.vanali@unipr.it

S. Manzoni · G. Busca
Dipartimento di Meccanica, Politecnico di Milano, Milan, Italy

E. Mola
ECSD Srl, Milan, Italy

S. Pakzad (ed.), *Dynamics of Civil Structures, Volume 2*, Conference Proceedings of the Society for Experimental Mechanics Series, https://doi.org/10.1007/978-3-319-74421-6_2

At first, the monitoring system installed is briefly described, then, the modal extraction algorithms and their performances are investigated. Two different approaches were tested: the polyreference-least squares complex-frequency domain (pLSCF) method and a method based on the complex mode indicator function (CMIF), which in turn is based on the singular value decomposition (SVD). The two methods are compared in terms of accuracy and bias effects on the estimated modal parameters using the Monte Carlo method. The analysis allowed to choose the best algorithm, as well as optimizing the parameters used in the analysis of the real data. Finally, the first data coming from the modal parameter extraction are presented.

2.2 Palazzo Lombardia and the Monitoring System

The "Palazzo Lombardia" building is the first in a series of high rise buildings which have been built in Milano in the last years. It is the current seat for the Regional Government and offices and therefore considered of strategic relevance. The complex is made up of five lower buildings (about 40 m high, called Cores 2, 3, 4, 5 and 6), surrounding the high-rise Tower (Core 1), which scored, at the time of construction, the new height record in Italy. The monitoring system is targeted to control core 1 tower and is capable of handling both dynamic vibration signals and static variables, as well as of the wind conditions. A full description of the system installation and performances can be found in [1]. A schematic representation of the system layout is given in Fig. 2.1.

Five floors are instrumented with clinometers and accelerometers according to the general layout given in Fig. 2.1. The setup was designed in accordance with the dynamic testing results [6, 8] in order to be able to identify at least the first three vibration modes and to asses wind comfort issues.

Data acquisition is performed 24 h per day with a final sampling frequency of 250 Hz on all channels enough to asses wind comfort, vibration disturbances and provide data to modal analysis purposes [9, 10]. A total of 24 high sensitivity piezo accelerometers and 10 clinometers are installed together with a wind measurement station on top of the building. A new data file is generated every 10 min ready to be analysed.

In order to be able to handle such an amount of data a continuous and automated modal extraction algorithm had to be implemented starting from the choice of the identification method and parameters.

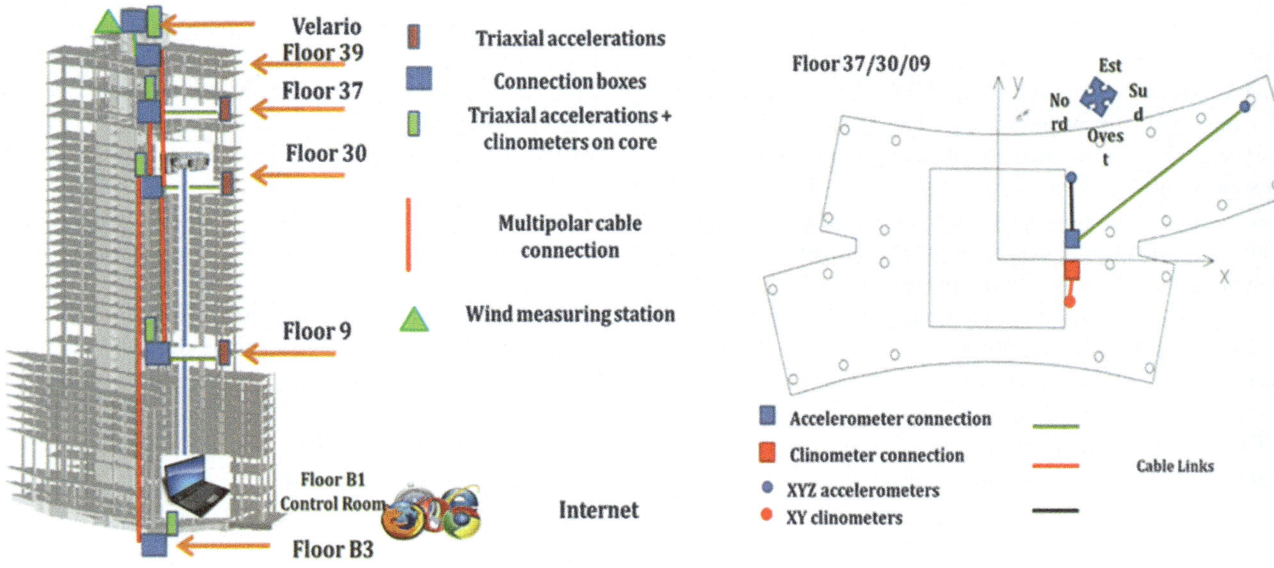

Fig. 2.1 CORE 1 Tower and a scheme of the installed monitoring system

2.3 Modal Identification Algorithm

The two algorithms which were taken into account are the pLSCF (Polyreference least square frequency domain) [11] and the FDD [12] (Frequency decomposition), as already mentioned. They are briefly discussed here to provide the most relevant information. More details can be found in the referenced works.

With regard to pLSCF, one of its main properties is that it provides clean and easy-to-interpret stabilization diagrams and this reduces the amount of complexity for its use as well as difficulties for getting reliable results. This has caused the pLSCF to become the industrial standard for modal analysis at the present time. pLSCF is a least-squares approach in frequency and can be used in OMA. In this case the inputs to the method are the positive power-spectra of the system responses.

The second algorithm taken into consideration is the FDD. As mentioned, it is based on SVD. SVD is a linear algebra technique that can achieve factorization of a complex matrix. FDD identification method works by decomposing the power spectral density (PSD) matrix in its principal components, at each spectral line. More details can be found in the wide referenced literature (e.g.).

There are two parameters which must be considered as inputs when using both mentioned OMA identification approaches: the frequency resolution R of the power-spectra and the number of averages N used to calculate them. It is known in the literature that a narrow frequency resolution allows to improve the accuracy of the modal identification. It is also easy to understand that the higher the number of averages is, the cleaner the power-spectra will be. This would suggest to increase the time-length of the acceleration signals used to calculate the power-spectra. Indeed, if T_t is the total time of the acceleration time-histories used for the modal extraction, and if the whole time-records are divided into N sub-records of time-length t_t (with no overlap), it results that:

$$R = \frac{1}{t_t} = \frac{N}{T_t} \Rightarrow T_t = \frac{N}{R} \tag{2.1}$$

Therefore, T_t must be increased in order to both increase N and decrease R.

However, T_t cannot be increased indefinitely because the structure changes its behaviour in time: e.g. the modal parameters of a structure change between day and night due to thermal shifts. According to these points, a maximum possible value for T_t was fixed equal to 10,800 s (i.e. 3 h), which is a time span over which an initial data check did not evidence any significant effect of the environmental conditions. Once this threshold was fixed, the effects of different values of R and N were studied by means of MCMC simulations.

2.4 Markov Chain Monte Carlo Simulations

The comparison of pLSCF and FDD was carried out by means of the MCMC method. Time-histories of accelerations were generated numerically with the goal to make them as close as possible to those collected by the accelerometers placed in the building. Then these numerical time-records were provided as inputs to the two algorithms.

These simulated signals were generated by means of a modal model of the structure, considering the first three modes Table 2.1. The PSDs of the generated signals were made as close as possible to the PDSs of the actual signals collected by the accelerometers. The reference actual PSDs were chosen from a day with very low wind in order to test the case with the poorest signal-to-noise ratio.

The effect of the electrical noise due to the transducers, cables, etc. was taken into account as well. Figure 2.2 shows a comparison between a PSD of a simulated signal and a real one. The agreement is satisfactory. This figure also shows that the value of the numerical PSD curve far from the resonances is close to the experimental one. This proves that the amount of random noise added to the numerical signals was correct.

Table 2.1 Reference parameters for the Monte Carlo simulations

First eigenfrequency	First mode shape	Second eigenfrequency	Second mode shape	Third eigenfrequency	Third mode shape
0.32 Hz	Bending (east-west direction)	0.40 Hz	Bending (north-south direction)	0.63 Hz	Torsion

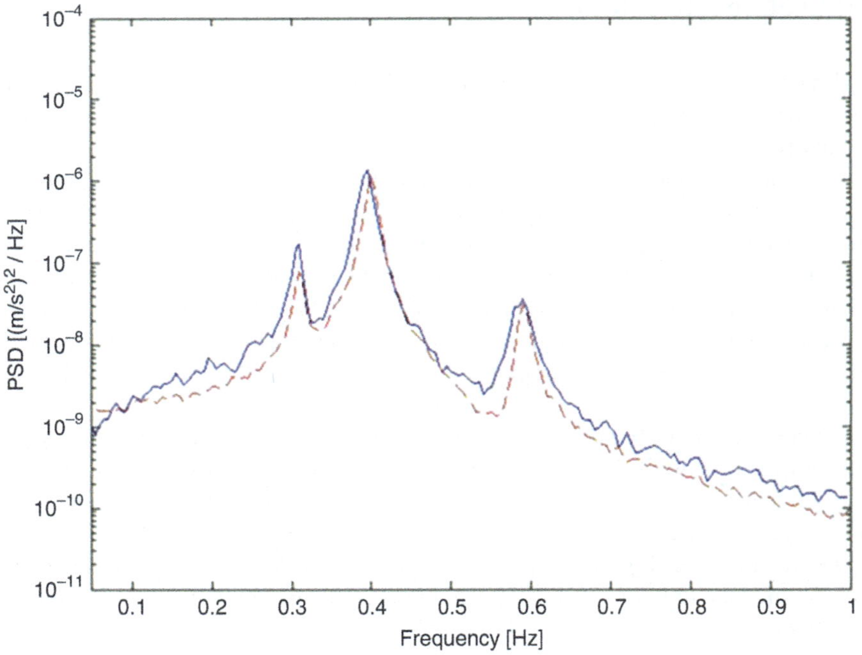

Fig. 2.2 Experimental (solid curve) and numerical (dashed curve) PSD for an accelerometer on floor 30 in case of low wind

Table 2.2 Values of T_t, R and N tested in the Monte Carlo simulations

Tested values of T_t [s]	Tested values of N	Tested values of R [mHz]
From 1000 to 10,800	From 13 to 72	From 3.3 to 13

The signal generation for each configuration tested (i.e. fixed values of N and R) was repeated 300 times. Then, modal identification was carried out for each iteration. The focus of the analysis was on the estimation of eigenfrequencies and mode shape components. The results related to mode shapes were described by a synthetic index: the modal assurance criterion (MAC) between the identified mode shapes and the reference ones used in the modal model.

This procedure allowed us to build statistical populations for the following indexes:

- errors between estimated and reference eigenfrequency values. These errors accounts for bias on the estimations by computing the population mean values μ and for the dispersion of the estimation by employing the standard deviation σ;
- MAC values. The mean MAC value μ accounts for bias, while again dispersion is related to the standard deviation σ of the MAC populations.

The trends of these mean and standard deviation values were investigated as function of two input variables: the number of averages N and the frequency resolution R employed to calculate the power-spectra provided as inputs to the modal identification algorithms. Table 2.2 shows the limit values of R and N tested in the MCMC simulations.

Figures 2.3, 2.4, 2.5 and 2.6 show the results in terms of μ and σ for the first three modes of the building as function of the frequency resolution and the number of averages. Particularly, Figs. 2.3 and 2.4 show the results for pLSCF, while Figs. 2.5 and 2.6 for FDD.

The analysis of these figures suggests the use of pLSCF for estimating the eigenfrequencies. Conversely, the FDD is more accurate for identifying mode shapes. Choosing a mixed approach allows to improve the accuracy associated to the estimated modal quantities. The MCMC simulation also allowed to find reliable data about the effect of frequency resolution and number of averages of the power-spectra and thus to choose properly their values (see Table 2.3).

The bias values associated to the identified eigenfrequencies are very low, as evidenced by Figs. 2.3a and 2.4a. Instead, the expected values of the dispersion associated to the identified eigenfrequencies are gathered in Table 2.4 and can be used as an estimation of the uncertainty of the identification method [13].

It is remarked that, for the configuration of values of Table 2.3, the MCMC test was repeated with a higher number of simulations (using an adaptive version of the MCMC) to assess statistical reliability. The results obtained are almost the same of those already shown with 300 simulations.

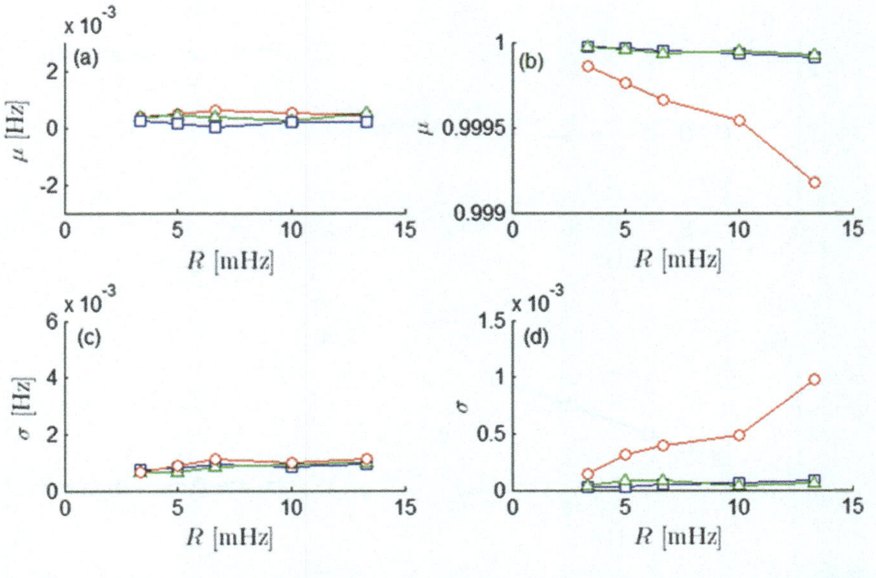

Fig. 2.3 Results of the MCMC simulations for pLSCF as function of R (N = 36): mean value of the error for the eigenfrequency estimation (**a**), mean value of the MAC (**b**), σ of the error for the eigenfrequency estimation (**c**), and σ for the MAC (**d**). Curves with squares (□) for the first mode, with circles (o) for the second mode, with triangles (Δ) for the third mode

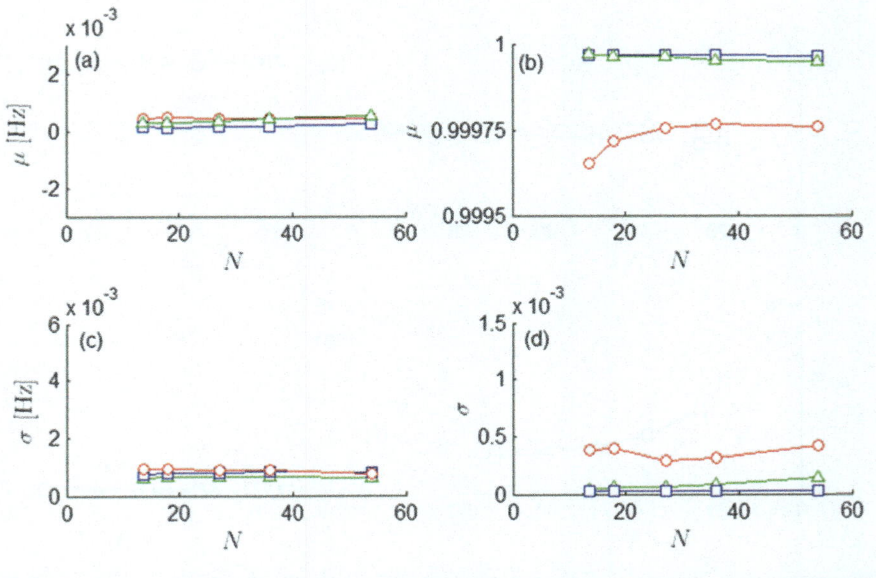

Fig. 2.4 Results of the MCMC simulations for pLSCF as function of N (R = 5 mHz): mean value of the error for the eigenfrequency estimation (**a**), mean value of the MAC (**b**), σ of the error for the eigenfrequency estimation (**c**), and σ for the MAC (**d**). Curves with squares (□) for the first mode, with circles (o) for the second mode, with triangles (Δ) for the third mode

2.5 First Results from the Real Data

The values of R and N chosen with the MCMC simulations, as well as the use of the mixed approach pLSCF/FDD, allowed to develop an automatic OMA identification system. To do this, an automated data-check procedure was also developed. Relying on the use of the Skewness coefficient and peak-peak values of the signals, corrupted time records (e.g. due to saturation, lightning, transducer damage) are automatically discarded before modal identification.

The automated extraction was applied to 8 months of vibration data coming from the installed monitoring system. At first the uncertainty interval equal to $\pm 2\sigma$ (see Table 2.4) around the identified eigenfrequency value for the first mode is shown in Fig. 2.7. This $\pm 2\sigma$ interval expresses a level of confidence of about 95%. The width of the uncertainty intervals is clearly overestimated. This is probably due the fact that the value of σ was estimated through MCMC considering a day with very low wind and consequently a very bad signal to noise ratio.

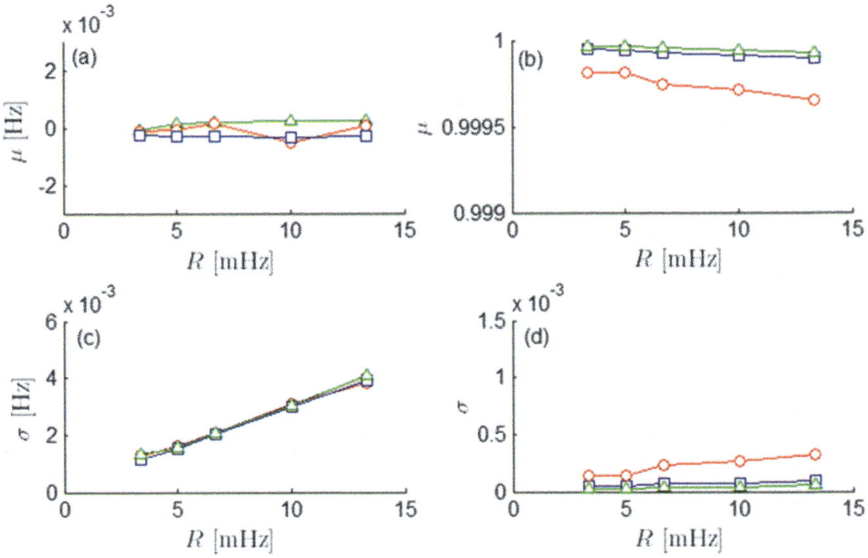

Fig. 2.5 Results of the MCMC simulations for FDD as function of R: mean value of the error for the eigenfrequency estimation (N = 48) (**a**), mean value of the MAC (N = 36) (**b**), σ of the error for the eigenfrequency estimation (N = 48) (**c**), and σ for the MAC (N = 36) (**d**). Curves with squares (□) for the first mode, with circles (o) for the second mode, with triangles (Δ) for the third mode

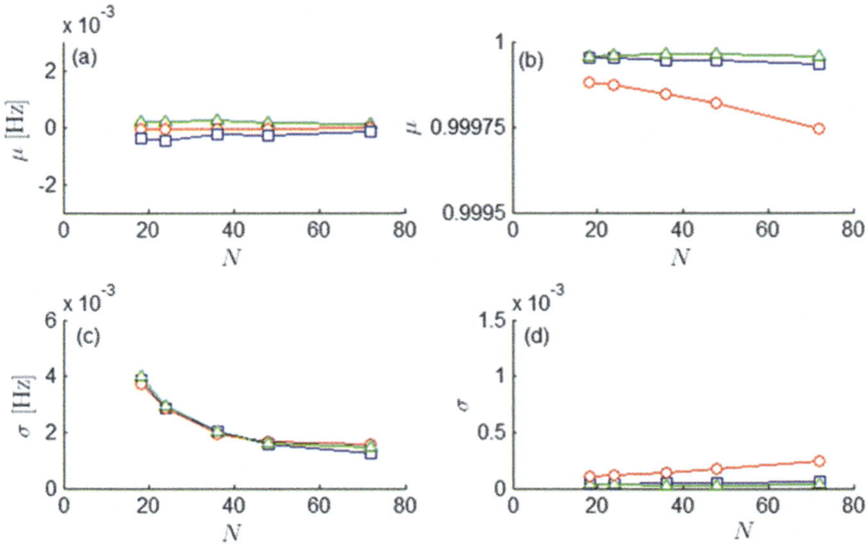

Fig. 2.6 Results of the MCMC simulations for FDD as function of N (R = 5 mHz): mean value of the error for the eigenfrequency estimation (**a**), mean value of the MAC (**b**), σ of the error for the eigenfrequency estimation (**c**), and σ for the MAC (**d**). Curves with squares (□) for the first mode, with circles (o) for the second mode, with triangles (Δ) for the third mode

Table 2.3 Chosen values of T_t, R and N to be used in OMA with real signals

Chosen value of T_t [s]	Chosen value of N	Chosen value of R [mHz]
7200	36	5

Figures 2.8, 2.9 and 2.10 show the trend of the first three eigenfrequencies of the structure. Instead, Fig. 2.11 shows the MAC trend for the first mode. The changes of the eigenfrequencies are always lower than 5%.

The wind speed is proved to be able to change the eigenfrequency values significantly. Indeed, the spikes towards zero in Figs. 2.8, 2.9 and Fig. 2.10 are always related to the presence of an increased value of the root mean square (RMS) of the acceleration signals (the RMS is calculated on the frequency band of the considered mode). This RMS is strongly correlated with the wind speed.

Table 2.4 Expected values of σ on the eigenfrequency estimations

σ on the first eigenfrequency	σ on the second eigenfrequency	σ on the third eigenfrequency
0.8 mHz	0.8 mHz	0.6 mHz

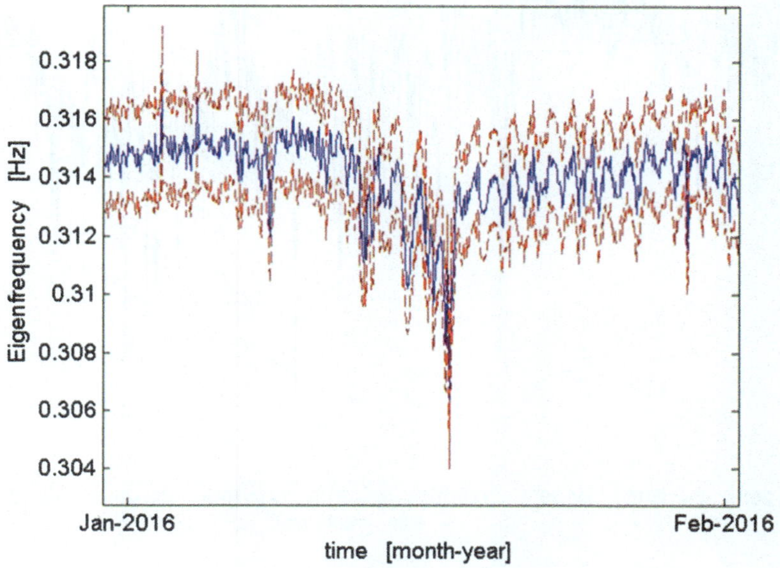

Fig. 2.7 Trend of the first eigenfrequency (solid blue curve) and ±2σ interval (dashed red curves)

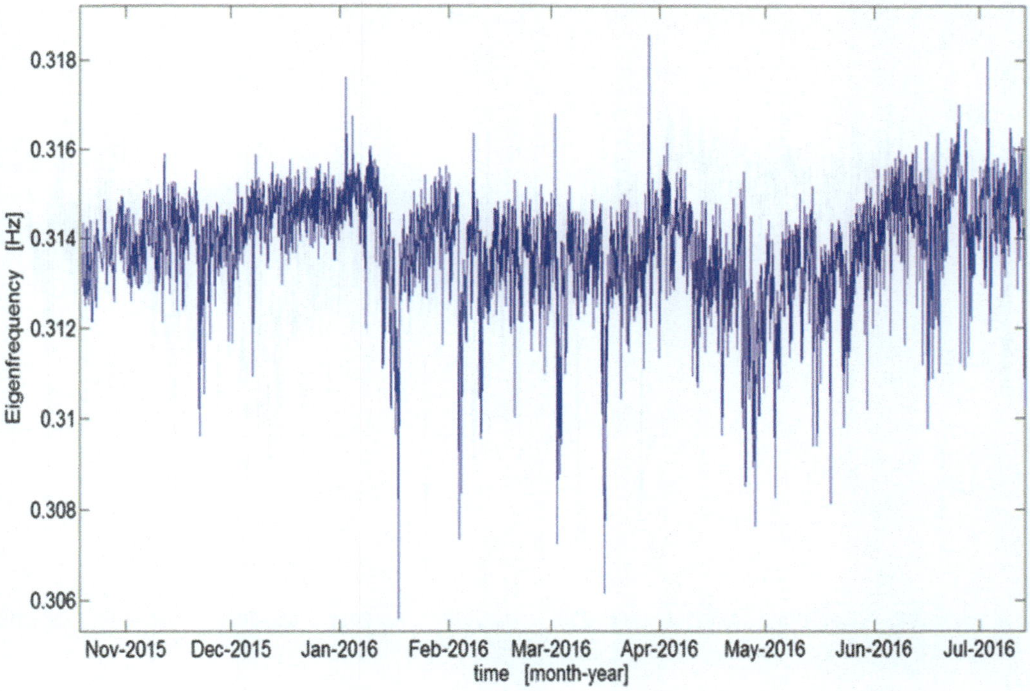

Fig. 2.8 Time trend of the first eigenfrequency

A slight descending trend is put into evidenze for the MAC coefficient even if looking at the scale in Fig. 2.11 it can be noticed that this variability is in the order of 1E-4, which is less than the identification ucertainty.

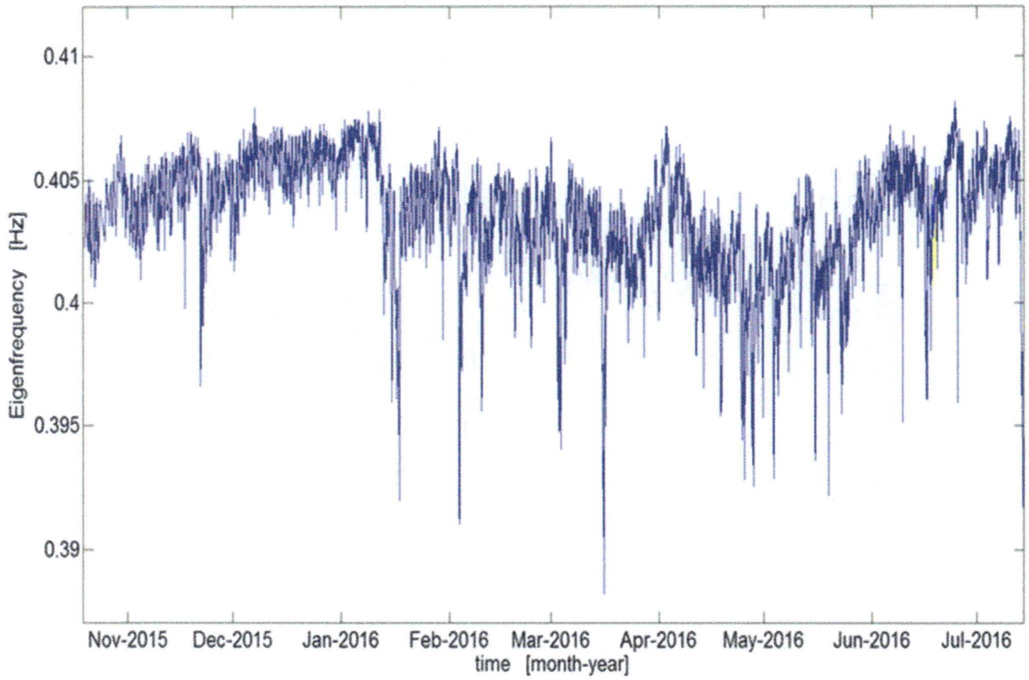

Fig. 2.9 Time trend of the second eigenfrequency

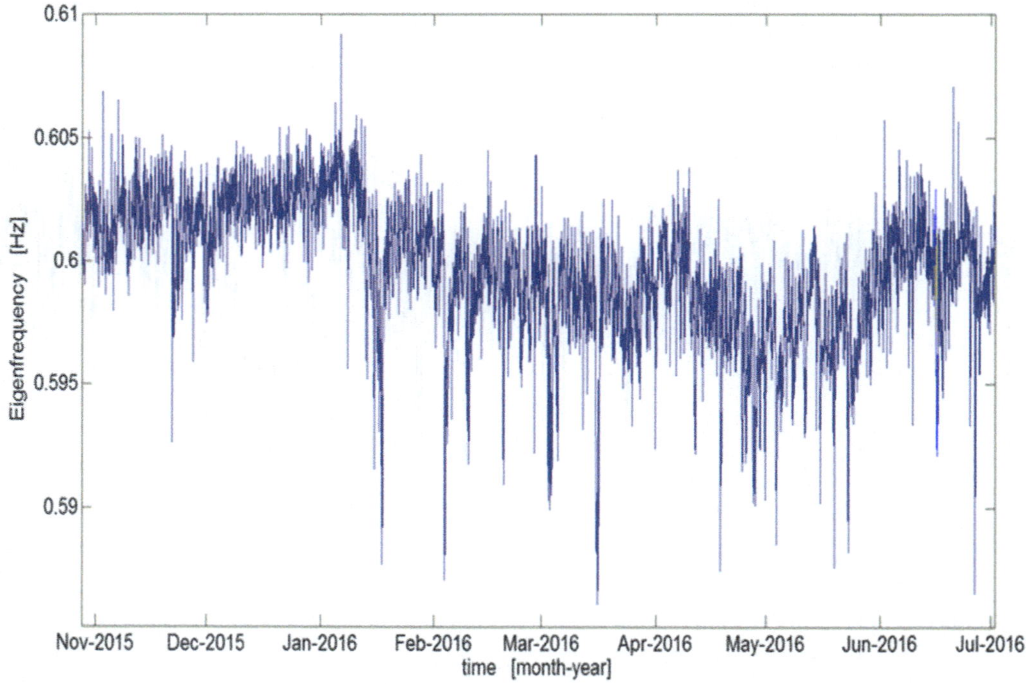

Fig. 2.10 Time trend of the third eigenfrequency

2.6 Conclusion

The paper has described the installed monitoring system and a proposed data analysis strategy for Palazzo Lombardia, one of the tallest building in Milano. The layout of the system has been presented, and the choices related to the used sensors and their locations are discussed.

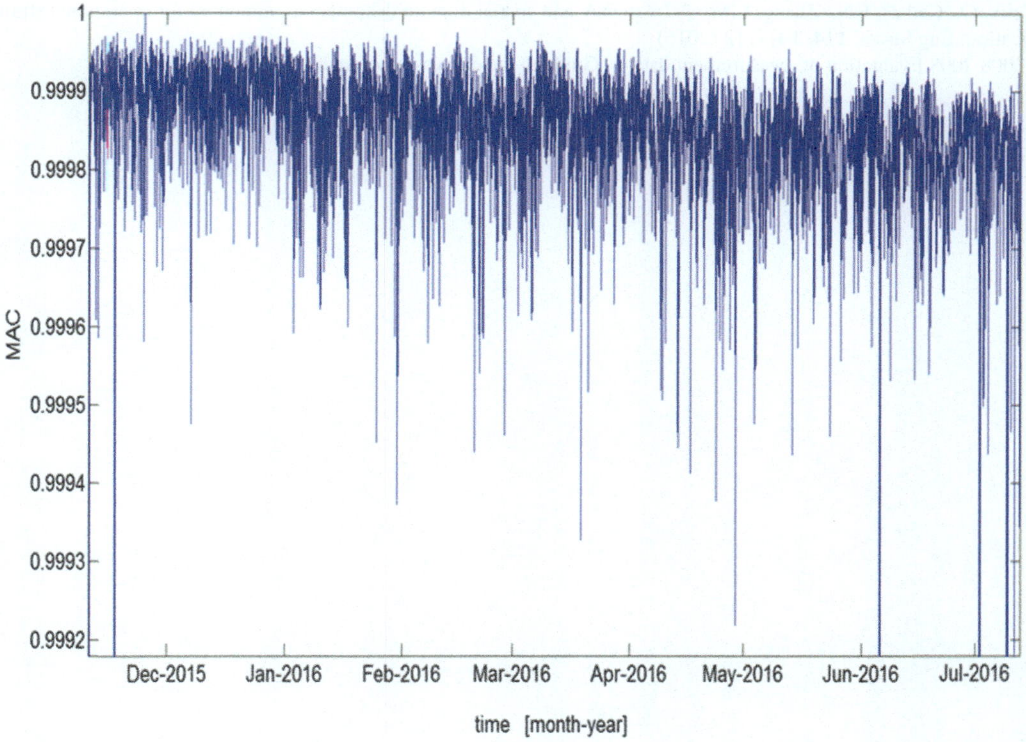

Fig. 2.11 Time trend of the MAC of the first mode

An approach based on automatic and continuous modal parameters extraction has then been presented for structural health monitoring purposes. The main features of the automatic modal analysis software used for modal identification are then provided, explaining how Monte Carlo simulations were used to optimise it.

This automatic identification system allowed to produce the first plots of the trends of the modal parameters of the building along 8 months of monitoring. These modal parameter trends will allow to develop a statistical model to describe the relationship between the first three eigenfrequencies of the building and a number of input variables representative of the environmental conditions. The statistical model will be used to detect anomalous trends indicating a change in the structure.

References

1. Berardengo, M., Cigada, A., Manzoni, S., Vanali, M.: Design and installation of a permanent monitoring system for Palazzo Lombardia in Milano, Italy VII Eur. Congr. Comput. Methods Appl. Sci. Eng. – ECCOMAS Congr. 2016 – Crete Isl. (Greece), 5–10 June 2016. **1**, 3640–51 (2016)
2. Farrar, C.R., Worden, K.: Structural Health Monitoring: A Machine Learning Perspective. Hoboken, NJ, Wiley, Hoboken (2012)
3. Mesquita, E., Antunes, P., Coelho, F., André, P., Arede, A., Varum, H.: Global overview on advances in structural health monitoring platforms. J. Civ. Struct. Heal. Monit. **6**, 461–475 (2016)
4. Yang, Z., Wang, L.: Structural damage detection by changes in natural frequencies. J. Intell. Mater. Syst. Struct. **21**, 309–319 (2010)
5. Rainieri, C., Fabbrocino, G.: Operational Modal Analysis of Civil Engineering Structures. Springer, New York (2014)
6. Busca, G., Cigada, A., Mola, E., Mola, F., Vanali, M.: Dynamic testing of a helicopter landing pad: comparison between operational and experimental approach. J. Civ. Struct. Heal. Monit. **4**, 133–147 (2014)
7. Ewins, D.J.: Modal Testing: Theory, Practice and Application. Research studies press Ltd., Baldock (2000)
8. Cigada, A., Mola, E., Mola, F., Stella, G., Vanali, M.: Dynamic behavior of the Palazzo Lombardia tower: comparison of numerical models and experimental results. J Perform Constr Facil. **28**(3), 491–501 (2014)
9. Mendis, P., Ngo, T., Haritos, N., Hira, A., Samali, B., Cheung, J.: Wind loading on tall buildings. EJSE Spec Issue Load Struct. **3**, 41–54 (2007)
10. Griffis, L.G.: Serviceability limit states under wind load. Eng J. **30**, 1–16 (1993)
11. Peeters, B., Van Der Auweraer, H., Guillaume, P., Leuridan, J.: The PolyMAX frequency-domain method: a new standard for modal parameter estimation? Shock Vib. **11**, 395–409 (2004)

12. Wang, T., Celik, O., Catbas, F.N., Zhang, L.M.: A frequency and spatial domain decomposition method for operational strain modal analysis and its application. Eng Struct. **114**, 104–112 (2016)
13. JCGM 100:2008 2008 Evaluation of measurement data — Guide to the expression of uncertainty in measurement

Chapter 3
Dynamic Characterization of the Little Belt Suspension Bridge by Operational Modal Analysis

Silas S. Christensen, Michael S. Andersen, and Anders Brandt

Abstract The (new) Little Belt Bridge, opened in 1970, is a Danish suspension bridge with largest span being 600 m, and a total length of 1700 m. During the design and construction phase, detailed analysis of the dynamic properties of the bridge were carried out both by hand calculations and by measurements on a scale model. Recently, the bridge was measured using a setup of 45 simultaneous responses, 30 vertical and 15 lateral, distributed over the main span. Operational modal analysis was carried out on the data set, and the first two vertical bending and torsional modes were compared to those of the model. It was found that the first vertical bending mode was 0.156 Hz, which is very near the frequency predicted by the original scale model. In total nine modes are reported, with frequencies from 0.156 to 0.808 Hz, and with damping values between 0.38% and 9.74%. This paper also demonstrates and discusses the use of inexpensive geophones and general measurement equipment for use in OMA applications based on the experience from this bridge measurement.

Keywords Dynamic characterization · Little belt suspension bridge · Operational modal analysis · Multi-reference Ibrahim time domain

3.1 Introduction

In the early 1965, the initial design of a suspension bridge connecting Jutland and Funen commenced. Five years later the Little Belt Suspension Bridge was officially open. The bridge itself was in its time a true masterpiece, and engineers have put a lot of thought into the final design. The bridge itself as seen in Fig. 3.1 consists of a 600 m main span and two 240 m side spans extended by approximately 300 m viaducts making the bridge almost 1700 m in length. The main and side spans are suspended by hanger cables for every 12 m which are upheld by two 1500 m main cables that are supported by two 120 m pylons and anchored on each side. The two main cables are fixed to the bridge deck at the center of the mid span. The main and side spans are separated in such a way that only lateral forces can be exchanged between the two bridge decks. Pendulum bearings mounted on the pylons are supporting the bridge deck in regard to vertical movement.

The Little Belt Suspension Bridge was one of the first long span bridges utilizing a closed box girder. The box shaped girder, shown in Fig. 3.2, was preferred instead of a truss because it required less material and was easier to manufacture. Due to the lower height, it has lower bending stiffness compared to a deep truss. However, the closed box had higher torsional stiffness which is favorable against wind instability. The box shaped girder bear some resemblance to the flat plate [2] which was used to model the aeroelastic effects of airfoils and bridge decks at that time. Due to the flat plate resemblance, torsional flutter, i.e. negative damping due to fluid-structural interaction in the torsional degree of freedoms, were not an issue contrary to the infamous first Tacoma Narrows which collapsed 25 years before the construction of the Little Belt Suspension Bridge began.

3.2 Theory

There are numerous modal parameter estimation techniques available [3]. Generally speaking there are time domain methods and frequency based alternatives. Depending on the application, one method is often preferred over the other. For output-only measurements, which is the case for operational modal analysis (OMA), time domain methods are often reported to perform

S. S. Christensen · M. S. Andersen · A. Brandt (✉)
Department of Technology and Innovation, University of Southern Denmark, Odense M, Denmark
e-mail: abra@iti.sdu.dk

S. Pakzad (ed.), *Dynamics of Civil Structures, Volume 2*, Conference Proceedings of the Society for Experimental Mechanics Series, https://doi.org/10.1007/978-3-319-74421-6_3

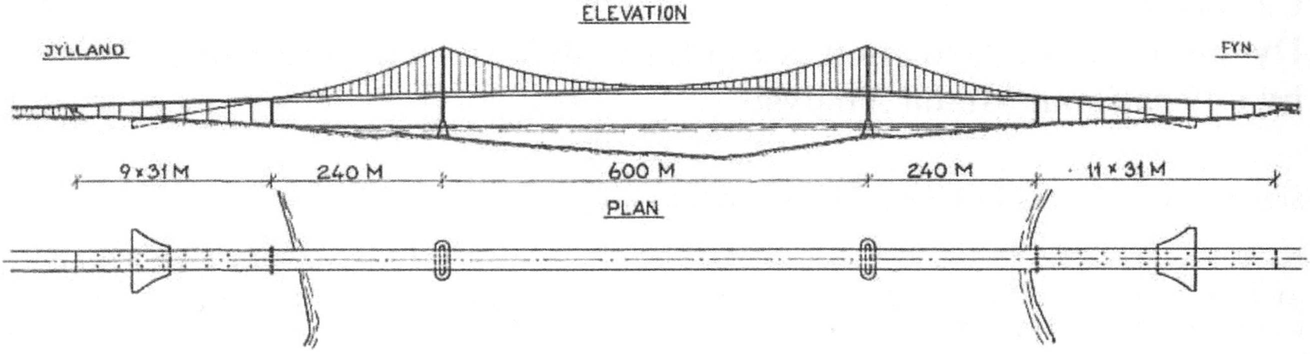

Fig. 3.1 Elevation and plan. (From [1])

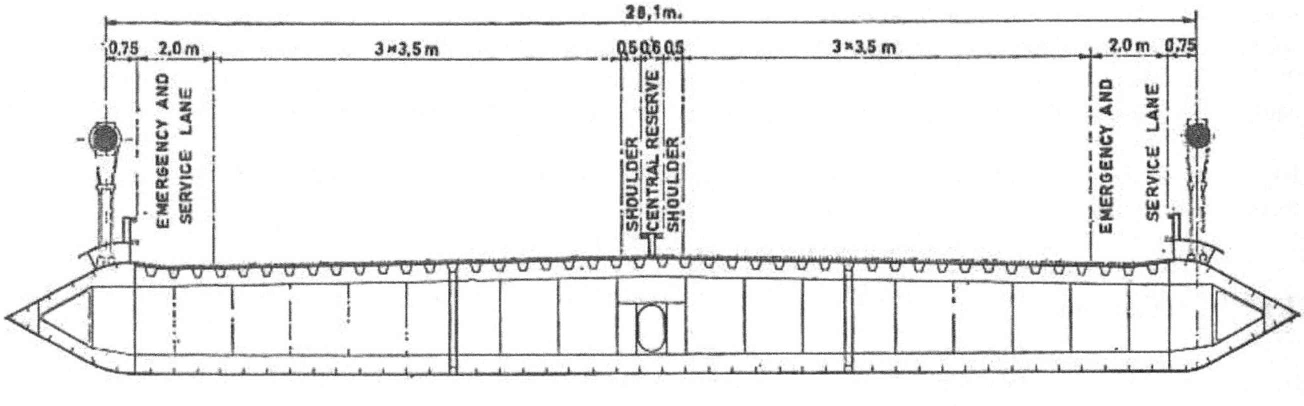

Fig. 3.2 Cross-section. (From [1])

best [4, 5]. In the present study, the multi-reference Ibrahim Time Domain (MITD) method [6, 7] for modal parameter estimation was utilized.

The total dynamic characteristics is best interpreted in a stabilization diagram, where physical poles tend to stabilize for increasing model order while spurious poles do not. From the stabilization diagram, a representative pole can be selected for each mode shape. To ensure that the selected mode shapes have been properly separated the Modal Assurance Criteria (MAC) is commonly used, see [8]. The MAC value between two mode shapes is evaluated by:

$$MAC_{rs} = \frac{\left|\{\psi\}_r^H\{\psi\}_s\right|^2}{\left(\{\psi\}_r^H\{\psi\}_r\right)\left(\{\psi\}_s^H\{\psi\}_s\right)} \tag{3.1}$$

where ^{H} is the Hermitian transpose (complex conjugate and transpose). The MAC value can take any value between 0 and 1. For independent mode shapes $\{\psi\}_r \neq \{\psi\}_s$, a value of 0 is common, whereas for identical mode shapes $\{\psi\}_r = \{\psi\}_s$, a value close to 1 is to be expected.

3.3 Experimental Setup

Measurements were taken on the mid span of the structure using 45 SS-4.5N geophones. Geophones are used within seismology and converts movement (velocity) into a voltage output using electromagnetic principles by having a moving coil in a magnetic field. For the vertical response, 30 geophones were installed in a double symmetric pattern, while 15 geophones were installed on one side of the bridge to measure the lateral response. The sensor layout is shown in Fig. 3.3.

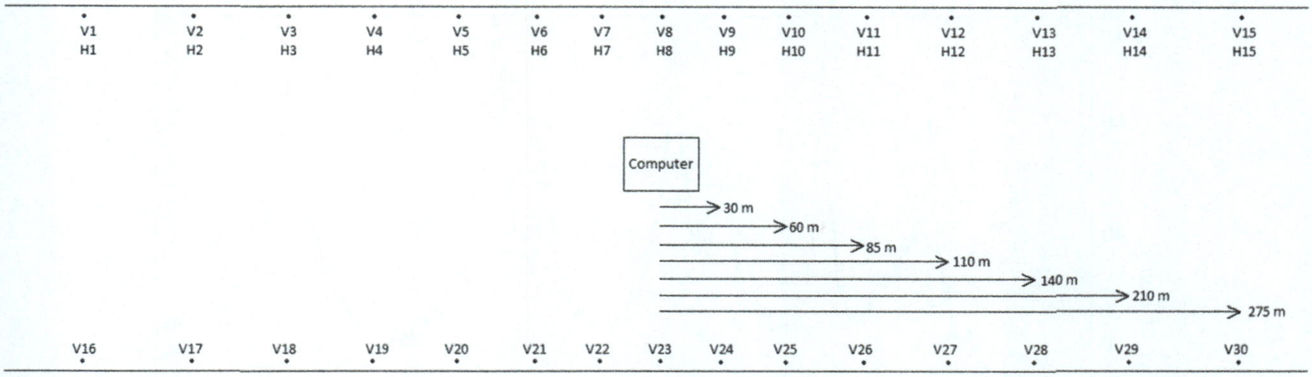

Fig. 3.3 Experimental setup. (From [9])

Along the bridge it is seen that the concentration of sensors are highest at the center of the mid span, while it decreases towards the pylons. Across the bridge the sensors were mounted as far from each other as possible. All geophones were mounted on wooden wedges and secured by adhesive tape. This mounting technique was tested in the laboratory prior to installation on the bridge. The geophones were connected to a National Instrument PXI system and measurements were stored on a computer as time data at a sampling frequency of 1000 Hz. Measurements were taken over several days in mid-April 2015.

3.4 Results and Discussion

All data processing was done using tools in [10]. A total of 1225 auto- and cross-correlation functions were produced with a blocksize of 512 samples for 60 min of data. Using 180 min of data, frequency and damping estimates were unchanged. For each correlation function, the first 5 lags were discarded and 50 samples were used to reduce the random error [11].

3.4.1 Frequencies Estimates and Mode Shapes

The stabilization diagram, seen in Fig. 3.4, looks clear and shows nine columns of stable poles indicating nine possible modes. The stabilization diagram has been overlaid by two mode identification functions (MIFs). The two MIFs are averages of the power spectral densities for all measurements taken in the vertical and lateral direction, respectively. The MIFs do not indicate that there are two closely spaced modes at frequencies of $f_n \approx 0.16$ Hz and $f_n \approx 0.17$ Hz. The lateral MIF shows two peaks at $f_n = 0.031$ Hz and $f_n = 0.448$ Hz that do no appear in the vertical MIF. The first of the two peaks is related to a lateral bending mode, but it is contaminated by some spurious motion. The second peak is a coupled torsional and lateral bending mode. These modes were not identified.

The Modal Assurance Criteria (MAC), as seen in Fig. 3.5, reveals that there are nine different modes, but some of the modes shapes do somewhat resemble one another. Particularly, the first, fourth and the fifth mode shapes have some similarities. It can be seen in Fig. 3.6 that these in fact all are symmetrical bending mode shapes. It is possible that DOFs outside the pylons would have to be included to produce lower MAC values.

The remaining four mode shapes are two additional vertical bending mode shapes, one antisymmetric and one symmetrical, along with the first and second torsional mode shapes.

The fourth mode shape looks like a symmetric version of the first antisymmetric vertical bending mode, and a similar mode was also reported in [12]. The mode shape has small movement at the center node, which may be related to the fact that the main cables are fixed to the bridge deck over the center of the mid span restraining the deck to the main cables.

From Fig. 3.6 it appears that the ends of the main span, at the pylons, are subject to some vertical motion. Particularly the part on the Funen side has higher vertical motion than the part towards Jutland.

The natural frequencies and damping ratios for each of the nine modes are seen in Table 3.1. There is reasonable agreement between the calculated natural frequencies before the bridge was built [1] and the frequencies estimated in the present study for the first two vertical and torsional modes. Some discrepancy is seen for the first antisymmetric vertical bending mode.

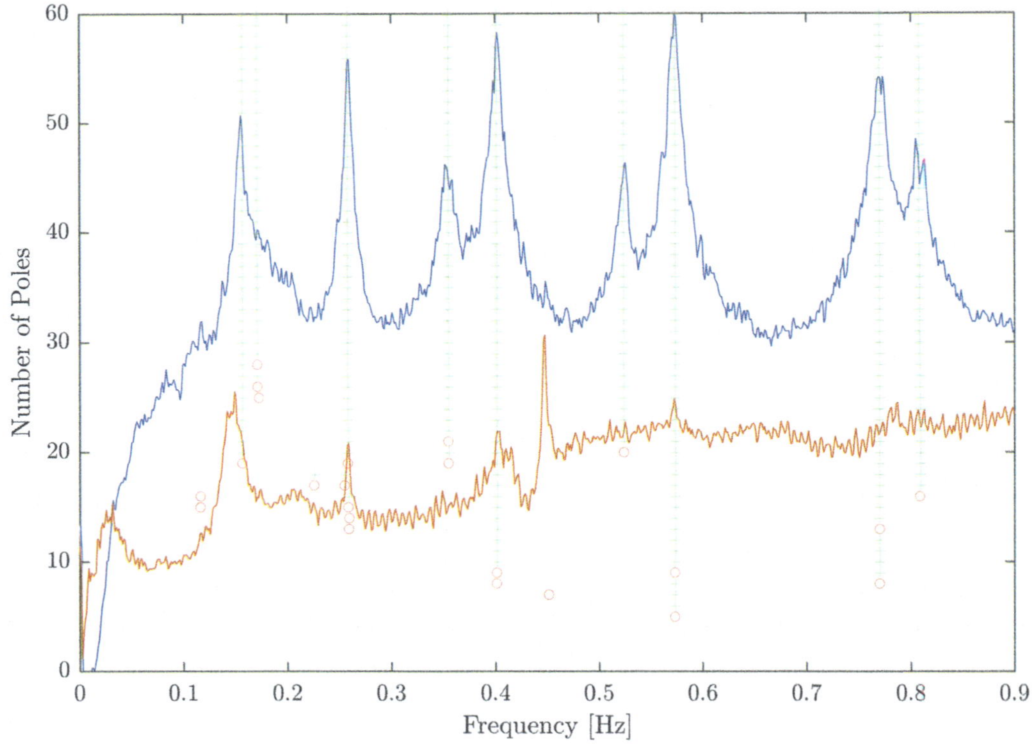

Fig. 3.4 Stabilization diagram. Red circle – unstable pole; green plus – stable pole; solid blue line – averaged PSD of vertical oriented sensors; solid red line – averaged PSD of lateral oriented sensors

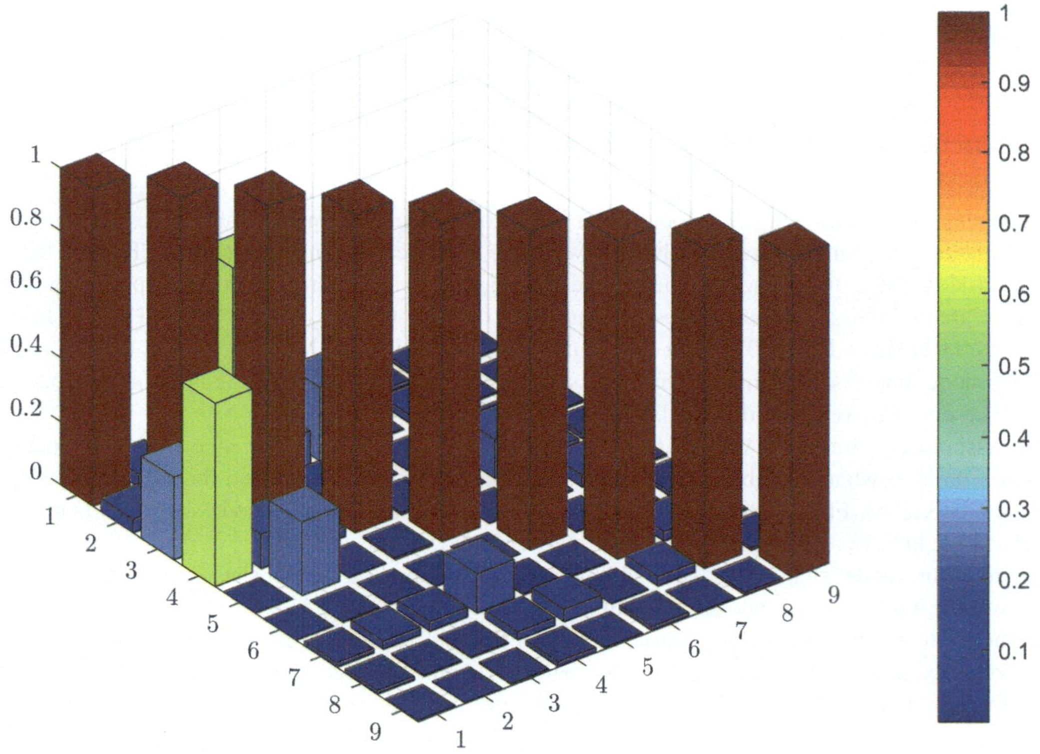

Fig. 3.5 Modal Assurance Criteria (MAC) of mode shapes for the Little Belt Suspension Bridge

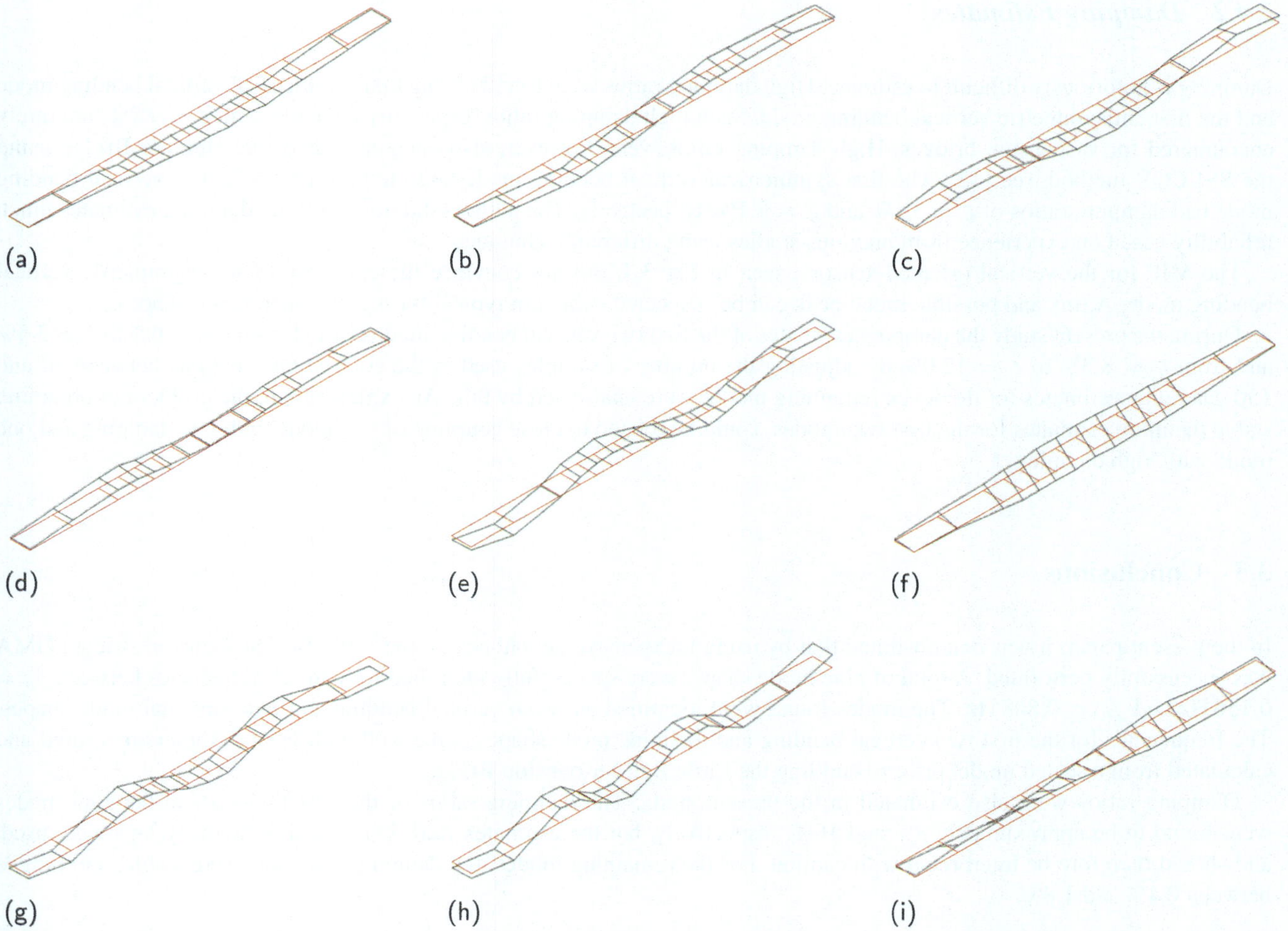

Fig. 3.6 Mode shapes with natural frequency and damping ratio identified by OMA on the Little Belt Suspension Bridge. The end of the bridge deck in the down left corner is pointing towards Jutland. (**a**) First symmetrical vertical bending mode, V1S $f_n = 0.156$ Hz; $\zeta = 3.157\%$. (**b**) First antisymmetric vertical bending mode, V1A $f_n = 0.171$ Hz; $\zeta = 9.741\%$. (**c**) Second symmetrical vertical bending mode, V2S $f_n = 0.258$ Hz; $\zeta = 0.786\%$. (**d**) Third symmetrical vertical bending mode, V3S $f_n = 0.355$ Hz; $\zeta = 0.1.777\%$. (**e**) Second antisymmetric vertical bending mode, V2A $f_n = 0.401$ Hz; $\zeta = 0.939\%$. (**f**) First torsional mode, T1S $f_n = 0.523$ Hz; $\zeta = 0.768\%$. (**g**) Fourth symmetrical vertical bending mode, V4S $f_n = 0.573$ Hz; $\zeta = 0.383\%$. (**h**) Third antisymmetric vertical bending mode, V3A $f_n = 0.769$ Hz; $\zeta = 0.608\%$. (**i**) Second torsional mode, T1A $f_n = 0.808$ Hz; $\zeta = 0.690\%$

Table 3.1 Natural frequencies and damping ratios for the present study and natural frequencies from [1]

Natural frequency [Hz], $\mathbf{f_n}$			Damping ratio [%], $\boldsymbol{\zeta}$		
Experimental – present	Experimental – from [1]	Computational – from [1]	Experimental – present	Experimental – from [1]	Computational – from [1]
0.156	0.156	0.147	3.157	–	–
0.171	0.153	0.145	9.741	–	–
0.258	–	–	0.786	–	–
0.355	–	–	1.777	–	–
0.401	–	–	0.939	–	–
0.523	0.500	0.480	0.768	–	–
0.573	–	–	0.383	–	–
0.769	–	–	0.608	–	–
0.808	–	0.800	0.690	–	–

3.4.2 Damping Estimates

Damping is notoriously difficult to estimate. High damping ratios were found for the first symmetrical vertical bending mode and the first antisymmetric vertical bending mode. Such high damping ratios, especially the latter with $\zeta = 9.7\%$, are rarely encountered for suspension bridges. High damping ratios were, however, also encountered for the Humber Bridge using the SSI-COV method, see [12]. The first symmetrical vertical bending mode and the first antisymmetric vertical bending mode had damping ratios of $\zeta = 3.1\%$ and $\zeta = 8.1\%$, respectively. The authors did not credit the damping estimates much reliability based on experience from previous studies using different techniques.

The MIF for the vertical oriented sensors, seen in Fig. 3.4, did not convince the existence of the asymmetric vertical bending mode. A tiny and possible broad peak can be suspected, which in typical for high damped mode shapes.

During the present study the damping estimates of the first two vertical bending modes varied from $\zeta = 1.6\%$ to $\zeta = 3.5\%$ and from $\zeta = 8.3\%$ to $\zeta = 12.0\%$ by adjusting the number of samples used in the correlation functions between 30 and 150. Damping estimates for the seven remaining modes were unaffected by this. An explanation to the problem of obtaining stable damping estimates for the first two modes, could be related to close coupling of one mode with low damping and one mode with high damping.

3.5 Conclusions

In the present paper, it was demonstrated that by using inexpensive geophones on the Little Belt Suspension Bridge, OMA was successfully performed. A total of nine mode shapes were successfully identified for natural frequencies between $f_n = 0.156$ Hz and $f_n = 0.808$ Hz. The modes found were identified as seven vertical bending and two torsional mode shapes. The frequencies for the first two vertical bending and torsional mode shapes agree well with what designers measured and calculated from a scaled model prior to building the Little Belt Suspension Bridge.

Damping ratios were also estimated in the present study. The damping ratios of the first two vertical bending modes were found to be approximately 3% and 10%, respectively, but the estimates were sensitive to how many lags were used, and should therefore be interpreted with caution. For the remaining modes, the damping estimates were stable, and ranged between 0.4% and 1.8%.

Acknowledgements We would like to thank The Danish Road Directorate for their help during the measurement campaign. We would also greatly want to thank Hilde Husby Knustad for producing good measurements.

References

1. Ostenfeld, C., Haas, G., Frandsen, A.G.: Motorway bridge across Lilleblt: model tests for the superstructure of the suspension bridge. Presented at Symposium on Supension Bridges, Lisbon (1966)
2. Theodorsen, T.: General theory of aerodynamic instability and the mechanism of Flutter. NACA report (1934)
3. Allemang, R.J., Brown, D.L.: A unified matrix polynomial approach to modal identification. J. Sound Vib. **221**(3), 301–322 (1998)
4. Peeters, B., Roeck, G.D.: Stochastic system identification for operational modal analysis: a review. J. Dyn. Syst. Meas. Control **123**(4), 659–667 (2001)
5. Brincker, R., Ventura, C.: Introduction to Operational Modal Analysis. Wiley, New York (2015)
6. Ibrahim, S.R., Mikulcik, E.C.: A method for the direct identification of vibration parameters from the free response. Shock Vib. Bull. **47**, 183–198 (1977)
7. Fukuzono, K.: Investigation of multiple-reference Ibrahim Time Domain modal parameter estimation technique. Master's thesis, University of Cincinnati (1986)
8. Brandt, A.: Noise and Vibration Analysis – Signal Analysis and Experimental Procedures. Wiley, New York (2011)
9. Knustad, H.H.: Operational modal analysis of the Little Belt Bridge using geophones. Master's thesis, University of Southern Denmark (2015)
10. Brandt, A.: ABRAVIBE A MATLAB toolbox for noise and vibration analysis and teaching. Department of Technology and Innovation, University of Southern Denmark (2011). http://www.abravibe.com
11. Orlowitz, E., Brandt, A.: Influence of Noise in Correlation Function Estimates for Operational Modal Analysis. XXXVI International Modal Analysis Conference (2018)
12. Brownjohn, J.M.W., Magalhaes, F., Caetano, E., Cunha, A.: Ambient vibration re-testing and operational modal analysis of the Humber Bridge. J. Eng. Struct. **32**, 2003–2018 (2010)

Chapter 4
The Prediction of Vibrations for Light Structures in Presence of Moving People

M. Berardengo, L. Drago, S. Manzoni, and M. Vanali

Abstract Recently, a model to describe the vibration of light structures (e.g. footbridges, staircases) was proposed by the authors of this paper. Such a model was developed with the aim of being accurate with a high number of people occupying the structure for long times. The present paper analyses the behaviour of the same model in the case of transient excitation of the structure. This allows to assess the accuracy of the model also in this further situation.

Keywords Human structure interaction · Ground reaction force · Dynamics · Vibration · Slender structure

4.1 Introduction

The dynamics of structures occupied by pedestrians is a widely studied topic. Many works treated the vibration serviceability issues of civil structures, with detailed focus on Human-Structure Interaction (HSI) and human induced vibrations [1–5]. Footbridges [6–12], grandstands of stadia [13–15], staircases [16–18], and other pedestrian structures [19–21] have been extensively investigated. Moreover, international standards and codes [22–26] have been developed with the aim of both designing and evaluating the structure dynamics under the crowd action and they are the usual reference when vibration serviceability is assessed. Unlike other standards, a recent guidance [27] (Joint Working Group, 2008) regarding dynamic performance requirements for permanent grandstands subjected to crowd action recommends to consider HSI.

There are many works which proposed models to describe HSI and/or predict the structure dynamics under the action of pedestrians (e.g., [7, 19, 20]). This work treats a model already presented in the literature [17, 18], which describes the action of each single person on the structure and considers the structure as a multi-degrees-of-freedom system. Therefore, even structures with coupled modes can be considered when computing the structural response due to people walking. The focus of the present paper is the analysis of the accuracy of this model in case of few people on the structure for a short time. Indeed, the model was originally developed for the case of many pedestrians on the structure for a long time.

The paper is structured as follows: Sect. 4.2 describes the theoretical model used to predict structural vibrations in presence of people; Sect. 4.3 describes the tests carried out for the aim of the paper; finally, Sect. 4.4 discusses the results.

4.2 The Model

When a pedestrian is in contact with a point of a structure, he/she produces a Ground Reaction Force (GRF). The GRF is the total force exchanged between the person and the structure. It is possible to see the GRF as the sum of a passive GRF (PGRF) and an active GRF (AGRF) [17, 18]. The PGRF is the force generated by structural movement. Indeed, when the structure vibrates, it excites the person. If the person is considered as a dynamic system, he/she starts to vibrate as well and thus exerts a force on the structure. This force is named PGRF. Conversely, the AGRF is generated by the person's active movement. The AGRFs do not depend on (and are not generated by) the vibration of the structure behind the person and are caused by the active movement of the person. The AGRF can be described as the force exerted by a moving person on a structure with an infinite stiffness.

M. Berardengo · M. Vanali
Dipartimento di Ingegneria e Architettura, Università degli studi di Parma, Parma, Italy

L. Drago · S. Manzoni (✉)
Dipartimento di Meccanica, Politecnico di Milano, Milan, Italy
e-mail: stefano.manzoni@polimi.it

S. Pakzad (ed.), *Dynamics of Civil Structures, Volume 2*, Conference Proceedings of the Society for Experimental Mechanics Series, https://doi.org/10.1007/978-3-319-74421-6_4

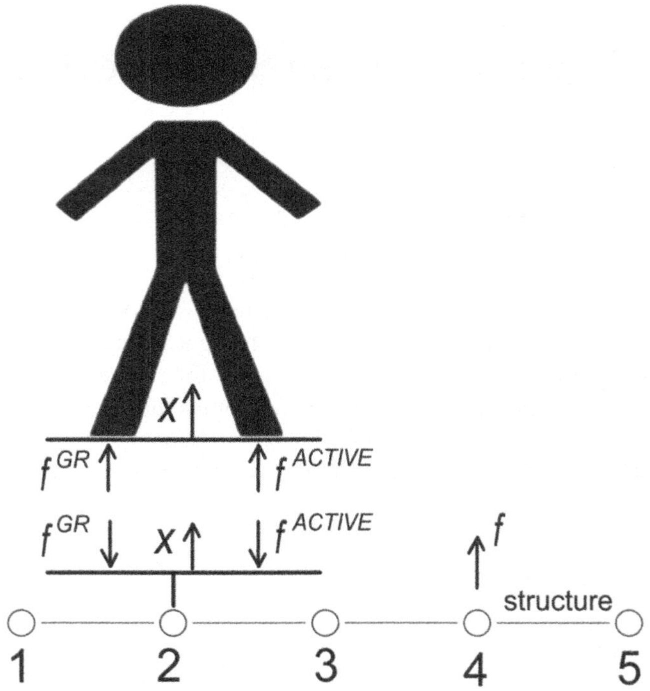

Fig. 4.1 Human-structure interaction: AGRFs ($\mathbf{f}^{\mathbf{ACTIVE}}$) and PGRFs ($\mathbf{f}^{\mathbf{GR}}$)

Therefore, the PGRF depends on the dynamic features of the person and on the motion of the structure. On the other hand, the AGRF is caused by an active movement of the pedestrian and is not related to the dynamics of the structure.

According to Fig. 4.1, the dynamics of the structure occupied by people can be described as:

$$\mathbf{x}(\omega) = \mathbf{G}(\omega)\left(-\mathbf{f}^{\mathrm{ACTIVE}}(\omega) - \mathbf{f}^{\mathrm{GR}}(\omega) + \mathbf{f}(\omega)\right) \tag{4.1}$$

Where $\mathbf{G}$ is the matrix containing the frequency response functions (FRF) of the empty structure, $\mathbf{x}$ is the vector of the displacements of the degrees-of-freedom in which the structure has been discretised, $\mathbf{f}$ is the vector of the external forces, $\mathbf{f}^{\mathrm{GR}}$ is the vector of the PGRFs, and $\mathbf{f}^{\mathrm{ACTIVE}}$ is the vector of the AGRFs. The expression of $\mathbf{G}$ is [28, 29]:

$$\mathbf{G}(\omega) = \sum_{j=1}^{n} \frac{\boldsymbol{\phi}_j \boldsymbol{\phi}_j^{\mathrm{T}}}{\omega_j^2 - \omega^2 + 2\mathrm{i}\zeta_j \omega \omega_{\mathrm{j}}} \tag{4.2}$$

where ω_{j} is the jth eigenfrequency, ζ_j is the associated non-dimensional damping ratio and $\boldsymbol{\phi}_j$ is the jth mode shape vector (scaled to the unit modal mass) evaluated/measured at discrete points. The superscript T indicates transposition. Finally, n is the (arbitrary) number of considered modes, i is the imaginary unit, and ω is the circular frequency. Since the eigenvector components are known at discrete (n_d) points, the matrix $\mathbf{G}(\omega)$ is an n_d x n_d matrix containing the FRFs for these degrees-of-freedom.

Equation 4.1 can be used to estimate the structural response due to people movement on the structure. To this purpose, the AGRFs and PGRFs have to be estimated.

As for the AGRFs, it is possible to build a database of forces for the considered kind of movement (e.g., as in [17] for people ascending and descending staircases).

As for PGRFs, we use the apparent mass of the pedestrians, which is the frequency response function (FRF) between the acceleration at the contact point between the pedestrian and the structure and the consequent force exerted by the pedestrian on the structure. This FRF can be measured as described in [17, 30–32].

However, we must consider that people change posture during motion and that they move over the structure so that the PGRFs change point of application in time. As for the former point, an equivalent apparent mass M_{eq}^* is defined considering many different postures of the subject during the motion. Therefore, each step is split in P postures and the apparent mass

$M^*_{a,i}$ is estimated (e.g., with experimental tests as in [17]) for each of them. Actually, each $M^*_{a,i}$ can be calculated as the average of the apparent mass of different people in the given posture. Hence, M^*_{eq} is calculated as:

$$M^*_{eq}(\omega) = \sum_{i=1}^{P} \alpha_i M^*_{a,i}(\omega) \tag{4.3}$$

where the weighting coefficients α_i are set in order to describe properly the amount of time spent by the pedestrians in the corresponding posture within the cycle time T (i.e., the time of a single step). In other words, P postures are frozen within the movement cycle and each of them is treated as a static posture.

Thus, each person produces a PGRF related to the apparent mass M^*_{eq}. Then, the PGRFs can be applied to the structure with two different approaches:

1. each PGRF (i.e., one for each pedestrian on the structure) is seen as a moving excitement. Therefore, the whole system is time-variant;
2. a fraction of the apparent mass $m*_{fr}(\omega)$ is calculated:

$$m^*_{fr}(\omega) = \frac{m}{n_d} M^*_{eq}(\omega) \tag{4.4}$$

Where n_d is the number of degrees-of-freedom in which the structure is discretised, and m is the number of pedestrians on the structure. Then, $m^*_{fr}(\omega)$ is applied to each of the n_d degrees-of-freedom. Therefore, the PGRF in each degree-of-freedom can be expressed as:

$$f_i^{GR}(\omega) = m^*_{fr}(\omega)\ddot{x}_i(\omega) = -m^*_{fr}(\omega)\omega^2 x_i(\omega) \tag{4.5}$$

In terms of the full displacement vector $\mathbf{x}(\omega)$, the $\mathbf{f}^{GR}$ can be described as [17, 18]:

$$\mathbf{f}^{GR}(\omega) = \mathbf{W_n H W_n^T x}(\omega) = -\omega^2 m^*_{fr}(\omega)\mathbf{W_n x}(\omega) \tag{4.6}$$

where $\mathbf{W_n}$ is a n_d x m matrix describing the connection of the m subjects with the structure degrees-of-freedom, $\mathbf{H}(\omega)$ is a m x m diagonal matrix containing the fractions of the equivalent apparent mass (i.e. $\mathbf{H}(\omega) = -\mathbf{W_n}\omega^2 m^*_{fr}(\omega)$). Substituting Eq. 4.6 into Eq. 4.1, we obtain (neglecting $\mathbf{f}$):

$$\left[\mathbf{G}^{-1}(\omega) + \omega^2 m^*_{fr}(\omega)\mathbf{W_n}\right]\mathbf{x}(\omega) = \mathbf{G}_{\mathrm{H}}^{-1}(\omega)\mathbf{x}(\omega) = -\mathbf{f}^{\mathrm{ACTIVE}}(\omega) \tag{4.7}$$

where $\mathbf{G_H}(\omega)$ is the n_d x n_d matrix representing the equivalent set of FRFs describing the dynamic behaviour of the joint system composed by the structure and the people. Clearly, the behaviour of this coupled system is an average behaviour because m^*_{fr} is employed.

The second approach of the previous list assumes a fixed form of $\mathbf{G}_{\mathrm{H}}(\omega)$ in time. Therefore, this assumption makes the simulation of the structure response fast and easy under the computational point of view.

The response of the structure to the movement of people can be finally calculated as the convolution between the AGRFs and the unit impulse response functions (IRF) of the coupled system. These IRFs can be found by applying the inverse Fourier transform to the FRFs composing $\mathbf{G}_{\mathrm{H}}(\omega)$. More details about this model can be found in [16–18].

When the number of people on the structure is high, and/or people occupy the structure for long times, the accuracy of this easy-to-apply approach is high [17]. Indeed, in this case the approximation due to the use of $m*_{fr}(\omega)$ results to be acceptable and does not introduce accuracy worsening in result estimation.

The next sections show the performance of this model when applied to the case of few people on the structure for a limited amount of time.

Table 4.1 Modal data identified for the test-structure in the frequency range 0–15 Hz (empty structure)

Mode number	$\omega_j/(2\pi)$ [Hz]	ζ_j [%]
1	7.84	0.33
2	8.89	0.43

Table 4.2 Tests description

Test ID	Type of test	Number of pedestrians
A	Test with pedestrians continuously walking on the stair in loop	1
B	Test with pedestrians crossing the stair once	3 (2 descending the staircase and 1 ascending)

4.3 Tests

This section describes the experimental campaign carried out to validate the model presented in Sect. 4.2. A staircase (made up of steel and marble, length 12.03 m, width 1.80 m, and height 5.22 m) was used as test-structure. The modal parameters of the empty structures were identified by means of experimental modal analysis [33] and the values of the eigenfrequencies and non-dimensional damping ratios are provided in Table 4.1. The modes taken into account are those in the frequency range 0–15 Hz, which is the frequency band where the pedestrians are mostly able to influence the structural dynamics.

The tests carried out with pedestrians crossing the stair were several. Here, we discuss two of them, which are gathered in Table 4.2, because their results are representative also of the results of the other tests. Each test was repeated several times with different pedestrians (for all of them apparent mass curves and AGRF time-histories were stored thanks to dedicated experimental tests). The structural response was measured by means of accelerometers.

The same tests were also simulated using the model described previously. Each test was simulated 100 times, extracting randomly the pedestrians and the AGRFs from the database. This allowed to take into account the natural dispersion of experimental results. Since the results showed Gaussian distributions, the results of the model are described by the mean value of the results of the 100 simulations plus/minus twice the standard deviation of the results (i.e. with a confidence level of about 95% [34]).

The results are expressed in terms of root-mean-square (RMS) of one of the acceleration signals showing the highest structural responses among all the accelerometers used to monitor the structural response. The same degree-of-freedom was considered in the simulation results. Moreover, a moving RMS was calculated as well (the moving RMS was calculated every 3 s). Its maximum value was also taken into account to express experimental and numerical results. This maximum RMS will be named here MRMS.

The RMSs and MRMSs were calculated in the frequency range 0–12 Hz.

4.4 Results

Figures 4.2 and 4.3 show the results for test A and test B, respectively. It is evident that when the staircase is occupied for long times (i.e. test A, Fig. 4.2), the model is still able to predict the vibration levels, both in terms of RMS and MRMS. Indeed, the order of magnitude of experimental and numerical result is the same (there are just few experimental results below the expected RMSs). Conversely, when the time length of the test is short (i.e. test B, Fig. 4.3), the numerical results clearly overestimate the experimental RMSs. As for the MRMSs, the model still overestimates the experiments because the interval provided by the model stretches over high MRMS values. However, the overestimation is not as much as in the case of the RMS.

Therefore, when long-time tests are taken into account, the model is able to correctly estimate RMSs and MRMSs. As for short tests, the model is still able to describe the vibration of the structure occupied by pedestrians. Indeed, the order of magnitude of RMSs and MRMSs values is correct and many experimental results fall into the intervals provided by the model. However, overestimations of both RMSs and MRMSs are evident.

This means that an accurate estimation of the vibration levels under operating conditions in case of transient excitation (i.e. situations comparable to the short test discussed here) would require the use of a more complex model able to properly describe the evolution of the PGRFs. This can be accomplished by developing time domain models where the PGRFs are described in time and space. Therefore, a model based on time integration is under development by the authors of the present paper. This model will be able to describe the position and the effect of each pedestrian on the structure as function of time. Such an approach is expected to produce results closer to the experimental evidence when transients are taken into account.

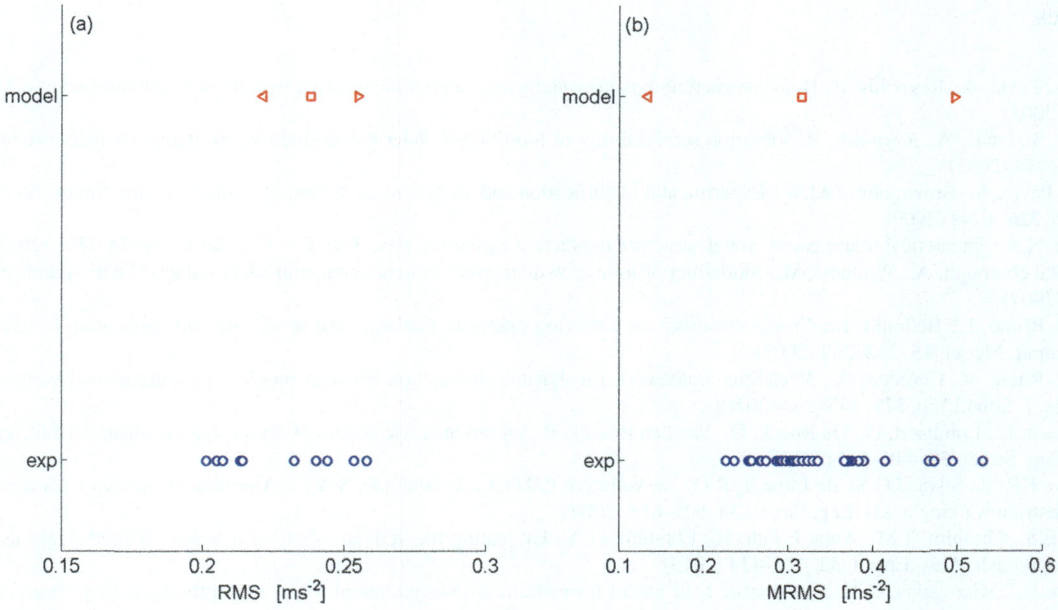

Fig. 4.2 Experimental and numerical results for test A. Numerical results are expressed in terms of mean value (square) plus/minus twice the standard deviation (triangles)

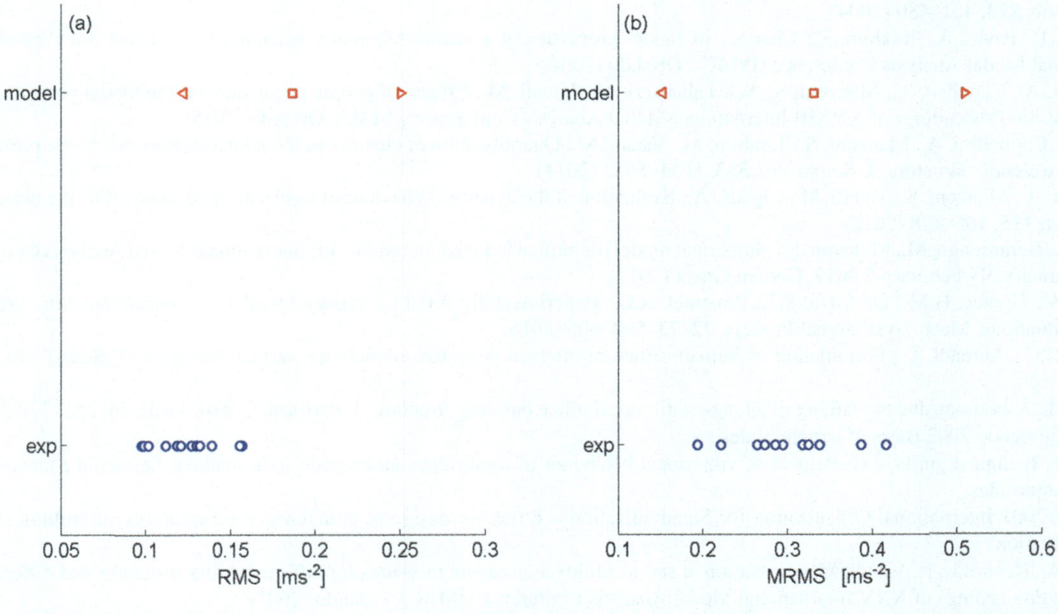

Fig. 4.3 Experimental and numerical results for test B. Numerical results are expressed in terms of mean value (square) plus/minus twice the standard deviation (triangles)

4.5 Conclusions

The paper has dealt with the problem of human-structure interaction. Particularly, the issue treated here has regarded a model to describe the vibration of light structures (e.g. staircases) proposed recently by the authors of this paper. This model was developed with the goal of being accurate in case of a high number of pedestrians on the structure for long times. The present paper has analysed the behaviour of the same model in the case of transients and few people on the structure. This has allowed to assess the accuracy of the model also in this further situation. The main outcome is the need of a more refined model for transient situations, which must be able to properly describe the contribution of the passive ground reaction forces in time and space.

References

1. Sachse, R., Pavic, A., Reynolds, P.: Human-structure dynamic interaction in civil engineering dynamics: a literature review. Shock Vib. Dig. **35**, 3–18 (2003)
2. Živanović, S., Pavic, A., Reynolds, P.: Vibration serviceability of footbridges under human-induced excitation: A literature review. J. Sound Vib. **279**, 1–74 (2005)
3. Racic, V., Pavic, A., Brownjohn, J.M.W.: Experimental identification and analytical modelling of human walking forces: literature review. J. Sound Vib. **326**, 1–49 (2009)
4. Alexander, N.A.: Theoretical treatment of crowd–structure interaction dynamics. Proc. Inst. Civ. Eng. Struct. Build. **159**, 329–338 (2006)
5. Sim, J., Blakeborough, A., Williams, M.: Modelling of joint crowd-structure system using equivalent reduced-DOF system. Shock. Vib. **14**, 261–270 (2007)
6. Venuti, F., Bruno, L., Bellomo, N.: Crowd dynamics on a moving platform: mathematical modelling and application to lively footbridges. Math. Comput. Model. **45**, 252–269 (2007)
7. Venuti, F., Racic, V., Corbetta, A.: Modelling framework for dynamic interaction between multiple pedestrians and vertical vibrations of footbridges. J. Sound Vib. **379**, 245–263 (2016)
8. Van Nimmen, K., Lombaert, G., De Roeck, G., Van den Broeck, P.: Vibration serviceability of footbridges: evaluation of the current codes of practice. Eng. Struct. **59**, 448–461 (2014)
9. Figueiredo, F.P., da Silva, J.G.S., de Lima, L.R.O., da Vellasco, P.C.G.S., de Andrade, S.A.L.: A parametric study of composite footbridges under pedestrian walking loads. Eng. Struct. **30**, 605–615 (2008)
10. Mashaly, E.S., Ebrahim, T.M., Abou-Elfath, H., Ebrahim, O.A.: Evaluating the vertical vibration response of footbridges using a response spectrum approach. Alex. Eng. J. **52**, 419–424 (2013)
11. Ingólfsson, E.T., Georgakis, C.T.: A stochastic load model for pedestrian-induced lateral forces on footbridges. Eng. Struct. **33**, 3454–3470 (2011)
12. Piccardo, G., Tubino, F.: Equivalent spectral model and maximum dynamic response for the serviceability analysis of footbridges. Eng. Struct. **40**, 445–456 (2012)
13. Sachse, R., Pavic, A., Reynolds, P.: Parametric study of modal properties of damped two-degree-of-freedom crowd-structure dynamic systems. J. Sound Vib. **274**, 461–480 (2004)
14. Reynolds, P., Pavic, A., Ibrahim, Z.: Changes of modal properties of a stadium structure occupied by a crowd. In: Proceedings of XXII International Modal Analysis Conference (IMAC), Orlando (2004)
15. Cappellini, A., Cattaneo, A., Manzoni, S., Scaccabarozzi, M., Vanali, M.: Effects of people occupancy on the modal properties of a stadium grandstand. In: Proceedings of XXXIII International Modal Analysis Conference (IMAC), Orlando (2015)
16. Busca, G., Cappellini, A., Manzoni, S., Tarabini, M., Vanali, M.: Quantification of changes in modal parameters due to the presence of passive people on a slender structure. J. Sound Vib. **333**, 5641–5652 (2014)
17. Cappellini, A., Manzoni, S., Vanali, M., Cigada, A.: Evaluation of the dynamic behaviour of steel staircases damped by the presence of people. Eng. Struct. **115**, 165–178 (2016)
18. Vanali, M., Berardengo, M., Manzoni, S.: Numerical model for human induced vibrations. In: International Modal Analysis Conference, IMAC XXXV, January 30–February 2 2017, Garden Grove (2017)
19. Toso, M.A., Gomes, H.M., Da Silva, F.T., Pimentel, R.L.: Experimentally fitted biodynamic models for pedestrian-structure interaction in walking situations. Mech. Syst. Signal Process. **72–73**, 590–606 (2016)
20. Caprani, C.C., Ahmadi, E.: Formulation of human–structure interaction system models for vertical vibration. J. Sound Vib. **377**, 346–367 (2016)
21. Setareh, M.: Vibrations due to walking in a long-cantilevered office building structure. J. Perform. Constr. Facil. **26**, 255–270 (2012)
22. EN1990-Eurocode 2002 Basis of structural design
23. Sètra 2006 Technical guide – assessment of vibrational behaviour of footbridges under pedestrian loading. Service d'Etudes techniques des routes et autoroutes
24. ISO10137 2007 International Organization for Standardization – Bases for design of structures – serviceability of buildings and walkways against vibration
25. Caprioli, A., Reynolds, P., Vanali, M.: Evaluation of serviceability assessment measures for different stadia structures and different live concert events. In: Proceedings of XXV International Modal Analysis Conference (IMAC), Orlando (2007)
26. Caprioli, A., Vanali, M.: Comparison of different serviceability assessment measures for different events held in the G. Meazza stadium in Milano. In: Proceedings of XXVII International Modal Analysis Conference (IMAC), Orlando (2009)
27. Institution of Structural Engineers 2008 Dynamic performance requirements for permanent grandstands subject to crowd action: recommendations for management, design and assessment
28. Ewins, D.J.: Modal Testing: Theory, Practice and Application. Research Studies Press Ltd., Baldock (2000)
29. Brandt, A.: Noise and Vibration Analysis – Signal Analysis and Experimental Procedures. Wiley, Chichester (2011)
30. Tarabini, M., Solbiati, S., Saggin, B., Scaccabarozzi, D.: Setup for the measurement of apparent mass matrix of standing subjects. IEEE Trans. Instrum. Meas. **65**, 1856–1864 (2016)
31. Matsumoto, Y., Griffin, M.J.: Mathematical models for the apparent masses of standing subjects exposed to vertical whole-body vibration. J. Sound Vib. **260**, 431–451 (2003)
32. Matsumoto, Y., Griffin, M.J.: Dynamic response of the standing human body exposed to vertical vibration: influence of posture and vibration magnitude. J. Sound Vib. **212**, 85–107 (1998)
33. Peeters, B., Van Der, A.H., Guillaume, P., Leuridan, J.: The PolyMAX frequency-domain method : a new standard for modal parameter estimation? Shock. Vib. **11**, 395–409 (2004)
34. JCGM 100:2008 2008 Evaluation of measurement data — Guide to the expression of uncertainty in measurement

Chapter 5
On Stationarity and the Interpretation of the ADF Statistic

K. Worden, I. Iakovidis, and E. J. Cross

Abstract The paper considers the nature of stationarity of a time series or signal, and how it may be quantified. It is argued that a subjective assessment is as effective as one based on mathematical definitions, if one actually has finite samples of data, and that such an assessment is fundamentally based on the number of cycles of the dominant periodic component visible in the sample. It is shown by dimensional analysis that one of the most often-used measures of stationarity – the Augmented Dickey-Fuller (ADF) statistic – supports this hypothesis. The paper should be of interest not just to engineers, but also to econometricians, or anyone concerned with time series analysis and the impact of nonstationarity.

Keywords Time series · Stationarity · Augmented Dickey-Fuller (ADF) statistic · Dimensional analysis

5.1 Introduction

The concept of *prediction* is central to science and engineering, also to economics and those of the social sciences that admit a mathematical description. However, as Neils Bohr is often quoted as saying: 'prediction is very difficult, especially about the future' [1]. Where prediction becomes extremely difficult, if not intractable, is when the physical (or otherwise) laws of the system of interest are changing with time. Determining if this is the case or not, is, therefore, an extremely important first step in deciding how one might model the system, in order to make predictions. In many situations of interest in engineering, one does not know the exact physical laws, and one is forced to try and infer some or all of them from observed data – this is essentially the problem of system identification [2]. The problem of deciding whether a system is changing with time, is thus 'reduced' to the problem of deciding if given *signals* are changing with time, and this leads directly to the concept of *stationarity* [3]. The idea of stationarity really originates in the study of random or stochastic processes, so the general definition of the term will be given in that context for now.

Suppose the set of observed signals is a vector time series $\underline{x}(t)$ governed by a stochastic process $\underline{X}_t$. The process is *strictly stationary* if the joint probability density $p_t(\underline{x}(t))$, is independent of time i.e.

$$p_t(\underline{x}(t)) = p_{t+\tau}(\underline{x}(t+\tau)), \qquad \forall \tau \tag{5.1}$$

where the variables (t, τ) can be discrete or continuous. This definition is beautiful from a mathematical point of view, but makes very little contact with engineering reality, as it is framed in terms of quantities which are not known, and in fact, cannot be known. This fact means that other indicators of stationarity are necessary, indicators that depend only on quantities that can be observed. One of the main objectives of this paper is to discuss the limitations of real (practical) measures of stationarity.

In the recent past, the idea of stationarity has become central to another engineering discipline, that of *Structural Health Monitoring* (SHM) [4]. The basic idea of SHM is to infer from measured data, whether a system or structure has become damaged during operation. For various reasons (discussed in detail in [4]), SHM is often implemented in practice by trying to determine if the measured data changes its nature at some point (i.e. when the structure moves from the undamaged state to a damaged state). Such a change is clearly an instance of nonstationarity; however, the problem is complicated by the fact that signals from structures may be nonstationary for entirely benign reasons. To give a concrete (no pun intended) example, consider a bridge to be the structure of interest. Suppose that the measurands/features of interest are the first few natural frequencies; these are known to be sensitive to damage because reduced structural stiffness (say, as the result of

K. Worden · I. Iakovidis (✉) · E. J. Cross
Dynamics Research Group, Department of Mechanical Engineering, University of Sheffield, Sheffield, UK
e-mail: iiakovidis1@sheffield.ac.uk

S. Pakzad (ed.), *Dynamics of Civil Structures, Volume 2*, Conference Proceedings of the Society for Experimental Mechanics Series, https://doi.org/10.1007/978-3-319-74421-6_5

cracks) will cause the natural frequencies to be reduced. Unfortunately, these same frequencies are also sensitive to the ambient temperature of the bridge, the traffic loading etc. These benign operational and environmental variations are known as *confounding influences* and represent one of the main technological problems in SHM implementation. Various methods have been proposed for the elimination of confounding influences [4]; one of the most powerful is based on the concept of *cointegration* [5, 6].

The theory of cointegration arose in the domain of econometrics, where it was mainly used in order to analyse and understand common trends in economic and financial time series. However, one application of the theory is to combine multiple nonstationary time series in such a way as to produce a stationary residual; this is the application that proves to be useful for SHM. The idea is that confounding influences in structural response measurements often manifest as common trends between signals i.e. the daily variations in temperature of a bridge will cause slowly-varying (compared to the dynamics) common trends in estimated natural frequencies; cointegration allows combination of the signals in such a way that the result is stationary, and sensitive only to changes due to structural damage (or previously unseen benign effects). One of the necessities of the cointegration technique, is a means of deciding if a signal is stationary or not, and this is often based on a hypothesis test using specialised statistics; the most common statistic being the *Augmented Dickey-Fuller* (ADF) statistic [7, 8]. In fact, the ADF statistic works in the abstract sense by detecting unit roots in the set of characteristic roots of the time series of interest. An important objective of this paper is to provide a simple interpretation of the ADF statistic and show that it makes contact with engineering intuition about the nature of stationarity. In fact, it will be argued that the decision on whether a time series is stationary or not is a matter of subjective judgement.

The layout of the paper is as follows: Sect. 5.2 will discuss the nature of stationarity and a hypothesis for how a human expert might judge stationarity. In Sect. 5.3, it is demonstrated that the ADF test statistic is a function of the dominant frequency of a mono-harmonic signal, its length and its signal to noise ratio, via a dimensional analysis. In Sect. 5.3.2 simulations are used to illustrate this point. Finally discussions and conclusions follow.

5.2 Practical Tests for Stationarity

As observed above, the mathematical definition of strict stationarity for a stochastic process is quite straightforward. However, it is framed in terms of multivariate density functions, which are notoriously difficult to estimate from data [9, 10]. To have any hope of practical success, a test for stationarity needs to be formulated in terms of quantities that are easier to estimate. Fortunately, this is possible. Suppose one is interested in detecting damage via a *control chart* [11], then one might be satisfied if the signal of interest is *weakly stationary*.[1] A signal is weakly stationary [12] if the mean of the process is constant i.e.,

$$\underline{m}(t) = E[\underline{X}_t] = \underline{m} \ \ (constant) \tag{5.2}$$

(where E denotes the expectation operator) and the auto-correlation function of process depends only on time differences, i.e.

$$E[\underline{X}_s \underline{X}_t] = \phi_{\underline{xx}}(s,t) = \phi_{\underline{xx}}(s-t) \ \ \forall s,t. \tag{5.3}$$

The second of these conditions is often replaced by the requirement that the covariance of the process, like the mean, is constant.

This is a step forward, but in general doesn't help a great deal because one does not know $\underline{m}(t)$ or $\phi_{\underline{xx}}$; one can only compute sample estimates from *finite* samples of data. (In fact, one also needs to assume *ergodicity* in order to equate ensemble statistics with time averages, but ergodicity is more difficult to establish than stationarity in practical terms.) Estimates of the mean and variance from a finite sample will of course be a constant, they can be nothing else. One can try and track the mean and variance over a sequence of moving windows, but the width of the window is then to be determined, again from the finite sample available. These considerations do not really matter if the signals of interest have been generated as stationary; however, the problem is that many of the signals of engineering interest are not. This is clearly an issue if one is interested in applying technology, like control charts for SHM, which require stationarity.

[1] Things are a little more complicated than this, detection using a control chart depends critically on the tails of the signal probability density, and these are independent/insensitive to the central statistics controlled in weak stationarity.

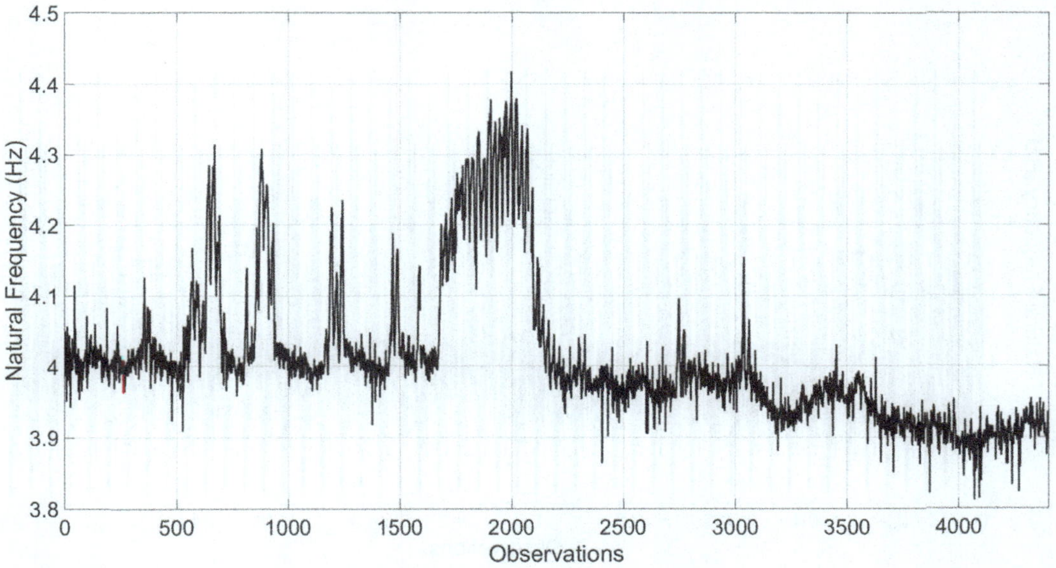

Fig. 5.1 First natural frequency of the Z24 Bridge, monitored over a single year

Typically, the signals of interest in SHM (i.e. damage-sensitive features polluted by confounding influences) are not stationary, but are at least *periodically-correlated* or *cyclostationary* [12]. A second-order stochastic process $\underline{X}_t$ is periodically correlated if for all s, t,

$$\underline{m}(t) = \underline{m}(t + T) \tag{5.4}$$

and,

$$\phi_{\underline{xx}}(s, t) = \phi_{\underline{xx}}(s + T, t + T) \tag{5.5}$$

for some *period* T, and there are no smaller values of $T > 0$ for which these conditions hold. Again, it is a difficult matter to apply these criteria directly to a finite sample of observed data.

At this point, it is useful to consider an illustration. Figure 5.1 shows the estimated (first) natural frequency of the well-known Z24 Bridge over an entire year of monitoring [13]. There are two main contributors to the signal. The first component is a high-frequency *noise* due to the variance in the frequency estimation; the second comes from large artefacts resulting from very low temperatures, during which the asphalt in the bridge deck stiffened [13]. Based on engineering judgement, an analyst would almost certainly conclude that the time series presented is nonstationary. Now, supposing that the cold periods in the region concerned occur at roughly the same points each year, one could simulate the effect of observing the bridge for 50 years, by repeating the one year data 50 times, as in Fig. 5.2. The analyst would almost certainly now say that the data are stationary; in fact, as the annual period is clearly visible, they would probably judge the series to be at least cyclostationary. Both of these judgments would clearly be subjective, but it is probable that they would be made by any expert.

The problem in SHM is that one wishes to detect damage from signals like the natural frequencies by detecting changes from a stationary response using tools like control charts. Our current best approach is to look at features/data measured over some finite window and to try and make a judgement as to their stationarity; if they are stationary, it is valid to use control charts etc.; if they are nonstationary, one needs to use cointegration first etc. It has already been noted that it is difficult to apply mathematical definitions of stationarity to finite samples of data, and it could be argued that the subjective judgements or engineers/experts might be just as reliable. However, engineers and scientists (rightly) distrust subjectivity as a rule, and they have produced summary statistics in order to try and render the judgement objective. One of the most commonly-used measures of stationarity is the ADF statistic [7, 8]; although this will be the main subject of this paper, alternatives include the Phillips-Perron [14] and KPSS [15] tests. All of these tests are based on a *unit root* hypothesis. The latter is a technical condition based on fitting an AR (auto-regressive) model of the time series of interest and computing the characteristic roots of the model [16]; if a unit root is found – indicating marginal stability of the model – the series is judged to be nonstationary.

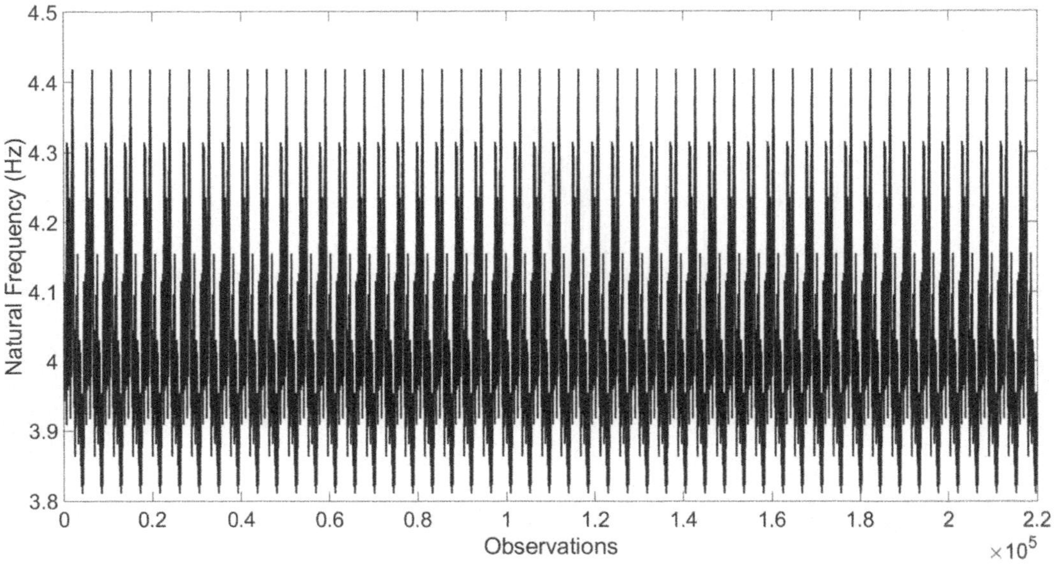

Fig. 5.2 First natural frequency of the Z24 Bridge, the single year repeated 50 times to simulate measurements over a 50-year period

Various criticisms of the unit root hypothesis can be found in the literature [15, 17, 18]. Criticisms are based on the fact that the unit root will never be found with complete precision and that a subjective judgement of closeness will be needed, and that there is usually too little information in the data to make the call with confidence.

Given that expert judgement will often be an excellent indicator of the nature of a signal, if scientists and engineers are to place their trust in proxies like the ADF statistic, they will do so with more confidence if the proxies agree with their intuitions about signals. So the question now is, what is the basis of expert intuition on the nature of a finite sample of a time series. It seems to the authors, that the main criterion would be: how may periods of the dominant frequency are present in the sample? At one extreme, an expert presented with data from one period or less of the dominant cyclostationary process (it will be assumed here that true noise is easily recognisable) will probably judge the process nonstationary. At the other extreme, an expert presented with many periods of the dominant frequency would probably judge it cyclostationary or stationary with confidence. If one accepts that this is the basis of expert judgement, one can now proceed to some analysis as to whether proxies like the ADF test respect this basis.

It has been observed in the literature that considerations affecting unit root tests include: the size of the sample, the sampling frequency and the overall time duration of the sample [19–24]. These observations will be captured in the simple dimensional argument presented in the following section.

5.3 Dimensional Analysis of the ADF Statistic

5.3.1 Analysis

In order to carry out the analysis of this section, a number of simplifying assumptions will be made about the signals under consideration. The basic signal studied will be,

$$x(t) = A \sin\left(\frac{2\pi t}{T}\right) + \epsilon(t) \tag{5.6}$$

where ϵ is a zero-mean white Gaussian noise sequence with variance σ_ϵ^2. Samples will be considered to have number of points N, and to be sampled every Δt seconds. The sampling frequency will thus be $f_s = 1/\Delta t$ and the dominant frequency will be $f = 1/T$. The *normalised frequency* corresponding to f will be defined by $f_n = f/f_s$.

Now, the details are not important here (the interested reader can consult [16]), but the ADF statistic is calculated by estimating a parameter θ of an AR model, and dividing by the standard error of that estimate, i.e.,

$$t_{ADF} = \frac{\overline{\theta}}{\text{se}(\overline{\theta})} \tag{5.7}$$

In order to judge stationarity, this statistic is compared to a critical value t_c; if the statistic is lower than the critical value, the original time series is judged to be stationary. The critical values are not standard for t-statistics, but were computed by Dickey and Fuller.

Now, the only thing that is needed for the following analysis, is that the t_{ADF} as defined, is *dimensionless*. This means that, by the Buckingham-Pi theorem [25], one has the relation,

$$t_{ADF} = \pi_1 = f(\pi_2, \dots, \pi_n) \tag{5.8}$$

where $\pi_2 \dots \pi_n$ are any other appropriate dimensionless groups defined for the problem of interest.

The quantities of interest in the problem under consideration are: T (the dominant period in the time series), Δt (the sampling interval), N (the number of points in the sample), σ_y^2 (the signal variance) and σ_ϵ^2 (the noise variance). Despite the fact that there are five variables of interest, only two dimensions are represented: time $[T]$ and 'length' (amplitude) $[L]$. This fact means that two primary variables for the dimensional analysis are needed; these are taken as noise variance σ_ϵ^2 ($[L]$) and sampling interval Δt ($[T]$). This, in turn, means that three secondary variables or dimensionless groups are possible, formed from T, N (already dimensionless) and σ_y^2. A valid set is immediately visible as: $\pi_2 = \Delta t/T = f/f_s$, $\pi_3 = N$ and $\pi_4 = \sigma_y^2/\sigma_\epsilon^2$. It follows from the Buckingham-Pi theorem that,

$$t_{ADF} = f(\frac{\Delta t}{T} = \frac{f}{f_s}, N, \frac{\sigma_y^2}{\sigma_\epsilon^2}) \tag{5.9}$$

where the first variable is simply the normalised frequency of the dominant periodic component and the third is a signal-to-noise ratio. Equation (5.9) identifies an exhaustive set of variables that the ADF statistic can depend on. The form of the dependence, in each case, can be investigated by simple simulations.

5.3.2 *Results*

In the first group of simulations, the SNR (signal-to-noise ratio) is set at 100 (a little noise is needed or there appear to be conditioning problems in fitting the AR models for the ADF statistic), T and ΔT are fixed by generating cycles of the sine wave with 32 points per cycle. This leaves only N to vary, and this is accomplished simply by concatenating cycles of the sine wave, from 1 to 20. Figure 5.3 shows the sample with 20 cycles, separated into the periodic component (black) and noise component (red).

Figure 5.4 shows the ADF statistic plotted against the number of cycles of the dominant periodic component. In agreement with the intuition that greater stationarity is supported for more periods of the sine wave, the ADF statistic falls monotonically with number of cycles. This simulation motivates two points:

1. Instead of using $\pi_3 = N$, it is more meaningful to use $\pi_3 = N\Delta t/T$ as this simply counts the number of cycles of the dominant period in the sample. This is the quantity which was identified above as the prime mover in intuition about stationarity. The ADF statistic falls monotonically with this quantity as expected.
2. Why is this happening? The ADF statistic is proportional to an AR coefficient estimate. However, adding the data in the form of more identical cycles does not change the underlying AR process. One needs to recall at this point, that the ADF statistic divided the parameter estimate by its standard error. Adding data, i.e. increasing N, means that the standard error will decrease monotonically (although it will eventually hit the Cramer-Rao bound and saturate). This fact means that the ADF will grow in magnitude i.e. become more negative, and judge the sample to be more stationary.

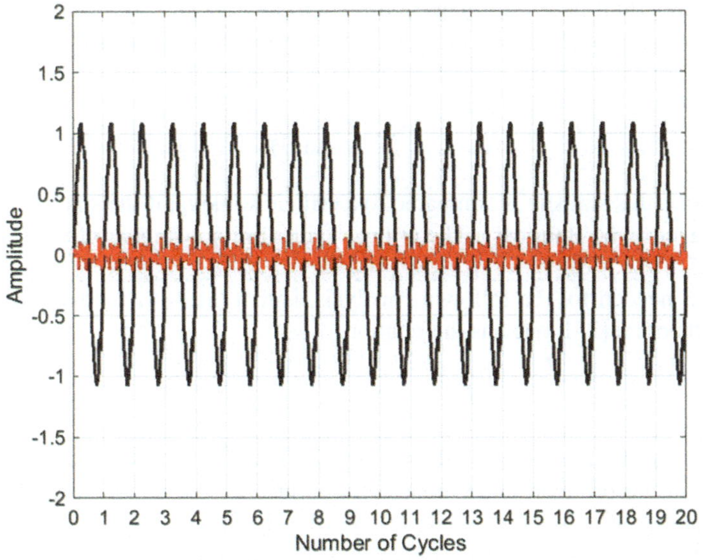

Fig. 5.3 Twenty cycles of the slightly noisy sine wave, separated into the periodic component (black) and the noise component (red)

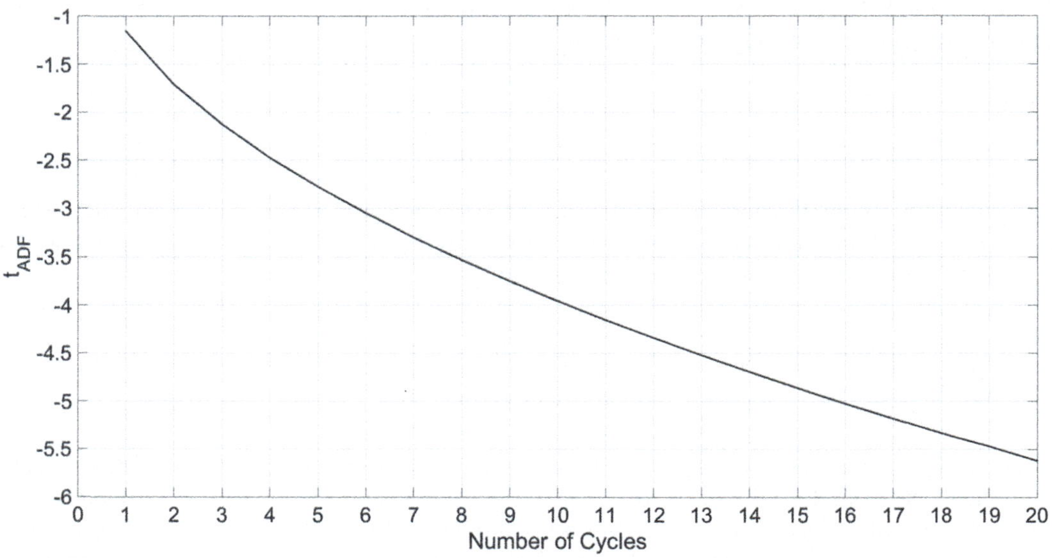

Fig. 5.4 ADF statistic as a function of the number of cycles of the dominant periodic component in a sample

So, now, the basic relation is,

$$t_{ADF} = f(\frac{f}{f_s}, \frac{N\Delta t}{T}, \frac{\sigma_y^2}{\sigma_\epsilon^2}) \tag{5.10}$$

The second set of simulations are concerned with varying f/f_s. Clearly one can increase the number of periods of the dominant frequency in a sample of fixed length by increasing that frequency or by decreasing the sampling frequency. Under the hypothesis expressed above, the signal would then appear more stationary and one would expect the ADF statistic to decrease. First, the SNR is fixed as before, and the number of points N, is fixed at 1024. The sampling frequency of the sample is varied by moving from one extreme when all 1024 points form one sine wave cycle, to the other extreme when 32 cycles are present, sampled at 32 points each; the extreme cases are depicted in Fig. 5.5. The sample with the highest sampling has an ADF statistic of -0.7702 which is judged nonstationary by comparison with the critical value $t_c = -1.94$ (at 95% confidence); the sample with the lowest sampling rate has an ADF statistic of -9.5011, and is thus judged stationary. Figure 5.6 shows that the ADF statistic decreases monotonically as the sampling rate decreases (or conversely as the dominant

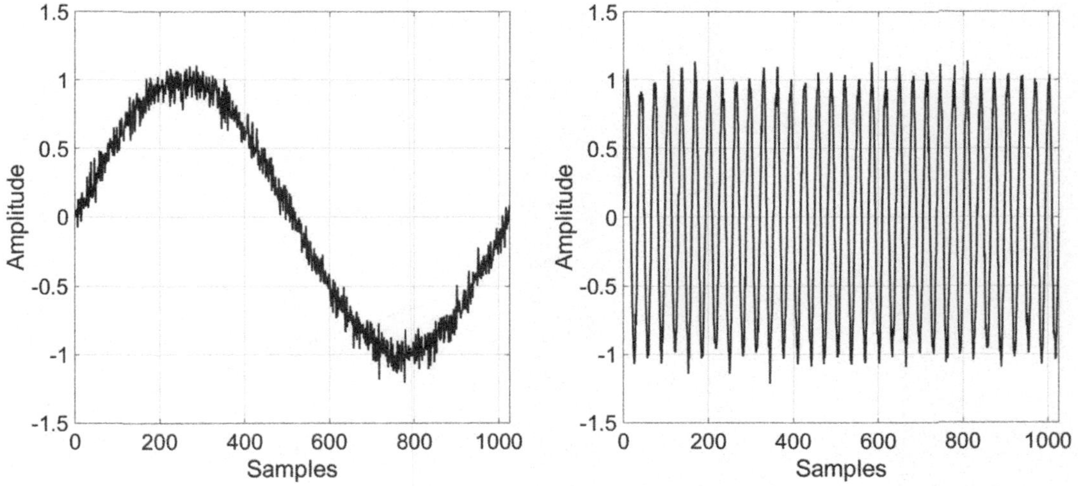

Fig. 5.5 Two extreme cases of sine waves sampled over 1024 points, representing one and 32 cycles respectively

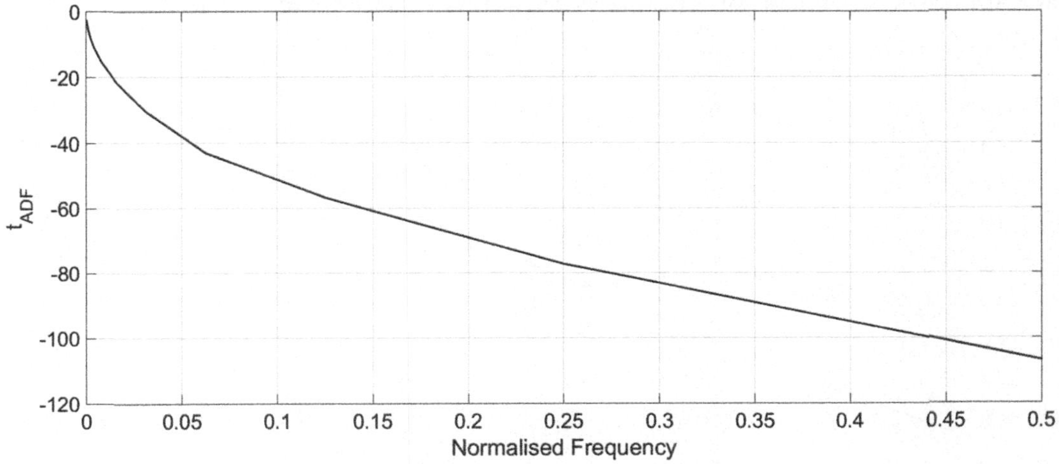

Fig. 5.6 ADF statistic versus normalised frequency

frequency increases) and this is once more consistent with the hypothesis that the ADF statistic essentially counts the number of cycles/periods of the dominant frequency component in a sample.

To illustrate in a little more detail, keeping the SNR equal to 100 and $N = 4096$, the number of cycles is set at 2^C for $C = 1, \ldots, 11$. Figure 5.6 shows that the ADF statistic decreases monotonically as the dominant frequency increases (or conversely as the sampling frequency decreases) and this is once more consistent with the hypothesis that the ADF statistic essentially counts the number of cycles/periods of the dominant frequency component in a sample.

The last two simulations reveal something rather interesting. For a fixed N, the ADF statistic falls monotonically with increased normalised frequency of the dominant component. This implies that, corresponding to a given critical ADF statistic t_c, there is a critical normalised frequency f_c such that, signals with normalised frequencies greater than f_c will be judged stationary, while those with $f < f_c$ will be judged nonstationary. For $N = 4096$ the critical ADF statistic is -1.94, and this corresponds to a critical normalised frequency of 0.00008 (Fig. 5.7). This frequency corresponds to less than one full cycle of a sine wave, and is indicated on Fig. 5.7 by linear extrapolation. Given the sampling frequency in any given situation, the critical frequency in Hz is easily derived from the critical normalised frequency, which is a function of N.

The final effect on the ADF statistic to consider is that of amplitude. As the statistic is dimensionless, it does not depend directly on amplitude, but only through the SNR, $\sigma_y^2/\sigma_\epsilon^2$. In the simplified signals presented here, this is the ratio of the periodic signal power to the noise power. The signal will thus be judged as stationary if the noise dominates and nonstationary if the periodic component dominates. Stationarity will thus increase monotonically with SNR, and this is supported by computation as shown in Fig. 5.8.

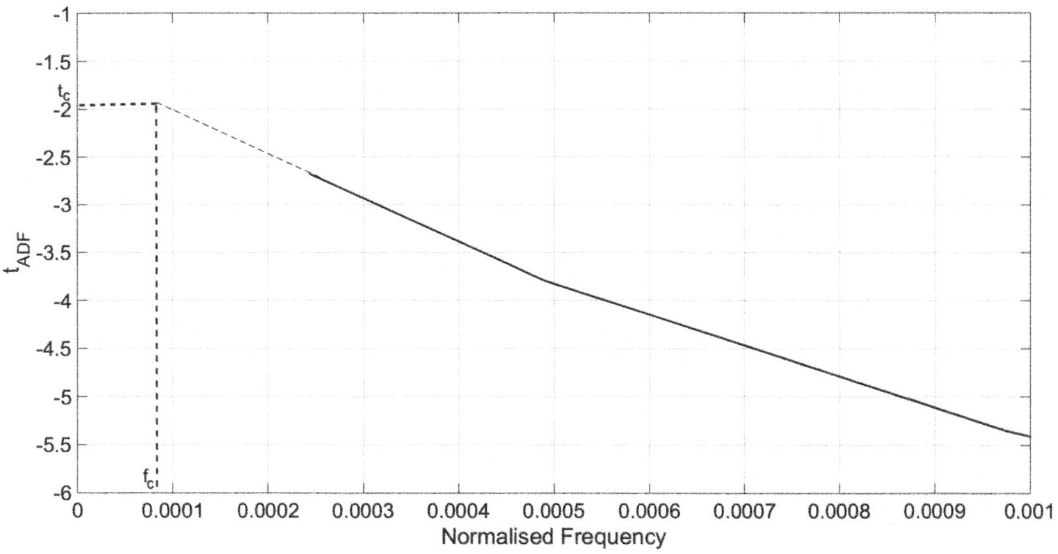

Fig. 5.7 Zoom of ADF statistic versus normalised frequency, indicating presence of critical normalised frequency

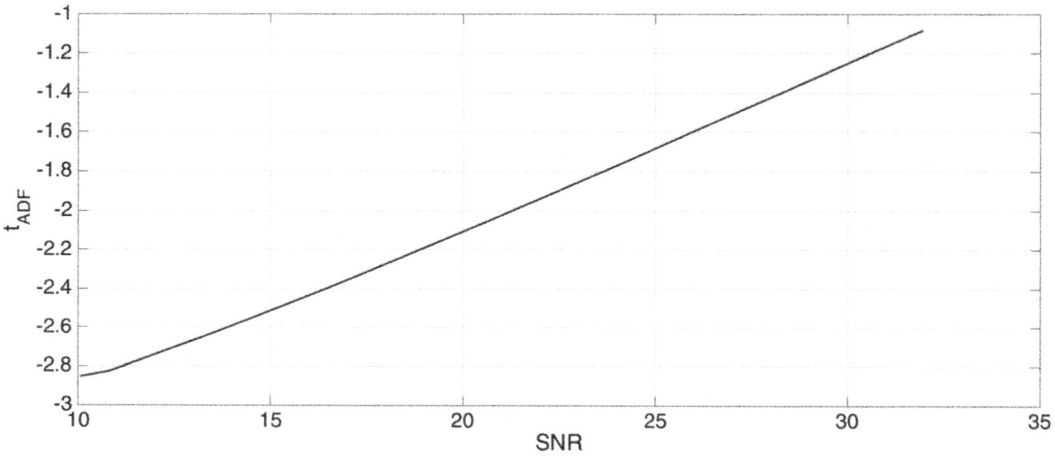

Fig. 5.8 ADF statistic versus signal-to-noise ratio

Of course, one could also define a critical normalised frequency with respect to SNR; however, this would be less interesting as it would apply only to the very restricted case of the signal being a noisy sine wave. In the general case, the ADF statistic will depend on the relative sizes of periodic components and the noise components. If periodic components with frequencies below the critical frequency dominate other components, one would expect the ADF statistic to judge the signal stationary. If the signal is a mixture of a narrowband random process and white noise, it might be possible to extend the arguments presented here by defining the 'dominant frequency' in terms of the peak in the spectrum or as the frequency centroid for example.

5.3.3 Discussion

The results so far support the hypothesis that the ADF statistic essentially counts the number of cycles of the dominant periodic component visible in the sample; it falls monotonically as this quantity increases. A second result of the paper is that there exists (at least for monocomponent signals + noise), a critical frequency f_c corresponding to the critical value of the ADF statistic for a fixed number of samples. This point requires a little more discussion, due to the fact that there is more than one critical ADF value. The critical value depends on the type of model assumed for the underlying process [16]. To simplify matters, the discussion here will be restricted to the Dickey-Fuller (DF) statistic, rather than the ADF, but the point is the same. In the DF test, three common models are considered:

1. No constant, no trend: this represents a random walk with no drift. the model structure is,

$$y_i = \rho y_{i-1} + \varepsilon_i \tag{5.11}$$

   where ε_i is the residual sequence.
2. Constant, no trend: the model is,

$$y_i = \rho y_{i-1} + \alpha + \varepsilon_i \tag{5.12}$$

   where α is a constant offset.
3. Constant and trend: the model is,

$$y_i = \rho y_{i-1} + \alpha + \beta t + \varepsilon_i \tag{5.13}$$

   where both α and β are constants.

Each of these models will generate a different t_c, and the same is true of the ADF test, which really only adds certain AR terms to DF model structures. So, for the situations above where the samples contain $N > 500$ points, and at 95% confidence, the ADF t_c values for the corresponding types of models (no constant, no trend etc.) are (1) −1.94, (2) −2.86, (3) −3.41. None of these models seem to completely fit the situation here where the deterministic components of the time series are periodic; however, model (1) has been chosen for its simplicity. The different choices do not affect the main conclusions of the paper relating to the monotonic behaviour of the ADF statistic, but they will have a small affect on the numerical value of f_c.

One can also make a general observation in the context of SHM. Consider the case when one is measuring the dynamic response of a structure. In the absence of any confounding influences, this would usually be stationary (assuming stationary excitation). If the structure suffers a damaging event, this usually has the effect of reducing the structural stiffness and thus the natural frequencies. In this case, the response would have marginally fewer periods of the dominant frequencies than for the undamaged structure measured over the same time window. With the interpretation of the ADF statistic in this paper, it would follow that the ADF statistic of the response would increase after damage, although this is likely to be a small effect unless there are major changes in the frequencies i.e. large damage. The use of the ADF statistic as a damage sensitive feature has actually already been proposed [26].

5.4 Conclusions

For various reasons, relating to applications in engineering like structural health monitoring (SHM) and system identification, this paper speculates on the nature of stationarity and how it is quantified. It is briefly argued that mathematical definitions are of little practical use on finite samples of data and that a subjective assessment by an expert may be as effective. A hypothesis is presented that experts would base their assessment on how many cycles of the dominant periodic component are present in the sample. If this hypothesis were true, then any statistics used for the quantitative assessment of stationarity would need to support it. Via a dimensional analysis argument, the paper shows that the this is the case specifically for the ADF statistic, which is one of the most commonly-used statistics used for assessing stationarity. It should be a simple matter to apply similar dimensional arguments to other measures e.g. the Phillips-Perron statistic. The results of this paper should be of interest, not only to engineers, but also to econometricians, or anyone serious concerned with time series analysis where stationarity is an issue.

Acknowledgements KW would like to acknowledge the support of the UK Engineering and Physical Sciences Research Council (EPSRC) through grant reference numbers EP/J016942/1 and EP/K003836/2.

References

1. Ellis, A.K.: Teaching and Learning Elementary Social Studies, 9th edn. Pearson, Boston (2010)
2. Worden, K., Tomlinson, G.R.: Nonlinearity in Structural Dynamics: Detection, Identification and Modelling. Institute of Physics Press, Philadelphia (2000)
3. Fuller, W.A.: Introduction to Statistical Time Series. Wiley Interscience, New York (1976)
4. Farrar, C.R., Worden, K.: Structural Health Monitoring: A Machine Learning Perspective. Wiley, New York (2012)
5. Cross, E.J., Worden, K., Chen, Q.: Cointegration: a novel approach for the removal of environmental trends in structural health monitoring data. Proc. R. Soc. Ser. A **467**, 2712–2732 (2011)
6. Cross, E.J., Manson, G., Worden, K., Pierce, S.G.: Features for damage detection with insensitivity to environmental and operational variations. Proc. R. Soc. Ser. A **468**, 4098–4122 (2012)
7. Dickey, D.A., Fuller, W.A.: Distribution of the estimators for autoregressive time series with a unit root. J. Am. Stat. Assoc. **74**, 427–431 (1979)
8. Dickey, D.A., Fuller, W.A.: Likelihood ratio statistics for autoregressive time series with a unit root. Econometrica J. Econ. Soc. **49**, 1057–1072 (1981)
9. Silverman, B.W.: Density Estimation for Statistics and Data Analysis. Monographs on Statistics and Applied Probability. Chapman and Hall, London (1986)
10. Scott, D.: Multivariate Density Estimation: Theory, Practice and Visualization. Wiley, New York (1992)
11. Montgomery, D.C.: Introduction to Statistical Quality Control. Wiley, Hoboken (2009)
12. Hurd, H.L., Miamee, A.: Periodically Correlated Random Sequences: Spectral Theory and Practice. Wiley, Hoboken (2007)
13. Kramer, C., de Smet, C.A.M., de Roeck, G.: Z24 Bridge damage detection tests. In: Proceedings of IMAC 17, the International Modal Analysis Conference, Kissimmee, pp. 1023–1029 (1999)
14. Phillips, P.C., Perron, P.: Testing for a unit root in time series regression. Biometrika **72**, 335–346 (1988)
15. Kwiatkowski, D., Phillips, P.C., Schmidt, P.C.: Testing the null hypothesis of stationarity against the alternative of a unit root: how sure are we that economic time series have a unit root? J. Econ. **54**, 159–178 (1992)
16. Perman, R.: Cointegration: an introduction to the literature. J. Econ. Stud. **18**, 3–30 (1993)
17. DeJong, D., Nankervis, J., Savin, N., Whiteman, C.: Integration versus trend-stationarity in macroeconomic time series. Technical report working paper, Department of Economics, University of Iowa (1988)
18. Nelson, C.R., Prosser, C.R.: Trends and random walks in macroeconomic time series: some evidence and implications. J. Monet. Econ. **10**, 139–162 (1982)
19. Hakkio, C.S., Rush, M.: Cointegration: how short is the long run? J. Int. Money Financ. **10**, 571–581 (1991)
20. Otero, J., Smith, J.: Testing for cointegration: power versus frequency of observation-further Monte Carlo results. Econ. Lett. **67**, 5–9 (2000)
21. Shiller, R.J., Perron, P.: Testing the random walk hypothesis: power versus frequency of observation. Technical report, National Bureau of Economic Research, Cambridge (1985)
22. van den Berg, H., Jayaretti, S.C.: A novel test of the monetary approach using black market exchange rates and the Johansen-Juselius cointegration method. Econ. Lett. **41**, 413–418 (1993)
23. Hooker, M.A.: Testing for cointegration: power versus frequency of observation. Econ. Lett. **41**, 359–362 (1993)
24. Lahiri, K., Mamingi, N.: Testing for cointegration: power versus frequency of observation – another view. Econ. Lett. **49**, 121–124 (1995)
25. Palmer, A.C.: Dimensional Analysis and Intelligent Experimentation. World Scientific Publishing Co., New Jersey (2008)
26. Dao, P.B., Staszewski, W.J.: Data normalisation for lamb wavebased damage detection using cointegration: a case study with single-and multiple-temperature trends. J. Intell. Mater. Syst. Struct. **25**, 845–857 (2014)

Chapter 6
Simulation of People's Movements on Floors Using Social Force Model

Ahmed S. Mohammed and Aleksandar Pavic

Abstract Vibration serviceability assessment of floors has been traditionally based on a scenario of a single person walking along a path which will generate maximum vibration level. This is due to the difficulty of predicting the real positions and paths of the walking people. With such a design scenario, it is possible to obtain calculated responses, which could be both over- or under-estimated, depending on the specifics. This could be due to considering only one person walking along one walking path in the simulations. This aspect in the design guidelines could be improved if realistic modelling of people's movements is utilised. Hence, this paper examines the performance of the social force model to simulate the behaviour of people's movements on floors. This method has been widely used to model a crowd of people in evacuation and panic situations. However, it has been reported in the literature that this approach could be used to model people's movements in normal situations as well. The simulation carried out in this paper focuses on the interaction between walking people themselves and between walking people and the surrounding boundaries in typical office floors. The results show that reasonable and realistic behaviour of the floor occupants could be obtained using the social force model. Furthermore, utilising the 'heatmap' can help the designers to visualise and obtain information about the proportion of time spent by walking individuals at various points on the floor. This approach can be adopted in a more realistic procedure for the vibration serviceability assessment of floors.

Keywords Social force model · people's movements · Walking path · Floors · Vibration serviceability

6.1 Introduction

Vibration serviceability assessment due to human walking excitation is increasingly governing the design of building floors [1]. The corresponding criterion is traditionally based on the assumption of a single person walking in a straight line related to the maximum vibration level [2–5]. In reality, people's movement on building floors is a complex process that involves multiple people where the walking behaviour of each person is affected by other people and the surrounding boundaries. Therefore, modelling natural people's movements on floors can add a significant improvement towards more realistic vibration serviceability assessment. This is required by the very nature of the serviceability limit state analysis which should take into account as realistically as possible the natural day-to-day operational condition of the structure.

Modelling people's movement in indoor environments is widely utilised in building evacuation and crowd safety analysis, planning of smart and energy efficient buildings and video games [6–8]. These models can be divided into two categories: macroscopic models, which are based on the similarity between crowd movements and the flow of fluids, and microscopic models, which focus on the behaviour of each person [7, 9]. The latter case is preferable for normal scenarios where low density of people is expected [10]. One of the most widely used models for this purpose is the *social force model* [10]. While the theory behind it is mostly based on 'panic scenarios', it can also be used for normal situations [10].

This paper summarises and examines how the social force model can simulate the behaviour of walking people on floors in normal situations. The model is used within an agent-based modelling (ABM) framework designed for this purpose. The performance is demonstrated using computer simulation of people walking on a typical office floor.

A. S. Mohammed (✉) · A. Pavic
Vibration Engineering Section, College of Engineering, Mathematics and Physical Sciences, University of Exeter, Exeter, UK
e-mail: asm221@exeter.ac.uk

S. Pakzad (ed.), *Dynamics of Civil Structures, Volume 2*, Conference Proceedings of the Society for Experimental Mechanics Series, https://doi.org/10.1007/978-3-319-74421-6_6

6.2 Methodology

One of the most efficient frameworks to simulate people's movements is the ABM approach. Using this method, each person thinks and behaves individually, but interacts with surrounding people and the environment. This interaction can be simulated using the social force model. The collective behaviour of walking people emerges from these individual interactions [10]. The key elements of modelling people's movements can be summarised as follows [6]:

- Decision-making. This governs the general behaviour of each person and it involves the "thinking" process and the acts the person decides to perform (e.g. making a decision to move from the current position to another position).
- Pathfinding. After a decision has been made, to move to another position, the person needs to find the shortest available path which s/he can follow to reach the destination.
- Steering. This guides the walking person to follow the pre-decided path.
- Perception. During walking, each person need to sense the surrounding people, obstacles and walls. This will guide the walking person during walking to prevent collision with the surrounding people or boundaries.

The design of the decision-making part can vary depending on the objectives of the simulations and the degree of detail needed. To design the pathfinding part, different strategies can be employed. A simple and efficient strategy is to subdivide the area and map it into nodes. These nodes are used to guide the pedestrians until they reach their destination [7]. Based on this, an algorithm for the shortest path can be implemented. During walking, the laws of the social force model control the steering and perception of individuals as explained in the next section.

6.3 Social Force Model

The social force model assumes that the internal and external stimuli acting on walking people can be represented by physical force vectors, which are known as social forces [11]. The resultant social force, which is the summation of all social forces, can be used to determine the walking acceleration, walking speed and the position of pedestrians. This is based on Newton's Second Law which states that the force (f) acting on an object is equal to the mass (m) of the object multiplied by its acceleration (a) (i.e. $f = m \cdot a$). Similarly, the vector of the walking acceleration ($\vec{a}_{p_1}(t_i)$) of a walking pedestrian p_1 at time step t_i can be computed if the corresponding body mass (m_{p_1}) and the vector of the resultant social force ($\vec{f}_{p_1}(t_i)$) are known, and as expressed in Eq. (6.1) [10, 12, 13].

$$\vec{a}_{p_1}(t_i) = \frac{\vec{f}_{p_1}(t_i)}{m_{p_1}}, \tag{6.1}$$

The walking acceleration can be integrated to obtain the walking speed ($\vec{v}_{p_1}(t_{i+1})$) and the position ($\vec{x}_{p_1}(t_{i+1})$) of the pedestrian at time step t_{i+1} as described in Eqs. (6.2) and (6.3), respectively.

$$\vec{v}_{p_1}(t_{i+1}) = \vec{v}_{p_1}(t_i) + \vec{a}_{p_1}(t_i)\ \Delta t\ , \tag{6.2}$$

$$\vec{x}_{p_1}(t_{i+1}) = \vec{x}_{p_1}(t_i) + \vec{v}_{p_1}(t_i)\ \Delta t + \frac{1}{2}\vec{a}_{p_1}(t_i)\ \Delta t^2\ , \tag{6.3}$$

where, Δt is the duration of the time step [s]. In Eq. (6.2) and Eq. (6.3), $\vec{v}_{p_1}(t_i)$ and $\vec{x}_{p_1}(t_i)$ are known from the previous steps or initial conditions when $t_i = 0$.

The resultant social force ($\vec{f}_{p_1}(t_i)$) can be obtained by computing the summation of three force vectors: driving force ($\vec{f}_{p_1,dri}(t_i)$), inter-pedestrian force ($\vec{f}_{p_1,ped}(t_i)$) and boundary force ($\vec{f}_{p_1,bound}(t_i)$) as described in Eq. (6.4) and shown in Fig. 6.1.

$$\vec{f}_{p_1}(t_i) = \vec{f}_{p_1,dri}(t_i) + \vec{f}_{p_1,ped}(t_i) + \vec{f}_{p_1,bound}(t_i)\,, \tag{6.4}$$

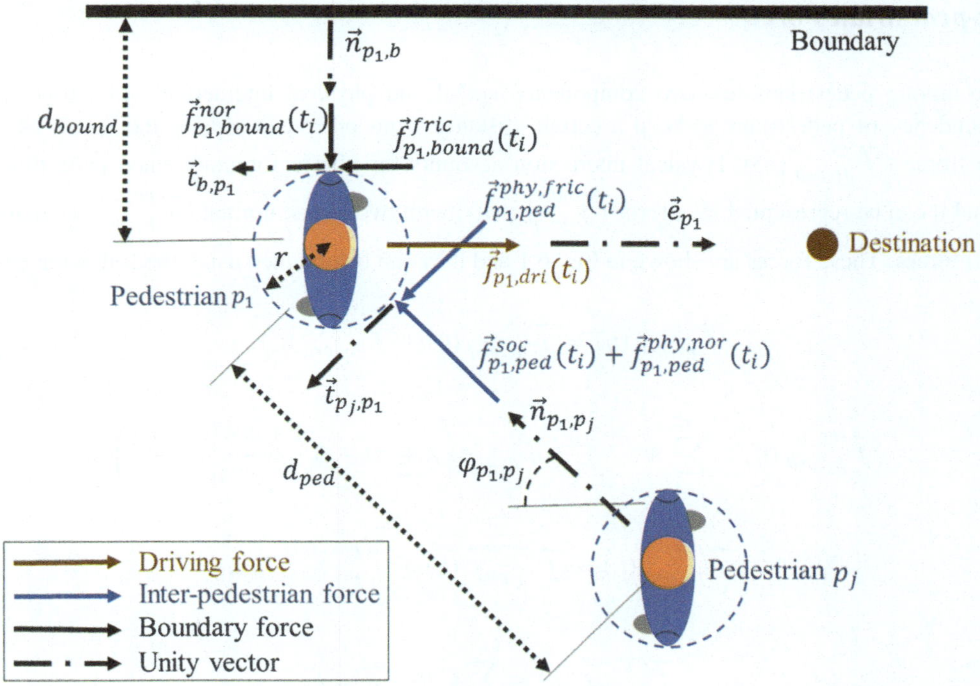

Fig. 6.1 Decomposition of the social forces acting on pedestrian p_1 [13, 14]

Table 6.1 Parameters of the social force model [11]

Parameter	Value
t_r	0.5 s
r	0.3 m
A	2000 N
B	0.08 N
λ	0.2
C	120,000 kg s^{-2}
D	240,000 kg s^{-2}

6.3.1 Driving Force

The driving force accounts for the desire of pedestrians to walk towards a target (Fig. 6.1). During walking, pedestrians try to maintain a desired walking speed. The actual speed may deviate from that due to the obstruction caused by other people and boundaries. Hence, the driving force stimulates the pedestrians to walk at the desired walking speed as described in Eq. (6.5) [10, 11].

$$\vec{f}_{p_1,dri}(t_i) = m_{p_1}\frac{v_d\,\vec{e}_{p_1} - \vec{v}_{p_1}(t_i)}{t_r}, \tag{6.5}$$

where, v_d is a scalar value describing the desired walking speed [m/s], $\vec{e}_{p_1}$ is a unit vector which represents the direction of movement (Fig. 6.1) and t_r is the relaxation time (Table 6.1). The relaxation time refers to the time needed by pedestrians to adapt from the actual ($\vec{v}_{p_1}(t_i)$) to the desired walking speed ($v_d\,\vec{e}_{p_1}$) [10].

6.3.2 *Inter-pedestrian Force*

The interaction among pedestrians has two components: social and physical interactions (Fig. 6.1). Social interaction describes the tendency of pedestrians to keep a certain distance from other pedestrians and it can be represented by a repulsive social force ($\vec{f}^{\,soc}_{p_1,ped}(t_i)$). Physical interaction accounts for physical contact when pedestrians are very close to each other and it can be represented by a force ($\vec{f}^{\,phy}_{p_1,ped}(t_i)$) with two terms: normal ($\vec{f}^{\,phy,nor}_{p_1,ped}(t_i)$) and sliding friction ($\vec{f}^{\,phy,fric}_{p_1,ped}(t_i)$) forces. These forces are shown in Fig. 6.1 and they can be obtained using the following eqs. [6.10].

$$\vec{f}_{p_1,ped}(t_i) = \vec{f}^{\,soc}_{p_1,ped}(t_i) + \vec{f}^{\,phy}_{p_1,ped}(t_i) \quad , \tag{6.6}$$

$$\vec{f}^{\,soc}_{p_1,ped}(t_i) = \sum_{j=1}^{n} A\, e^{\frac{2r-d_{ped}}{B}}\, \vec{n}_{p_1,p_j} \left(\lambda + (1-\lambda)\frac{1+\cos\left(\varphi_{p_1,p_j}\right)}{2}\right) \quad , \tag{6.7}$$

$$\vec{f}^{\,phy}_{p_1,ped}(t_i) = \vec{f}^{\,phy,nor}_{p_1,ped}(t_i) + \vec{f}^{\,phy,fric}_{p_1,ped}(t_i) \,, \tag{6.8}$$

$$\vec{f}^{\,phy,nor}_{p_1,ped}(t_i) = \sum_{j=1}^{n} C\, \theta_{ped}\, \vec{n}_{p_1,p_j} \quad , \tag{6.9}$$

$$\vec{f}^{\,phy,fric}_{p_1,ped}(t_i) = \sum_{j=1}^{n} D\, \theta_{ped} \left\langle \overrightarrow{\Delta v}_{p_1,p_j} . \vec{t}_{p_j,p_1} \right\rangle \vec{t}_{p_j,p_1} , \tag{6.10}$$

$$\theta_{ped} = \begin{cases} 2r - d_{ped} & if \quad 2r - d_{ped} > 0 \\ 0 & if \quad 2r - d_{ped} \le 0 \end{cases} \quad , \tag{6.11}$$

$$\overrightarrow{\Delta v}_{p_1,p_j} = \vec{v}_{p_1}(t_i) - \vec{v}_{p_j}(t_i) \,, \tag{6.12}$$

where, A is the interaction strength (Table 6.1), r is the radius of the pedestrian (Table 6.1), d_{ped} is the distance between the centre of masses of pedestrians p_1 and p_j [m] (where, j denotes the number of another pedestrian), n is the number of pedestrians excluding the pedestrian p_1, B is the range of repulsive interaction (Table 6.1), $\vec{n}_{p_1,p_j}$ is a unit vector pointing from an pedestrian p_j to pedestrian p_1 (Fig. 6.1), φ_{p_1,p_j} is the angle between the direction of motion and the direction of the force exerted by pedestrian p_j on pedestrian p_1 (Fig. 6.1), λ is the potential factor which is related to the fact that the pedestrians in front of pedestrian p_1 have more effect than those behind this pedestrian (Table 6.1), C is the body force strength due to contact (Table 6.1), θ_{ped} is a parameter which depends on the existence of physical contact between pedestrians p_1 and p_j as calculated in Eq. (6.11) [m], D is the sliding friction strength (Table 6.1), $\overrightarrow{\Delta v}_{p_1,p_j}$ is the relative walking speed between pedestrians p_1 and p_j and it can be calculated using Eq. (6.12) [m/s], $\vec{t}_{p_j,p_1}$ is a unit vector perpendicular to $\vec{n}_{p_1,p_j}$ (Fig. 6.1) and $\langle\rangle$ denotes a scalar product of two vectors.

6.3.3 *Boundary Force*

The interaction between pedestrians and boundaries can be treated in a similar way to the interaction among pedestrians (Fig. 6.1). Hence, the force corresponding to this interaction has two components, normal ($\vec{f}^{\,nor}_{p_1,bound}(t_i)$) and sliding friction ($\vec{f}^{\,fric}_{p_1,bound}(t_i)$) forces as shown in Fig. 6.1 and described in the following Eq. (6.10).

$$\vec{f}_{p_1,bound}(t_i) = \vec{f}^{\,nor}_{p_1,bound}(t_i) + \vec{f}^{\,fric}_{p_1,bound}(t_i) \quad , \tag{6.13}$$

$$\overrightarrow{f}^{nor}_{p_1,bound}(t_i) = \sum_{j=1}^{n} A\, e^{\frac{2r-d_{bound}}{B}}\, \overrightarrow{n}_{p_1,b} + C\, \theta_{bound}\, \overrightarrow{n}_{p_1,b}\ , \tag{6.14}$$

$$\overrightarrow{f}^{fric}_{p_1,bound}(t_i) = \sum_{j=1}^{n} D\, \theta_{bound} \left\langle \overrightarrow{v}_{p_1}(t_i)\,.\,\overrightarrow{t}_{b,p_1} \right\rangle \overrightarrow{t}_{b,p_1}, \tag{6.15}$$

$$\theta_{bound} = \begin{cases} 2r - d_{bound} & if \quad r - d_{bound} > 0 \\ 0 & if \quad r - d_{bound} \leq 0 \end{cases}, \tag{6.16}$$

where, d_{bound} is the distance between the centre of pedestrians and the boundary, $\overrightarrow{n}_{p_1,b}$ is a unit vector perpendicular to the boundary and pointing towards pedestrian p_1 (Fig. 6.1), θ_{bound} is a parameter which depends on the existence of a physical contact between pedestrian p_1 and the boundary as explained in Eq. (6.16) [m] and $\overrightarrow{t}_{b,p_1}$ is a unit vector perpendicular to $\overrightarrow{n}_{p_1,b}$ (Fig. 6.1).

6.4 Demonstration

6.4.1 Description

To demonstrate the performance of the social force model to simulate the behaviour of people's movements in normal situations, a typical office floor (21 m long and 14 m wide) is used (Fig. 6.2).

The floor comprises eight zones: two office rooms (two zones), a conference room, printing and storage space, three desks (one zone), a kitchen, a reception and a WC. The entrance/exit is situated near the reception (Fig. 6.2). One hour of the office floor usage by 20 people is simulated and, for simplicity, the following assumptions were made:

- The floor was divided into a grid of nodes with a spacing of 0.5 m between each of two horizontal or vertical nodes. This is to assist the pedestrians finding the shortest walking path and to guide them during walking.
- All floor occupants entered the floor within 5 min after the beginning of the simulation and they left the floor within the last 5 min. The exact entrance/exit times were assigned to each person randomly.
- During the simulation, the floor occupants moved independently to random nodes (positions) in any of the floor zones in Fig. 6.2. They spent 5 min at each node before moving to another random node in a random zone.
- The floor occupants interacted with each other and with the boundaries (walls and desks) during walking based on the social force model explained in the previous section.
- For simplicity, stationary people were assumed to have no effect in the social force model.
- A desired velocity of 1.5 m/s and a body mass of 70 kg were assigned for all floor occupants.
- The time step of the simulation was 0.1 s.

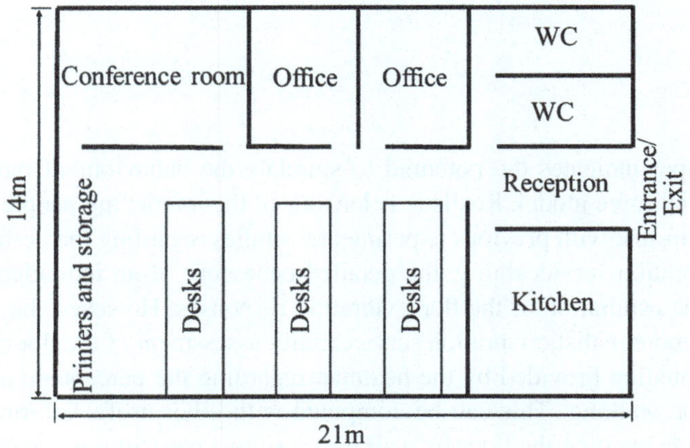

Fig. 6.2 Plan of a typical office floor used in the simulation

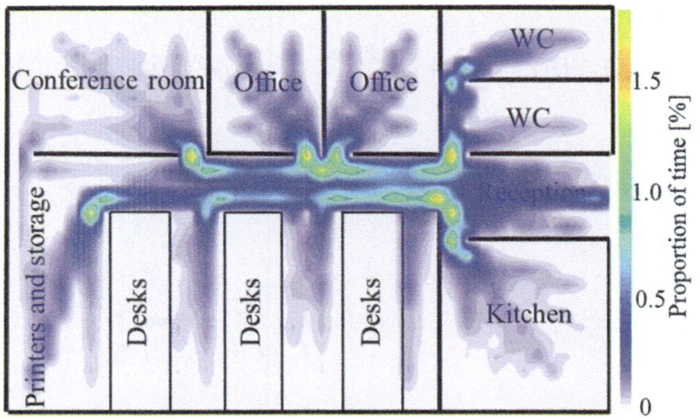

Fig. 6.3 The heatmap corresponding to the positions of the walking floor occupants

These assumptions were adopted to allow for a relatively high interaction among the floor occupants and between the floor occupants and the boundaries. A Matlab script was written to conduct this simulation and the results are shown below.

6.4.2 Results

There are different ways to present the distribution of walking people on the floor for the duration of the simulation. In this paper, this has been carried out as follows. The proportion of time spent by each walking person at each node in the floor is obtained. The summation of the proportions from all floor occupants is calculated and presented as a heatmap in Fig. 6.3. Therefore, this heatmap provides an indication of the positions where individuals are spending most of their time during walking. Similarly, other useful information can be obtained, for example: the probability of 1, 2 or more people walking close together.

In this figure, it is obvious how the boundary force has shaped the area where the floor occupants are walking and prevented them from walking into the walls or desks. A higher proportion of time spent by the walking people can be observed near the corners. This can be explained by the pedestrians reducing their walking speed when they approach corners (Fig. 6.3). A closer view on the behaviour of the floor occupants can be obtained by focusing on specific parts in the simulation. Figures 6.4 and 6.5 show how the natural behaviour of walking people emerges when they interact with each other.

In Fig. 6.4, one person is entering while another person is leaving a room through the same door opening. Due to the space limit of the entrance, one person had to 'wait' until the other person went past before moving on. This waiting period is a result of that inter-pedestrian and boundary forces are opposing the driving force resulting in zero or low walking speed.

Another example of similar behaviour is given in Fig. 6.5, when one person had to make a temporary change in the walking path due to the entrance of another person into the corridor. This behaviour replicates the natural manoeuvres of walking people in indoor environments. The results indicate that the social force model can be used to simulate the natural behaviour of walking people in normal situations, as previously reported in the literature [10].

6.5 Conclusions

The work presented in this paper indicates the potential to simulate the behaviour of multiple pedestrian movements in normal situations using the social force model. Realistic behaviour of the interacting people emerges from the realistic rules applied to each person. This is in-line with previous experimental studies regarding the performance of this model.

From the perspective of vibration serviceability, the detailed behaviour of an individual person traversing a floor may have no significant effect on the estimation of the floor vibration responses. However, the collective behaviour of walking people can be used for a much more realistic vibration serviceability assessment of the floors they walk on. For this purpose, the designers can use the information provided by the heatmap regarding the percentage of time various parts of the floor are occupied by a single person walking. This can be computed with other useful information, such as the frequency of occurrence of one or more people exciting the floor by walking close to a particular mode of vibration.

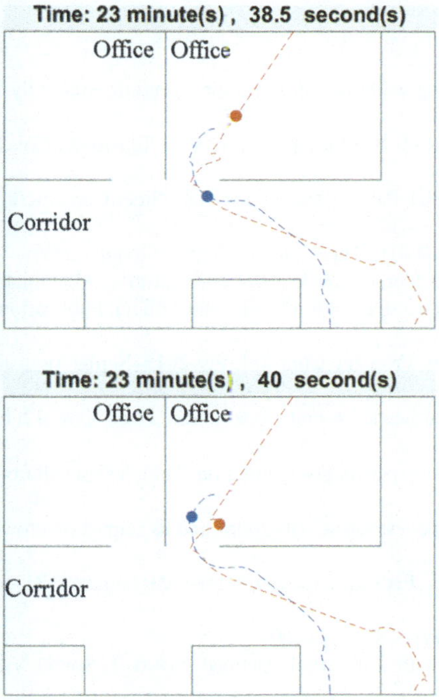

Fig. 6.4 Time-lapse snapshots of two floor occupants walking in opposite directions. One of them (blue circle) is entering while the other one (red circle) is leaving a room. The dashed lines represent their full walking paths

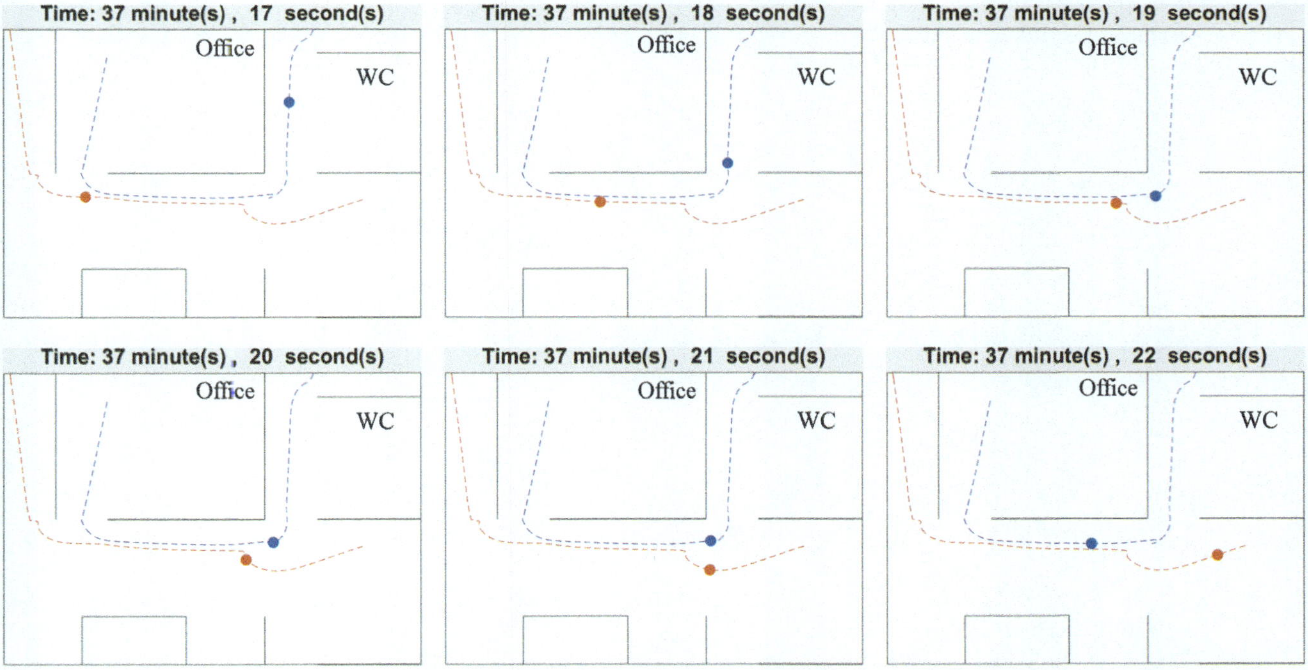

Fig. 6.5 Time-lapse snapshots of two floor occupants, one of them (red circle) is walking through the corridor while the other one (blue circle) is entering the corridor near the WC zone. The dashed lines represent their full walking paths

Acknowledgements The authors are grateful for the College of Engineering, Mathematics and Physical Sciences in the University of Exeter for the financial support they provided for the first author and his PhD program. The authors would also like to acknowledge the financial support provided by the UK Engineering and Physical Sciences Research Council (EPSRC) for grant reference EP/K03877X/1 ('Modelling complex and partially identified engineering problems- Application to the individualised multiscale simulation of the musculoskeletal system').

References

1. Brownjohn, J.M.W., Racic, V., Chen, J.: Universal response spectrum procedure for predicting walking-induced floor vibration. Mech. Syst. Signal Process. **70-71**, 741–755 (2016)
2. Pavic, A, Willford, M.R.: Vibration serviceability of post-tensioned concrete floors – CSTR43 App G. Append. G Post-Tensioned Concr. Floors Des. Handb. – Tech. Rep. **43**, 99–107 (2005)
3. Smith, A.L., Hicks, S.J., Devine, P.J.: Design of Floors for Vibration – a New Approach SCI P354, Revised edn. The Steel Construction Institute, Ascot (2009)
4. Willford, M.R., Young, P.: A Design Guide for Footfall Induced Vibration of Structures – CCIP-016. The Concrete Centre, Slough (2006)
5. Feldmann, M., Heinemeyer, C., Butz, C., Caetano, E., Cunha, Á., Galanti, F., Goldack, A., Hechler, O., Hicks, S., Keil, A., Lukic, M., Obiala, R., Schlaich, M., Sedlacek, G., Smith, A., Waarts, P.: Design of Floor Structures for Human Induced Vibrations. Office for Official Publications of the European Communities., ISBN: 9789279140945, Luxembourg (2009)
6. O'Connor, S., Liarokapis, F., Jayne, C.: Perceived realism of crowd behaviour with social forces. Proc. Int. Conf. Inf. Vis. **2015–Septe**(August), 494–499 (2015)
7. Farina, F., Fontanelli, D., Garulli, A., Giannitrapani, A., Prattichizzo, D.: Walking ahead: the headed social force model. PLoS One. **12**(1), 1–23 (2017)
8. Helbing, D., Buzna, B., Johansson, A., Werner, T.: Self-organised pedestrian crowd dynamics: experiments, simulations, and design solution. Transp. Sci. **39**(1), 1–24 (2005)
9. Moussaïd, M., Perozo, N., Garnier, S., Helbing, D., Theraulaz, G.: The walking behaviour of pedestrian social groups and its impact on crowd dynamics. PLoS One. **5**(4), e10047 (2010)
10. Helbing, D., Farkas, I.: Simulation of pedestrian crowds in normal and evacuation situations. Pedestr. Evacuation Dyn. **21**(January), 21–58 (2002)
11. Helbing, D., Farkas, I., Vicsek, T.: Simulating Dynamical Features of Escape Panic. Nature. **407**, 487–490 (2000)
12. Carroll, S.P., Owen, J.S., Hussein, M.F.M.: Modelling crowd–bridge dynamic interaction with a discretely defined crowd. J. Sound Vib. **331**(11), 2685–2709 (2012)
13. Jimenez-Alonso, J.F., Saez, A., Caetano, E., Magalhães, F.: Vertical crowd – structure interaction model to analyze the change of the modal properties of a footbridge. J. Bridg. Eng. **21**, C4015004 (2016)
14. Helbing, D., Molnár, P.: Social force model for pedestrian dynamics. Phys. Rev. E Stat. Phys. Plasmas Fluids Relat Interdiscip. **51**(5), 4282–4286 (1995)

Chapter 7
Using Correlation Functions as Free Decays

Rune Brincker, Sandro Amador, and Martin Juul

Abstract It is a general assumption in OMA that correlation functions are free decays. In multiple input OMA this assumption also implies that any column in the correlation function matrix is to be considered as multiple output free decays. This assumption is discussed in this paper together with issues concerning estimation and application of correlations functions in OMA.

Keywords Operational modal analysis · Correlation functions · Unbiased estimation · Modal parameters

7.1 Introduction

In operational modal analysis (OMA), a normal procedure to estimate modal parameters is to fit a parametric model to the estimated correlation functions when using time domain identification, and vice versa when using spectral densities in frequency domain identification, Brincker and Ventura [1] and Rainieri and Fabbrocino [2]. However, since the spectral densities can be computed as the Fourier transforms of the correlation functions, it all boils more or less down to getting the correlation functions right.

In the Data-driven Stochastic Subspace Identification (SSI-DATA), Overschee and De Moor [3] also it can be proven that the so-called projection matrix is a linear combination of the correlation functions, Brincker and Andersen [4]. Even in techniques where a parametric model is fitted directly to the operating response – like and ARMA model – it is central to make sure that correlation functions are modelled right. In the ARMA case this is done by securing the covariance equivalence of the ARMA model, Kozin and Natke [5].

In this paper we will consider the basic problems of "getting the right correlation functions", with the aim of minimizing estimation errors on the modal parameters. First, we will discuss under what conditions the correlation functions can be considered free decays of the system. Then we will consider problems of over fitting (when we have too many inputs), and why we might estimate modal participation in order to help us to distinguish between physical modes and noise modes. Afterwards, we will consider different ways of estimating the correlation functions from operational data, and finally, how we can minimize the influence of estimation and measurement noise on the modal parameter estimates.

7.2 Free Decays and Correlations Functions

In Brincker and Ventura [1] it is shown that for a white noise input, the correlation function matrix for positive time lags $\tau \geq 0$ is given by,

$$\mathbf{R}(\tau) = 2\pi \sum_{n=1}^{M} \left(\boldsymbol{\gamma}_n \mathbf{b}_n^T e^{\lambda_n \tau} + \boldsymbol{\gamma}_n^* \mathbf{b}_n^H e^{\lambda_n^* \tau} \right) \tag{7.1}$$

R. Brincker (✉) · S. Amador
Department of Civil Engineering, DTU, Lyngby, Denmark
e-mail: runeb@byg.dtu.dk

M. Juul
Department of Engineering, Aarhus University, Aarhus, Denmark

S. Pakzad (ed.), *Dynamics of Civil Structures, Volume 2*, Conference Proceedings of the Society for Experimental Mechanics Series, https://doi.org/10.1007/978-3-319-74421-6_7

Where $\mathbf{b}_n$ are the mode shapes, λ_n are the poles, and $\boldsymbol{\gamma}_n$ are the modal participation vectors. This is when we use the classical definition of the correlation function matrix $\mathbf{R}(\tau) = \mathrm{E}[\mathbf{y}(t)\mathbf{y}^T(t+\tau)]$. If we use the transposed version of the matrix, that will swap the two terms $\mathbf{y}(t)\mathbf{y}^T(t+\tau)$ in the definition to be $\mathbf{y}(t+\tau)\mathbf{y}^T(t)$, and the outer products $\boldsymbol{\gamma}_n\mathbf{b}_n^T$ in the resulting correlation function matrix to be $\mathbf{b}_n\boldsymbol{\gamma}_n^T$. We then realize that only in this case the correlation function matrix have the general form of a free decay. If we do not take the transpose, we will estimate the modal participation vectors instead of mode shapes.

However, if we take the transpose of the correlation function matrix, then every column in the matrix is indeed a free decay, and all free decay techniques can then be used for identification, like Ibrahim time domain [6] Polyreference time domain, Vold et al. [7], the ERA technique, Juang and Pappa [8], Polyreference frequency domain, Guillaume et al. [9].

The main difference between the free decays used in classical EMA (like impulse response functions) and the correlation functions used in OMA are the following: (1) the correlation functions has typically more noise, and that (2) in OMA it is a normal practice to use more sensors. The reason for the praxis of using more sensors in OMA is partly because we have more noise and partly because we are looking for the modal parameters due to a certain operational condition.

Thus, in order to capture the modal behavior for a certain operating condition, we might want to use enough degrees of freedoms (DOF)'s to get all information in one test. For instance, if we use 50 sensors we will have 50 free decays, with 50 DOF's in each decay. Using a polyreference with 2 AR matrices, this will give us 50 modes in the model, which might easily lead to over-fitting. Therefore, it might be useful to restrict the number of DOF's in the identification process, for instance by using condensation, Brincker et al. [10].

Even with a reasonable condensation, it is often difficult to get rid of the noise modes. In these circumstances, botaining the estimated modal participation vectors $\widehat{\boldsymbol{\gamma}}_n$ in the identification process might be useful to distinguish between the physical and noise mode, since it is well known that physical modes tend to have higher scalar modal participation p_n which is calculated as

$$p_n = \sqrt{\widehat{\boldsymbol{\gamma}}_n^H \widehat{\boldsymbol{\gamma}}_n} \tag{7.2}$$

see Brincker and Ventura [1].

7.3 Correlation Function Estimation and Application

Classically the correlation function is often estimated by Welch averaging in the frequency domain, and then later taking the spectral density back to time domain by inverse FFT to obtain the correlation functions. However, this will introduce the well-known leakage bias that will typically result in over estimation of the damping. This FFT-driven method was originally introduced to save time in the estimation process, but nowadays it makes more sense to take the time needed to calculate the direct estimate

$$\widehat{\mathbf{R}}(k) = \frac{1}{N-k}\sum_{n=1}^{N-k} \mathbf{y}(n)\mathbf{y}^T(n+k) \tag{7.3}$$

The estimation noise on the estimate is well-known. Influence of this noise can be minimized by discarding the noise tail of the correlation function in the identification process, see Tarpø et al. [11].

The measurement noise is concentrated at the first time lags of the correlation function. One way of reducing its influence is to discard the first values of the correlation function. Another way to reduce such influence is to estimate the correlation functions from the spectral density using only the upper singular values of the SVD of the spectral density matrices.

If we want to use the correlation functions to get to the spectral densities, then we need to discard the noise tail, and/or the final values of correlation when not totally decayed. This can be done using some kind of window that suppress the outer part of the correlation functions. In this case, the well-known exponential window or a more innovative solution like the flat-triangular window proposed by Brincker and Ventura [1] can be used to reduce the influence of a non-decayed correlation function on spectral density estimates.

Acknowledgements The authors acknowledge the funding received from Centre for Oil and Gas – DTU/Danish Hydrocarbon Research and Technology Centre (DHRTC).

References

1. Brincker, R., Ventura, C.: Introduction to Operational Modal Analysis. Wiley, Hoboken (2015)
2. Rainieri, C., Fabbrocino, G.: Operational Modal Analysis of Civil Engineering Structures. An Introduction and Guide for Applizations. Springer, New York (2014)
3. Van Overschee, P., De Moor, B.: Subspace Identification for Linear Systems. Kluwer Academic Publishers, Boston/London/Dordrecht (1996)
4. Brincker, R., Andersen, P.: Understanding stochastic subspace identification. In: Proceedings of IMAC-XXIV (2006)
5. Kozin, F., Natke, H.G.: System identification techniques. Struct Saf. **3**, 269–316 (1986)
6. Ibrahim, S.R., Milkulcik, E.C.: A method for direct identification of vibration parameters from the free response. Shock Vib. Bull. **47**, 183–196 (1977)
7. Vold, H., Kundrat, J., Rocklin, G.T., Russell, R.: A multi-input modal estimation algorithm for mini-computers. SAE Technical Paper 820194 (1982)
8. Juang, J.N., Pappa, R.S.: An eigen system realization algorithm for modal parameter identification and modal reduction. J. Guid. Control. Dyn. **8**(5), 620–627 (1985)
9. Guillaume, P., Verboven, P., Vanlanduit, S., Van der Auweraer, H., Peeters, B.: A poly-reference implementation of the least-squares complex frequency domain-estimator. In: Proceedings of the 21th International Modal Analysis Conference, Kissimmee Florida, February (2003)
10. Brincker, R., Camargo, E.A., Skafte, A.: An example of correlation matrix based mode shape expansion in OMA. In: Proceedings of the IMAC Congerence (2015)
11. Tarpø, M., Olsen, P., Amador, S., Juul, M., Brincker, R.: On minimizing the influence of the noise tail of correlation functions in operational modal analysis. In: Proceedings of the X Internation Conference on structural dynamics, EURODYN 2017

Chapter 8
Footbridge Vibrations Predicted by Stochastic Load Model

Lars Pedersen and Christian Frier

Abstract Actions of humans on footbridges may result in structural vibrations that may be annoying to bridge users potentially rendering footbridges unfit for their intended use. Hence, it is useful to make predictions of footbridge vibrational performance already at the design stage involving estimation of levels of vibrations in the footbridge. Nowadays both deterministic and stochastic approaches are available for such evaluations. The have primary focus on probability-based approaches for predicting levels of floor vibrations. The predictions involve employing Monte Carlo simulations and the initial setting up of a stochastic framework describing the action of a walking person. The paper investigates the influence of selected decisions made by the engineer when setting up the basis for the prediction of levels of vibration in the footbridge.

Keywords Footbridge vibrations · Walking loads · Walking parameters · Stochastic load models · Serviceability-limit-state

Nomenclature

a	Bridge acceleration
F	Walking load
f_1	Bridge fundamental frequency
f_s	Step frequency
i	Integer
L	Bridge length
l_s	Step length
m_1	Bridge modal mass
θ	Phase
Q	Modal load
t	Time
v	Pacing speed
W	Weight of pedestrian
Φ	Mode shape
α	Dynamic load factor
μ	Mean value
σ	Standard deviation
ζ_1	Bridge damping ratio

8.1 Introduction

It is useful to erect footbridges as they provide new corridors for pedestrians. A new corridor for pedestrians was provided in London by erecting the Millennium Bridge. However, the response of the bridge to pedestrian traffic was found to be unacceptable [1]. The incident has resulted in an increased focus on how to model the load generated by pedestrians and how to evaluate the response of footbridges.

L. Pedersen (✉) · C. Frier
Department of Civil Engineering, Aalborg University, Aalborg, Denmark
e-mail: lp@civil.aau.dk

S. Pakzad (ed.), *Dynamics of Civil Structures, Volume 2*, Conference Proceedings of the Society for Experimental Mechanics Series, https://doi.org/10.1007/978-3-319-74421-6_8

Unlike some of the existing codes of practice on the matter (such as [2, 3]), the present paper employs a probability-based approach when attempting to predict footbridge response induced by pedestrians. By this approach the main characteristics of a pedestrian (the walking parameters covering step frequency, step length and dynamic load factors) are treated as random variables. This is in line with the fundamental nature of walking as will appear by reviewing [4–7].

Adapting such approach, however, involves a number of decisions to be made by the engineer in charge of computations of the load to be employed for computing footbridge response (more decisions than had a purely deterministic approach for modelling the action been employed).

The paper outlines some of the decisions to be made and explores how sensitive the computed stochastic nature of footbridge response (represented by selected quantiles of the computed probability distributions of peak midspan vertical accelerations) is to some of the choices made by the engineer.

For such evaluations, it is considered useful to take offset in a simple structural configuration and to consider the most simple form of walking-induced excitation possible. Hence, the studies of this paper considers single-person pedestrian traffic on a single-span pin-supported footbridge. Different assumptions are made for the random variables (the walking parameters) and the impact this have on the stochastic nature of bridge vibrational response will be examined.

Section 8.2 introduces the mathematical framework for computing the modal load induced by walking and identifies a set of options found in the literature for modelling walking parameters. Section 8.3 outlines the study assumptions in terms of the specific models and the setting of values for walking parameters employed for computing bridge response, and Sect. 8.4 presents the results in terms of the stochastic nature of footbridge response computed with different assumptions.

8.2 Framework for Computing Bridge Response

As mentioned single-span pin-supported bridges will be assumed for the investigations of this paper. When computing the bridge response to vertical walking loads only the vertical bridge motion associated with the fundamental mode (first vertical bending mode) of the bridges will be considered. Hence, the bridges are modelled as SDOF systems excited by vertical loads from single-person traffic. The modal load generated by a pedestrian on such system may be computed using Eq. (8.1).

$$Q(t) = \Phi(t)F(t) \tag{8.1}$$

In this equation, $F(t)$ represents the vertical load acting on the bridge at the position of the pedestrian and $\Phi(t)$ represents the value of the mode shape function of the first vertical bending mode of the bridge at that position in time.

Assuming that the pedestrian walks with a constant pacing speed, v, the mode shape function, $\Phi(t)$, may be computed using Eq. (8.2):

$$\Phi(t) = \sin(\pi vt/L) \tag{8.2}$$

where L represents the length of the single-span bridge.

It appears that the pacing speed, v, is a parameter influencing the modal load, and the paper considers different ways of determining this parameter.

Basically, the pacing speed of a pedestrian would depend on the step length, l_s, of the pedestrian and the step frequency, f_s, of the pedestrian. Given that both parameters (l_s and f_s) are constants during locomotion, the pacing speed can be determined using Eq. (8.3).

$$v = f_s l_s \tag{8.3}$$

For the two parameters used for computing v, different models have been proposed in the literature.

Table 8.1 presents two stochastic models for the step frequency of a pedestrian. Both models rely on a Gaussian distribution, and the table lists the mean value, μ, and the standard deviation, σ, obtained by fitting this distribution to two different sets of experimental data with outcomes of step frequencies observed for different pedestrians.

Table 8.1 Mean values and standard deviations for step frequency, f_s

Model	μ	σ	Reference
A1	1.87 Hz	0.186 Hz	[5]
A2	1.99 Hz	0.173 Hz	[4]

Table 8.2 Mean values and standard deviations for step length, l_s

Model	μ	σ	Reference
B1	0.71 m	0.071 m	[5]

Table 8.3 Mean values and standard deviations [8]

–	α_2	α_3	α_4	α_5
μ	0.07	0.05	0.05	0.03
σ	0.030	0.020	0.020	0.015

Table 8.2 lists the mean value and the standard deviation of a model for the step length, l_s, in which the parameter is modelled as a random variable following a Gaussian distribution.

Returning to Eq. (8.1) it is apparent that a model for computing the modal load, $Q(t)$, requires settling on a model for $F(t)$ representing the vertical load acting on the bridge at the position of the pedestrian. To this end different proposals are available in the literature.

This paper will employ the quite advanced load model for single-person traffic suggested in [8]. The model recognises that a pedestrian enforces a load on the bridge which is not fully periodic resulting not only in main harmonic components of the action occurring at the frequencies if_s ($i = 1, 2, ..., 5$) but also in subharmonic load components appearing at frequencies "in between" the main harmonics.

Equations (8.4), (8.5) and (8.6) gives the overall mathematical expressions for the load.

$$F(t) = \sum_{i=1}^{5} F_i(t) + \sum_{i=1}^{5} F_i^S(t) \tag{8.4}$$

$$F_i(t) = W\alpha_i \sum_{\bar{f}_j = i-0.25}^{i+0.25} \bar{\alpha}_i(\bar{f}_j) \cos\left(2\pi \bar{f}_j f_s t + \theta(\bar{f}_j)\right) \tag{8.5}$$

$$F_i^S(t) = W\alpha_i^S \sum_{\bar{f}_j^S = i-0.75}^{i-0.25} \bar{\alpha}_j^S\left(\bar{f}_j^S\right) \cos\left(2\pi \bar{f}_j^S f_s t + \theta\left(\bar{f}_j^S\right)\right) \tag{8.6}$$

The model is also capable of modelling the leakage of energy known to occur around the center frequencies of main and subharmonics.

The load model allows for modelling the stochastic nature of walking parameters such as step frequency, step length and a set of dynamic load factors, denoted α_i.

In terms of dynamic load factors, the model, to a large extend, rely on findings in [6] suggesting that the mean value of the first dynamic load factor, α_1, and the adjoining standard deviation corresponds to:

$$\mu = -0.2649 f_S^3 + 1.3206 f_S^2 - 1.7597 f_S + 0.7613; \quad \sigma = 0.16\mu \tag{8.7}$$

and that dynamic load factors in general follow a Gaussian distribution. For higher load harmonics (α_i, $i = 2, 3, \ldots, 5$), the values listed in Table 8.3 apply when making use of this load model.

In the model, phases between load harmonics, θ, are assumed to follow a uniform distribution in the range $[-\pi,\pi]$.

It is beyond the scope of the present paper to go into further details about this load model but reference is made to [8] for more details. The model is somewhat more refined than the walking load models traditionally employed such as the models suggested in [9–11]. A central difference (but not the only one) is that the model considered for the present paper includes subharmonic load components. For instance, this has the effect that should a pedestrian walk using a step frequency of 1.9 Hz, this pedestrian would also be capable of bringing a bridge with a natural frequency of 2.85 Hz into resonant vibration.

8.3 The Studies of This Paper

8.3.1 *The Overall Methodology*

After having identified models for the walking parameters, it can be seen that there, for some walking parameters, are different proposals in the literature for describing their stochastic nature. The engineer performing design stage computations for prediction of the stochastic nature of bridge response might be interested in knowing whether they result in identical predictions of bridge behaviour. Further, he may be interested in knowing whether he, instead of stochastic models, can employ simpler deterministic models for some of the walking parameters without loosing information about the vibrational behaviour of the bridge. Hence, in the paper, bridge response is computed with different assumptions for modelling the walking parameters in order to examine potential divergence in the stochastic nature of bridge response. The response in focus is the midspan vertical acceleration of the bridge and particularly the peak value of this response occurring while a pedestrian crosses the bridge. The stochastic nature of this parameter is determined numerically by having numerous pedestrians crossing the bridge one by one. Each pedestrian is set to take on walking parameters accommodating the predefined/assumed stochastic or deterministic nature. The pedestrian is assumed to walk with at a constant pacing speed while crossing the bridge (using also a constant step frequency and step length). Of cause, in reality, this may prove quit difficult.

Monte Carlo simulations accommodates such investigations (100.000 runs were made for each study assumption) and the result section of this paper will present selected quantiles of bridge response (the acceleration quantiles a_{95}, a_{90} and a_{75}). The bridge responses were computed using standard Newmark-time-integration procedures.

The following sections outline the different assumptions made for modelling the walking parameters.

8.3.2 *Modelling of Walking Parameters*

For the determination of modal load from walking using Eq. (8.1) and for the determination of pacing speed using Eq. (8.3), three different approaches will be considered. They are listed and described in Table 8.4.

Possibly, approach 1 represents the most advanced approach, also being the approach employed in [8]. At least, approach 2 and 3 represent simplifications of matters particularly related to the modelling of step length.

When deciding on a scheme for computing bridge response, a decision as to how to model the stochastic nature of step frequency also needs to be made. The paper has introduced two options (model A1 and model A2). The result section of this paper will also address implications of this decision.

For all computations, the value of W (the weight of the pedestrian) is set to 750 N, as this corresponds to the assumption for W employed in [8]. Dynamic load factors are modelled as described in Sect. 8.2.

8.3.3 *The Bridges Considered for the Study*

Bridge modal properties are outlined in Table 8.5 giving information about natural frequency, f_1, damping ratio, ζ_1, and modal mass, m_1, of the first vertical bending mode of the bridge. The table also provides information about the bridge length, L.

Table 8.4 Study assumptions for step frequency, f_s, and step length, l_s

Approach	Description
1	f_s and l_s are treated as independent random variables using model A1 and B1, respectively.
2	f_s is treated as a random variable using model A1 and l_s is treated as a deterministic property $l_s = 0.71$ m (being the mean value for model B1).
3	f_s is treated as a random variable using model A1, however the pacing speed is determined as $v = 1.87$ Hz • 0.71 m ≈ 1.33 m/s, regardless of the outcome of f_s. I.e. the pacing speed is for all bridge crossings determined as the mean value for model A1 multiplied by the mean value for model B1.

Table 8.5 Modal properties of bridges and bridge length

Bridge	f_1 [Hz]	ζ_1 [%]	m_1 [10^3 kg]	L [m]
I	1.90	0.5	43.8	45.3
II	2.85	0.5	19.5	30.2

Table 8.6 Acceleration quantiles derived for bridge I (the 1.9 Hz-bridge)

–	Approach		
Acc. quantile	1	2	3
a_{95}	0.33654 m/s^2	0.33333 m/s^2	0.33809 m/s^2
a_{90}	0.26354 m/s^2	0.26119 m/s^2	0.26622 m/s^2
a_{75}	0.12787 m/s^2	0.12616 m/s^2	0.12963 m/s^2

Table 8.7 Acceleration quantiles derived for bridge II (the 2.85 Hz-bridge)

–	Approach		
Acc. quantile	1	2	3
a_{95}	0.07398 m/s^2	0.07421 m/s^2	0.07418 m/s^2
a_{90}	0.06648 m/s^2	0.06655 m/s^2	0.06677 m/s^2
a_{75}	0.05572 m/s^2	0.05570 m/s^2	0.05589 m/s^2

It appears that two bridges are considered for the studies of this paper (bridge I and bridge II). They are not representative of any particular bridges but the model properties are selected to describe realistic bridges. The motivation for addressing the response of two different bridges is that the load model employed for the studies of this paper potentially may cause both bridges to resonate. Bridge I may resonate because the basic locomotion frequency of the pedestrian very well may be close to 1.9 Hz. Bridge II is considered because a subharmonic load component is predicted to occur at a frequency in the vicinity of 1.5 multiplied by the basic locomotion frequency (1.5 times 1.9 Hz is equal to 2.85 Hz).

8.4 Results

Sections 8.4.1 and 8.4.2 presents results of calculations in terms of the computed stochastic nature of bridge response.

8.4.1 Influence of Choice of Approach for Modelling Pacing Speed

From computed probability distributions of bridge acceleration response, selected acceleration quantiles have been derived. Table 8.6 presents results obtained for the three different approaches outlined in Sect. 8.3.2 for modelling the pacing speed of a pedestrian. For the computations, the stochastic nature of step frequency was modelled using model A1.

The results in Table 8.6 are those obtained for the 1.9 Hz bridge. It is recalled that approach 1 is the approach in which the step frequency as well as the step length are modelled as random variables. It can be seen that systematically, approach 3 provides the highest values for the three acceleration quantiles, and approach 2 provides the lowest values. For the quantile a_{95}, the results differ from the value of a_{95} obtained using approach 1 by less than 1%. For the quantile a_{75}, the differences are the highest but still the difference is less than 1.5%. Hence for the 1.9 Hz bridge is does not seem to matter much which of the three approaches (1, 2, or 3) that is used for determining the walking velocity of the pedestrian. In fact the deterministic value of $v = 1.33$ m/s used in approach 3 is seen to provide quite accurate estimates of the computed acceleration quantiles.

Table 8.7 provides results for the 2.85 Hz bridge.

For this bridge, the acceleration quantiles are lower than for the 1.9 Hz bridge. This is attributed to the fact that the dynamic load at the natural frequency of this bridge is cannot reach as high values as at the natural frequency of bridge II. It can be seen that there is very limited difference between the results obtained for the three approaches. Hence, the conclusions are similar to those drawn from inspecting Table 8.6.

Table 8.8 Acceleration quantiles derived for the bridges using different step frequency models

–	Bridge I (1.9 Hz-bridge)		Bridge II (2.85 Hz-bridge)	
–	Step frequency model		Step frequency model	
Acc. quantile	A1	A2	A1	A2
a_{95}	0.33654 m/s^2	0.33169 m/s^2	0.07398 m/s^2	0.07396 m/s^2
a_{90}	0.26354 m/s^2	0.25671 m/s^2	0.06648 m/s^2	0.06751 m/s^2
a_{75}	0.12787 m/s^2	0.12544 m/s^2	0.05572 m/s^2	0.05776 m/s^2

8.4.2 Influence of Choice of Model for Step Frequency

This section focuses on the sensitivity of the stochastic nature of bridge response to choices made in terms of the stochastic nature of step frequency modelled for computations. Two different models were introduced (model A1 and A2). Both models relied on a Gaussian distribution; however mean values and standard deviations were different.

Table 8.8 shows acceleration quantiles computed for the bridges I and II. For the computations, the pacing speed of the pedestrian was computed using approach 1.

For all three computed quantiles there is only a very small difference between accelerations derived employing model A1 and model A2 for the step frequency. It is recalled that both models relies on a Gaussian distribution. The difference is the value of mean values (1.87 Hz and 1.99 Hz for model A1 and A2, respectively) and standard deviations (0.186 Hz and 0.173) of the distributions. The natural frequency of footbridge I was 1.90 Hz, thus a value close to and in between the two means values. This might explain why almost identical results are obtained for each of the three computed acceleration quantiles. It cannot be ruled out that had a footbridge with a natural frequency of (perhaps) 2.1 Hz been excited by the two different models for step frequency, the acceleration quantiles calculated with the two different assumptions would give somewhat different results. Futhermore, there might be other proposals in the literature in terms of the mean values and the standard deviation of step frequency than those suggested in [4, 5]. Hence, it would be too drastic, on the basis of the results obtained and presented in this paper, to conclude that the stochastic nature of bridge acceleration response is insensitive to the stochastic parameters defining the stochastic distribution for step frequency. Further, it would be going too far to conclude that the two models for step frequency would suggest identical results for any single-span footbridge.

8.5 Conclusion

With offset in a mathematical framework (relying on a load model modelling 5 main harmonic and 5 subharmonic load components) for computing the modal load induced by single-person traffic on a single-span pin-supported bridge, it was found that:

The stochastic nature of bridge response represented by its peak-acceleration determined at floor midspan (having modelled numerous pedestrians crossing the bridge one by one) to a large extend is independent of the choices made related to how to model the pacing speed of the pedestrian. The findings seem to suggest that it is not worth the effort to model the pacing speed as a random variable as long as realistic value for the average pacing speed is assumed for computations of bridge response. This was found for two bridges with different natural frequencies.

Two possibilities related to how to model the stochastic nature of step frequency were introduced and investigated in the paper. Only very small differences in the computed stochastic nature of footbridge response were found for the two bridges.

However, as explained in the paper, it would be going too far to conclude that the stochastic nature of bridge acceleration response is insensitive to the stochastic parameters used to define the stochastic distribution of step frequency. Further, that two models employed for step frequency (used in the studies of this paper) would end up suggesting almost identical stochastic natures of bridge response for any single-span pin-supported footbridge.

More calculations in which a larger number of single-span footbridges (bridges with other natural frequencies that the two bridges examined in the paper) loaded by pedestrian traffic are recommended.

Acknowledgements This research was carried out in the framework of the project "UrbanTranquility" under the Intereg V program and the authors of this work gratefully acknowledge the European Regional Development Fund for the financial support.

References

1. Dallard, P., Fitzpatrick, A.J., Flint, A., Le Bourva, S., Low, A., Ridsdill-Smith, R.M., Wilford, M.: The London Millennium Bridge. Struct. Eng. **79**, 17–33 (2001)
2. Ontario Highway Bridge Design Code, Highway Engineering Division; Ministry of Transportation and Communication, Ontario (1983)
3. British Standard Institution: Steel, Concrete and Composite Bridges. Specification for Loads, BS 5400: Part 2. British Standards Institution, London (1978)
4. Matsumoto, Y., Nishioka, T., Shiojiri, H., Matsuzaki, K.: Dynamic design of footbridges. In: IABSE Proceedings, No. P-17/78: pp. 1–15 (1978)
5. Živanovic, S.: Probability-based estimation of vibration for pedestrian structures due to walking. PhD Thesis, Department of Civil and Structural Engineering, University of Sheffield, UK (2006)
6. Kerr, S.C., Bishop, N.W.M.: Human induced loading on flexible staircases. Eng. Struct. **23**, 37–45 (2001)
7. Pedersen, L., Frier, C.: Sensitivity of footbridge vibrations to stochastic walking parameters. J. Sound Vib. (2009). https://doi.org/10.1016/j.jsv.2009.12.022
8. Živanovic, S., Pavic, A., Reynolds, P.: Probability-based prediction of multi-mode vibration response to walking excitation. Eng. Struct. **29**, 942–954 (2007). https://doi.org/10.1016/j.engstruct.2006.07.004
9. Ellis, B.R.: On the response of long-span floors to walking loads generated by individuals and crowds. Struct Eng. **78**, 1–25 (2000)
10. Bachmann, H., Ammann, W.: Vibrations in Structures – Induced by Man and Machines. IABSE Structural Engineering Documents 3e, Zürich (1987)
11. Rainer, J.H., Pernica, G., Allen, D.E.: Dynamic loading and response of footbridges. Can. J. Civ. Eng. **15**, 66–78 (1998)

Chapter 9
Non-structural Masses and Their Influence on Floor Natural Frequencies

Christian Frier, Lars Pedersen, and Lars Vabbersgaard Andersen

Abstract Excessive floor vibrations are problematic and may potentially render a floor unfit for its intended use. A design-stage check of vibrational performance of a floor design would encompass design-stage estimates of floor dynamic characteristics such as floor natural frequencies. Non-structural masses such as furniture might be present on the in-service floor. For a prediction of floor dynamic characteristics it is not common to account for the fact that non-structural masses elevated above the floor plane may contribute with inertial energy as a result of their horizontal motion occurring during vertical floor vibration. The paper addresses this subject by setting up a finite element model for the floor, which also accounts for an elevation of the non-structural masses. It is shown how different configurations of non-structural masses influence floor natural frequencies. For the investigations, the elevations and weights of the masses are modelled as random variables and Monte Carlo simulations are used for setting up the random configurations of non-structural masses across the floor area.

Keywords Modal properties of floors · Floor dynamics · Numerical prediction · Serviceability-limit-state · Estimation accuracy

9.1 Introduction

Vibration serviceability evaluation of a floor design is sometimes required. Excitations that may result in floor vibrations can be many. Examples of potentially problematic excitation sources encompass external actions such as nearby road or rail traffic [1–3] or pile driving occurring close to the building containing the floor. Internal actions such as human activity on floors or operating machinery within the building also constitute potentially problematic sources of vibration [4]. The actions may result in annoyance for people or vibration sensitive equipment operating on the floor.

Regardless of the source of dynamic excitation of a floor, dynamic amplification of vibrations can occur at floor natural frequencies. Hence, design-stage predictions of floor natural frequencies are central input parameters for predicting in-service floor behavior in terms of floor vibrations.

It is evident that floor natural frequencies will depend on the imposed load on the in-service floor. For an office floor, the imposed load would consist of items such as decks, bookshelves etc. Such items are referred to as non-structural masses as they do not contribute with either strength or stiffness to the floor. For a specific floor design there would be numerous possible ways of placing these non-structural masses on the floor (in space and time) and the masses of the individual items present on the floor are not likely to be equal. This explains why it generally is difficult to predict natural frequencies of an in-service floor.

Recognizing this, it is considered useful, for design-stage predictions of floor natural frequencies, to consider input parameters for computation of floor frequencies as being random variables. The paper adapts this line of thinking, which allows for modelling and displaying the stochastic nature of predictions of floor frequencies.

The input parameters modelled as random variables in this paper are defined in subsequent sections of this paper, but they all relate to defining the configuration of the non-structural masses present on the floor covering positioning and sizing of the non-structural masses. Studies reported in [5] have revealed that accounting for the possible elevated position of non-structural masses on floors can have significant impact on estimates of floor natural frequencies. Hence, it is considered

C. Frier · L. Pedersen (✉)
Department of Civil Engineering, Aalborg University, Aalborg, Denmark
e-mail: lp@civil.aau.dk

L. V. Andersen
Department of Engineering, Aarhus University, Aarhus C, Denmark

S. Pakzad (ed.), *Dynamics of Civil Structures, Volume 2*, Conference Proceedings of the Society for Experimental Mechanics Series, https://doi.org/10.1007/978-3-319-74421-6_9

useful to account for this possibility for the investigations of the paper with the implication that the elevated position of each individual non-structural mass will be one of the parameters modelled as a random variable. The basis for mapping the stochastic nature of natural frequencies involves settling on a stochastic model for the considered random variables. To this end, the paper considers different models in order to widen the basis for discussion of results.

All investigations will be based on computations of natural frequencies of a rectangular concrete floor representative of a floor in an office building. Results will be presented for the first five modes of floor vibration.

Section 9.2 outlines the methodology involving a description of the floor and a very brief introduction to the finite-element (FE) model of the floor is provided. The section also outlines the assumed usage of the floor (defining the considered possible configurations of non-structural masses present on the floor) and defines the parameters associated with the configuration that are modelled as random variables. Further, Sect. 9.2 introduces the models assumed for random variables and introduces the different stochastic frameworks assumed for mapping the stochastic nature of floor natural frequencies.

Section 9.3 presents the results, and Sect. 9.4 presents and discuses conclusions.

9.2 Methodology

9.2.1 The Empty Floor

The floor is assumed made of reinforced concrete and the floor spans in two directions between pin supports. The slap thickness of the solid 8 m by 9 m rectangular floor (pin-supported along all four edges) is set to 180 mm. The floor is modelled as being homogenous, isotropic and linear elastic. The latter assumption is useful for many vibration problems. For the material, a value for Young's modulus, *E*, of 30 GPa is assumed. Poisson's ratio, ν, is set to 0.15, and the mass density is defined as 2400 kg/m^3.

The FE model of the floor is built employing biquadratic quadrilateral shell elements [6]. For more details about the FE model of the floor, reference is made to [5]. Here it suffices to mention that a 12-by-12-element grid has been used.

9.2.2 Assumed Usage of the Floor

The floor is assumed occupied by an imposed load consisting of nine non-structural masses placed on the floor. These masses (m_k, $k = 1, 2, ..., 9$) are assumed located at predefined positions in the horizontal floor plane, and each individual mass is modelled with the possibility of being elevated above the floor as defined by the parameter z_k. The positions of the non-structural masses in the horizontal plane of the floor are shown in Fig. 9.1. The positions coincide with nodal points of the FE model of the floor.

For calculations, non-structural masses are modelled as a lumped masses.

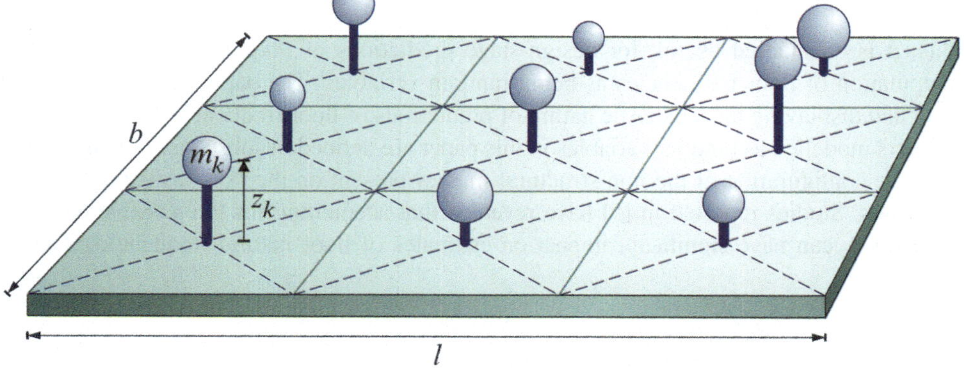

Fig. 9.1 Definition of points of attachment of non-structural masses to the horizontal plane of the floor, and definition of the parameters m_k and z_k. $b = 8$ m and $l = 9$ m

For modelling the effect of the presence of elevated non-structural masses on the floor, a FE model equipped with massless, axial and bending stiff elements placed in between the FE floor-plane and each of the non-structural masses could be used. The effect is implemented differently in the FE model, however results in terms of the dynamic behavior of the floor have been found to be the same.

9.2.3 *Random Variables*

Randomness in usage of the floor is implemented in the studies of the paper for the parameters listed below.

A. The total magnitude of the non-structural mass present on the floor (denoted total non-structural mass, M)
B. The relative distribution of the size of the individual non-structural mass (m_k, $k = 1, 2, ..., 9$)
C. The elevated position of each individual non-structural mass (z_k, $k = 1, 2, ..., 9$)

Having decided on a certain total non-structural mass to be present on the floor, M, there would still be different possible ways of distributing the masses m_k in the predefined positions in the horizontal plane of the floor. For the relative distribution of the individual masses a uniform distribution is employed (for each mass) with the constraint that the sum of the nine lumped masses equals the assumed total imposed load on the floor, M.

9.2.4 *Stochastic Models Assumed for the Random Variables*

The stochastic models assumed for the random variables are outlined below.

In general the magnitude of the imposed load on a floor would depend on the assumed usage of the floor. Eurocode 1 [7] defines categories of imposed loads and adjoining magnitudes of distributed loads in the unit kN/m^2. The present floor will be assumed to be an office floor. For such a floor, Eurocode 1 [7] specifies a distributed imposed load of 2.5 kN/m^2. This value (the characteristic value) represents the 98% quantile of the load. In [8] a Gumbel distribution is proposed for the imposed load with a coefficient of variation of 0.2.

For the calculations of this paper a similar approach and a similar distribution is adapted for modelling the magnitude of the total non-structural load on the floor, which in this paper is defined in terms of the total non-structural mass M. The value of M is simply determined by multiplying the distributed load with the area of the floor and dividing the total load by the gravitational acceleration, $g = 9.81$ m/s^2.

The stochastic nature of the magnitude of imposed load on a floor is by nature quite complex. In some cases it is useful to distinguish between sustained (long-term) components and intermittent (short-term) components of the imposed load and their individual rate of occurrence, such as described in [9].

Such advanced approaches to modelling the magnitude of the total mass, M, is not adapted for the studies of this paper. However, M will be modelled as a random variable, which in itself is considered useful for the investigations of this paper.

The elevated position of each non-structural mass added to the floor, z_k, is assigned a value dictated by a uniform distribution that can take on values between 0 and 1.5 m. Employing this range implies that the non-structural mass needs to be located at a position above the upper side of the floor but that there is also a maximum value of its elevated position. This value is set to be 1.5 m above the upper side of the floor, which is considered a fair upper limit for non-structural masses on an office floor, given that the value represents the mass midpoint.

9.2.5 *Study Approaches and Simulations*

Employing Monte Carlo simulations allows for mapping probability distributions for floor natural frequencies with offset in a predefined stochastic framework describing the loading scenario with the nine non-structural masses.

For the studies of this paper the following stochastic frameworks are considered.

- Approach 1: The stochastic framework described in the previous section (the reference scenario).
- Approach 2: As approach 1, however shifting the distribution for the value of M to a uniform distribution.
- Approach 3: As approach 1, however bonding all non-structural masses to the floor (i.e. assuming $z_k = 0$, for every non-structural mass attached to the floor).

Approach 3 is introduced because it probably is quite common for the prediction of floor natural frequencies not to account for the possibility of elevating the non-structural masses present on the floor. Hence, the results obtained assuming Approach 3 provide a central reference when discussing results obtained with other assumptions.

Approach 2 is introduced to investigate outcomes of results in terms of floor frequencies by another assumption than the assumed Gumbel distribution. As an alternative to the Gumbel distribution, a uniform distribution was chosen. As was the case for the Gumbel distribution, the uniform distribution is calibrated such that the 98% quantile corresponds to a distributed load of 2.5 kN/m^2. The lower limit of the distribution is set to 0 kN/m^2 (corresponding to $M = 0$ kg), and as a result of having 2.5 kN/m^2 to represent the 98% quantile, the upper limit of the distribution is at a value slightly higher than 2.5 kN/m^2. It is noted that this approach will not imply a coefficient of variation of 0.2 for the load. For the employed uniform distribution the coefficient of variation is close to 3 times higher.

It proved sufficient to conduct 1,000,000 simulations for obtaining converged results for the first five natural frequencies of the floor.

9.3 Results

This section presents results obtained in terms of predictions of the stochastic nature of floor natural frequencies. It is considered useful to initiate the presentation of results by addressing the fundamental frequency of the floor.

9.3.1 Results for Fundamental Floor Frequency

Figure 9.2 shows probability distribution functions for the fundamental floor frequency computed under three different study assumptions describing the stochastic nature of the loading scenario with nine non-structural masses present on the floor (i.e. results obtained for Approaches 1, 2, and 3).

For ease of interpreting the results, it is useful to recognize that for a specific probability distribution function, the higher values of floor frequency would be those occurring for scenarios in which the total non-structural masses M. In the limit where the value of M approaches zero, the empty floor condition is reached, and this is the scenario that results in the uppermost possible value for the floor frequency for a specific probability distribution function.

For the present floor eigenmode, the empty floor natural frequency is 8.15 Hz. From Fig. 9.2 it can be seen that this frequency is close to being reached for Approach 2, in which a uniform distribution is assumed for the total non-structural mass. For Approaches 1 and 3 (in which a Gumbel distribution is assumed for the total non-structural mass) it is not that obvious that the distributions approach 8.15 Hz. This is a result of the fact that the Gumbel distribution does not foresee many outcomes of the value of M being close to zero in combination with the fact that an infinite number of simulations has

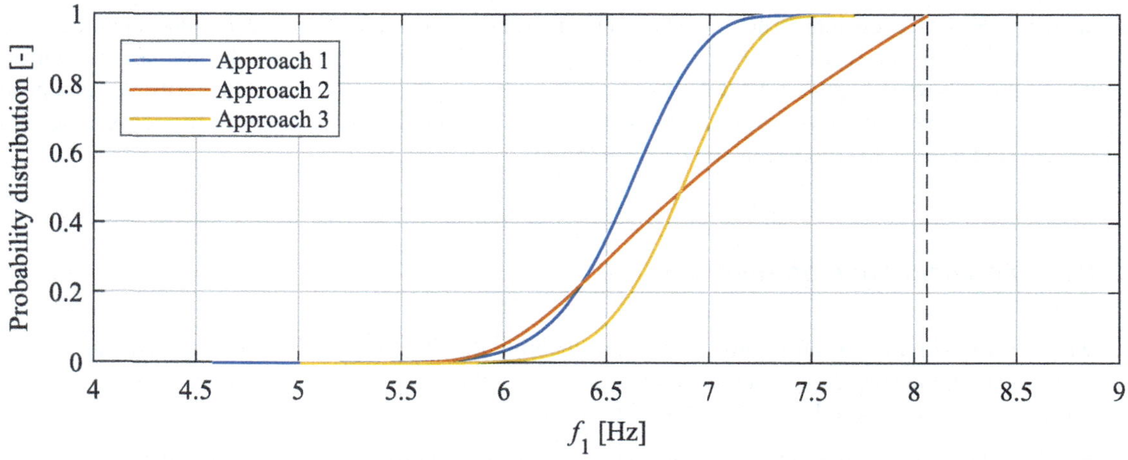

Fig. 9.2 Probability distribution functions for the 1st natural frequency of the floor obtained using Approaches 1, 2, and 3. The dashed line represents empty floor

not been conducted. For the uniform distribution of non-structural mass, a larger number of realizations will assume values of M close to zero, which explains why close to empty floor conditions are reached in simulations.

Another useful recognition is that when outcomes of the value of M increases, the floor frequency drops. For Approach 3, all individual non-structural masses are bounded to the floor, whereas for Approach 1, elevation of these masses are possible. Hence, comparing the probability distribution functions derived by the two approaches provides an idea of the implications of accounting for the possibility of having elevated non-structural masses present on the floor.

A study of Fig. 9.2 reveals that there is some difference between the results obtained for Approaches 1 and 3. However, there is also some difference between results obtained for Approaches 1 and 2, where the difference in study assumption is the distribution type for the total non-structural mass present on the floor.

It proves useful to present more results (probability distribution functions for more modes of vibration) prior to discussing the results further.

9.3.2 Results for Floor Frequencies in Higher Modes of Vibration

Figure 9.3 shows probability distribution functions obtained for the natural frequencies associated with Modes 2–5 of vibration. The dashed lines represent the empty floor natural frequencies.

It can be seen that for these modes of floor vibration, generally Approach 1 end up with the lowest estimate of floor natural frequencies.

It is of interest to compare the results obtained employing Approaches 1 and 3. The following discussion of results takes offset in the results obtained for Mode 4 (i.e. for f_4 with the results presented in Fig. 9.3c). The highest floor frequencies obtained by the two approaches are below the empty floor frequency for Mode 4 being 32.46 Hz. The reason for this has already been addressed.

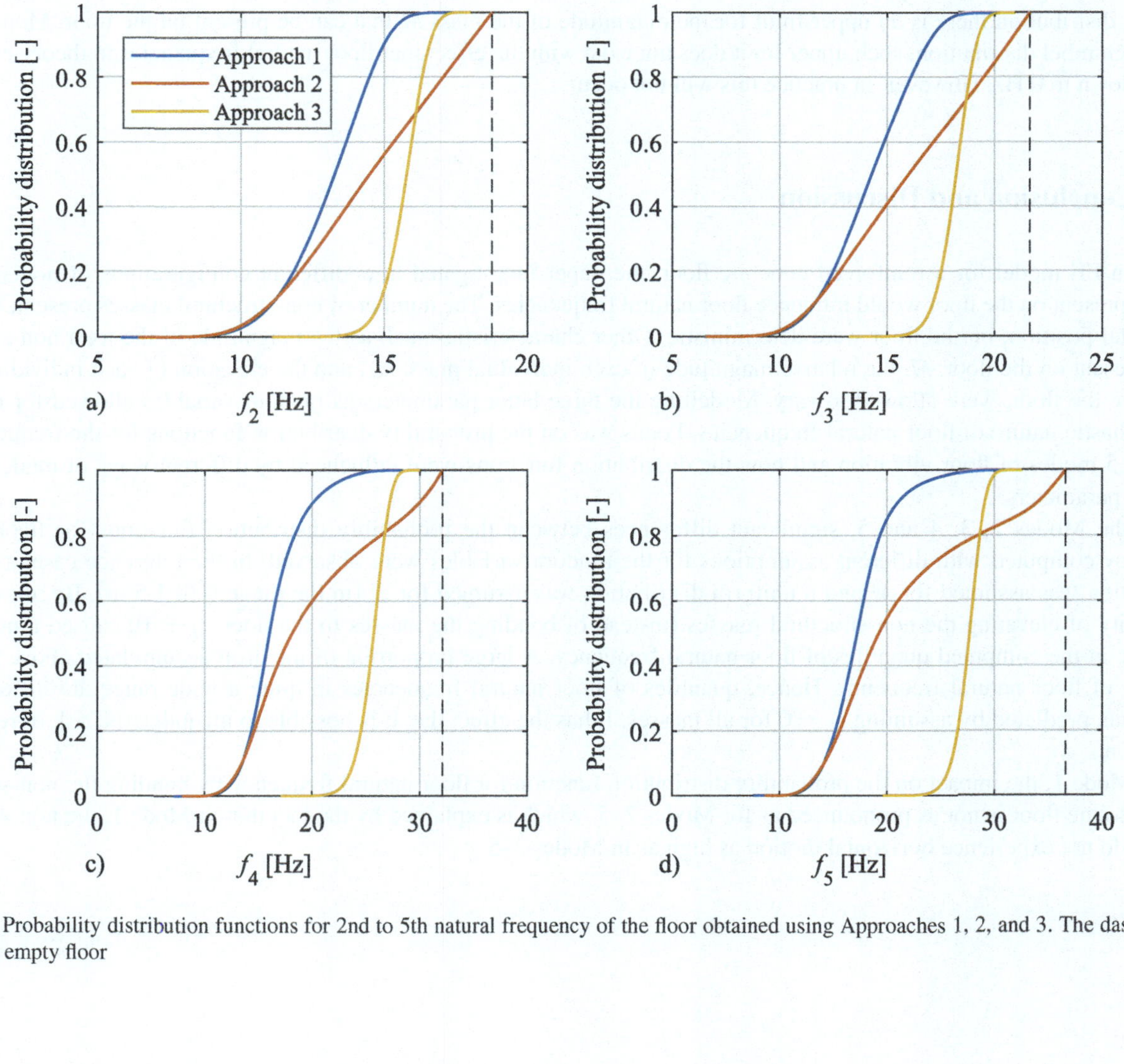

Fig. 9.3 Probability distribution functions for 2nd to 5th natural frequency of the floor obtained using Approaches 1, 2, and 3. The dashed lines represent empty floor

Adding more non-structural mass to the floor, however, quickly results in quite different paths for the probability distribution functions for floor frequency. Approach 1 suggests that a steep decline in the higher quantiles of floor frequency will occur when increasing the total non-structural mass from the "empty floor condition" (moving downwards from the 100% quantile of the floor frequency). For Approach 1, the 80% quantile of the floor frequency is (already) at a value of (only) 17.5 Hz. The 80% quantile for Approach 3 is at 27 Hz. The accumulated difference of approximately 10 Hz is solely a result of the fact that Approach 1 accounts for the possibility of elevated positions of the non-structural masses (which Approach 3 does not). The 10 Hz difference appears to prevail for any quantile of floor natural frequency lower than 80%.

The fact that the elevation of non-structural mass has a significant impact on the floor frequency for this particular mode of vibration is that some of the elevated non-structural masses are positioned on neutral lines of eigenmodes of vertical floor vibration. At these positions, the masses will move horizontally when mode shape functions of vertical floor motions are activated for these modes. This causes a rotational energy to enter into the system, hereby increasing the modal mass and bringing down the floor natural frequency. This matter has been addressed in [5]. When bounding the non-structural masses to the floor (as assumed in Approach 3), this impact on floor frequency would be very limited.

A drop in floor frequency when elevating non-structural masses is also observed for Modes 2, 3, and 5 (see Fig. 9.3a–d), and it is also believed to be attributed to horizontal motion of elevated non-structural masses.

For the fundamental mode of the floor studied in Sect. 9.3.1, the neutral lines of vibration are at the boundaries of the floor (at which no non-structural mass is positioned). For this mode, the horizontal motion of the elevated masses would not be as pronounced, which explains why the effect of elevating the masses is not that significant for the fundamental mode of vibration.

Turning to a comparison of results obtained assuming a Gumbel distribution for the total non-structural mass (Approach 1) and a uniform distribution (Approach 2), there are also systematic differences in the computed probability distribution functions that can be noticed in Fig. 9.3 for all modes of vibration. The frequency difference that can be observed at the 100% quantile of floor frequency has already been discussed. Moving downwards from the 100% quantile, the difference in frequency quantiles increases (possibly most different results are observed at the 90% quantile) but the two approaches appear to end up suggesting almost identical values for the lower quantiles for the floor natural frequency. There is the slight difference though that for Approach 2, there will be a lower limit for the frequency that can be reached as by assuming a uniform distribution there is an upper limit for the magnitude of the mass M that can be present on the floor. Modelling M using a Gumbel distribution, such upper limit does not exist with the effect that floor natural frequencies in theory can attain values down to 0 Hz. However, in practice this will not occur.

9.4 Conclusion and Discussion

Using an FE model for a reinforced concrete floor, the paper investigated how different configurations of non-structural masses present on the floor would influence floor natural frequencies. The number of non-structural masses present, and their horizontal position, on the floor were deterministic. Other characteristics such as the magnitude of the total non-structural mass present on the floor, M, the relative magnitude of each individual mass, m_k, and the elevation of each individual mass, z_k, above the floor, were allowed to vary. Modelling the three latter parameters as random variables allowed for mapping the stochastic nature of floor natural frequencies. Focus was on the probability distribution functions for the frequencies of the first 5 modes of floor vibration and how the distribution functions were influenced by different ways of modelling the random parameters.

For the Modes 2, 3, 4 and 5, significant differences between the probability distribution functions for floor natural frequency computed with different assumptions for the random variables were observed. In the reference case, a Gumbel distribution was assumed for M and a uniform distribution was assumed for z_k (in the range 0 til 1.5 m). By allowing the possibility of elevating the non-structural masses (instead of bonding the masses to the floor, $z_k = 0$) caused a noticeable decrease in the computed quantiles of floor natural frequency. A large proportion of the drop accumulated above the 90% quantile of floor natural frequency. Hence, quantiles of floor natural frequencies in quite a wide range are different and lower than predicted by assuming $z_k = 0$ for all masses. It has the effect that it is possible to misjudge the risk of resonance excitation.

For Mode 1, the impact on the probability distribution function for floor natural frequency by bonding the non-structural masses to the floor is not as pronounced as for Modes 2–5, which is explained by the fact that in Mode 1, the non-structural masses do not experience horizontal motion as high as in Modes 2–5.

It was also found that there is some difference in the probability distribution functions for floor natural frequency found assuming a Gumbel distribution and assuming a uniform distribution for describing the stochastic nature of M. For more details about the observed differences, reference is given to the Sect. 9.3.

The stochastic nature of imposed load on a floor is by nature quite complex. For the studies of this paper quite simplistic stochastic models were employed for modelling the imposed load. Improvements at this end may well be possible. For instance the studies assumed a Gumbel distribution calibrated to ensure that the 98% quantile of imposed load corresponded to 2.5 kN/m^2 (the characteristic value for an office floor suggested in Eurocode 1 [7]). It might be more appropriate to employ a stochastic model which generally predicts lower values of the non-structural masses as this probably would reflect the stochastic nature of imposed load in the serviceability-limit-state better. However, it is believed that the studies reported in the paper have provided a good initial picture of the mechanisms controlling the stochastic nature of floor natural frequencies considering the presence of non-structural masses on the floor. It is, in itself, considered useful when judging the risk of resonant excitation to adapt the thinking that floor natural frequencies are not deterministic properties.

Acknowledgements This research was carried out in the framework of the project "UrbanTranquility" under the Intereg V program and the authors of this work gratefully acknowledge the European Regional Development Fund for the financial support.

References

1. Nagy, A.B., Fiala, P., Márki, F., Augusztinovicz, F., Degrande, G., Jacobs, S., Brassenx, D.: Prediction of interior noise in buildings generated by underground rail traffic. J. Sound Vib. **293**, 680–690 (2006)
2. Fiala, P., Degrande, G., Augusztinovicz, F.: Numerical modelling of ground-borne noise and vibration in buildings due to surface rail traffic. J. Sound Vib. **301**, 718–738 (2007)
3. Lombaert, G., Degrande, G., Cloutaeu, D.: Road traffic induced free field vibrations: numerical modelling and in situ measurements. In: Proceedings of the International Workshop Wave 2000, pp. 195–207 (2000)
4. Bachmann, H., Ammann, W.: Vibrations in Structures—Induced by Man and Machines, IABSE Structural Engineering Documents 3e, Zürich (1987)
5. Pedersen, L., Frier, C., Andersen, L.V.: Flooring systems and their interaction with usage of the floor. In: Dynamics of Civil Structures, Volume 2: Proceedings of the 35th IMAC, A Conference and Exposition on Structural Dynamics 2017, pp. 205–211 (2017)
6. Ahmad, S., Irons, B., Zienkiewicz, O.: Analysis of thick and thin shell structures by curved finite elements. Int. J. Numer. Methods Eng. **2**, 419–451 (1970)
7. EN 1991-1-1.: Eurocode 1 Actions on structures. Part 1-1 General actions. Densities, selfweight, imposed loads for buildings, CEN (2002)
8. NKB/SAKO.: Basis of design of structures, proposal for modification of partial safety factors in Eurocodes. 1999:01 E. Nordic Council of Ministers, Oslo (1999)
9. Leonardo da Vinci Pilot Project.: Handbook 3, Action Effects for Buildings. Aachen (2005)

Chapter 10
Probabilistic Analysis of Modal Properties for Floor Systems with Uncertain Support Conditions

Lars Vabbersgaard Andersen, Christian Frier, and Lars Pedersen

Abstract Traffic and construction work as well as internal sources may cause vibration of floors in buildings. Potentially, this leads to annoyance for people living or working in the buildings—especially when resonance occurs as a result of excitation frequencies coinciding with eigenfrequencies of the floors. Hence, proper design of floors requires insight into the dynamic properties of the system in order to avoid resonance. In this context, the boundary conditions for the floor—or the connections to the main structure—play an important role. A floor clamped along the entire edge reacts differently than a floor which is simply supported. However, whereas the floor system may well be described in terms of material and geometry, an assessment of the supports can be difficult. Often, calculated eigenmodes and eigenfrequencies do not match those identified for a real floor system and this is, to a great extent, due to uncertain and poorly described supports. Hence, the paper suggests a probabilistic approach focussing on the dynamic properties of the floor given uncertain support conditions. Especially, a rectangular concrete floor, representative of a floor in an office or residential building, is assessed regarding its eigenfrequencies. A stochastic model is introduced for the rotational stiffness of the supports, and a numerical analysis is performed in order to quantify how uncertainty related to the supports for the floor system transfers into uncertainty of its eigenfrequencies.

Keywords Floor · vibration · finite-element model · partial fixity · uncertainty

10.1 Introduction

As discussed by, among others, Bernard [1] and Negreira et al. [2], vibration of floors potentially cause annoyance to occupants. In particular, this occurs when sources of vibration excite the floor at its eigenfrequencies, leading to resonance. The sources may be external (e.g. roadway or railway traffic, or construction work), or vibration can be caused by internal sources such as appliances and physical activity of the occupants.

In order to assess whether problems with resonance may potentially occur, the modal properties of the floor system must be known. However, the modal properties of real floors are uncertain and may vary over time as a result of changing use. Pedersen et al. [3] and Frier et al. [4] studied the influence of non-structural mass placed on a floor, thus simulating the presence of furniture. They found that not only the amount of mass but also the height above floor level at which it is attached have significant influence on the eigenfrequencies.

Another important source of uncertainty is the properties of the materials and structural members used within the floor system. In this context, for example, Foschi and Gupta [5] analysed dynamic impact on timber floors, studying the influence of uncertain joist stiffness. Further, different ways of connecting elements within a floor system, e.g. by glue or screws, can have significant impact on the stiffness. A study of such effects related to wooden walls was conducted by Domadiya et al. [6], who found that glued connections of plates and studs provide entirely different dynamic response compared to screwed connection.

Finally, the support conditions influence the dynamic properties of the floor system. Eigenfrequencies of a simply supported floor are lower than those of a clamped floor, given that all other properties are identical. Depending on support geometry and properties of the adjacent structures, partial clamping may occur in cases where the floor was designed to be simply supported or fully clamped. Further, the rotational stiffness of the supports at low deformation levels associated with

L. V. Andersen
Department of Engineering, Aarhus University, Aarhus C, Denmark

C. Frier · L. Pedersen (✉)
Department of Civil Engineering, Aalborg University, Aalborg, Denmark
e-mail: lp@civil.aau.dk

S. Pakzad (ed.), *Dynamics of Civil Structures, Volume 2*, Conference Proceedings of the Society for Experimental Mechanics Series, https://doi.org/10.1007/978-3-319-74421-6_10

man-made vibration may be influenced by pressure, e.g. from walls standing on top of the floor. In this context, Andersen and Kirkegaard [7] studied a multi-storey building in which the floor-to-wall coupling rigidity increased with the number of overlying storeys.

This paper presents the continuation of the work by Andersen and Frier [8]. Thus, a rectangular concrete floor has been studied, especially regarding the influence of uncertain variations in the boundary conditions on variations in the eigenfrequencies related to the first five eigenmodes of the floor system. Section 10.2 describes the finite-element model of the floor and the applied approach to modelling uncertain supports. The results of the analyses are presented and discussed in Sect. 10.3. Finally, Sect. 10.4 provides the conclusions and some additional discussion.

10.2 Computational Model for Concrete Floor with Partial Fixities

In this section, the computational model used for analysis of the concrete floor will be presented. Firstly, the finite-element model of the concrete floor is shortly described, after which the approach to modelling partial fixities will be discussed. Finally, the stochastic model for the supports of the floor system is defined.

10.2.1 Finite-Element Model of Rectangular Concrete Floor

The concrete floor is assumed to behave as a linear elastic material, given the small deformation levels relevant to vibration induced by external sources, e.g. traffic, or internal sources, e.g. walking. In the present case, Young's modulus is assumed to be 30 GPa, Poisson's ratio is 0.15, and the mass density of the concrete is 2400 kg/m^3. Reinforcement is not modelled explicitly, but the material parameters are assumed to be representative of reinforced concrete. A real concrete floor is orthotropic, which will have some influence on the eigenmodes and the associated eigenfrequencies. However, since the primary focus of the present analysis is to examine the influence of uncertain boundary conditions, this orthotropic behaviour has been disregarded.

A solid floor with dimensions 8×9 m^2 and a slab thickness of 0.180 m has been analysed. The floor is divided into 12×12 shell finite elements. Thick-plate theory (i.e. Mindlin theory) has been employed to obtain better accuracy of the response in higher modes than can be expected with thin-plate Kirchhoff elements. The quadrilateral elements have nine nodes and employ biquadratic interpolation of the displacement and cross-sectional rotation fields. No drilling degrees of freedom are introduced. However, a small artificial stiffness is introduced to ensure well-conditioning. Further, shear locking is avoided by using full integration of the bending and membrane stiffness but reduced integration for the transverse shear components. The mass matrix is established by full integration.

Figure 10.1 shows the finite-element model of the concrete floor with blue lines indicating the elements whereas the red lines indicate the position of the reference plane. Nodes are present at the crossings of the red lines, i.e. full biquadratic Lagrange interpolation is utilized with a node placed at the centre of each element.

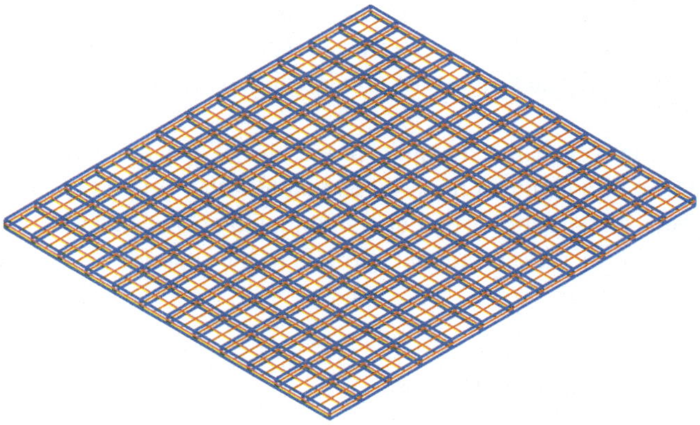

Fig. 10.1 Three-dimensional shell finite-element model of concrete floor

10.2.2 Modelling Partial Fixities Along the Edges of the Floor

The main objective of the analysis is to examine the influence of uncertain support conditions in terms of the rigidity related to rotation around the edges. This leads to the necessity of introducing a quantitative measure of rotational stiffness. The following assumptions and definitions have been applied:

1. The floor is supported along all four edges at the reference plane, i.e. at the mid-plane of the shell.
2. All three displacement components are fixed along the entire edge.
3. The rotations around the local axes orthogonal to the edge of the mid-plane (i.e. the vertical axes and one of the horizontal axes) are not fixed.
4. The rotations around each of the four edges are treated as partially fixed, identified by the relative fixity, κ:
   (a) A relative fixity of $\kappa = 0$ provides the same eigenfrequency as a simply supported edge.
   (b) A relative fixity of $\kappa = 1$ provides the same eigenfrequency as a clamped edge.
   (c) Between the two extremes, a linear scaling is used, i.e. a relative fixity of $\kappa = 0.5$ provides an eigenfrequency halfway between the values corresponding to simple supports and fully clamped conditions.

Partial fixity is obtained by introducing rotational boundary springs along the edges of the floor. Regarding items 1 and 2 in the list above, supporting the edges at the mid-plane of the slab ensures decoupling of rotation and translation of the boundary. Conversely, supporting the floor at bottom of the slab would imply partial fixity even if no rotational springs were implemented.

In order to obtain the relative fixity as function of rotational boundary spring stiffness, a number of calculations have been done in which the rotational boundary spring stiffness per unit length of the edge has been assumed constant along the entire edge, employing consistent nodal representations of the rotational boundary spring stiffness. A logarithmic spacing of the stiffness has been used, and the results are presented in Fig. 10.2 with cubic-spline interpolation between the discrete values of the stiffness considered in the numerical analysis. For reference, the eigenfrequencies f_s and f_c of the simply supported and fully clamped floors, respectively, are listed in Table 10.1. Evidently, the transition from simply supported to fully clamped conditions takes place on the interval 10^4 to 10^{10} Nm/m with the main part of the transition occurring between 10^6 and 10^8 Nm/m. The relative fixity only approaches the limits 0 and 1 asymptotically for the logarithm of rotational boundary spring stiffness going towards minus and plus infinity, respectively. However, for any practical purposes, a floor with rotational boundary springs having a stiffness outside the range 10^3 to 10^{11} Nm/m can be regarded as being either simply supported or fully clamped.

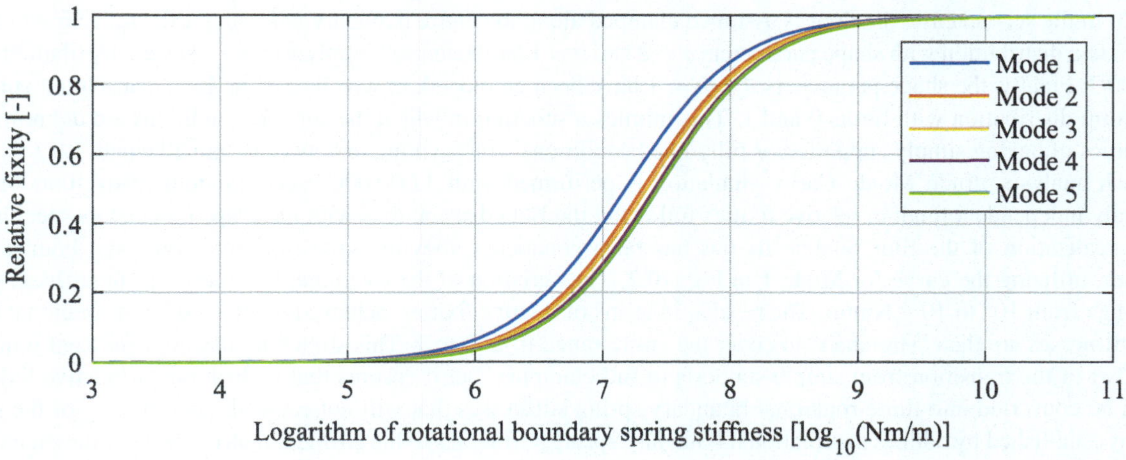

Fig. 10.2 Relative fixity for eigenmodes 1–5 versus logarithm of rotational boundary spring stiffness

Table 10.1 Eigenfrequencies (Hz) for Modes 1–5 for simply supported or fully clamped floor

Boundary conditions	Frequency	Mode 1	Mode 2	Mode 3	Mode 4	Mode 5
Simply supported	f_s	8.06	18.76	21.63	32.11	36.60
Fully clamped	f_c	14.87	28.03	32.30	44.35	48.90

As an important finding, Fig. 10.2 shows that different transitions of the relative fixity occur for different eigenmodes. Thus, Mode 1 approaches a relative fixity of 1 at lower rotational boundary stiffnesses than Modes 2 and 3, and even larger rotational boundary stiffnesses are required in order to reach nearly clamped conditions in Modes 4 and 5. Note that the discrepancy between the eigenfrequencies related to Modes 2 and 3, and again between the eigenfrequencies associated with Modes 4 and 5, are a result of the floor being rectangular. Hence, a relative fixity cannot be specified in a unique manner. In the following analyses, unless otherwise stated, the curve obtained for Mode 1 will be applied. The rotational boundary stiffness applied on a given edge is therefore defined as the value that would provide a given target value of the relative fixity in Mode 1, given that all four edges were supported in the same manner.

As a further note, it is observed from Table 10.1 that the absolute difference between the eigenfrequencies f_s and f_c generally increases with increasing eigenmode number. For Mode 1, the difference is $f_c - f_s = 6.81$ Hz, whereas a difference of 12.30 Hz is related to Mode 5. However, the ratio f_c/f_s decreases with increasing mode number, thus providing a smaller relative difference between the eigenfrequencies of the simply supported and fully clamped floors in Mode 5 compared to Mode 1, for example. This implies that variations of the boundary conditions have relatively less impact on the higher eigenmodes compared to lower eigenmodes. In the higher eigenmodes, a large central part of the structure does not "see" the boundaries. It can be expected that this will result in a smaller coefficient of variation (CV) for the eigenfrequencies related to a higher eigenmode compared to the CV for the eigenfrequencies associated with a lower eigenmode.

10.2.3 Stochastic Model of Boundary Conditions for the Floor

A stochastic model with four random variables is considered. Thus, the relative fixities along the four sides of the rectangular floor are introduced as independent random variables following the Beta distribution

$$p = F(\kappa|a,b) = \frac{1}{B(a,b)} \int_0^\kappa k^{a-1}(1-k)^{b-1} dk, \quad 0 \le \kappa \le 1, \tag{10.1}$$

where $B(a,b)$ is the Beta function with shape parameters a and b. The Beta distribution is chosen due to its versatility and since the probability of having a relative fixity on the interval from 0 to 1 is 1, independently of the shape parameters. However, the Beta distribution can provide left or right skewed distributions as well as distributions with more or less variation.

In the present paper, three basic cases are considered regarding the relative fixities. The first case assumes that the floor was designed to be simply supported along all four edges but that partial fixity occurs. In this case the Beta distribution with shape parameters $a = 1$ and $b = 2$ has been employed, leading to a right skewed triangular distribution as depicted in Fig. 10.3. In the second case, the floor is assumed clamped along the entire boundary but with partial reduction of the fixity. Here the Beta distribution with shape parameters $a = 2$ and $b = 1$ has been used, implying a left skewed distribution as shown in Fig. 10.3. Finally, the shape parameters $a = b = 1$ have been employed, in which case the Beta distribution simplifies to the Uniform distribution with limits 0 and 1. This mimics a situation in which the support conditions are unknown, i.e. the probabilities of having simply supported or fully fixed conditions—or anything in between—are all equal.

In each analysis, crude Monte Carlo simulation is performed with 1,000,000 pseudorandom realizations of the four statistically independent random relative fixities following the Beta distributions with the shape parameters specified above. Once, a realization of the four relative fixities has been conducted, they are converted into rotational boundary spring stiffnesses, utilizing the curve for Mode 1 in Fig. 10.2. The variation of the rotational boundary spring stiffness is limited to the range from 10^0 to 10^{25} Nm/m. The results, in terms of relative fixities, achieved for this range of rotational boundary spring stiffnesses are then "stretched" to cover the entire range from 0 to 1. This stretching has no significant impact in the central part of the transition from simple supports to full clamping, but it ensures that realizations of relative fixities close to 1 will be converted into finite rotational boundary spring stiffnesses that will not cause ill conditioning of the system of equations established by adding the rotational boundary spring stiffnesses to the stiffness matrix for the finite-element model of the floor system.

10.3 Analysis of a Rectangular Floor

For the three cases identified above, the probability distributions of the first five eigenfrequencies are determined. Section 10.3.1 provides the results for the intended simply supported floor, and Sect. 10.3.2 provides the results for the

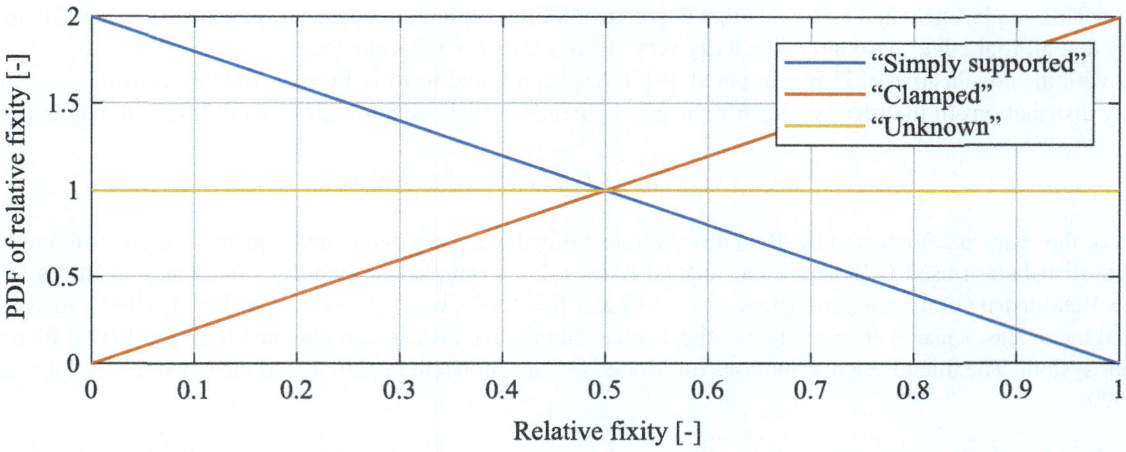

Fig. 10.3 Beta probability density functions used for modelling "simply supported" and "clamped" floors as well as floors with "unknown" uniformly distributed support conditions

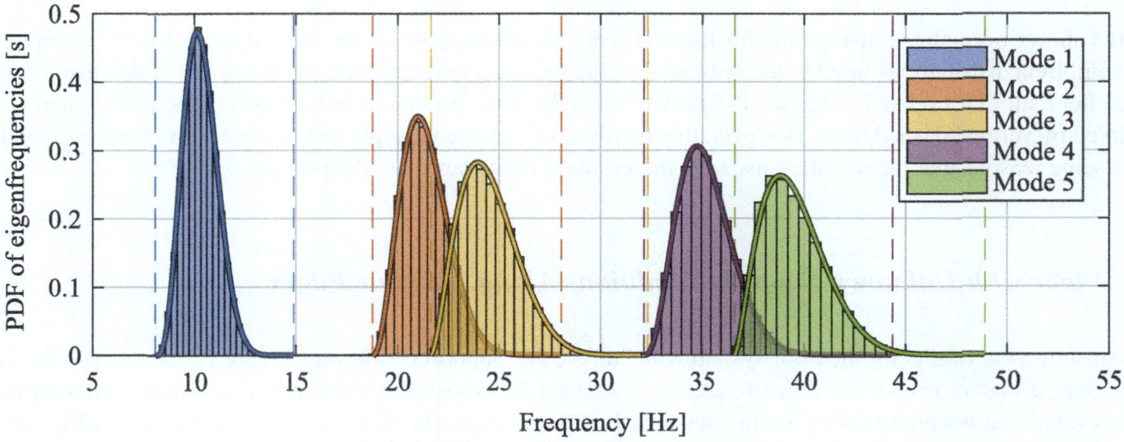

Fig. 10.4 Probability density functions for the first five eigenfrequencies of the "simply supported" floor

intended clamped floor. Finally, Sect. 10.3.3 presents the results for the floor with unknown support conditions, represented by uniformly distributed relative fixities. Here the influence of using another definition of relative fixity, based on another eigenmode, will be assessed.

10.3.1 Floor Designed to Be Simply Supported Along All Four Edges

Based on Monte Carlo simulation with 1,000,000 realizations, Fig. 10.4 shows histograms for the first five eigenfrequencies of the floor designed to be simply supported along all edges. The histograms have been normalized to provide approximate probability density functions (PDFs) for the sample. For reference, the vertical dashed lines indicate the limits provided by the eigenfrequencies f_s and f_c of the simply supported and fully clamped floors, respectively, cf. Table 10.1. The standard deviation of the eigenfrequencies in Modes 4 and 5 are greater than those of Modes 2 and 3—and significantly greater than that of Mode 1. However, as expected, the CV of the eigenfrequencies is smaller in the higher modes compared to the lower modes.

Evidently, the mean values of all eigenfrequencies lie to the left of the respective interval midpoints, and the probability distributions are increasingly more right skewed with increasing mode number. This is caused by the trend observed in Fig. 10.2 and discussed already in Sect. 10.2.2, namely that a given rotational boundary stiffness provides a higher relative fixity in Mode 1 compared to Modes 2 and 3, and the difference is even larger when compared to Modes 4 and 5. The probabilities of eigenfrequencies close to the limits provided by f_s and f_c are small—not only at the upper limit but also at the lower limit. This can be attributed to the fact that the relative fixities of the four edges are statistically independent. Hence, it is unlikely that all four edges have a relative fixity very close to 0 or 1 at the same time.

In line with the conclusions of Damgaard et al. [9], it has been found that the Beta distribution provides a good fit to the probability distributions of the first five eigenfrequencies, given that they are normalized according to the definition

$$\phi = (f - f_s) / (f_c - f_s), \quad 0 \leq \phi \leq 1. \tag{10.2}$$

However, the shape parameters of the Beta distributions fitting the normalized eigenfrequencies ϕ are different from those of the Beta distribution used to generate the relative fixities. For example, the first eigenfrequency of the sample can be fitted by a Beta distribution with parameters $a = 4.819$ and $b = 9.692$ with a Goodness of Fit of 0.9998, measured as the normalized mean least squares error relative to the discrete cumulative distribution obtained from the 1,000,000 realizations of the floor system. The fitted Beta distributions for Modes 1–5 are shown in Fig. 10.4, and the Goodness of Fit is in all cases above 0.999.

10.3.2 Floor Designed to Be Clamped Along All Four Edges

Figure 10.5 shows normalized histograms for the first five eigenfrequencies of the floor designed to be clamped along all edges. Again, Beta distributions fitted to the normalized eigenfrequencies have been shown on top of the histograms. For the "clamped" floor, all PDFs related to Modes 1–5 peak to the right of the midpoints in the intervals between f_s and f_c which are again marked by vertical dashed lines. However, the distributions are decreasingly left skewed with increasing mode number due to the same reasons that the distributions were increasingly right skewed in the previous section.

10.3.3 Floor with Unknown Support Conditions Along All Four Edges

Figure 10.6 shows the results for the floor system with "unknown" support conditions. Figure 10.6a contains the histograms and fitted Beta distributions for the eigenfrequencies obtained by converting uniformly distributed values of the relative fixities into rotational boundary spring stiffnesses based on the response in Mode 1. As two alternatives, Fig. 10.6b shows the results for a sample with conversion from relative fixities into rotational boundary spring stiffnesses based on the response in Mode 2, while Fig. 10.6c shows the results with conversion based on the response in Mode 5.

The effect of using one or another mode as reference when converting relative stiffness into rotational boundary spring stiffness is clear. In Fig. 10.6a (based on Mode 1) all distributions are right skewed, even though the distribution of eigenfrequency 1 is nearly symmetrical. However, when Mode 2 is used as reference, the probability distribution for

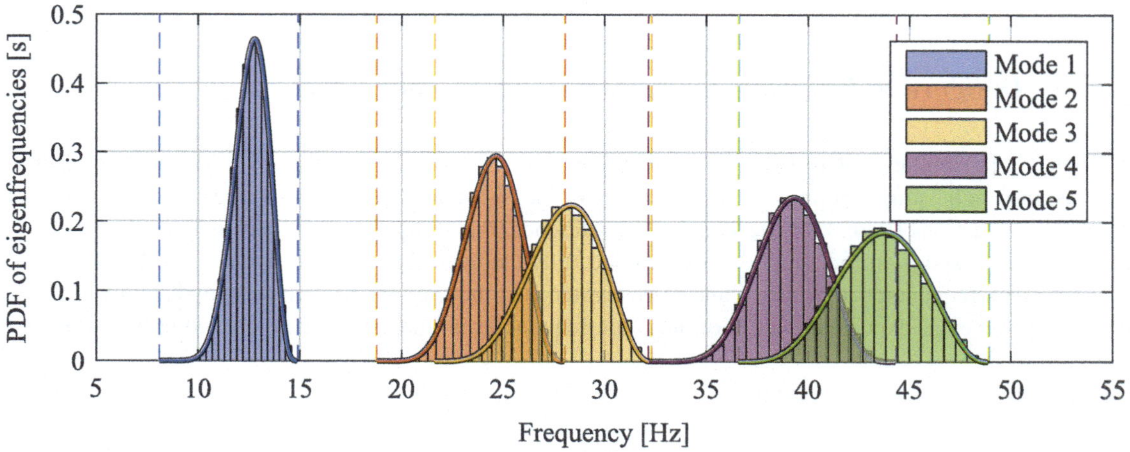

Fig. 10.5 Probability density functions for the first five eigenfrequencies of the "clamped" floor

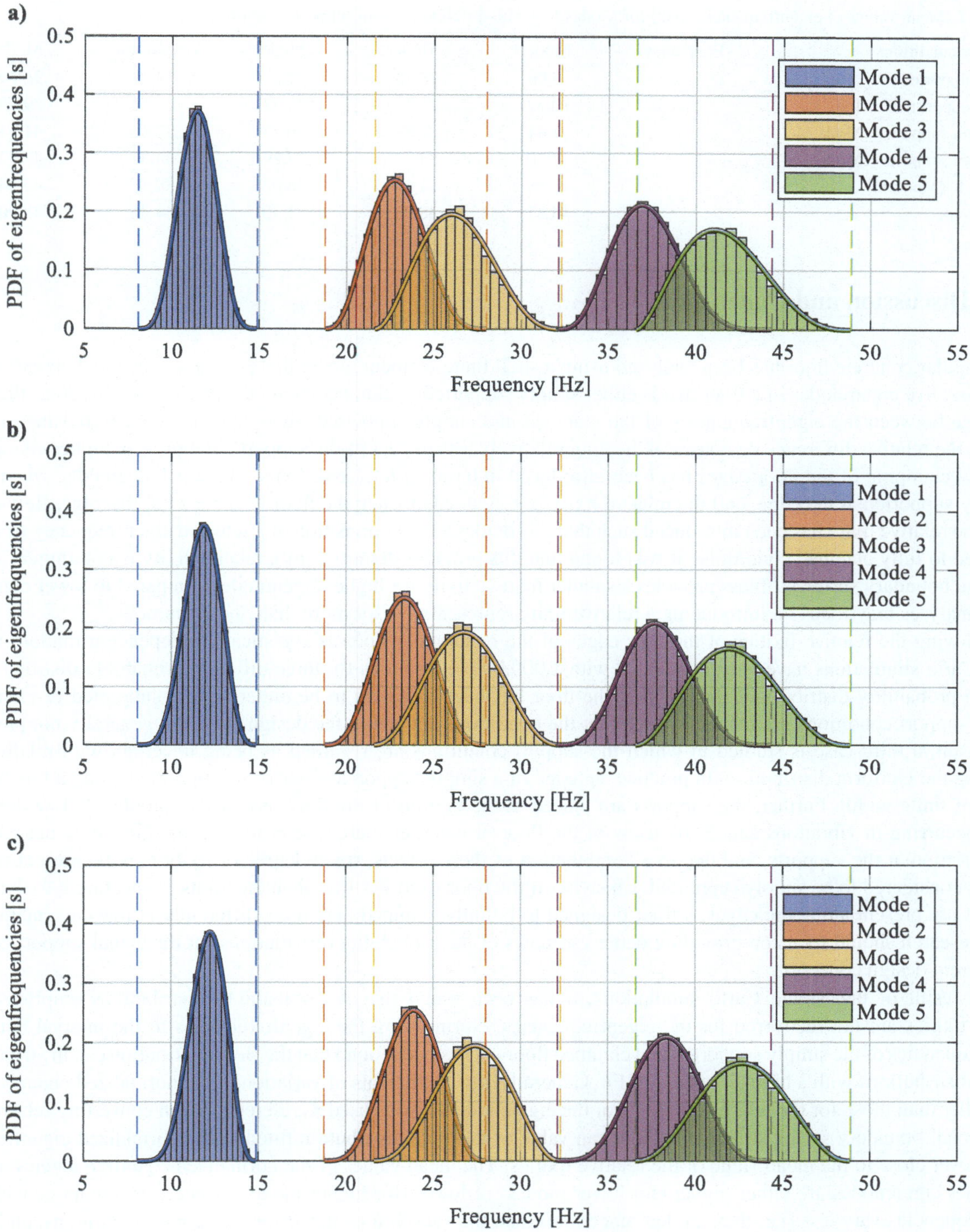

Fig. 10.6 Probability density functions for the first five eigenfrequencies of the floor with uniformly distributed relative fixities: (**a**) based on response in Mode 1; (**b**) based on response in Mode 2; (**c**) based on response in Mode 5

eigenfrequency 1 becomes slightly left skewed, cf. Fig. 10.6b, and when Mode 5 is used as reference (see Fig. 10.6c) the distributions for Modes 1–3 are left skewed while the distribution for Mode 4 is nearly symmetrical and the distribution for Mode 5 is slightly right skewed. Mean values of eigenfrequencies are shown in Table 10.2.

Table 10.2 Mean values of eigenfrequencies (Hz) for Modes 1–5 with different random support conditions

Boundary conditions	Frequency	Mode 1	Mode 2	Mode 3	Mode 4	Mode 5
Simply supported	f_s	8.06	18.76	21.63	32.11	36.60
"Simply supported" (Mode 1)		10.32	21.37	24.52	35.18	39.55
"Unknown supports" (Mode 1)		11.44	22.89	26.27	37.16	41.54
"Unknown supports" (Mode 5)		12.10	23.79	27.29	38.33	42.72
"Clamped" (Mode 1)		12.58	24.44	28.02	39.15	43.54
Fully clamped	f_c	14.87	28.03	32.30	44.35	48.90

10.4 Discussion and Conclusion

A rectangular concrete floor has been analysed using a shell finite-element model that provides converged eigenfrequencies for the first five eigenmodes of a floor that is either simply supported or clamped along the entire edge. Whereas the absolute difference between the eigenfrequencies of the clamped and simply supported floors increase with increasing eigenmode number, the relative difference decreases, indicating relatively less impact of the support conditions in higher eigenmodes.

Partial clamping of the floor edges has been introduced in terms of a relative fixity which is 0 at an edge when the floor is simply supported at the edge, and the relative fixity is 1 at an edge when the floor is clamped at the edge. Between these limits, the relative fixity has been introduced such that it provides a linear transition of a selected eigenfrequency from simply supported to fully clamped conditions. It was found that this provides definition of a relative fixity is not unique. The same rotational boundary spring stiffness provides a smaller relative fixity for higher eigenmodes compared to lower eigenmodes. Future analyses may focus on introducing a relative fixity representative for more than one eigenmode.

Employing the relative fixities of the four edges of the rectangular floor as statistically independent random variables, Monte Carlo simulations have been conducted with 1,000,000 realizations for three different probability distributions. The first two probability distributions assume that the floor has been designed to be either simply supported or clamped, but that the support conditions are uncertain such that the conditions defined in the design stage are generally not present. For comparison, a third case is studied in which the support conditions are regarded as being uncertain with relative fixities following the Uniform distribution. In practice, an edge of a simply supported floor is not supported along a line but within a zone of finite width. Further, the supports are placed at the bottom of the floor, not at the middle. At low deformation levels, occurring in vibrations caused by users of the floor or external sources (e.g. traffic), the floor may not slide freely due to friction at the supports, and due to a combination of these effects, some degree of fixity may be present for a floor that was considered to be simply supported. Likewise, if the floor itself is stiffer than the joints connecting it to the walls, or if cracks are present along the edges, a floor designed to be fully clamped may react differently at low deformation levels. Future research should focus on providing better estimates of the probability distributions for the actual support conditions for different design concepts.

As a result of the Monte Carlo simulations, it has been found that the probability distributions employed for the relative fixities are not recovered for the eigenfrequencies. Normalizing the eigenfrequencies to the interval between the eigenfrequencies of the simply supported and clamped floors, it has been found that the Beta distribution can in all cases fit the sample distributions with a high Goodness of Fit. Generally, the coefficients of variation of the normalized eigenfrequencies are smaller than those for the relative fixities. For the eigenmode that was used as reference when converting relative fixities to rotational boundary spring stiffnesses, the mean value of the Beta distribution fitted to the normalized eigenfrequencies are however close to the mean value of the relative fixities. The mean values of the normalized eigenfrequencies associated with other eigenmodes are either higher (for lower modes) or lower (for higher modes). Hence, as a main conclusion, the eigenfrequencies vary less (i.e. they are less uncertain) than the variation assumed for the support conditions, and the Beta distribution is able to fit the sample distributions of the normalized eigenfrequencies for all considered cases of uncertain support conditions.

Acknowledgement The research was carried out in the framework of the project "Urban Tranquility" under the Interreg V programme. The authors of this work gratefully acknowledge the European Regional Development Fund for the financial support.

References

1. Bernard, E.S.: Dynamic serviceability in lightweight engineered timber floors. J. Struct. Eng. **134**(2), 258–268 (2008)
2. Negreira, J., Trollé, A., Jarnerö, K., Sjökvist, L.-G., Bard, D.: Psycho-vibratory evaluation of timber floors–towards the determination of design indicators of vibration acceptability and vibration annoyance. J. Sound Vib. **340**, 383–408 (2015)
3. Pedersen, L., Frier, C., Andersen, L.: Flooring-systems and their interaction with usage of the floor. In: Caicedo, J., Pakzad, S. (eds.) Dynamics of Civil Structures, Volume 2: Proceedings of the 35th IMAC, A Conference and Exposition on Structural Dynamics 2017, pp. 205–211. Springer, Cham (2017)
4. Frier, C., Pedersen, L., Andersen, L.V., Persson, P.: Flooring-systems and their interaction with furniture and humans. In: Procedia Engineering X International Conference on Structural Dynamics, EURODYN 2017. Elsevier, Rome (2017)
5. Foschi, R.O., Gupta, A.: Reliability of floors under impact vibration. Can. J. Civ. Eng. **14**(5), 683–689 (1987)
6. Domadiya, P.G., Dickow, K.A., Andersen, L.V., Sorokin, S.V.: Evaluation of various joints between studs and plates on flanking noise transmission within lightweight periodic structures. In: Topping, B.H.V. (ed.) Proceedings of the 11th International Conference on Computational Structures Technology, CST2012 ECT2012. Civil-Comp Press, Glasgow (2012)
7. Andersen, L.V., Kirkegaard, P.H.: Vibrations in a multi-storey lightweight building structure: influence of connections and nonstructural mass. In: Research and Applications in Structural Engineering, Mechanics and Computation: Proceedings of the 5th International Conference on Structural Engineering, Mechanics and Computation, SEMC 2013 (2013)
8. Andersen, L.V., Frier, C.: Impact of support uncertainties on the modal properties of flooring systems. In: Høgsberg, J., Pedersen, N.L. (eds.) Proceedings of NSCM 30: The 30th Nordic Seminar on Computational Mechanics, 25–27 Oct 2017, pp. 35–38. Deparment of Mechanical Engineering, Technical University of Denmark, Lyngby (2017)
9. Damgaard, M., Andersen, L.V., Ibsen, L.B., Toft, H.S., Sørensen, J.D.: A probabilistic analysis of the dynamic response of monopile foundations: soil variability and its consequences. Probab. Eng. Mech. **41**, 46–59 (2015)

Chapter 11
Usage of MEMS Capacitive Acceleration Sensors for Structural Monitoring

Marine Dumont and Dan Wolf

Abstract Bridge Structural Monitoring has been a recurring topic over the last few years. A focus has been placed on avoiding any future catastrophic bridge collapses by having a better understanding of the mechanical reactions caused by traffic and environmental conditions. This desire for a better understanding has caused a rise in the usage of sensor technologies which measure the vibrations of a structure.

The IEPE (Integrated Electronic Piezoelectric) or MEMS Capacitive technologies are widely used accelerometer families for structural monitoring. The IEPE technology is ideal for smaller structures where higher frequencies are being focused on. While the MEMS capacitive technology is best used for larger structures which experience DC or low frequency signals.

In the first part of this paper, we will detail the technical specificity of MEMS Capacitive technologies. Some application oriented technical considerations will be made such as driving long cables, EMC protection and thermal stability. In the second part of this paper, three applications will be described where Kistler MEMS Capacitive Technologies have been used for Modal Investigation and Structural Monitoring. Details on sensor selection, implementation and signal computing relevant to those applications will be provided.

Keywords Accelerometers · MEMS · Capacitive · SHM · Structural health monitoring

11.1 MEMS Capacitive Technology

Variable capacitance accelerometers utilize a micro electro-mechanical system (MEMS) that is fabricated from silicon using a bulk micro machining process. Within the structure is a small proof mass suspended by a beam acting as a spring that deflects as a function of applied acceleration (Fig. 11.1). The deflection causes a capacitance change between the proof mass and two electrodes that are fixed on either side separated by a small air gap.

An application specific integrated circuit (ASIC) is used to interface with the MEMS sensor to detect these capacitance changes. The ASIC contains clocks, a demodulator, and other signal conditioning to convert the capacitance changes to an output voltage proportional to the applied acceleration. Changes in the MEMS geometry allow it to be scaled for various acceleration input ranges. A gas enclosed inside the structure provides damping. Open loop devices strictly measure applied acceleration by converting the change in capacitance seen by the sensing element to a voltage that is proportional to it. The proof mass is not constrained from movement inside the MEMS structure.

Variable capacitances accelerometers differ from piezoelectric (or IEPE) type sensors in that they can sense acceleration down to 0 Hz, or static levels. This makes them useful for applications such as inertial measurement, tilt, and low frequency vibration measurement while insuring a very high linearity of the sensor sensitivity together with a very high stability while subjected to temperature changes. Newer designs such as the single axis 8316A . . . allows for measurement down to 100 μg level with a linearity of 0.1%. Now that we have a better understanding on the sensor technology itself, it is important to look into some application related considerations for optimal usage of the sensors.

M. Dumont (✉) · D. Wolf
Kistler Instrument Corporation, Amherst, NY, USA
e-mail: marine.dumont@kistler.com

S. Pakzad (ed.), *Dynamics of Civil Structures, Volume 2*, Conference Proceedings of the Society for Experimental Mechanics Series, https://doi.org/10.1007/978-3-319-74421-6_11

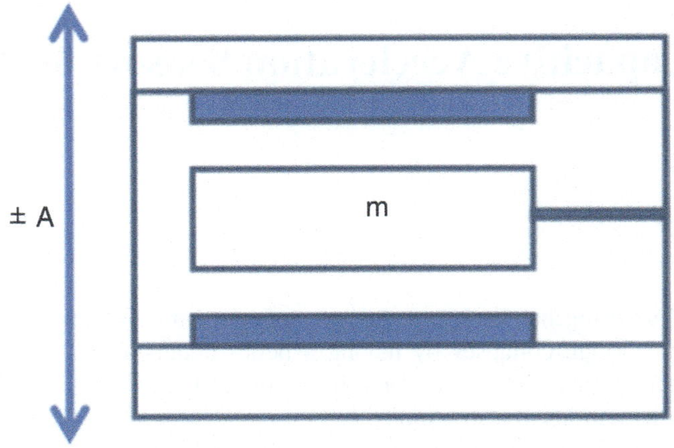

Fig. 11.1 Schematic representation of a MEMS Capacitive sensor

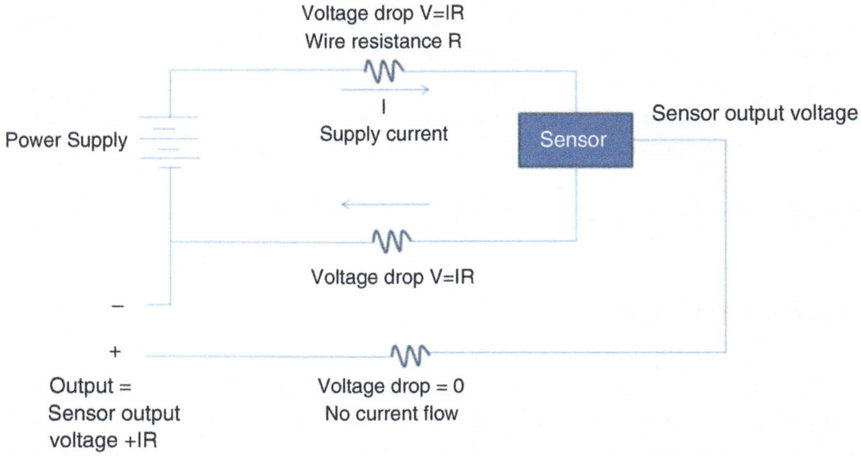

Fig. 11.2 Electrical diagram of a measuring chain including MEMS Capacitive sensor, power supply and cabling

11.2 Driving Long Cables: Effects on Required Supply Voltage, Offset Changes and Cable Capacitance Considerations

When using accelerometers in applications requiring long cables between the power source and data acquisition system as it might be the case on bridges, it is importance to consider the resistance of the wires along with the supply voltage and current draw of the sensor. As cable length is increased the voltage drop between the power source and sensor will increase. In order to adjust for this the supply voltage should be set so that the minimum required voltage at the sensor can be maintained to properly power the sensor.

Knowing the resistance of the supply wire and the power return wire along with the supply current of the sensor the voltage drop can then be calculated. As can be seen in Fig. 11.2, the voltage drop is simply then:

$$V = IR$$

Where I is the supply current and R is the sum of both wire resistances. The sensor supply voltage should then be adjusted so that it is set for at least the minimum sensor supply voltage plus the calculated voltage drop or higher. Use of larger diameter wire gauges and sensors with lower operating currents will also help to reduce the voltage drop.

There can also be changes in the measured acceleration signal as cable length is changed. When the power supply returned is used as the sensor output voltage measurement reference, an increase in cable length causes the measured output voltage to also increase. Because the power supply return line is carrying current the voltage increase will be due to the product of the cable resistance and the current draw of the sensor.

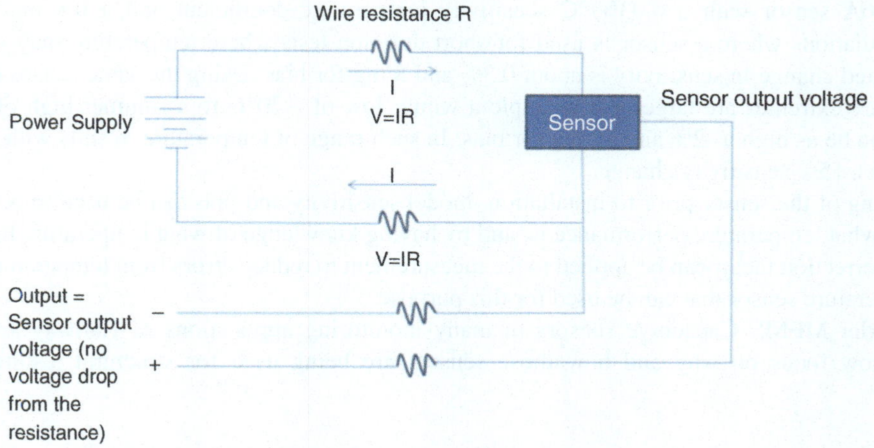

Fig. 11.3 Electrical diagram of a measuring chain using differential output MEMS Capacitive sensor

To alleviate this issue an additional wire at the sensor power return can be used as the signal reference. If this is done there will be no voltage drop since the data acquisition input impedance is high and the offset will not change. Differential output sensors will not show an increase in offset with increased cable length, however consideration of minimum supply voltage at sensor is still required (Fig. 11.3).

It is also important to be aware of cable capacitance and its effect on sensor frequency response. The sensor output impedance and cable capacitance create a low pass filter whose −3 dB response follows the equation:

$$f = \frac{1}{2\pi RC}$$

So as cable length is increased the capacitance of the cable increases causing the frequency response to decrease. To alleviate this, select cable that has a lower capacitance. Variable capacitance sensors have a low frequency response, so depending on the application cable capacitance may or may not be an issue.

11.3 EMC Protection

The accelerometer type 8316A has been EMC tested per EN 61326-1:2006 and CISPR 11:2004/EN 55011:2007 at an accredited test facility and carry the CE mark. Electromagnetic compatibility testing verifies that if exposed to certain electromagnetic environments sensors will operate properly (immunity). The tests also ensure that the sensors do not generate unwanted interference (emissions) that may affect operation of other electronics. This verifies operation through exposure to levels of electrostatic discharge (ESD), radiated radio frequency interference, electrical fast transients, surge, conducted radio frequency interference, magnetic fields, and that conducted and radiated emissions are within acceptable limits. The sensor design for protection in these environments includes filtering chokes, capacitors, and transient voltage protection diodes on all interface lines including power, power return, all sensor output lines, and proper shielding. The test levels applied for emissions are for industrial, scientific and medical equipment. The test levels applied for immunity are for electrical equipment for measurement, control, and laboratory use.

11.4 Thermal Stability

As with most sensors, temperature changes will cause small performance changes. Those changes are much lower than in case of IEPE technologies based on Ceramic. In case of MEMS Capacitive sensors, those changes are mostly seen with sensitivity and bias. Those sensors have specifications for sensitivity and bias temperature coefficients which detail the expected changes over temperature. Depending on the installation location and how much the ambient temperature changes this may or may not be a concern, it however is always something to consider.

Consider an 8316A sensor with a 0.03%/°C sensitivity temperature coefficient and a 0.8 mg/°C bias temperature coefficient. For installations where a sensor is used for short duration tests where temperature may vary during 1 day by only 10°C the expected change in sensitivity is about 0.3% and 8 mg for bias. Using the same sensor for a long term study where the temperature extremes are large (For example a winter low of −30°C to a summer high of 38°C) the expected sensitivity change can be as high as 2% and 54 mg for bias. In such range of temperature, results with Ceramic based IEPE sensors could lead to a 15% sensitivity change.

Temperature testing of the sensor prior to installation, model sensitivity and bias can be used to correct for temperature errors. By knowing what temperature performance is, and by having knowledge of what temperature the accelerometer is at when taking data a correction factor can be applied to the measurement to reduce errors from temperature. The 8316A sensor has an integral temperature sensor that can be used for this purpose.

We can find Kistler MEMS Capacitive sensors in many monitoring applications of Railway and Civil Engineering industries. Let us now focus on why and how those sensors are being used for structural testing and monitoring of infrastructures.

11.5 Structural Testing and Monitoring. Why?

On August 1, 2007, the I-35W Bridge over the Mississippi River in Minneapolis, MN collapsed during rush hour, plunging dozens of cars and their occupants into the river. The calamity disrupted transportation and aimed a spotlight on public infrastructure, specifically methods of inspection. In the aftermath of this event, the Minnesota Department of Transportation decided to design a replacement bridge that could easily and accurately detect structural issues.

The US has 578,000 highway bridges, which are the lifelines of US commerce. The average life span of highway bridges is about 70 years and the majority of bridges currently in use were built after 1945. However, significant environmental damage requiring repair typically occurs before the average bridge reaches mid-life. Corrosion, cracking and other damage can all affect a bridge's load carrying capacity. Therefore, all of the elements that directly affect performance of the bridge including the footing, substructure, deck, and superstructure must be periodically inspected or monitored. Such an issue is encountered worldwide.

11.6 Modal/Structural Testing and Analysis

Experimental Modal Analysis or Operational Modal Analysis (OMA) are methods, strategies and techniques which allow assessing of the structural state of an infrastructure through natural frequency, modal damping ratios and mode shape studies. They are often carried on medium and long span-bridges. It provides global and summary information on the dynamic behavior of the structure whose tracking might inform about potential damage presence. This data is usually correlated with stiffness, damping and mass properties of the structure for assessment or further actions to be taken.

Operational Modal Analysis is based on ambient response data induced by traffic and wind, but not by artificially controlled inputs. Such investigations are possible thanks to new generations of MEMS capacitive accelerometers allowing for accurately measured of low level signals induced by ambient excitation. Such investigations usually include an analysis of the structural state using acceleration and strain measurements, environmental state with air temperature, air humidity, wind speed and direction. In addition it could also include looking into the operation state with gross vehicle weights/per vehicle, axle-load distributions.

11.7 Bridge Structural Monitoring

Once Modal/Structural Testing and Analysis is completed, it is common that a structural monitoring system is set in place. Bridge Structural Monitoring would avoid the catastrophe of a bridge collapsing, such as in the case of the Minnesota Bridge. We are talking here about state of the art monitoring systems looking into strain, movement vibrations and temperature. The most popular instrumentations here would be accelerometers for vibration, linear potentiometer for distance, strain gage for tensile and compressive load. Usually preliminary testing and analysis are performed to identify the weakest points. Also worst case measurements are performed to optimize the monitoring system settings.

In the civil engineering infrastructure world, we can identify three purposes for monitoring. First is Design Verification Monitoring, it allows for verification of the assumptions made by designers during executive design stages. The monitoring is done by controlling major design values during the construction process, especially for unconventional erection stages. It would then allow for detection of default conditions and anomalies of the new structure. Second is the Serviceability Limit State Monitoring, this allows for evaluating the serviceability of a structure under operating conditions. It compares some quantities with threshold values (ex: If the wind is too high, traffic closes). Last but not least, the Structural Health Monitoring (SHM) has been the subject in the last decade of many international research subjects. Its goal is to monitor the health state of a structure which may be subjected, during its life span, to "sudden" and/or "gradual" events. Examples of sudden events can be earthquakes, blasts, fatigue failures, breaks due to corrosion of stay cables. While an example of gradual events can be normal aging of structural materials.

The main purpose of the SHM is the real time assessment of structural conditions if a certain structural state jeopardizes the structures overall safety due to both unpredictable events (like earthquake) and sudden event happening during life span (manmade disasters, blasts, etc.). The system should be able to evaluate if the structure can be used or if remedial actions should be taken.

The key characteristics for accelerometers to be used in such applications are high sensitivity, stability, durability, reliability robustness, isolation and protection. The technology used (IEPE or MEMS Capacitive) will be mostly depending on the bridge Span and traffic. As mentioned earlier, accelerometers are not the only instrumentation needed for a proper SHM system. Here is a non-exhaustive list of instrumentation often encountered in monitoring projects. Some of these are illustrated in Fig. 11.4.

- Accelerometers: IEPE or MEMS capacitive depending of the bridge span and the natural frequency of the bridge
- Strain sensor: Dynamic (vibrating wire gage) and Static (Piezoresistive). Optical fiber could cover both but long term stability is unfortunately not very good.
- Displacement sensors
- GPS to monitor displacement of bridge towers
- Temperature and humidity sensors
- Wind speed measurements with anemometers
- WIM (Weight in motion measurement) is located at the entrance of the bridge and is used in combination with a high speed camera. If a displacement sensor located in the middle of the bridge is detecting a big shift, data from WIM are analyzed in order to see if displacement is coming from car load or from something else.
- Many Data Acquisition Systems located along the bridge. They are all, most of the time, linked together by which typically go to the monitor center.

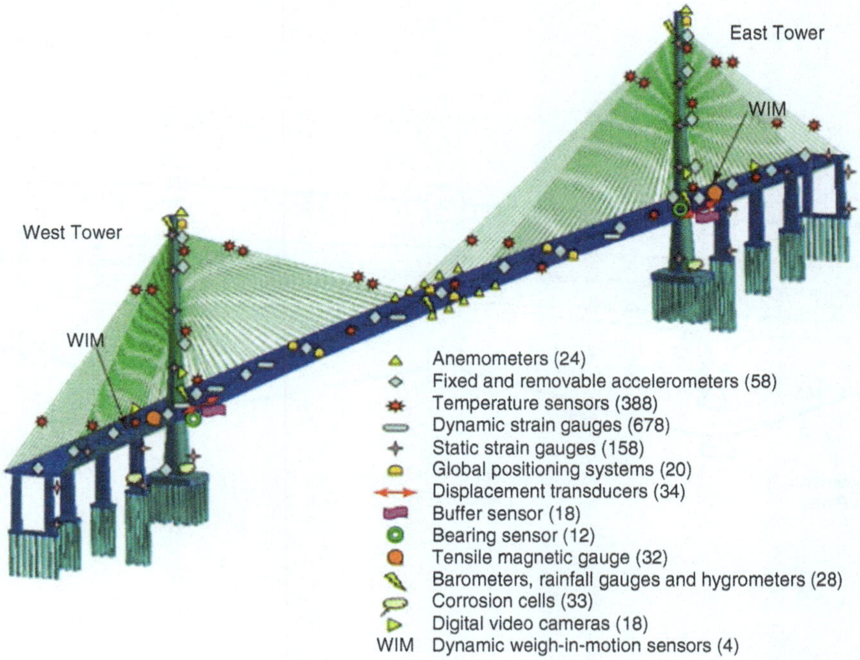

Fig. 11.4 Monitoring system used on the Stonecutters Bridge [1]

11.8 I-35W Bridge, Mississippi River, Minneapolis MN [2, 3]

As mentioned in the introduction, I-35W collapsing was a disaster that the US country will always remember. When building the new bridge, a huge focus has been placed on safety. Mere visual inspection is the primary method used to evaluate the condition of the majority of the nation's highway bridges. Inspectors periodically (about every 2 years) pay each bridge a visit to assess its condition.

The goal for this new project was to support construction processes, record structural behavior (structural monitoring) and bridge security. When the new bridge project started, the bridge contractors started to look for accelerometers and data acquisition systems as well as linear potentiometers. Kistler MEMS Capacitive (former 8316A...) were then chosen – primarily for their low noise, high sensitivity, ability to measure down to DC levels, thermal stability and high linearity. They are used to measure how much the bridge moves or compresses from traffic, wind and air temperature.

The replacement bridge consists of separate north and south lanes supported by common piers. Each lane has three main spans composed of two hollow concrete box girders. Twelve accelerometer sensors are used to monitor the vibration at the midpoint of each concrete box girder with accelerometer cables ranging in length from 10 to 39 m. The remaining 14 accelerometers are used for periodic modal studies. They allow measurement of traffic-induced vibrations and modal frequencies (Eigen frequencies). They address the following risk/uncertainty: Excessive vibrations and dynamic amplifications caused by the traffic induced vibrations. Modal Frequencies allow for correspondence with Finite Element Method (FEM). Dynamic damping studies allow for stuck joints and anomalous global behavior to be identified. Nodes were established on the bridge at strategic locations to ensure worst case measurements. Sensors with indicated output levels were allocated to the nodes depending upon measurement requirements.

To identify specific node locations the bridge is divided into northbound and southbound lanes. The major structural supports are denoted as abutments and piers. Beginning at the south end of the bridge and moving in a northbound direction, the order is abutment 1, pier 2, pier 3, pier 4, and abutment 5. Figure 11.5 provides a schematic view of the bridge that includes structural and node (Nx) locations as well as the distances between them. Nodes not located directly on a pier are placed at the centerline between structural supports (the theoretical weakest point) for a worst-case measurement.

When the new Interstate 35W bridge opened on September 18, 2008 it included safety features exclusive to this project. The Smart Bridge technology has 300 sensors that monitor the structure of the bridge. The various sensors are buried in the concrete, across the expansion joints, attached to bridge girders and attached to wires routed to a central computer. Engineers from the Minnesota Department of Transportation and the University of Minnesota will monitor the information to detect any early warning signs of problems and advance the art of bridge design.

Figure 11.6 is showing the progression of a large vehicle across the bridge. Starting at the bottom of the screen, moving toward the top, and correlating it with the bridge schematic of Fig. 11.5, we can clearly see that the vehicle is first detected

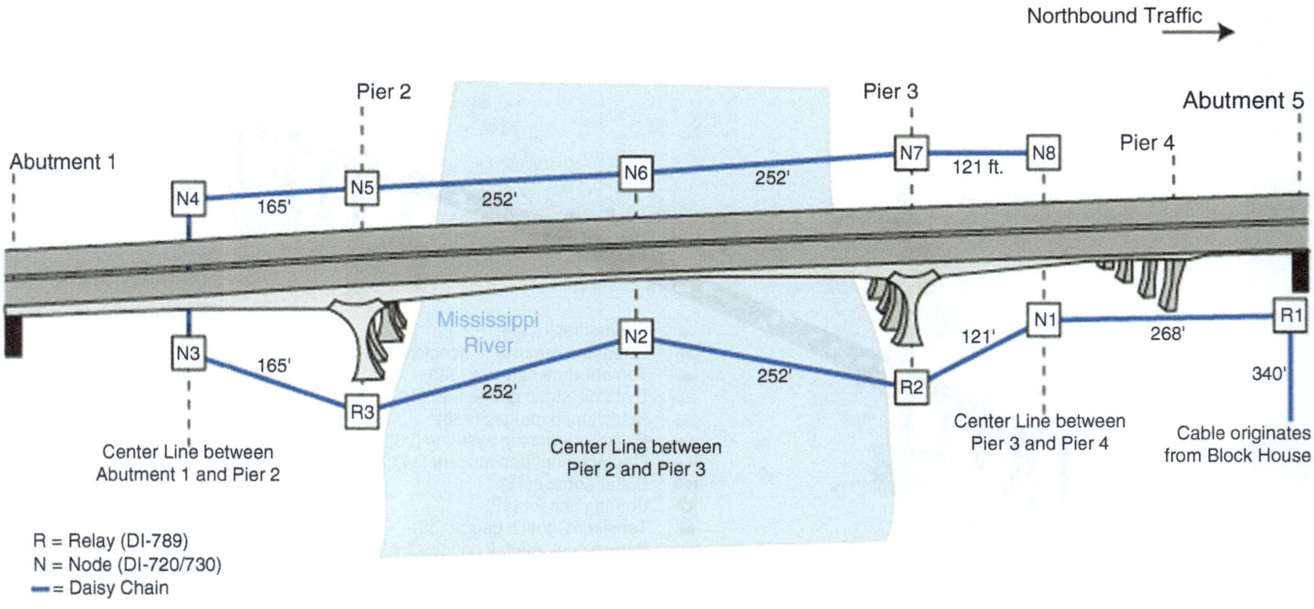

Fig. 11.5 Representation of instrumentation nodes (Nx) and relays (Rx location) [2, 3]

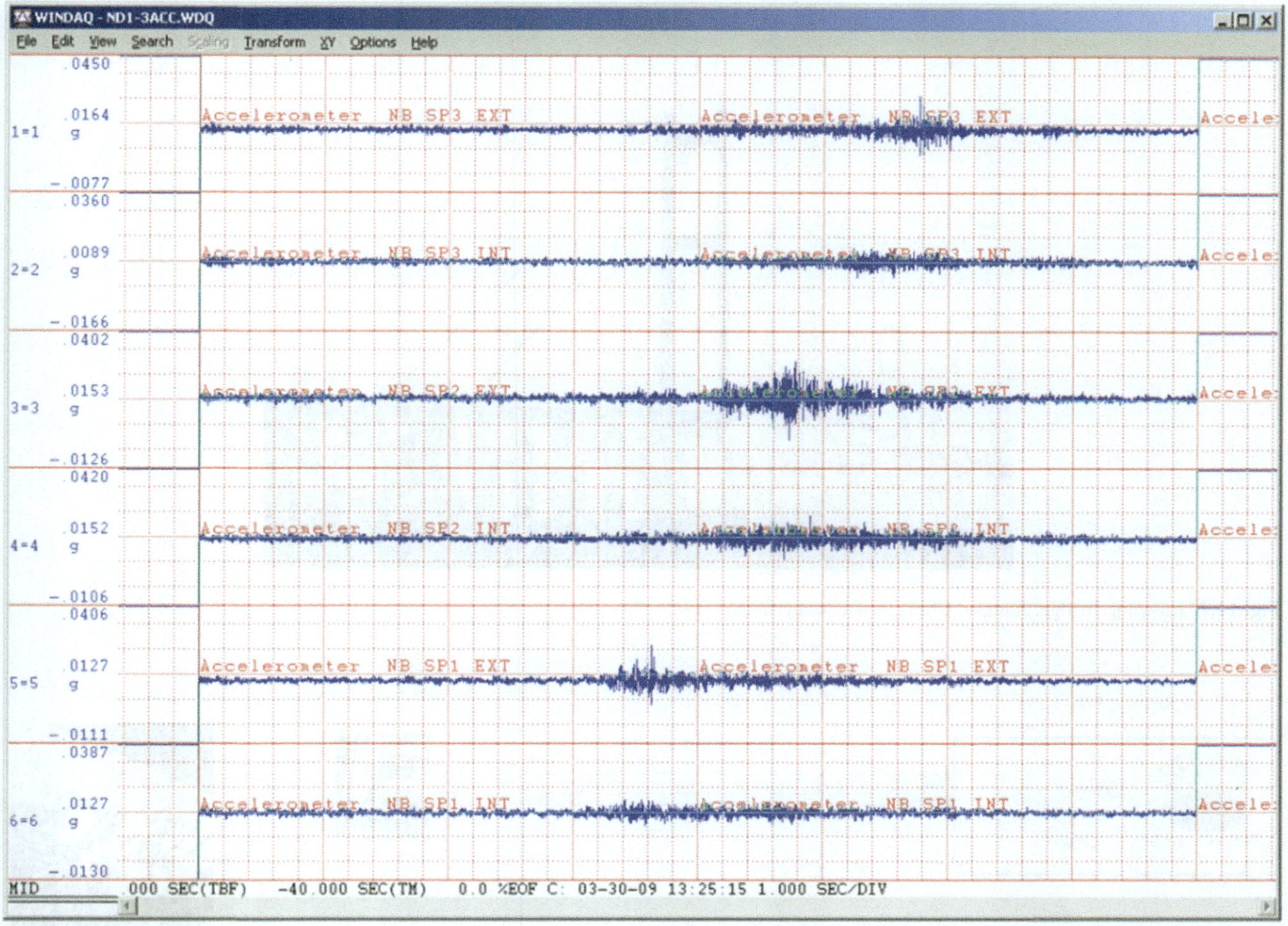

Fig. 11.6 Instrumented node 1, 2 and 3 acceleration signal [2, 3]

by node 3 accelerometers on the first bridge span, then by node 2 accelerometers on the second bridge span, and finally by accelerometers of node 1 on the third and final bridge span.

11.9 Runyang Bridge of Jiangsu, China

Chinese infrastructures, being newer construction, usually already include such measurement and monitoring systems. An example can be shown with the Jiangsu Transportation Institute of Nanjing in China. Their role is to perform surveillance of more than 10 bridges around the Nanjing Area in China. Each bridge being instrumented with many sensors, for example . . . , the Runyang Bridge of Jiangsu (Fig. 11.7).

This project has been done in collaboration with Kistler for Bridge Monitoring using 8330B3 Servo MEMS Capacitive sensors. Those sensor type were using a closed loop technology that was at the time the only solution to reach noise levels allowing for measurements around 100 μg. The new generation of open loop MEMS capacitive type 8316A allows to measure those levels. This allows a much more economical, rugged and compact solution.

Along the bridge, five stand alone DAQ stations PXIe from National Instruments were powering and receiving data from the accelerometers. Two sensors were mounted orthogonally in an IP68 box to protect them from the environmental conditions and to measure acceleration on two orthogonal axes (Fig. 11.8). The IP68 boxes were located in the main tower and along the side of the bridge. Cables of about 200 m were used to connect the sensors to five data acquisition systems from National Instruments. They were connected to a main system at the very end of the bridge providing for power and data collection. Those data were then sent by optical fiber or wireless to a Monitoring Center located several kilometers away.

Fig. 11.7 Runyang Bridge, Jiangsu, China (www.jscd.gov.cn)

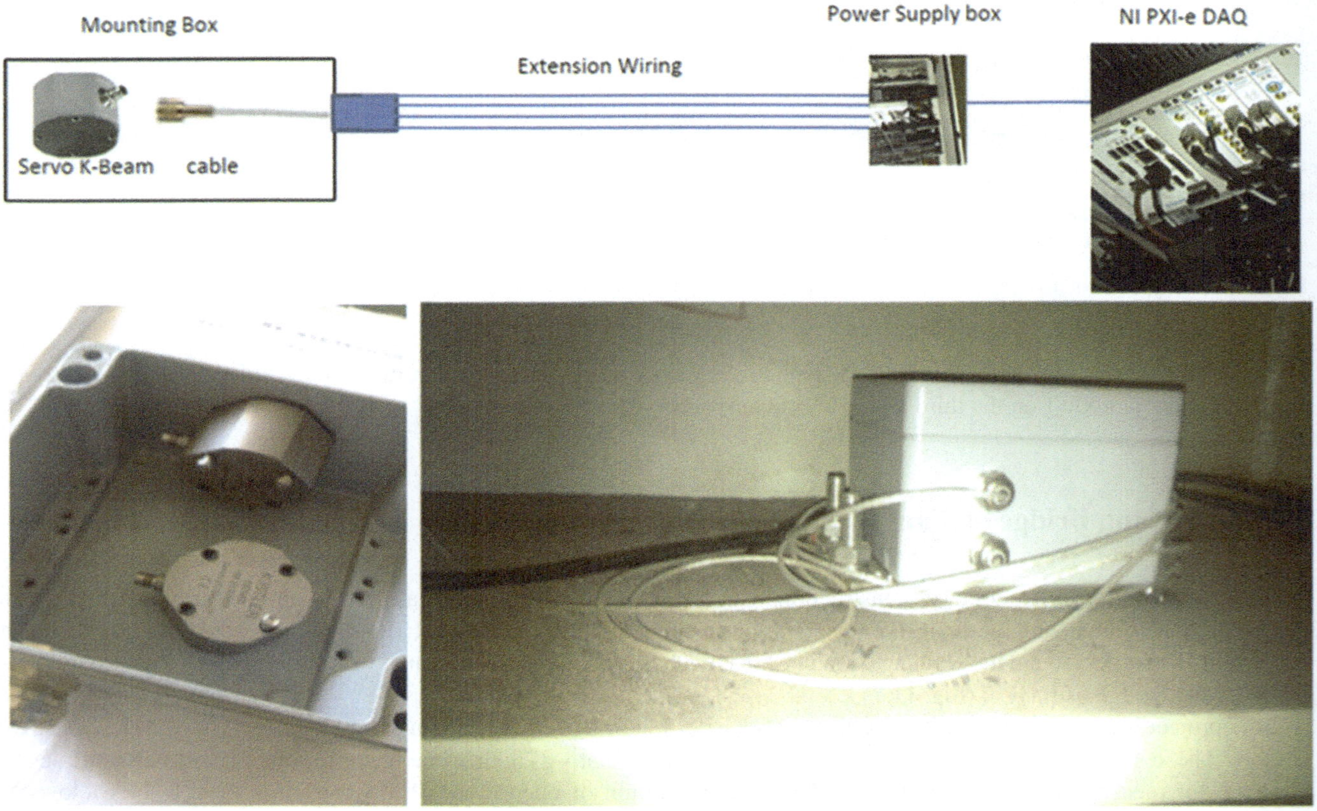

Fig. 11.8 Monitoring measuring chain using MEMS Capacitive sensors and NI PXI-e data acquisition system

11.10 1889 Paerni Railway Iron Bridge, Italy [4]

Some European cities have another issue. Some of their bridges are being considered as monuments and cannot be rebuilt. Their structural monitoring or regular Modal Analysis has become a must to ensure safety of the users. The Paerni railway bridge in Italy is an Iron bridge from 1889. It has been subjected through time to very poor maintenance and significantly damaged by corrosion. It is still however used for road and rail traffic.

The goal of their project was to better understand the dynamic characteristic of this old bridge, describe its modal behavior and use the data to develop a structural monitoring system. Investigations were performed in operational conditions with road traffic and wind. Vertical accelerations were measured on downstream and upstream side of the bridge and the lateral acceleration at different cross sections. Transversal behavior has been identified in a frequency range between 0...5 Hz. Usage of MEMS Capacitive sensors from KISTLER allow getting displacement data from acceleration measurement and correlating those measurements to two thermocouples. They provide timely information on the actual structural state over time.

In a first step, the main target of the modal estimate investigation was to get information on very slight variations of the natural frequency through time. In a second phase, the work was to provide the automated estimation of the modal parameters and investigate their variations in order to identify possible dependencies on either temperature or traffic intensity and allow for surveillance of the health state of the bridge. The monitoring system developed and shown in Fig. 11.9 is in usage since November 2011. It consists of a total 21 Servo MEMS 8330A3 (closed loop MEMS capacitive that could now be replaced with 8316A2D0... considering the noise level requirements) plus seven DAQ Units, two thermocouples, two Ethernet switch devices and one industrial PC.

Among all the products available on the market, Kistler sensor has been selected here for several reasons:

1. During train passages the acceleration experienced by the girder usually ranges between 0.4 and 0.5 g. As the vertical MEMS is subjected to gravity, one needs at least a ±2 g range to avoid sensor overload.
2. Good compromise between measuring range and sensitivity, noise level and threshold values to accurately collect the signals during the different monitoring scenarios.
3. DC response sensor and linearity allow for accurate calculation of displacement from acceleration data.

Each bridge cross section has been equipped with three sensors: vertical acceleration on downstream, upstream and lateral acceleration. The thermocouple is measuring the air temperature nearby the structure. One DAQ unit is available at each instrumented cross-section to minimize local wiring from sensors to A/D converter. Two switch devices collect the Ethernet cables from each group of channels to transmit the data to an industrial PC (Fig. 11.10). Sampling is made at 200 Hz.

One of the signals investigated and later used for monitoring is the average RMS of the vertical acceleration over all the channels. Figure 11.11 is an example of a series with four train passages. Peaks and valleys correspond to the train passages and temporary stops of alternate traffic. We can see major peaks induced by the train passage. An RMS value of the acceleration is being used as a local measure of traffic effects at a certain vertical/horizontal instrumented position. The mean value of all the RMS acceleration is used as global measures of traffic intensity on all the vertical/horizontal instrumented positions.

A though signal processing allows them to provide an hourly dataset of ambient vibration signals related to road traffic, wind and microtremors as well as several datasets corresponding to a train passage and temperature time histories. Figure 11.12 shows a typical daily pattern of the RMS vertical acceleration at channel 11 (blue) and horizontal acceleration at channel 19 (red). We can see that horizontal acceleration is generally greater than the vertical one. We can also identify two rush hours at 7 a.m. and 5 p.m.

Tracking of natural frequencies of critical modes and at critical nodes allow for a continuous monitoring of the structural state of the bridge. In Fig. 11.13, we can see that during the ice period the natural frequency of modes VB1 and VB2 shows a strong variation in the curve below. The signal is clearly exhibiting dependency on traffic intensity or on temperature or both and eventually on the vibration level. Understanding those interdependencies allow for structural health monitoring. Of course, modes sensitive to traffic induced vibrations but less to temperature are the ones that will be used.

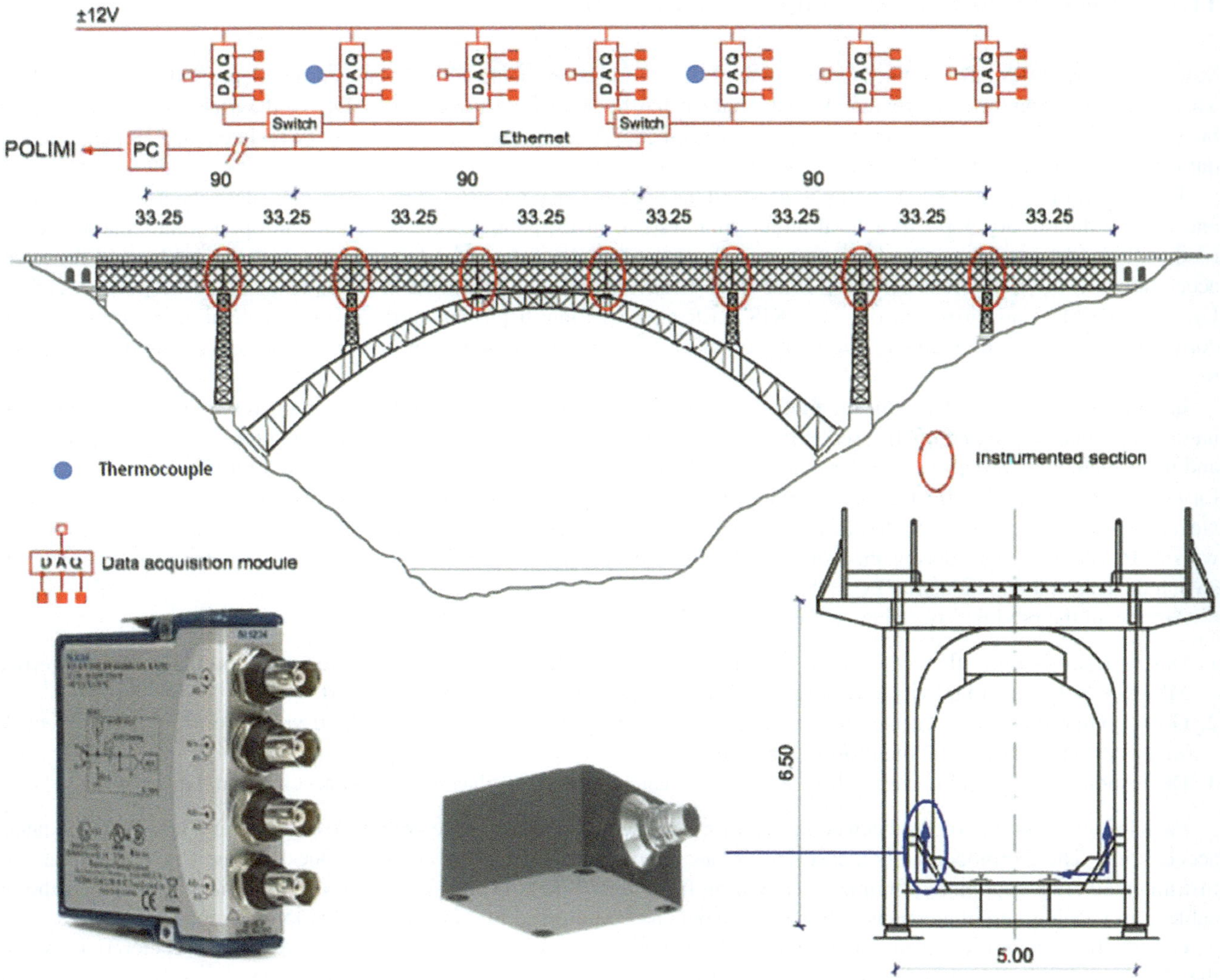

Fig. 11.9 Monitoring system using MEMS capacitive accelerometer and NI data acquisition system [4]

11.11 Conclusion

In the aftermath of the Minneapolis disaster, the official enquiries into the causes of the bridge's collapse soon disclosed hints that some vital parts of the structure, dating from the early sixties, had been grossly undersized due to some obvious design flaws. It is not a secret that the I-35W bridge collapse was not an isolated incident, but much of the US' infrastructure is in a deplorable state. With a large number of bridges, dams, and power supply lines being nearly 100 years old.

An MIT estimate says that, over the next few years, costs for repairs and refits will amount to roughly 600 billion dollars a year. Investing more money into structural testing as well as structural monitoring tools will for sure prevent such high costs to happen for rehabilitated or brand new structures worldwide. One of the exceptional tools worth investing in is the Capacitive Sensor Technology.

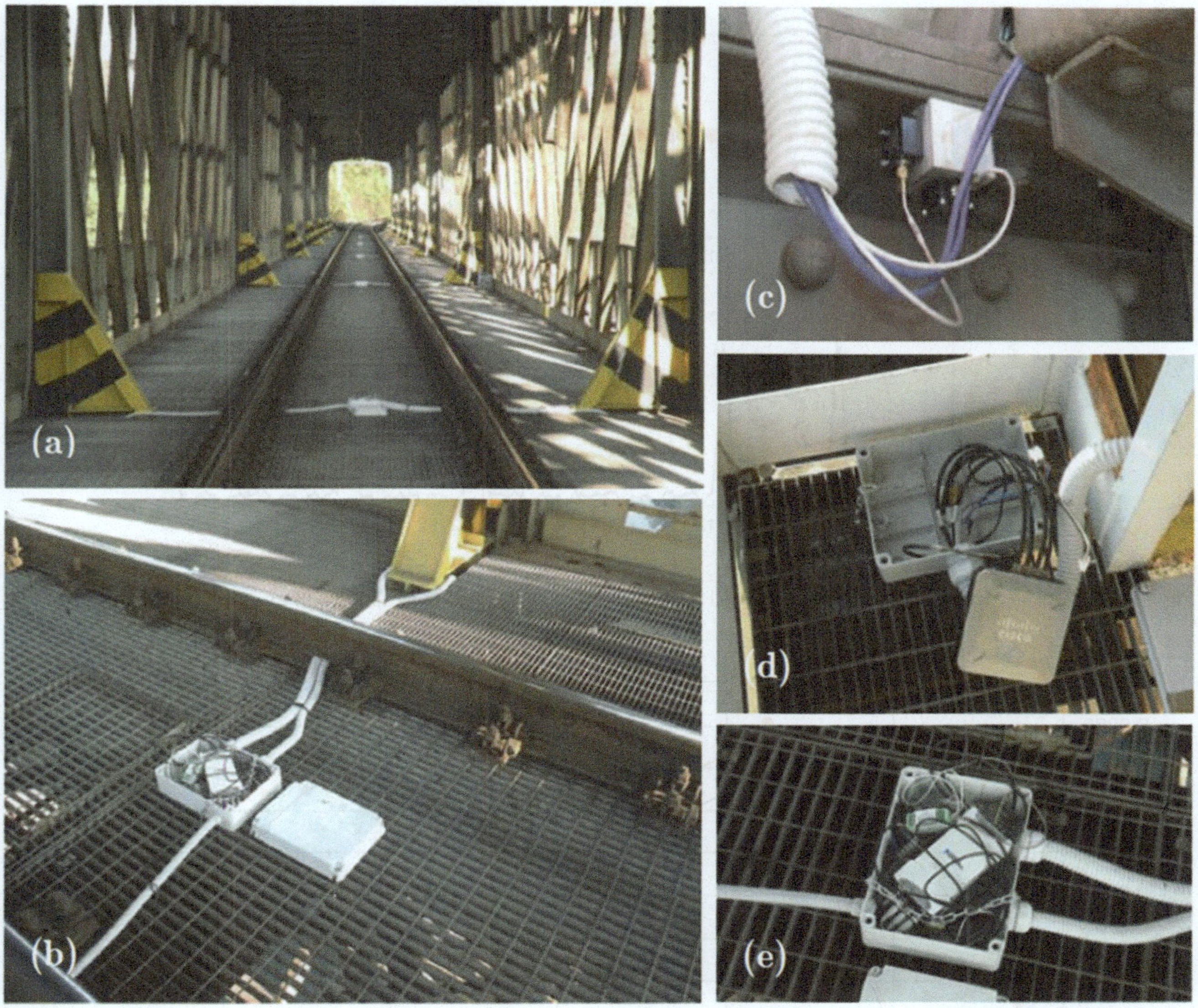

Fig. 11.10 Monitoring system installation [4] – (**a**) and (**b**) Distributed architecture along the bridge; (**c**) Kistler Servo MEMS capacitive accelerometer; (**d**) Switch; (**e**) National Instruments data acquisition system

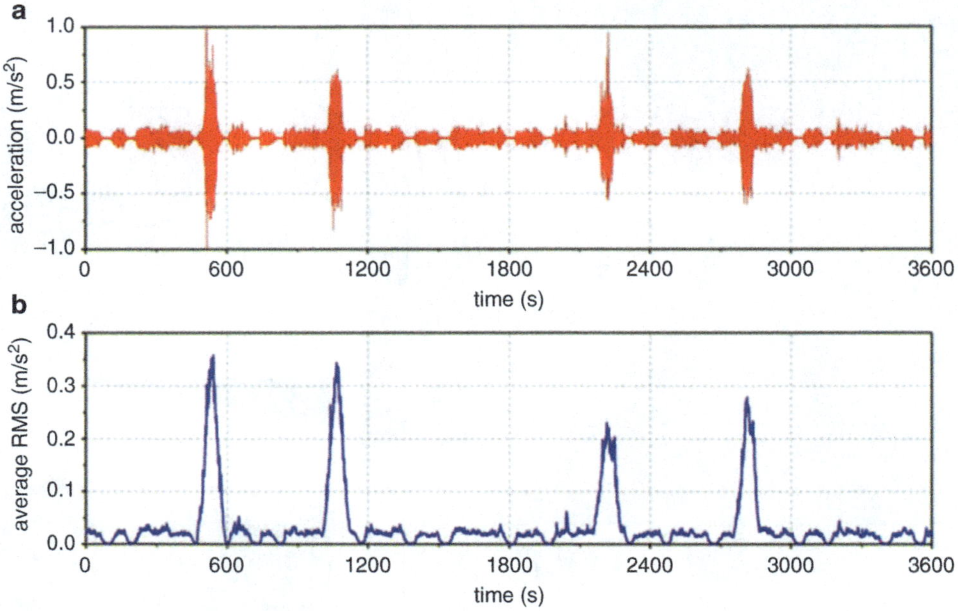

Fig. 11.11 Rush hour data set acceleration induced by rail and road traffic; (**a**) Acceleration time history; (**b**) Averaged RMS function [4]

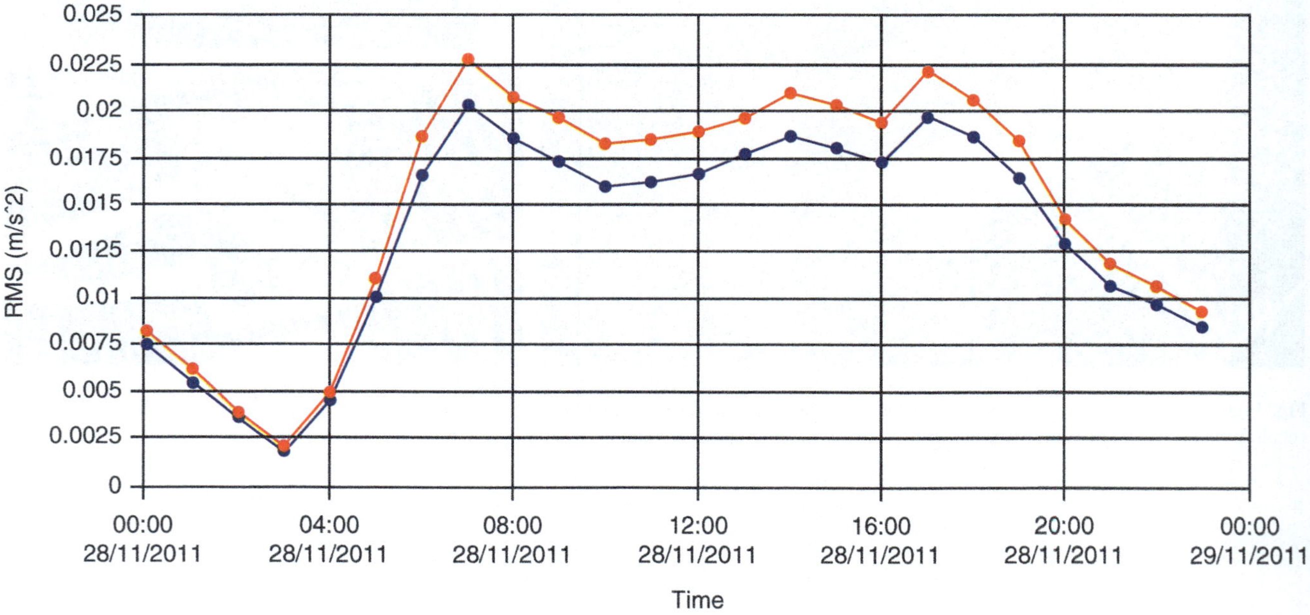

Fig. 11.12 RMS function of acceleration at channel 11 (vertical, blue) and 19 (horizontal, red) [4]

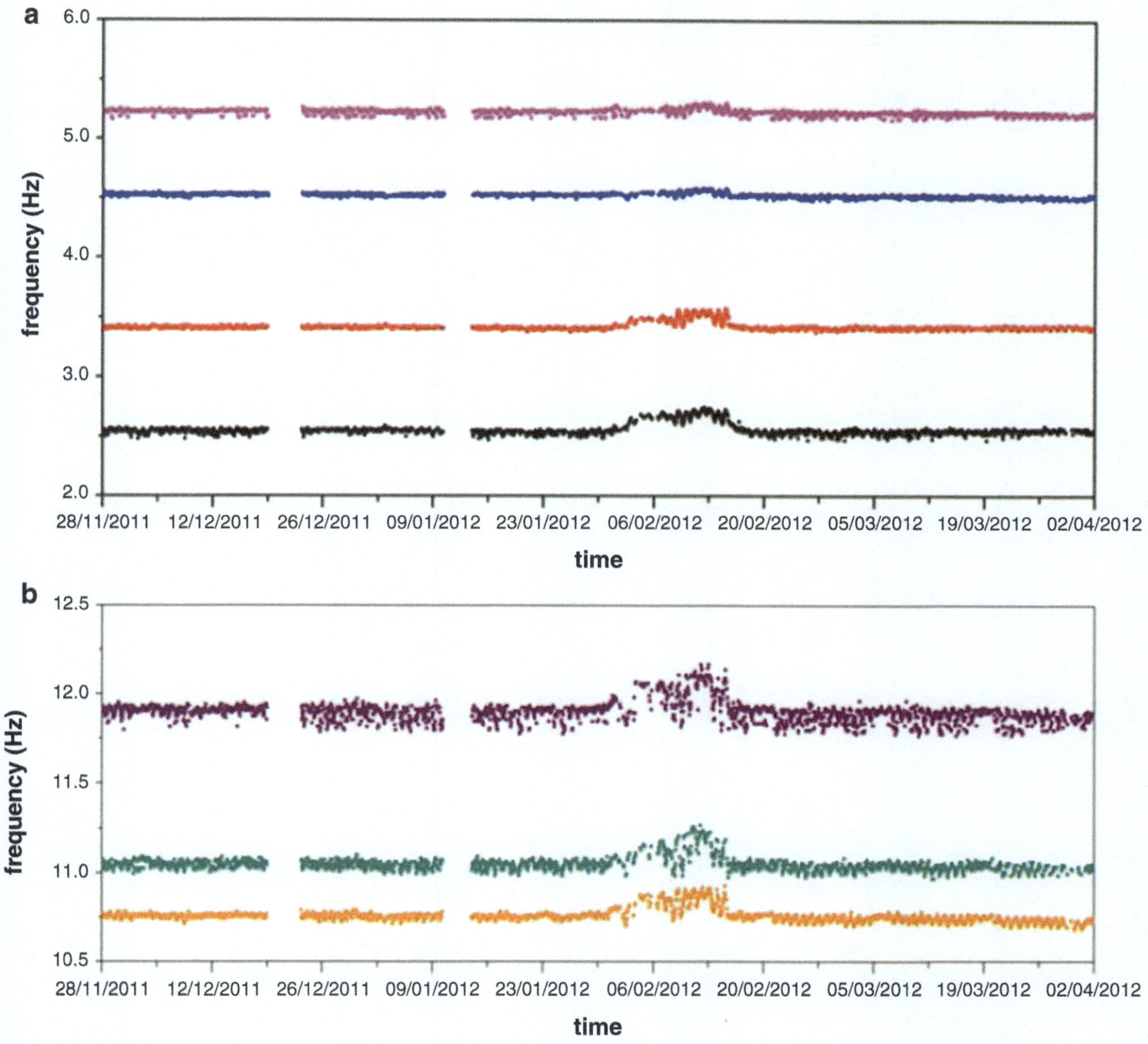

Fig. 11.13 Natural frequency modes including ice period; (**a**) VB1–VB4; (**b**) VB5–VB7 [4]

References

1. Wong, K.-Y., et al.: Modular architecture of SHM system for cable-supported bridges (2009)
2. Inaudi, D., et al.: Structural health monitoring system for the new I-35W St. Anthony Falls Bridge (2009)
3. Lockhart, R.W.: Bridge structural monitoring system, distributed synchronous Ethernet data acquisition systems satisfy demanding I35W St. Anthony Falls Bridge health measurements
4. Busatta, F.: Dynamic monitoring and automated modal identification of large structures: methodological aspects and application to a historic iron bridge (2012)

Chapter 12
Using the Random Decrement Technique on Short Records with Varying Signal-to-Noise Ratios

Ronwaldo Emmanuel R. Aquino and Yukio Tamura

Abstract An earlier paper by the authors discusses potential issues in the practical implementation of the Random Decrement Technique (RDT), particularly when the data has a low signal-to-noise ratio (SNR), and when filtering, which distorts the data and affects the results, is necessary. This is the case for practically every measurement of full-scale buildings, typically done at very low ambient response amplitudes. The RDT procedure used also enables a form of error quantification which allows the engineer or data analyst to judge how reliable the result is. In one of the examples in the earlier paper, it was shown that for a certain SNR, an improvement in the damping estimate is evident when the recording length is significantly longer. For the studied structure, this meant that for better results the recording had to be 72 h (or about 46,000*T where T is the building period of interest) instead of the 4 h initially considered. Even then it required a much longer record duration, estimated to be equivalent to a month of recording, to arrive at an acceptably accurate result. However, such a very long record length is impractical. The current study thus firstly aims to determine what is the minimum SNR (maximum noise level) given say just a 1-h recording, to achieve a certain level of accuracy. This is done by varying the noise level and therefore the SNRs, and applying the mentioned RDT procedure and comparing SNR against the calculated error. The SNRs of ambient response measurements of a full-scale building with and without a supplementary damping system are then viewed considering the finding on minimum SNR.

Keywords Free decay curve fitting technique · Random decrement technique · Short-duration records · Signal-to-noise ratio · System identification

12.1 Introduction

An earlier paper by the same authors [1] discusses potential issues in the practical implementation of the Random Decrement Technique (RDT), particularly when the data has a low signal-to-noise ratio (SNR), and when filtering, which distorts the data and affects the results, is necessary. This is the case for practically every measurement of full-scale buildings, typically done at very low ambient response amplitudes. The system identification procedure (SIP) used by the authors in conjunction with RDT also enables a form of error quantification allowing an engineer or data analyst to judge how reliable the result is.

In one of the examples in the mentioned earlier paper [1], it was shown that for a certain SNR, an improvement in the damping estimate is evident when the recording length is significantly longer. For the studied system, this meant that for better results the recording length had to be about 72 h (or about 46,000T_n where T_n is the building period of interest) instead of the 4 h initially considered. Even then, it required a much longer record duration, estimated to be equivalent to a month of continuous recording, to arrive at an acceptably accurate result. However, such a very long record length would have been impractical for most structures.

The current study thus aims to determine what is an acceptable SNR given a short recording (e.g. about 2500T_n), to achieve a certain level of accuracy using the RDT.

R. E. R. Aquino (✉)
RWDI, Guelph, ON, Canada
e-mail: ron.aquino@rwdi.com

Y. Tamura
Chongqing University, Chongqing, China

S. Pakzad (ed.), *Dynamics of Civil Structures, Volume 2*, Conference Proceedings of the Society for Experimental Mechanics Series, https://doi.org/10.1007/978-3-319-74421-6_12

12.2 Literature Review

The RDT is attributed to research work in the aerospace industry [2], but it has since found popular application in civil engineering structures [1, 3–7]. In the earlier paper mentioned [1], RDT is viewed as something more like a data conditioning technique, akin to a filter. The principle behind the RDT is simple: take several short segments from a data recording and any noise may be filtered out by effectively cancelling out after superimposing (adding) enough data segments. The selection of the data segments to be superimposed are an important feature of the RDT. After the RDT is performed, the result is what is called a Random Decrement Signature (RDS). With the proper RDT parameters (including recording length) and given sufficient data quality (SNR level), among other considerations, this RDS can be a good representation of the damped free vibration response (herein referred to simply as "free decay" or FD) of the system. A system identification technique then needs to be applied to the RDS, treating it like FD.

Past research has focused on improving the results of RDT by selecting the most appropriate triggering condition [3], which is evident in the work in Refs. [4, 6, 7]. There is usually a minimum quantity of superimposed segments (e.g. [6]), or a bandpass filter is required (e.g. [5]). As mentioned in Ref. [1], however, filtering only distorts the data and consequently the results, and the quantity of superimposed segments or the length of data recording therefore becomes a more important variable. There is no mention of minimum SNR in the literature that the authors are aware of, whereas typically, research in RDT come up with minimum record length or number of superimposed segments. As shown in Ref. [1], in some cases it would need weeks of measurements for RDT to work, which is impractical, and so in this study, the SNR is treated as a condition for performing RDT instead of record length or number of superimposed segments.

12.3 Methodology

For the current study, the response of a linear one-degree-of-freedom (1DOF) system to random excitation is first simulated. The 1DOF system represents the first fundamental mode of a super tall building, with undamped natural frequency $f_n = 0.18$ Hz (or undamped natural period $T_n = 5.56$ s). It is assumed that the higher order modes are not captured in the same recording. A damping ratio of $\zeta = 2\%$ is assumed for the system. The generated random response represents a measurement under ambient conditions. The response simulation has a duration of 4 h, representing about $2500{*}T_n$. (Thus, if a 5 Hz structure was simulated, the duration would have only been under 9 min.) The sampling rate used is 50 Hz, or about $250{*}f_n$.

The next step is to simulate a separate set of Gaussian white noise of the same duration and sampling rate as the simulated random structural response. The noise is multiplied by various scalar numbers and added to the simulated random response arriving at the six different records shown in full in Fig. 12.1, with corresponding spectra in Fig. 12.2.

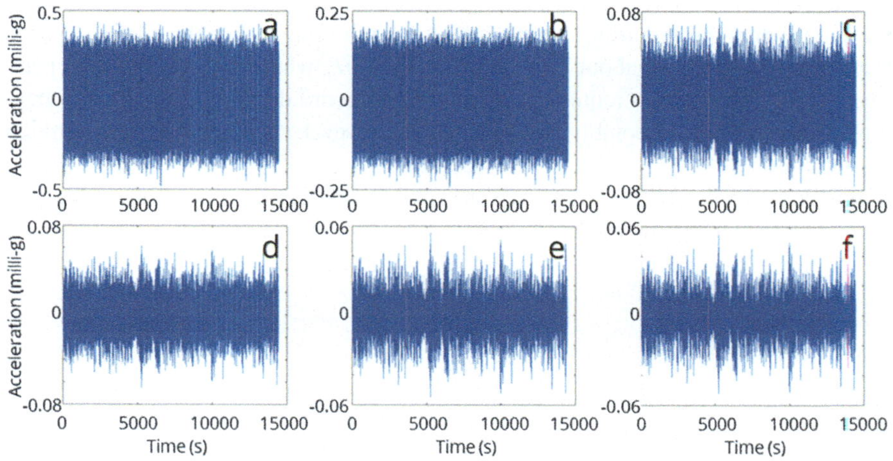

Fig. 12.1 Full time-histories of simulated response with various levels of noise and various SNR values. SNR = (**a**) −24.812 dB, (**b**) −20.560 dB, (**c**) −10.433 dB, (**d**) −8.940 dB, (**e**) −8.326 dB, and (**f**) −8.298 dB

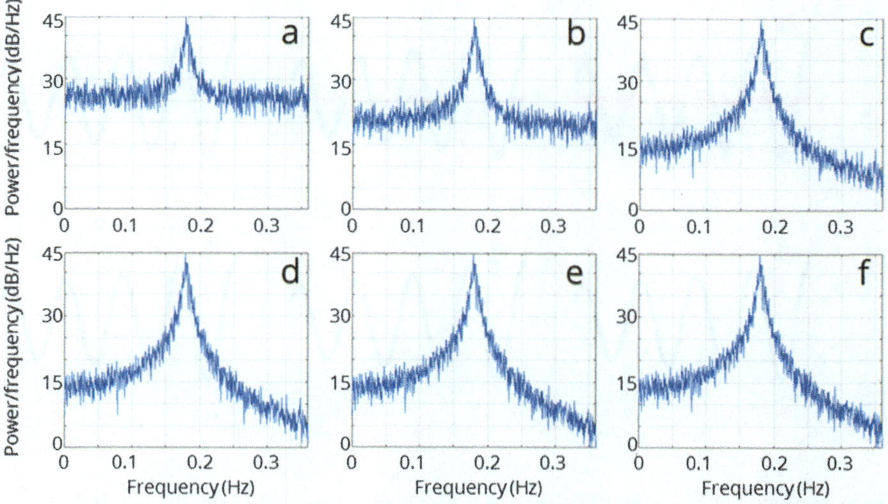

Fig. 12.2 Power spectral density plots of simulated response with various levels of noise and various SNR values. SNR = (**a**) −24.812 dB, (**b**) −20.560 dB, (**c**) −10.433 dB, (**d**) −8.940 dB, (**e**) −8.326 dB, and (**f**) −8.298 dB

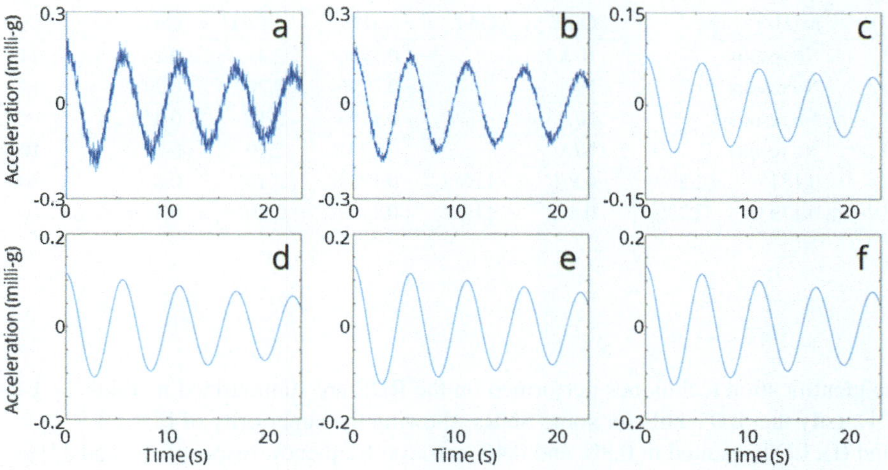

Fig. 12.3 RDS generated for records with SNR = (**a**) −24.812 dB, (**b**) −20.560 dB, (**c**) −10.433 dB, (**d**) −8.940 dB, (**e**) −8.326 dB, and (**f**) −8.298 dB

As indicated in Figs. 12.1 and 12.2, (a) would be the recording with the highest added noise level and (e) is the one with the lowest noise level, while (f) does not have any added noise. Note how visible the white noise is in the first two spectra which correspond to the recordings with the highest SNRs.

The next step is to perform RDT on these records, arriving at the corresponding RDSs shown in Fig. 12.3. After the RDSs are generated, finally, the two system identification techniques are used: the Logarithmic Decrement Technique (LDT) which has been traditionally used with RDT, and a Free Decay Curve Fitting Technique (FDCFT) which was first introduced in Ref. [7]. The resulting fitted damped 1DOF response curves as well as the residuals are shown in Fig. 12.4.

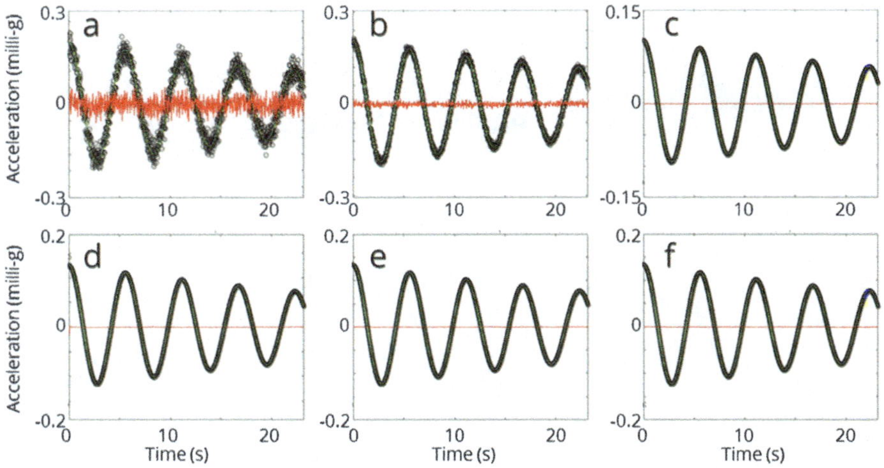

Fig. 12.4 Fitted curves and residuals plotted against RDS for records with SNR = (**a**) −24.812 dB, (**b**) −20.560 dB, (**c**) −10.433 dB, (**d**) −8.940 dB, (**e**) −8.326 dB, and (**f**) −8.298 dB

Table 12.1 Results of system identification on RDSs

		LDT results		LDT errors		FDCFT results		FDCFT errors		
Record	SNR (dB)	f_n (Hz)	ζ	On f_n	On ζ	f_n (Hz)	ζ (%)	On f_n (%)	On ζ (%)	LSE (%)
(a)	−24.812	No results		N/A		0.1796	2.47	0.2	24	20.0
(b)	−20.560	No results		N/A		0.1796	2.26	0.2	13	5.3
(c)	−10.433	No results		N/A		0.1793	2.16	0.4	8	0.5
(d)	−8.940	No results		N/A		0.1791	2.19	0.5	10	0.8
(e)	−8.326	0.1815	2.22%	0.8%	11%	0.1789	2.19	0.6	10	0.3
(f)	−8.298	0.1793	2.22%	0.4%	11%	0.1793	2.20	0.4	10	0.3

12.4 Results

The results from the identification techniques performed on the RDS are summarized in Table 12.1 and in Fig. 12.5. It can be seen in Table 12.1 firstly that LDT fails for some SNRs, showing the superiority of FDCFT.

For records (e) and (f), LDT resulted in 0.8% and 0.4% error on frequency, respectively, and 11% error on damping ratio for both. For the FDCFT results, the error on frequency and damping ranges from 0.2% to 0.6% and 8% to 24%, respectively. The least squares error defined in Ref. [1] is also listed for the FDCFT results, generally showing very low least squares error or LSE (i.e., very good curve fitting) for higher SNRs.

Figure 12.5 further summarizes the results in Table 12.1. It shows a good linear relationship on a logarithmic plot (except for apparently one outlier) between SNR and LSE (shown in the plot as LSQerror), that ideally the LSE is below 1%, and finally that at a minimum SNR of say −15 dB, we may be able to obtain the best estimates of damping (with under 10% error) and frequency (with under 0.4% error).

12.5 Application

Two short duration measurements on a super tall building were obtained: first with its mass damper disabled, and second with it enabled. The records needed to be filtered to obtain SNRs of −4.26 and −14.21 dB for the without- and with-damper case, respectively. These are better than our earlier estimated minimum SNR of −15 dB. RDT is performed at multiple amplitudes (i.e. "RDma") to arrive at various damping and frequency estimates for each record such as shown in Fig. 12.6 for the without-and with-damper cases. Note that the with-damper case required a 2DOF curve fit.

The results show average values of 1.3% damping ratio and 0.146 Hz frequency for the building without damper over a range of amplitudes, and about 4.0% and 4.1%, and 0.137 and 0.156 Hz frequency for the building with damper enabled, which are in line with predictions for this building and damper.

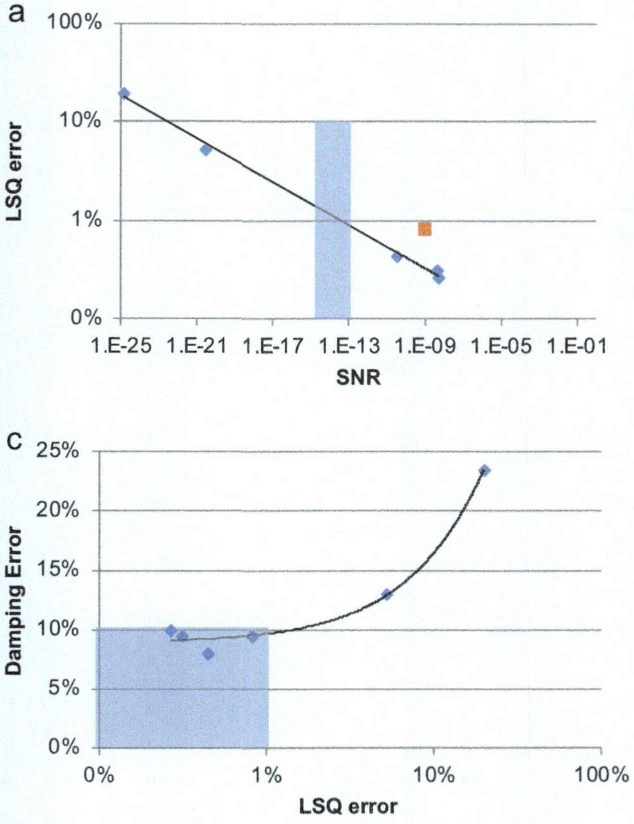

Fig. 12.5 Summary of FDCFT results: (**a**) SNR vs. LSQ error, (**b**) SNR vs. absolute frequency error, (**c**) LSQ error vs. damping error, and (**d**) SNR vs. damping error

12.6 Conclusion

By way of a simple numerical experiment, a SNR of about −15 dB or maximum LSE on FDCFT of under 1% was found necessary to arrive at acceptable frequency and damping results from RDT. This is illustrated by analysis of data from a building before and after its tuned mass damper was enabled, which unfortunately needed to be filtered to have SNR above −15 dB.

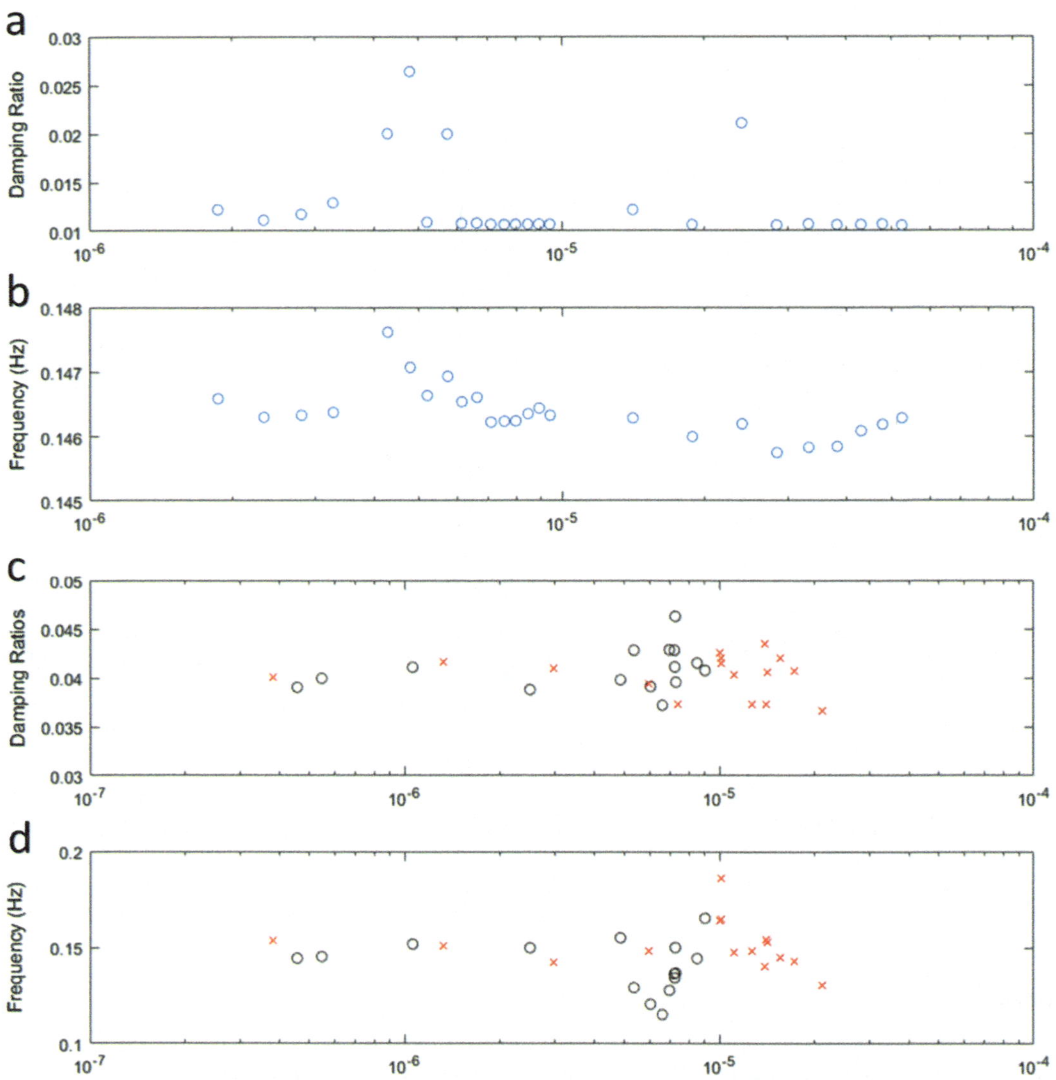

Fig. 12.6 Summary of RDma + FDCFT (**a**) damping and (**b**) frequency results on tall building without a damper and (**c**) damping and (**d**) frequency results on tall building with damper

References

1. Aquino, R.E.R., Tamura, Y.: Potential pitfalls in the practical application of the random decrement technique. In: Proceedings of the 5th International Structural Specialty Conference, 2016 CSCE Annual Conference, Canadian Society of Civil Engineering, London, ON, Canada (2016)
2. Cole, H.A.: On-Line Failure Detection and Damping Measurement of Aerospace Structures by Random Decrement Signatures. NASA, Mountain View, CA (1973)
3. Cole, H.A.: Randomdec in Retrospect. CreateSpace Independent Publishing Platform, Los Altos, CA (2014)
4. Asmussen, J.C.: Modal analysis based on the random decrement technique – application to civil engineering structures. Ph.D. dissertation, University of Aalborg, Denmark (1997)
5. Jeary, A.P.: Damping in tall buildings – a mechanism and a predictor. Earthq. Eng. Struct. Dyn. **14**, 733–750 (1986)
6. Tamura, Y., Suganuma, S.: Evaluation of amplitude-dependent damping and natural frequency of buildings during strong winds. J. Wind Eng. Ind. Aerodyn. **59**(2), 115–130 (1996)
7. Kijewski, T., Kareem, A.: Reliability of random decrement technique for estimates of structural damping. In: Proceedings of the 8th ASCE Specialty Conference on Probabilistic Mechanics and Structural Reliability, Notre Dame, IN, USA (2000)

Chapter 13
Why a Curb Shouldn't Be Kicked to the Curb: The Importance of Non-Structural Elements in Dynamic Modelling

Michael J. Wesolowsky, Melissa Wong, Allan Raun, and John C. Swallow

Abstract Occupant footfalls are often the most critical sources of floor vibration on the elevated floors of buildings, especially if rhythmic activity is expected (e.g. dancing or aerobics). Achieving reasonable vibration levels on these floors requires sufficiently stiff and massive floor structures to effectively resist the forces exerted by larger groups of people. In many cases, further vibration control is provided by Tuned Mass Dampers (TMDs). A difficulty for engineers in modelling buildings for these scenarios can be exacerbated due to the uncertainty provided by non-structural elements (e.g. non-load bearing partitions, floor toppings, curbs or railings).

In this paper, three case studies are presented of modelling structures in order to predict vibrations due to rhythmic activity. The first structure is a sports arena which features a large cantilevered balcony upon which dancing by 600 people was expected to occur. The structural design included TMDs to control these expected vibrations. Validation testing conducted once construction was complete indicated that the balcony was significantly stiffer than expected, and a complete redesign of the TMDs was required. The second structure is a long-span floor office tower that was designed with light steel trusses. Modelling predicted excessive vibration from aerobic activity on the amenity floor, which was proposed to be mitigated with TMDs. Validation testing indicated that the measured frequencies were almost 250% higher than those in the model, completely removing the need to implement TMDs. The third case study is a project consisting of two hospital towers that were nearing completion. Peer-review modelling indicated expected marginal exceedance of the required criteria, so the decision was made to measure the as-built floors. Measurements showed that frequencies were considerably higher than predicted, and extraordinarily high damping.

In all three case studies, it was concluded that non-structural elements were the cause of the large discrepancies between modelled and measured dynamic properties.

Keywords Vibration serviceability · Footfall vibration · Vibration measurements · Model validation · Sensitive floors · Dynamic loading

13.1 Introduction

The primary sources of vibration in most buildings are due to human activity, especially footfalls from walking or aerobic activity. As people walk or jump, the impact from each footfall induces floor motions that may also easily transmit to nearby spaces. Quantifying vibration from walking, whether through measurement of existing spaces or numerical predictions for guiding the design of a new building, is a complex task. This task is complicated in cases where unexpectedly stiff non-structural elements can significantly modify the as-built dynamic properties of the structures.

This paper presents the three case studies to compare measurements to conventionally modelling structures. The first structure is a sports arena which features a large cantilevered balcony upon which dancing by 600 people was expected to occur. The second structure is a long-span floor office tower that was designed with light steel trusses. The third case study is a project consisting of two hospital towers that were nearing completion.

M. J. Wesolowsky (✉) · M. Wong · A. Raun · J. C. Swallow
Swallow Acoustic Consultants Ltd./Thornton Tomasetti, Mississauga, ON, Canada
e-mail: mwesolowsky@thorntontomasetti.com

S. Pakzad (ed.), *Dynamics of Civil Structures, Volume 2*, Conference Proceedings of the Society for Experimental Mechanics Series, https://doi.org/10.1007/978-3-319-74421-6_13

13.2 Vibration Criteria

For human comfort, vibration criteria are normally expressed as the root mean square (RMS) response of each one-third octave band from 1 to 80 Hz [1]. For sensitive equipment, the criteria may be expressed in one-third octave bands, or other formats, including power spectral densities, peak-to-peak levels, etc. Over the past 25 years, generic vibration limits have been developed which provide frequency-dependent sensitivities for wide classes of equipment, and are used extensively in design of healthcare and research facilities [2]. These vibration criterion (VC) curves are internationally accepted as a basis for designing and evaluating the performance of vibration sensitive equipment and the structures that support them. The VC curves range between Workshop (least stringent) through VC-G (most stringent). Most laboratories target at least VC-A as a maximum baseline for the majority of the facility, with certain areas requiring more stringent criteria, depending on the expected use of the space.

These curves were originally based on the ISO 2631-2 [3] base curve for human response to whole body vibration, which is considered the threshold of human perception, but have since evolved. The ISO base curve is often referred to as the ISO-Operating Room criteria. The above noted criteria are also specified as RMS velocities in one-third octave bands. The VC curves should not be used to replace manufacturers' specifications for vibration requirements, but are beneficial where manufacturers' specifications are non-existent, incomplete, or where specific equipment has not yet been selected.

The VC curves will be used as a basis of evaluating the performance of the second and third structures in this paper.

13.3 Case Study #1: Sports Arena

The building used for Case Study #1 is a sports arena, and includes an exterior cantilevered balcony of variable length, ranging from 11 to 14 m (see Fig. 13.1). The expected use of the space is to be as a restaurant or club, with an occupancy of approximately 600 people. Typical construction of the cantilevered trusses can be seen in Fig. 13.2. A structural finite element model (FEM) of a portion of the arena was provided by the structural engineer, which was used for the predictive analysis. The model was modified for the purpose of the dynamic analysis using SAP2000 Advanced v16 analysis software package. The balcony had both a concrete curb to which was attached a glass railing, as well as a concrete topping that was placed over a resilient material. Both of these non-structural elements were modelled as distributed dead loads over the area of the balcony. The predicted natural frequencies of the balcony ranged between 3 and 3.3 Hz.

The following parameters were considered for the dynamic analysis:

- Activity participants weighing 76 kg (167 lb), as recommended by the SCI P354 [4];
- Modal damping of 1.1%, as recommended by the SCI for walking scenarios;

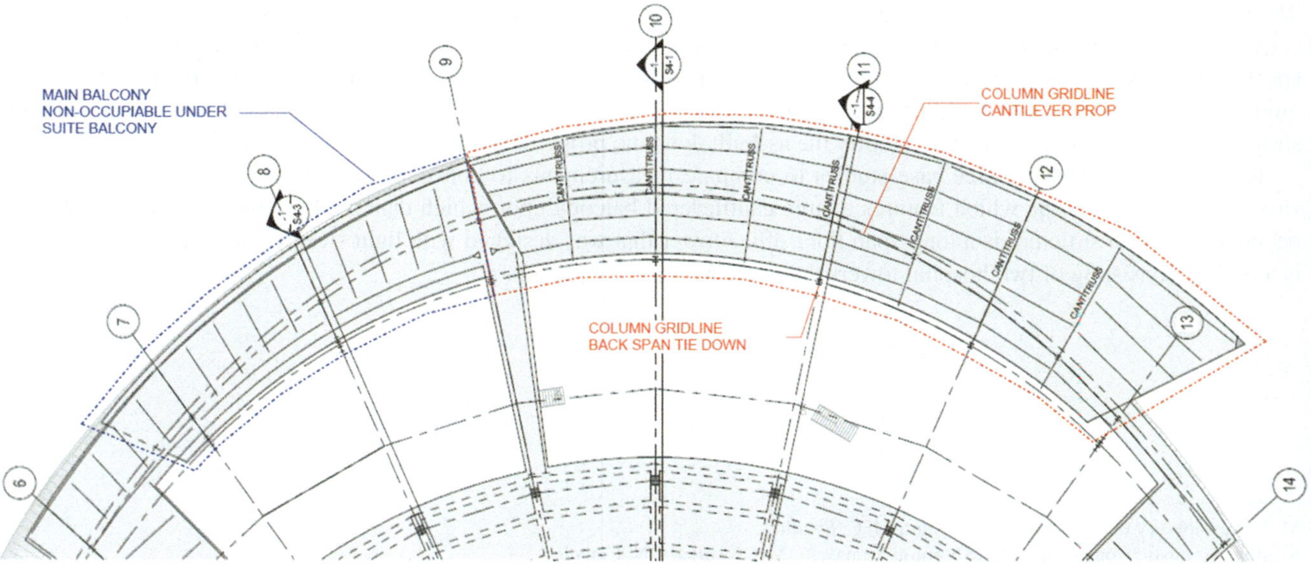

Fig. 13.1 Plan view of the cantilevered balcony

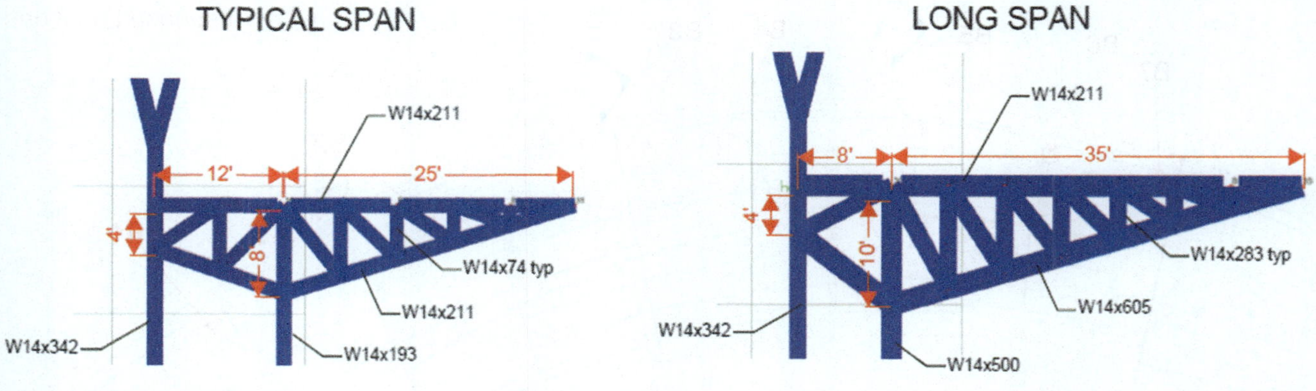

Fig. 13.2 Typical cantilevered trusses

Table 13.1 Predicted peak accelerations along the external cantilevered balcony (at the locations shown in Fig. 13.3 – note B5 and B7 are intentionally not listed in Table 13.1)

Location	General socializing (%g)	Moderate dancing (%g)	Heavy dancing (%g)
B1	9.7	58.1	130.3
B2	2.2	12.7	31.6
B3	0.7	7.0	17.6
B4	0.7	7.2	28.5
B6	1	6.7	12.4

- Modal damping of 6.0%, as recommended by the AISC [5] for rhythmic activity scenarios;
- Rhythmic activities having dancing (jumping/bouncing) speeds between 90 and 168 steps per minute (spm) (1.5–2.8 Hz), as recommended by the SCI; and,
- A target vibration criterion of 7%g, as per the recommendation of the AISC.

The following occupant activity schemes were considered, assuming participants are arrayed evenly over the accessible portions of the cantilevers (i.e. between the railings and the main supporting structure to which the cantilevers are affixed):

1. 'General socializing' – this scheme involves people milling about and swaying to the music without significant synchronization;
2. 'Moderate dancing' – this scheme involves the occupants as a group dancing moderately to the music with significant synchronization, and with 1 ft or another occasionally leaving the ground; and,
3. 'Heavy dancing' – this scheme is more similar to aerobic activities, where occupants are jumping up and down to the beat of the music. It can also be described as 'moshing' or 'slamdancing'.

The maximum peak acceleration predictions for each activity for several locations along the balcony were tabulated, and can be found in Table 13.1. The locations considered can be found in Fig. 13.3.

From Table 13.1 it can be seen that the predicted vibrations at location B1 exceeds the recommended criterion for general socializing, however all other areas are expected to have perceptible, though acceptable vibration levels. When moderate dancing is considered, most areas were expected to either marginally or excessively exceed the criterion. The predicted levels at B2 would be considered disturbing, and at B1 would likely cause panic. For most areas, heavy dancing is expected to result in vibrations that would likely cause panic.

Based on this analysis, it was decided to proceed with the design and installation of Tuned Mass Dampers (TMDs) at locations B1–B7, as shown in Fig. 13.3. The TMDs were expected to range in mass between 1.5 and 3 tonnes. The locations were chosen based on the predicted mode shapes of the cantilevered balcony. As a side note, the alternative non-TMD design to control vibration would have required an increase in structural steel by a factor of 10, which was determined to be uneconomical.

Once the structure was substantially complete, the as-built cantilever vibration characteristics were determined by applying a broadband vibration input force using an electromagnetic shaker. The floor's response was measured at the seven locations indicated in Fig. 13.3, and natural frequencies were identified. Each natural frequency was investigated to determine the associated mode shape, modal mass, and damping ratio. Testing identified that the expected frequencies of 3–3.3 Hz were measured at 3.9–5.6 Hz. These frequencies are tabulated in Table 13.2. Further, the measured mode shapes were significantly different than those predicted by the model.

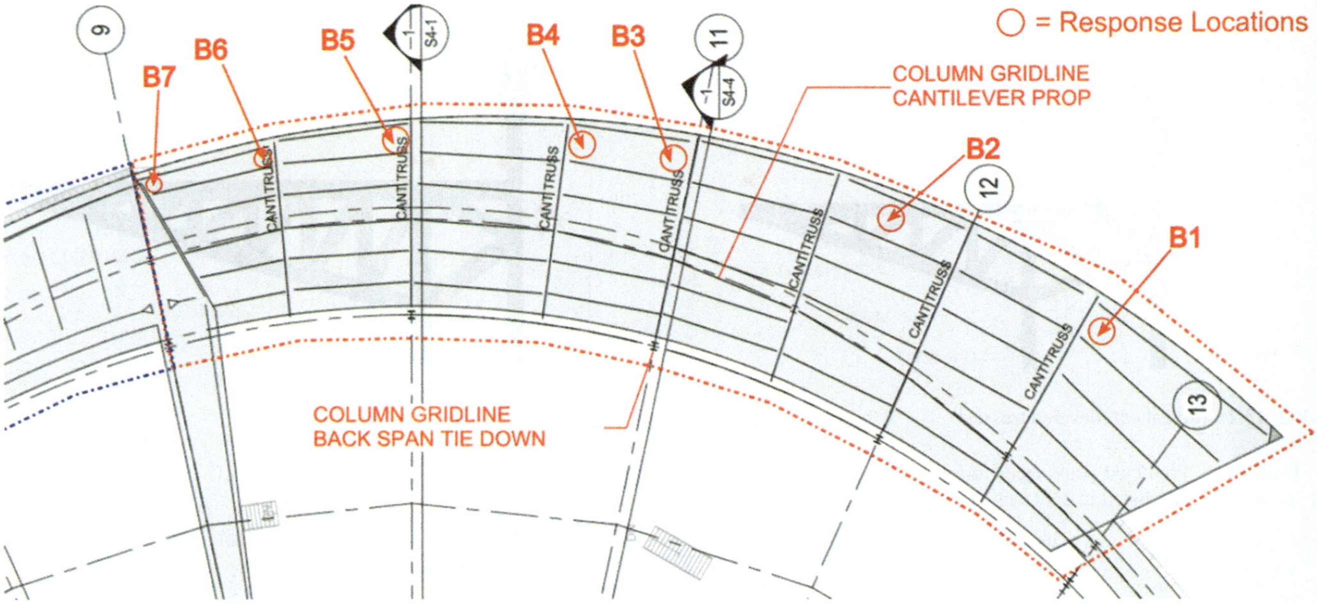

Fig. 13.3 Locations considered for acceleration predictions on the external balcony

Table 13.2 Predicted vs. measured frequencies on the cantilevered balcony

Mode	Predicted frequency (Hz)	Measured frequency (Hz)	Frequency increase (%)	Stiffness increase (%)
1	3.0	3.9	30	69
2	3.1	5.0	61	160
3	3.3	5.6	70	188

These significant differences required a substantial re-design and reduction in the number of TMDs, as compared to the initial designs developed during the predictive phase of the design. After the investigation, it was determined that the differences were a result of the concrete curb and toppings both being inadvertently structurally linked (by the contractor during construction) to the main supporting structure of the cantilevered balcony.

13.4 Case Study #2: Long-Span Office Floor

The building used for Case Study #2 is a high-rise office tower, constructed of structural steel with a composite concrete/steel deck. The main steel supports are long span (>13 m) deep joist structures. The fourth floor of the tower was designed as an amenity floor, which included a wellness centre, meditation rooms, prayer rooms and an aerobics studio. The aerobics studio (Wellness Studio 'A'), where 20 participants were expected to attend classes, was located within the same structural bay as the meditation rooms (see Fig. 13.4). Preliminary modelling by the structural engineer indicated that a bay frequency of 4.7 Hz was expected, which was considered a cause for concern. Further modelling using the SCI methodology indicated that vibrations were expected to be in excess of 14%g within the studio, and only slightly less in the prayer rooms. In consultation with the design team and the owners, it was decided that a criterion of 5%g in the aerobics studio, and Residential Night (ISO) in the prayer rooms, would be targeted.

As a result of the expected excessive vibrations, it was decided to design and install two Tuned Mass Dampers (TMDs) to control the vibrations in this area. The first one (2 tonnes) was to be placed under the aerobics studio, while the second one (3 tonnes) was to be placed in the bay directly adjacent to the west. Even with these two TMDs, it was expected that the vibrations in the prayer rooms to the north of the aerobics studio would significantly exceed the criterion of Residential Night (ISO), although would be deemed acceptable (<5%g) in the aerobics studio itself. Further, based on this analysis it was decided to move the prayer rooms to a location far from the aerobics studio, and to repurpose those rooms with storage areas.

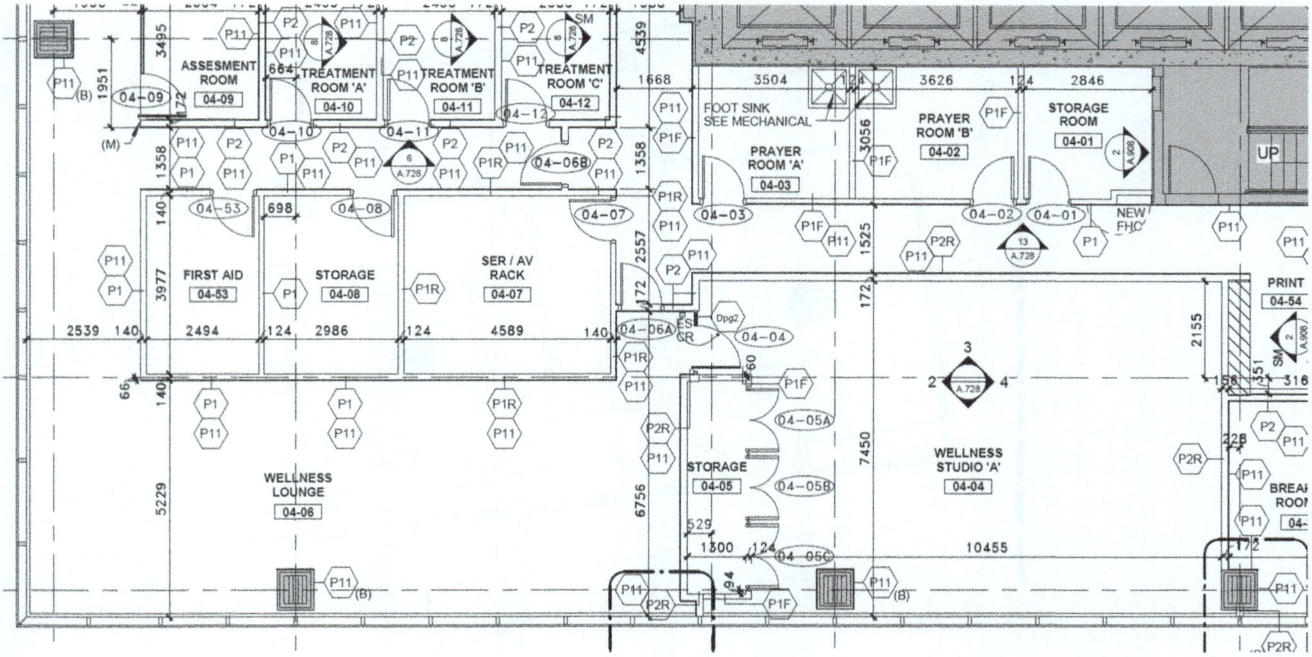

Fig. 13.4 Architectural layout of a portion of the amenity floor

After substantial completion was achieved, floor vibration measurements were performed using an APS 400 electromagnetic shaker located in Wellness Studio A, and the floor's motion was measured in multiple locations, using varying input frequencies and amplitudes. The results indicated that the natural frequency of the first mode of vibration of the structural bay in which Wellness Studio A was located was 10 Hz, which was significantly higher than the 4.7 Hz as predicted by the models. This mode had approximately 4% critical damping. After the investigation, it was determined that the partitions directly below the floor were installed from floor to ceiling, which was not the design indicated on the architectural drawings. Further, the partitions were supposed to have slip joints on the headers of the wall, which had been generally over-tightened by the contractor, inadvertently causing the walls to act as structural supports under serviceability loading. As a result, the decision to proceed with the construction and installation of TMDs was suspended, pending feedback by the users of the studio.

13.5 Case Study #3: Hospital Towers

The third case study is a project consisting of two hospital towers that were nearing completion. Construction deficiencies initiated a peer review process, during which our analysis indicated that the expected vibrations in the operating rooms, patient rooms and diagnostic imaging suites would marginally exceed the criteria. Since the construction was almost complete, it was further decided to perform a series of sample testing throughout each of these three room types in both towers of the hospital.

The following criteria were used for this assessment:

1. Operating Rooms (ORs) – Operating Theatre (ISO)
2. Patient Rooms (PRs) – Residential Night (ISO)
3. Imaging Rooms (IRs) – VC-C

A sampling of IRs, ORs and PRs were pre-selected in order to provide a range of measured responses over a wide area of the buildings. Time-history responses were recorded using PCB 393B31 accelerometers and an 8-channel LMS SCADAS Mobile real-time analyzer. Heel-drop tests were conducted to determine the natural frequencies of the floor. Based on the natural frequencies found, the walking speeds at which the maximum floor excitation would occur were determined. These worst-case walking speeds were then used for the walking scenario load cases. These heel-drop tests also allowed the damping of the floor to be estimated.

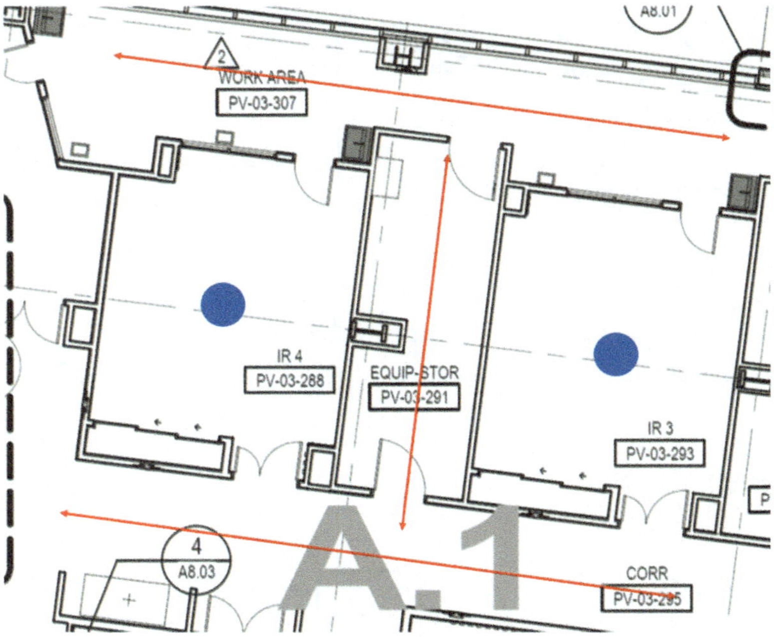

Fig. 13.5 IR accelerometer (blue dots) and walking (red arrows) locations

Table 13.3 Predicted and measured floor frequencies

Testing area	Predicted floor natural frequency (Hz)	Measured floor natural frequency (Hz)
IRs	8–9	14–15
ORs	8–9	16
PRs	6	12–13

The load cases were conducted by having a walker walk at a specified walking speed along pre-determined paths. The walking paths were chosen to run along the lengths of the corridor around the areas being tested, as well as in the rooms, as appropriate. The same procedure was repeated for all tested areas. The placement of the accelerometers for an example test area can be seen in Fig. 13.5. This figures also indicates the walking pathways used.

Finite Element Modelling had been conducted before the testing in order to predict the expected natural frequencies of the floors. These predicted values, as well as the range of floor natural frequencies as determined by the heel-drop tests are shown in Table 13.3.

In most rooms, the test data showed indications of natural frequencies that were close to the predicted values, but the dominant frequencies were much higher (as seen in Table 13.3). It is believed that the placement/construction of the full height partitions provided enough stiffness at the centre of each bay, under service loading, to force the floor system to vibrate predominately at the higher frequency in each area. A sample of this behavior can be seen in Fig. 13.6.

Typical damping for a fully fitted out steel/concrete floor is in the range of 3–4.5% [4]. Inherent damping was calculated from the heel-drop measurements, and was found to be in the range of 12–25% for most areas. In the authors' experience, this level of inherent damping is extremely high, and it is assumed that it is due to the heavy double-wall partitions that extend from floor to ceiling throughout much of the construction. A typical heel-drop time-history decay is shown in Fig. 13.7.

13.6 Conclusions

Three case studies have been presented to compare measurements to modelling of structures that had significant non-structural elements (e.g. curbs, partitions, floor toppings). In all cases, the floors exhibited significantly higher than expected frequencies and damping. The following conclusions have been made:

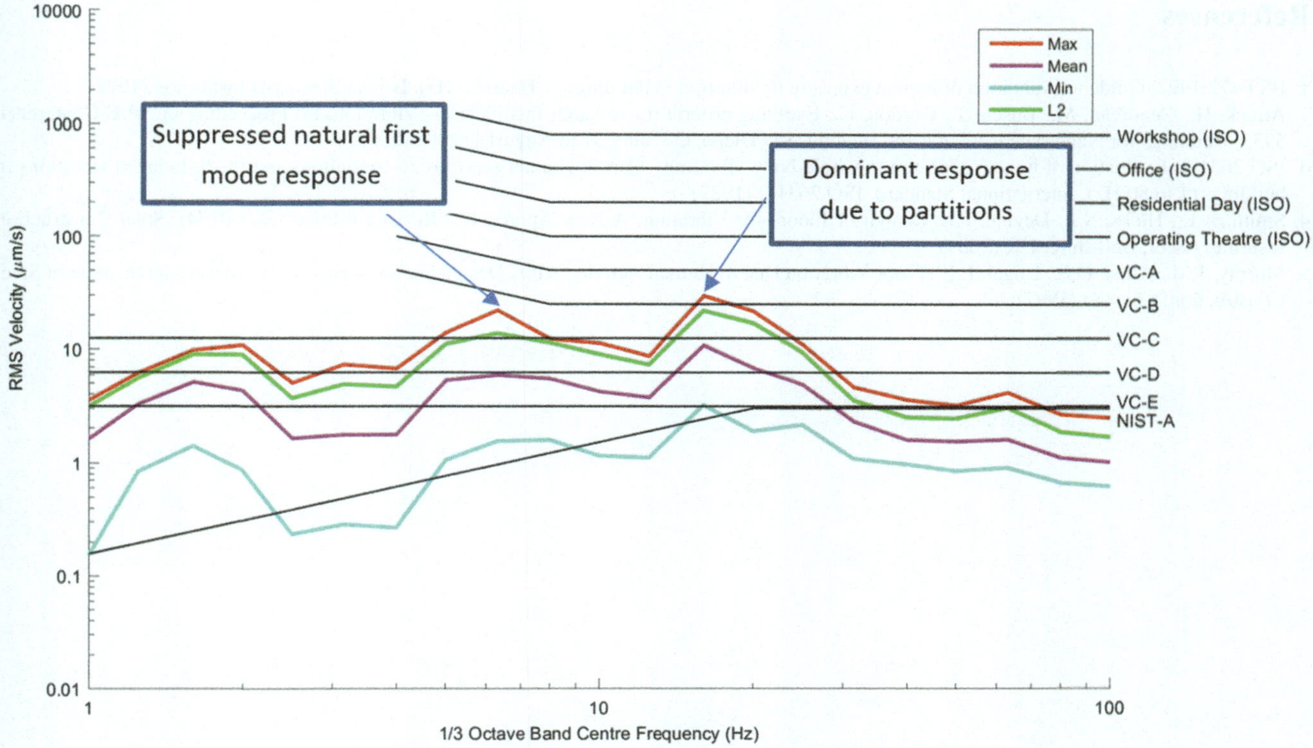

Fig. 13.6 Typical frequency response due to walking

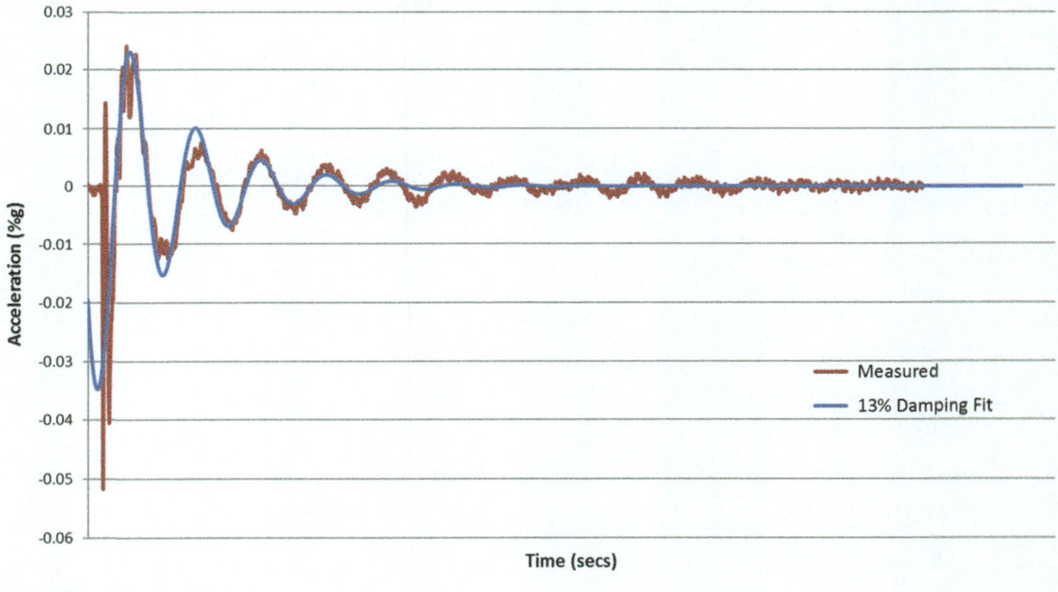

Fig. 13.7 Typical heel-drop time-history

1. When modelling structures that have significant non-structural elements, care must be taken to appropriately account for these elements in order to appropriately capture the expected dynamic properties of the as-built structure.
2. The impact of partitions, which have been traditionally treated as having the effect of only increasing inherent damping, also appear to significantly increase the stiffness and frequency of as-built floors.
3. Further research should be conducted in order to provide appropriate guidance to the design community in order to appropriately account for these non-structural elements during the predictive modelling of structures.

References

1. BS 6472:1992: Guide to evaluation of human exposure to vibration in buildings (1 Hz to 80 Hz). British Standards Institution (1992)
2. Amick, H., Gendreau, M., Busch, T., Gordon, C.: Evolving criteria for research facilities: I – Vibration. In: Proceeding of SPIE Conference 5933: Buildings for Nanoscale Research and Beyond, San Diego, CA, July 31 to August 1, 2005 (2005)
3. ISO 2631-2: Evaluation of human exposure to whole-body vibration—Part 2: Human exposure to continuous and shock-induced vibrations in buildings (1 to 80 Hz). International Standard, ISO 2631-2 (1989)
4. Smith, A.L., Hicks, S.J., Devine, P.J.: Design of Floors for Vibration: A New Approach – Revised Edition (SCI P354). Steel Construction Institute, Ascot, Berkshire, UK (2009)
5. Murray, T.M., Allen, D.E., Ungar, E.E.: Floor Vibration Due to Human Activity, AISC Design Guide, Series No. 11. American Institute of Steel Construction, Chicago (1997)

Chapter 14
Dynamic Behavior of a 130 Years Old Building Under Excessive Sound Pressure

Reto Cantieni

Abstract The Casineum in the City of Luzern, Switzerland, has been built in 1883. In 1911 a major reconstruction added a floor on top of the building and significantly changed the roof load bearing structure. The Casineum serves as a Disco since about 2004. The admissible sound pressure was increased from 98 dBA to 100 dBA recently. As a consequence of respective complaints the structural response to such excitation was measured. In contrary to the façade the roof vibrations were found as being not acceptable. The critical frequency region is f = 30..0.80 Hz where the bass speakers are powering. Acting on either vibration source or transfer path found not being feasible the receiver (the roof) was given a closer look at. The result: The roof structural details are so complex that a straightforward structural solution could not be estimated without experimental validation. Therefore, an Experimental Modal Analysis was performed using ambient as well as white noise excitations.

Keywords Historical building · Disco · Dynamic sound pressure · Excessive roof vibrations · Dynamic load test, experimental modal analysis

14.1 The Structure

The Casineum is part of the Grand Casino Luzern, built in 1883 (Fig. 14.1). It is used as a Disco since 2004. Figures 14.2 and 14.3 give an impression of the music hall and the roof as seen from inside. The 1911 reconstruction is illustrated in Fig. 14.4. The original roof timber truss upper half was cut off and a steel truss was added on top. This added an extra floor located between the steel truss lower and upper chords. The Casineum plan view dimensions are about 20 x 17 m, its height is about 15 m (Fig. 14.5).

14.2 Structural Behavior in Normal Operation

14.2.1 The Problem

Soon after having increased the maximum admissible sound power from 98 dBA to 100 dBA, guests complained about excessive vibrations of the windows on the lake-side balcony. This is the only place where smoking is allowed (Fig. 14.5). Subsequently, the Casineum Management initiated the investigations presented here. Their main concern was the risk of baroque angels falling down on guests (Fig. 14.3).

14.2.2 The Tests

Structural vibrations were measured in the night of March 24/25, 2016, before and during the Disco Night as presented by "Voilà Special". Instrumentation consisted of 8 PCB 393A03 (1 V/g) and 2 PCB 393B31 (10 V/g) piezo-type acceleration

R. Cantieni (✉)
rci dynamics, Structural Dynamics Consultants, Duebendorf, Switzerland
e-mail: reto.cantieni@rcidynamics.ch

S. Pakzad (ed.), *Dynamics of Civil Structures, Volume 2*, Conference Proceedings of the Society for Experimental Mechanics Series, https://doi.org/10.1007/978-3-319-74421-6_14

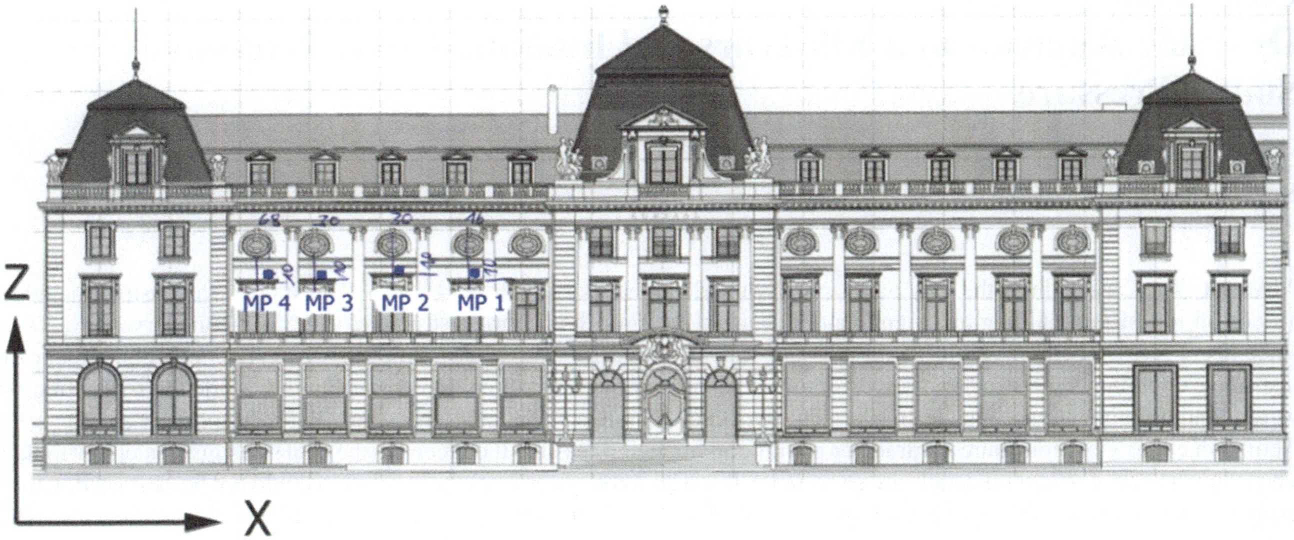

Fig. 14.1 The Grand Casino Luzern as seen from the lake. The Casineum is in the lefthand wing of the building. Also indicated is the location of the four sensors MP 1 to MP 4 as used for the tests described in Chap. 2

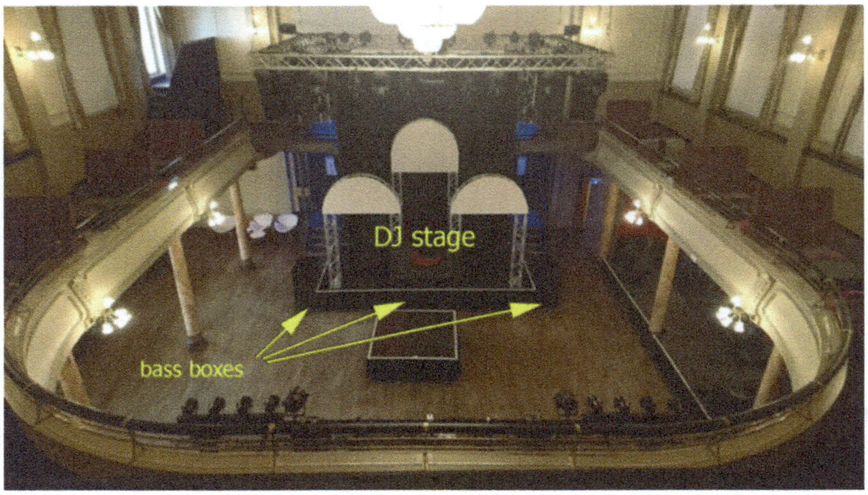

Fig. 14.2 The Casineum

sensors. Four 1D A03 sensors were located at the lake-side façade where the complaints had come from (MP 1..0.4, Figs. 14.1, 14.5 and 14.6). Two 3D–measurement points were located on the roof (MP 5 and MP 6). They consisted of 2 A03 and 1 B31 each. Since the roof bottom view was not easily accessible, these measurement points were located on the lowest level of the roof construction: Directly on the layer of plastered reed (Fig. 14.7). The signals were acquired through an LMS SCM05 frontend (Fig. 14.17) with a sampling rate sR = 512 Hz and a resolution of 24 bit and were stored on a laptop. Time window length was T = 5..0.60 min. Data processing was performed using the LMS TestXpress software package.

A total of 53 tests were performed. Parameters varied were the type of sound (pink noise, DJ action), the power level (98 and 100 dBA) and the intensity of some third-octave bands of interest (40, 50, 75, 80 Hz). Summarizing the results very shortly here: (a) Vibrations at the façade in MP 1 to MP 4 are acceptable: <1 mm/s; (b) Vibrations measured at roof level (MP 5) are not acceptable: 11 mm/s for 98 dBA, 16 mm/s for 100 dBA ([2, 3]). Frequency analysis revealed that f = 30..0.80 Hz is the range of concern. Figure 14.8 shows a time signal and PSD for MP 5 and a 100 dBA DJ action. f = 60..0.80 Hz was found being significant for other excitation types.

Fig. 14.3 The Casineum roof

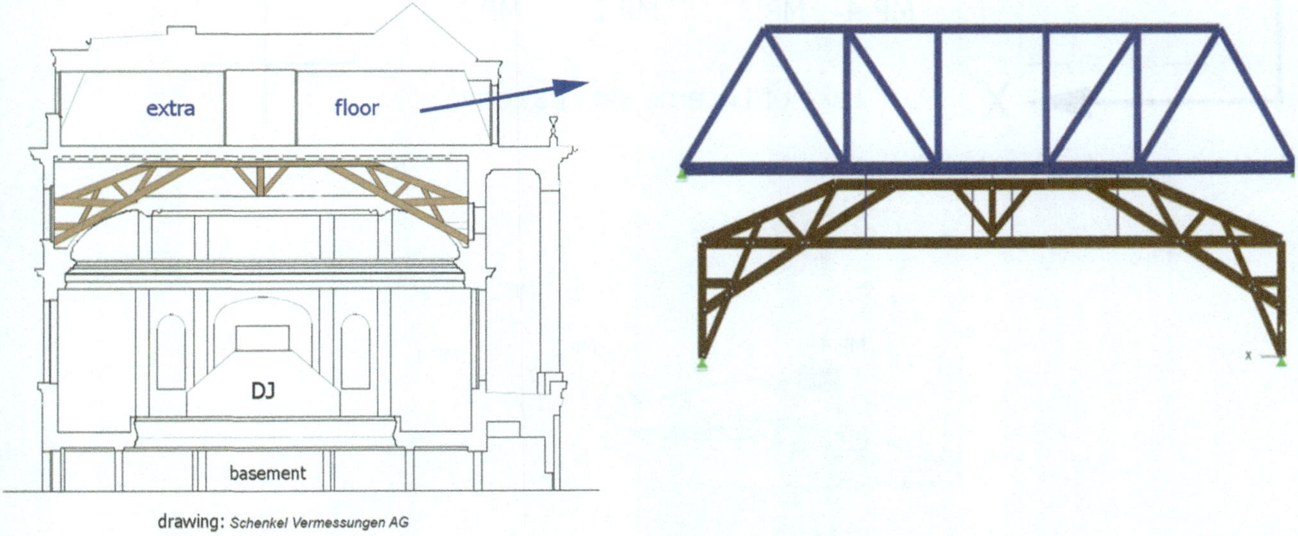

Fig. 14.4 The Casineum roof construction as of today ([1]). Brown: remaining part of the original timber truss. Blue: steel truss (new extra floor). Brown timber truss today's function: transfer the roof load from plastered reed to steel hangers. Thin blue vertical lines: steel hangers transferring the global loads from timber to steel

14.3 Experimental Modal Analysis

14.3.1 Test Procedures

Since remedial measures on the source of the excitation or the transfer path were found not to being feasible, an attempt was undertaken to determine the part of the roof construction being most susceptible to bass speaker power. First of all the roof structure had to be analyzed in detail. The "primary" structure is a steel truss, indicated in blue in Figs. 14.4 and 14.9). This had been added in 1911 after the top part of the original timber truss had been cut off. No longer being able to carry the roof loads, the timber truss had been connected to the new steel truss with four hangers (Fig. 14.9). This wooden truss is called "secondary" here. The space between upper and lower chords of the secondary timber truss is accessible today. This is possible through a wooden beam/plank system having been added on top of the secondary timber structure lower chord. This has a local load carrying function only, is adding mass to the roof system only and is not discussed here. Further important elements are the "local" timber cross girders connecting the secondary structure lower chords and the lathing structure being

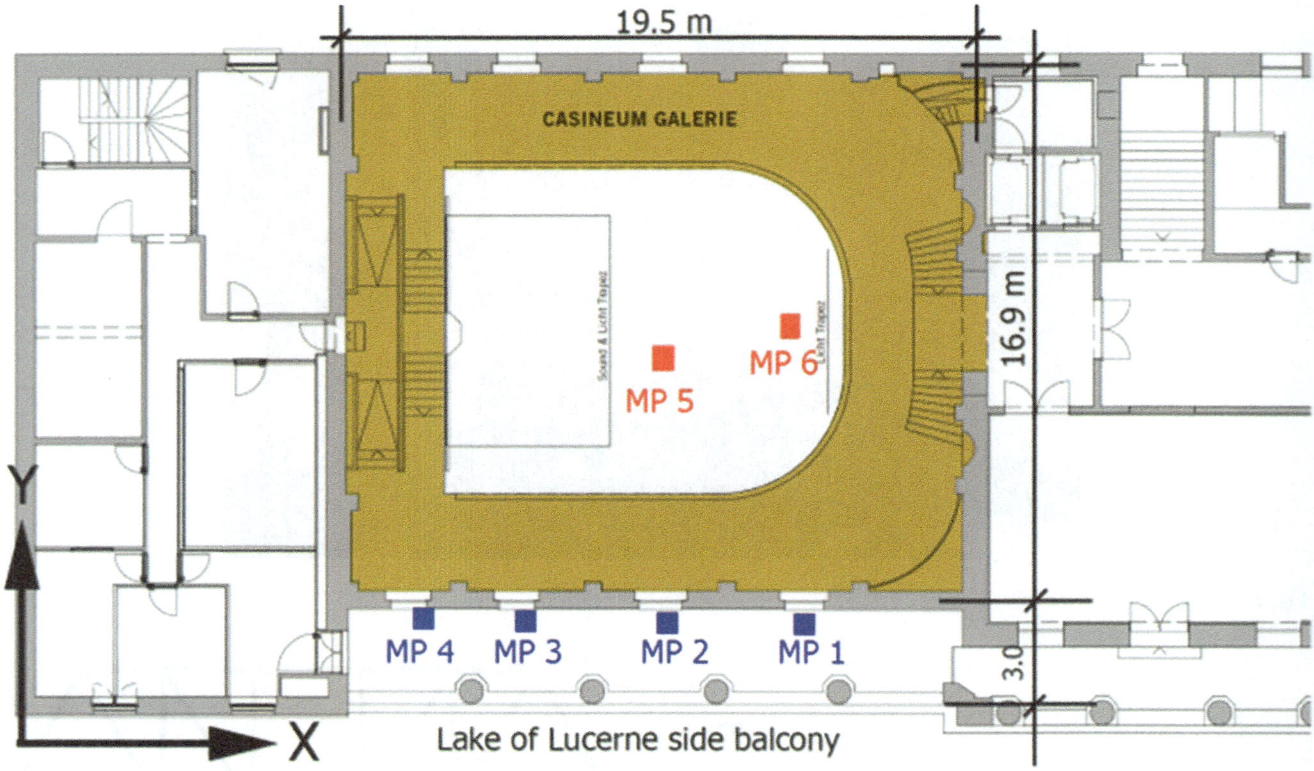

Fig. 14.5 Casineum plan view and instrumentation

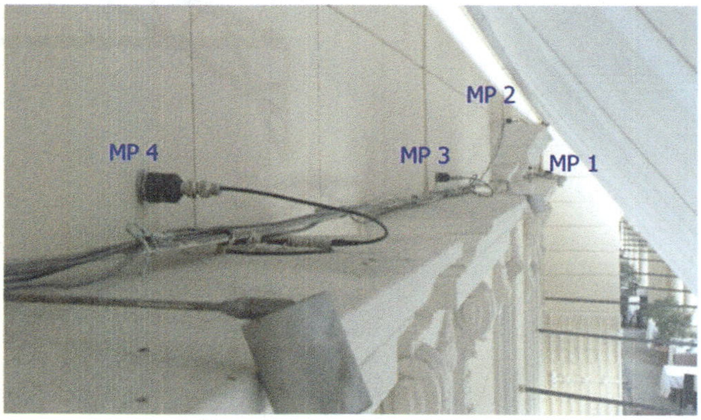

Fig. 14.6 Sensors at the lake side façade

nailed to the cross girders bottom fiber. This finally is carrying the plastered reed layer which is seen from the Casineum inside as "the roof".

Planning the instrumentation setup was based on this detailed investigation of the roof load carrying system. The four primary and secondary main truss' axes A to D located at a distance of 3.84 m each are indicated in Fig. 14.10. Firstly it was decided to concentrate on main truss axes A and B. In a next step, the two cross axes 10 and 20, distant at 1.75 m, were chosen. This considered the fact of the secondary timber truss global stiffness being high in the vicinity of two "closely spaced" hangers and low in a region with the hangers being relatively distant (Fig. 14.9).

The remaining two load carrying elements are the cross girders connecting the timber truss' lower chords and the lathing being nailed to their bottom fiber. The distance between cross girders is 1.75 m and between lathing elements 0.65 m. Each cross girder is carrying the load transferred by 5 lathing beams.

Instrumentation thus covered the primary and secondary elements in two main sections A and B and in two cross sections 10 and 20. In addition, the cross girders in sections 10 and 20 in three points (two end and one mid-point) plus the middle lathing beam between the two in two additional points (Figs. 14.11 and 14.12).

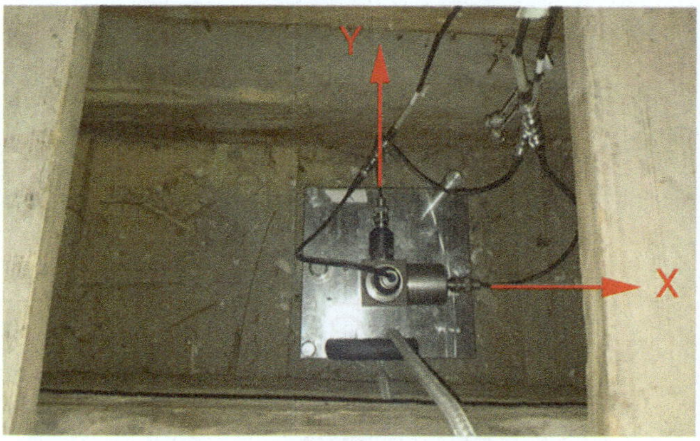

Fig. 14.7 MP 5 on the roof. The sensor support plate is located directly on the plastered reed layer

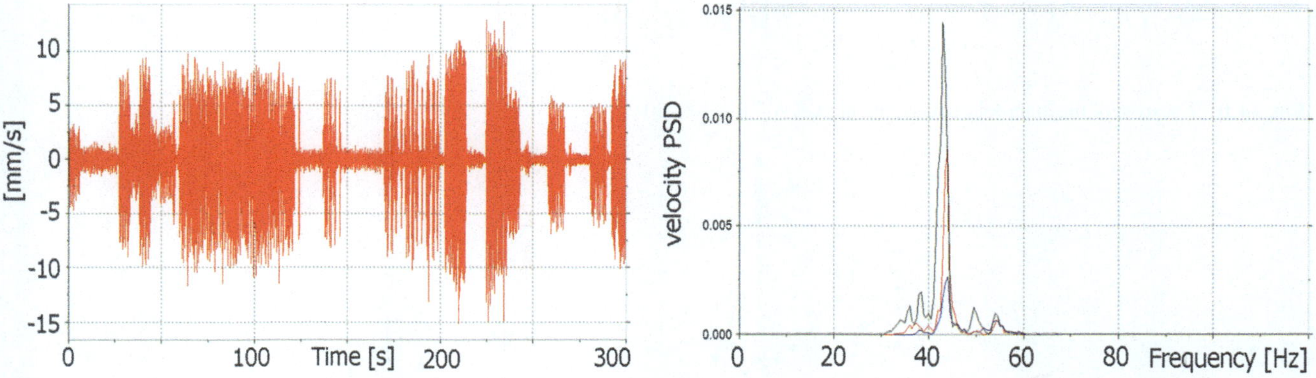

Fig. 14.8 Z Time Signal (left) and PSD Frequency spectra (right) for MP 5, 100 dBA DJ action. X (blue), Y (red) and Z (black)

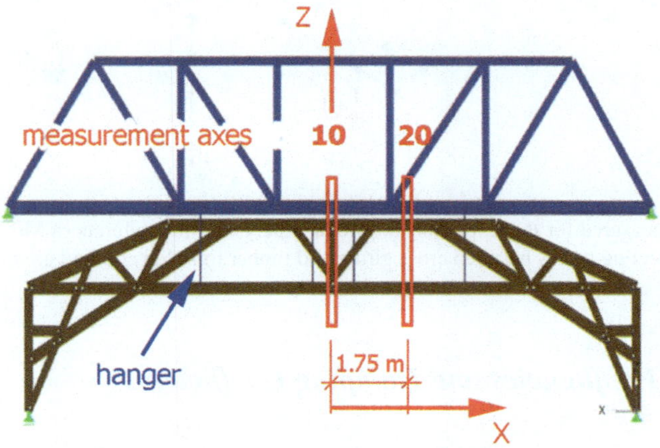

Fig. 14.9 Primary (blue) and secondary (brown) roof structures

Four 3D accelerometers were chosen for the modal analysis (Dytran 3093B, 100 mV/g and PCB 356B18, 1 V/g). Two of them were kept as reference points located in the middle of the cross beams (points 10R and 20R), the remaining two were roved over the measurement point grid as given in Fig. 14.11. A total of 10 Setups was necessary to cover this grid. The sampling rate was chosen to sR = 512 Hz, the time window length to T = 250 s. Considering the lowest frequency of interest being some f = 20 Hz, these parameters are reasonable and include some reserve. Ambient as well as white noise with a power of 98 dBA provided by the DJ equipment free of charge were used as system excitations. A total of 25 tests (including 5 preliminary tests) were performed June 20, 2016 (Figs. 14.13, 14.14, 14.15, 14.16 and 14.17).

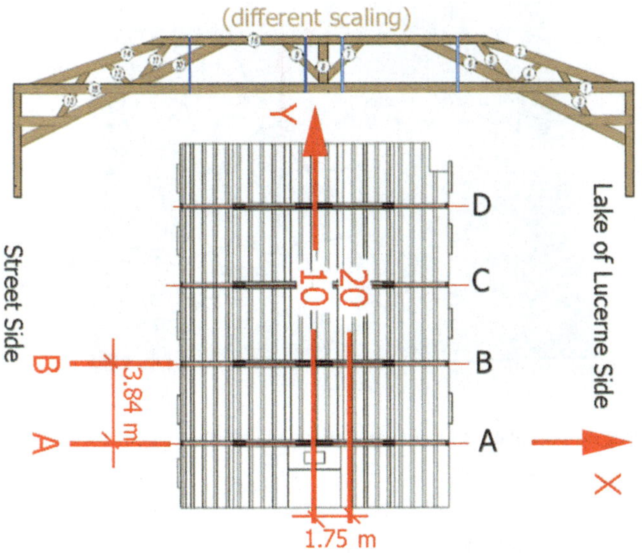

Fig. 14.10 Location of the main four measurement axes A – B – 10 – 20

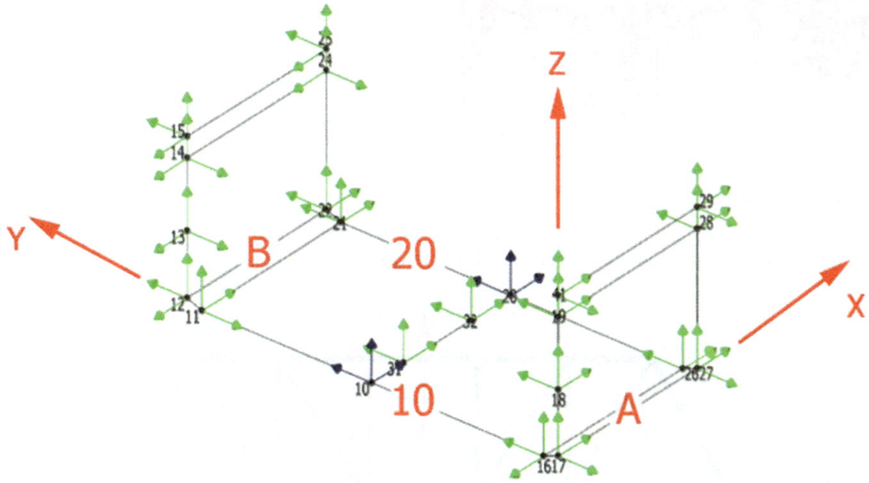

Fig. 14.11 Degrees of freedom as measured for the Experimental Modal Analysis. Blue: references MP 10R and MP 20R at cross girder's mid span points. Green: rovers. The connecting points between cross girder and timber truss lower chord are measured separately (e.g. MP 11 and 12, or MP 21 and 22; see Fig. 14.14)

14.3.2 Results, Natural Frequencies and Damping Coefficients

EFDD technology as offered through the ARTeMIS software package was used to process the data. No decimation was applied (dec = 1), the FFT resolution was chosen to 2 K and the frequency range displayed to f = 0..0.100 Hz. SVD diagrams for ambient and white noise excitation are given in Figs. 14.18 and 14.19. The modal parameters as derived from the modal tests are given in Table 14.1.

14.3.3 Results, Mode Shapes

Due to space restrictions it is not possible to discuss the shapes of all 21 modes identified here (Table 14.1). However, the main result can be illustrated with the two shapes shown in Figs. 14.20 and 14.21.

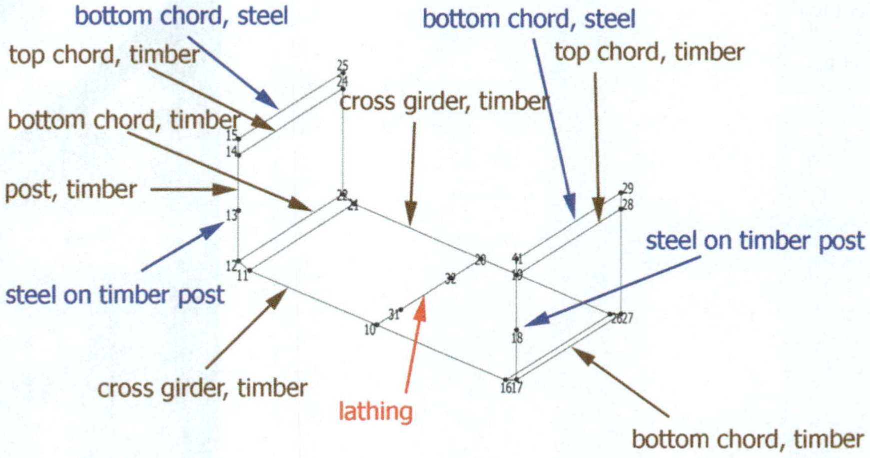

Fig. 14.12 Details of the roof structural elements. Blue: steel, brown: timber, red: lathing

Fig. 14.13 MP 20R, cross girder mid point reference point instrumented. Special supporting steel bars have been developed to allow quick sensor roving

Fig. 14.14 MP 26, cross girder end point (left), and MP 27, timber truss lower chord (right), the latter consists of two beams in parallel

The shape of Mode 1 is representative for all modes with f < 20 Hz: Primary and secondary structural elements are determining: Significant (vertical) modal amplitudes in the steel and timber truss main elements. Cross girders and lathing exhibit an “own life” to a very limited extent only.

Fig. 14.15 MP 14, timber truss upper chord. Also visible: Bottom flange of the steel truss lower chord (MP 15)

Fig. 14.16 MP 31 on top of the (central) lathing between cross girders 10 and 20. The state of health of the plastered reed layer being nailed to the lathing can also be seen

Fig. 14.17 Measurement Center

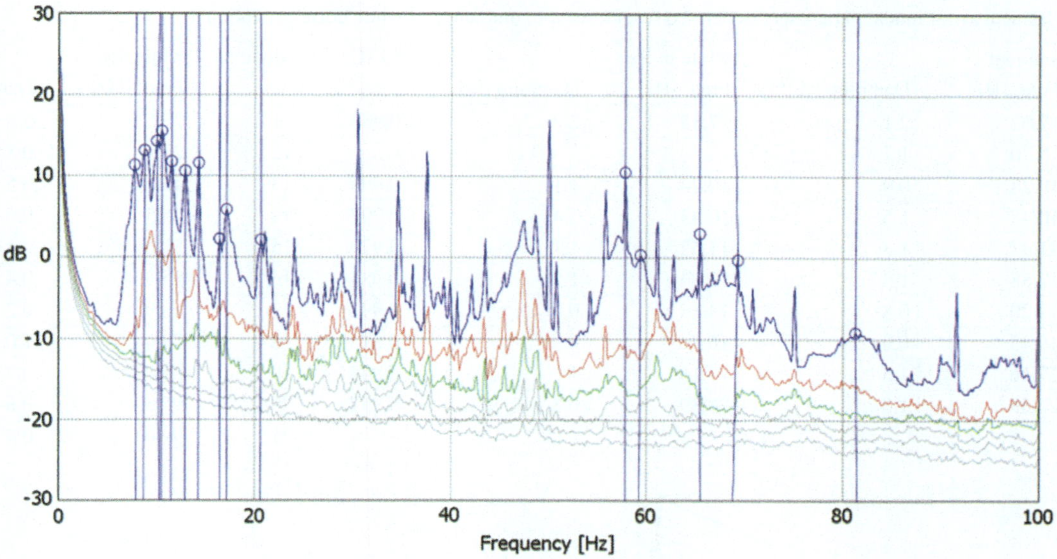

Fig. 14.18 SVD diagram for ambient excitation. This was used for EFDD analysis in the f = 0..0.20 Hz range

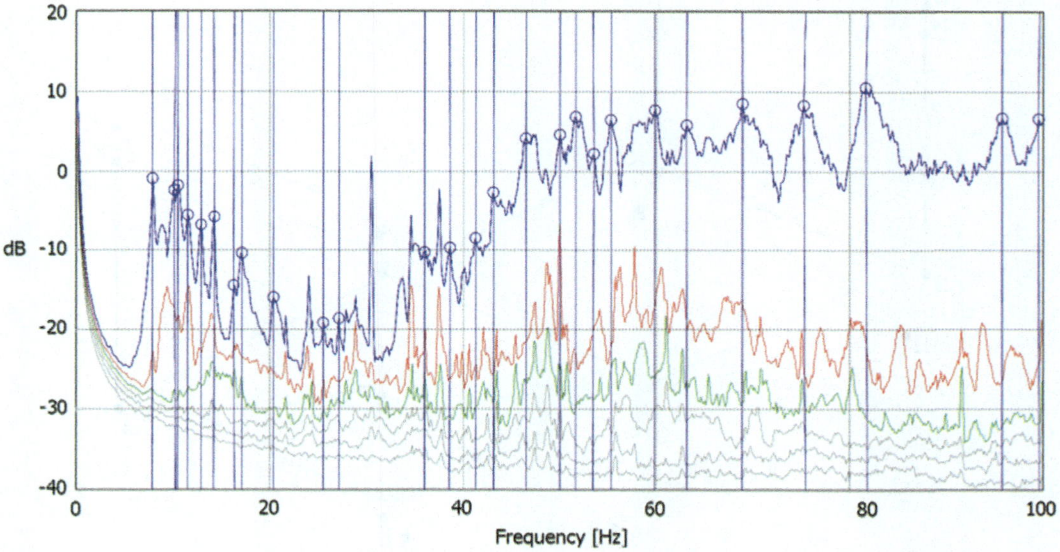

Fig. 14.19 SVD diagram for 98 dBA white noise excitation. This was used for analysis in the f = 0..0.100 Hz range

The shape of Mode 14 is representative for all modes with f > 30 Hz: The modal amplitudes in the primary and secondary structural elements are mainly equal to zero. As an exception, the timber truss lower chord shows some modal motion in MP 20 because this is in the middle between two distant hangers. What is important: Cross girders and lathing exhibit significant local modal deformations.

14.4 Summary, Conclusion

Experimentally determined roof vibrations in a 130 years old building being used as a Disco proved not to be acceptable for a historical structure. This applied to the state with the sound devices being limited to an overall power of 98 dBA as well as for 100 dBA. Analysis of the measured acceleration time signals in the frequency domain revealed the action of bass boxes acting in a frequency range f = 30..0.80 Hz being responsible for this.

Table 14.1 Casineum roof modal parameters as derived from tests with ambient and white noise excitation

Mode	ambient		white noise		MAC	Mode	white noise	
	Freq. [Hz]	Damping [%]	Freq. [Hz]	Damping [%]			Freq. [Hz]	Damping [%]
1	7.92	2.1	7.85	2.2	0.988	11	25.5	0.6
2	8.70	1.7	–	–	–	12	27.1	0.5
3	10.28	0.9	10.24	1.1	0.998	13	43.2	0.5
4	10.48	1.5	10.43	1.4	0.997	14	46.5	0.6
5	11.47	1.3	11.52	1.1	0.979	15	51.6	0.4
6	12.88	1.5	12.84	1.5	0.985	16	55.3	0.4
7	14.20	0.8	14.18	0.8	0.992	17	68.9	0.2
8	16.43	0.9	16.34	0.9	0.815	18	75.4	1.0
9	17.16	1.3	17.13	–	0.945	19	81.8	0.7
10	20.59	1.0	20.39	0.8	0.957	20	95.8	0.6
						21	99.6	0.4

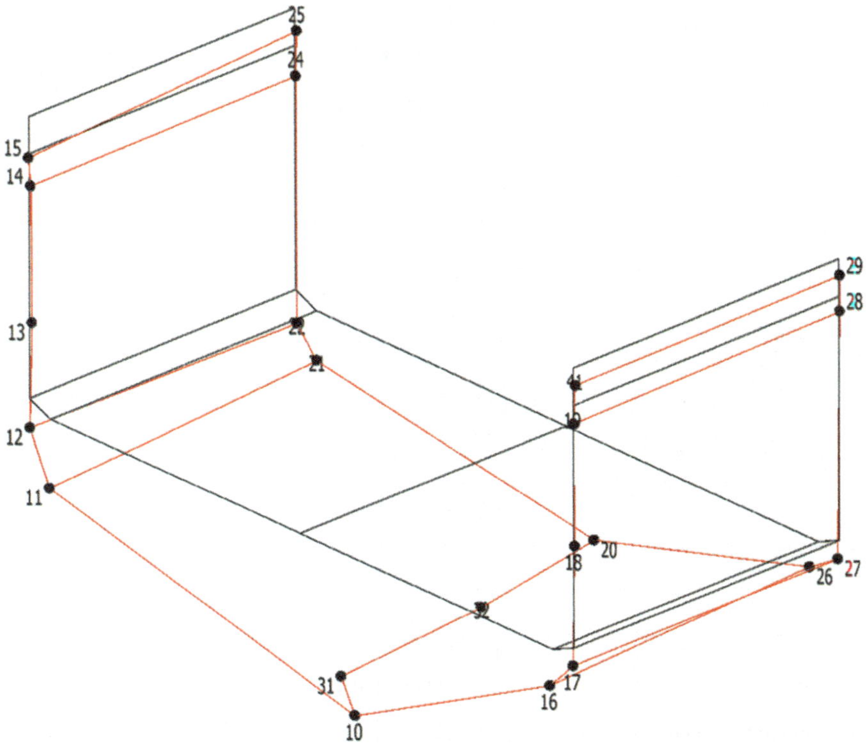

Fig. 14.20 Mode 1, f = 7.92 Hz, $\zeta = 2.2\%$

An Experimental Modal Analysis was performed in an attempt to identifying the element of the quite complicated roof structure being most susceptible to bass loudspeaker's excitation. Extending the instrumentation over four structural element levels and four main structural axes allowed identification of the local structural elements exhibiting maximum modal amplitudes in the frequency range of concern: Cross girders, lathing carrying the bottom plastered reed layer. The main structural elements, the trusses, are not susceptible to dynamic sound pressure.

The building owners are looking for a reasonable way of increasing the structure's serviceability as asked for under full power DJ action at the moment. This is not easy with the weak points being the connection of the cross girders to the timber truss lower chords and the lathing element bending stiffness.

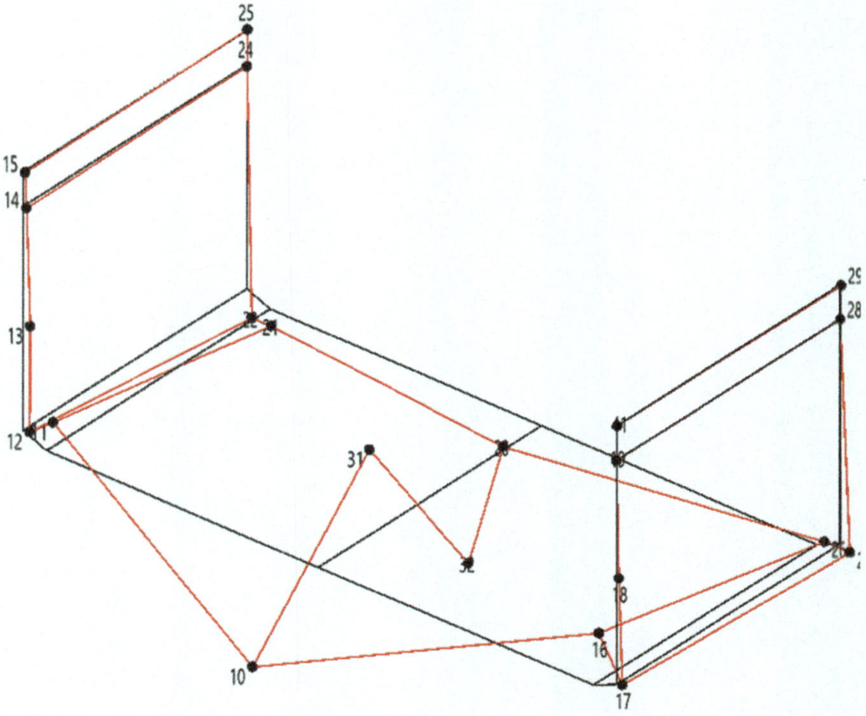

Fig. 14.21 Mode 14, f = 46.5 Hz, $\zeta = 0.6\%$

References

1. Lauber Essential Timber Engineering, Bestandesaufnahme Zwischendecke Casineum, Luzern (2013)
2. Erschütterungen – Erschütterungseinwirkungen auf Bauwerke. Schweizer Norm SN 640′312. VSS, Verein Schweizerischer Strassenfachleute, Zürich (2013)
3. Erschütterungen im Bauwesen, Teil 3: Einwirkungen auf bauliche Anlagen. DIN 4150-3, Deutsches Institut für Normung, Berlin (1999)

Chapter 15
Scenario Based Approach for Load Identification

Michael Vigsø, Marius Tarpø, Jannick B. Hansen, Rune Brincker, and Christos T. Georgakis

Abstract In output only analysis the load identification has been a puzzle for several years. Different techniques have been purposed to cope with the inversion problem that lies within this field. However it has been shown, that most methods struggle to obtain robust and consistent results in cases of modal truncation and noise contaminated signals. In the light of these challenges, a scenario based method is proposed. This approach utilizes model updating along with mode shape expansion to obtain a reliable numerical model of the given structure. Then, by evaluating a series of rational load scenarios, it is possible to obtain a reasonable input identification – both the spatial distribution and the temporal variation of the load. The method is demonstrated numerically and experimentally.

Keywords System identification · Operational modal analysis · Response estimation · Modal truncation · FE updating

Nomenclature

Vectors are noted by a single underline and matrix by a double underline. Superscript* indicates complex conjugated while superscript ^ is an approximation.

†	Moore Penrose pseudo inverse
ω_r	Natural frequency of mode r
ζ_r	Damping ratio of mode r
$\underline{\underline{\Phi}}$	Mode shapes, arranged in columns
λ_r	Pole of mode r
ε	RMS error
Q_r	Modal scaling constant for mode r
$\underline{f}(t)$	Force
$\underline{f_0}$	Spatial distribution of load
$g(t)$	Scaling function of load
$\underline{\underline{h}}(t)$	Impulse response function
$\underline{y}(t)$	Displacements
$\underline{q}(t)$	Modal coordinates
$\underline{Y}(\omega)$	Fourier transformed of $\underline{y}(t)$
$\underline{C}(\omega)$	Fourier transformed of $\underline{c}(t)$
$G(\omega)$	Fourier transformed of $g(t)$
$\underline{\underline{H}}(\omega)$	Frequency response function

M. Vigsø (✉) · C. T. Georgakis · M. Tarpø · J. B. Hansen
Aarhus University, Aarhus, Denmark
e-mail: mvigso@eng.au.dk

R. Brincker
Technical University of Denmark, Kongens Lyngby, Denmark

S. Pakzad (ed.), *Dynamics of Civil Structures, Volume 2*, Conference Proceedings of the Society for Experimental Mechanics Series, https://doi.org/10.1007/978-3-319-74421-6_15

15.1 Introduction

Being able to determine the dynamic loading of a mechanical system is of interest in many contexts. In most civil engineering structures, however, it is not feasible to get direct measurements of the dynamic loading and one must rely on indirect methods by measuring the response. Often however, either due to inaccessibility or to limited amount of sensors, the response of the structure is hardly ever available at all desired locations which increases the complexity of the problem [1].

Utilizing the concepts of Operational Modal Analysis, OMA, on a structure, it is possible to extract key parameters of the system, which can be used to reconstruct a model representation [2]. Then by carefully expanding the mathematical formulation of the system, it is possible to estimate the response at locations whereas it is not feasible to get measurements. By using the expanded mode shapes along with the frequencies and damping ratios obtained from the OMA, one can establish a frequency response function of the system. This yields an inverse identification problem where the number of unknowns exceeds the number of equations and a unique solution is not possible to achieve. Methods to cope with this inversion instability have been proposed either through least square schemes [3] or filtering using singular value decomposition [4]. Since complete dynamic characteristics of the system are rarely available, truncation errors and noise can cause the input estimates to be rather non-physical.

This paper presents a way to improve the load estimate by constraining the inversion problem by evaluating a set of pre-defined load scenarios. The hypothesis is that when evaluating the "correct" load scenario this yields a minimum error. The mode shape expansion provides the option to detect loads at locations where the response is not originally recorded. The method is developed for time invariant systems with linear assumption and small damping.

15.2 Concept

The main idea in this approach is to assume the load can be written as a product between a spatial distribution $\underline{f_0}$ and a time dependent scaling function $g(t) \in \mathbb{R}$.

$$\underline{f}(t) = \underline{f_0}\, g(t) \tag{15.1}$$

There is nothing novel in making this separation of variables as several authors have already published papers including this approach, e.g. [5–10]. However the assumption proves to yield robust solutions for load identification. In the time domain, the response of a multi-degree-of-freedom system can be found as a convolution between the impulse response function and the load.

$$\underline{y}(t) = \underline{\underline{h}}(t) * \underline{f}(t) \tag{15.2}$$

Given that the load $\underline{f}(t)$ can be written as in Eq. (15.1), the response can then be approximated as

$$\hat{\underline{y}}(t) = \left(\underline{\underline{h}}(t)\ \underline{f_0}\right) * g(t) \tag{15.3}$$

$$= \underline{c}(t) * g(t) \tag{15.4}$$

The easiest way to evaluate this convolution is in the frequency domain, so by means of the Fourier transform equation (15.3) becomes:

$$\hat{\underline{Y}}(\omega) = \left(\underline{\underline{H}}(\omega)\ \underline{f_0}\right)\ G(\omega) \tag{15.5}$$

$$= \underline{C}(\omega)\ G(\omega) \tag{15.6}$$

It is then possible to obtain a least square solution for the scaling function $G(\omega)$.

$$\hat{G}(\omega) = \underline{C}^{\dagger}(\omega)\underline{Y}(\omega) \tag{15.7}$$

By making this formulation, one is omitting the challenges associated with inverting the frequency response function $\underline{\underline{H}}(\omega)$, which very well may be ill posed [1, 4].

Then by evaluating Eq. (15.7) and back substituting into Eq. (15.5), it is possible to obtain an estimate of the response of the system given the load distribution, $\underline{f_0}$, which was chosen initially. This can then be compared to the recorded response and $\underline{f_0}$ can be changed until the error has been minimized.

15.3 Simulation Case

A simulation case is made on a plane cantilever beam. Bernoulli-Euler beam element formulation is used with two degrees of freedom at each node and a distributed mass. Assuming, that the cantilever beam is subjected to loading approaching Gaussian white noise then the modal parameters can be extracted using common OMA approaches [2]. Mode shapes, $\underline{\underline{\Phi}}$, and associated frequencies, ω_r, and damping rations, ζ_r, are extracted from the model and are assumed to be of high quality. For the sake of convenience, the damping ratio is assumed to be the same for all modes. Table 15.1 shows the extracted modal parameters for the first eight modes and their associated damping ratio.

In the current simulation case the model is subjected to a load at a single node as shown in Fig. 15.1. Responses are extracted from a limited number of nodes of the model which are considered "measured" or "experimental" responses. Noise is added to the "measured" response in order to stress the algorithm. The level of noise in this case is 100 dBW and no filtering is performed.

Often, the load source will be located at a point where no response is recorded. This is also chosen to be the case in this simulation, where the load is acting exactly between two experimental nodes, see Fig. 15.1. In order to estimate the loads at locations where data is not available, the measured response must be expanded to the unknown locations. This is done by a transformation to modal coordinates as follows:

$$\underline{\hat{q}}(t) = \underline{\underline{\Phi}}^{\dagger}\ \underline{y}(t) \tag{15.8}$$

Table 15.1 Extracted modal parameters

Mode	1	2	3	4	5	6	7	8
Frequency [Hz]	8.32	52.1	146	286	473	707	988	1320
Damping ratio [%]	1.0	1.0	1.0	1.0	1.0	1.0	1.0	1.0

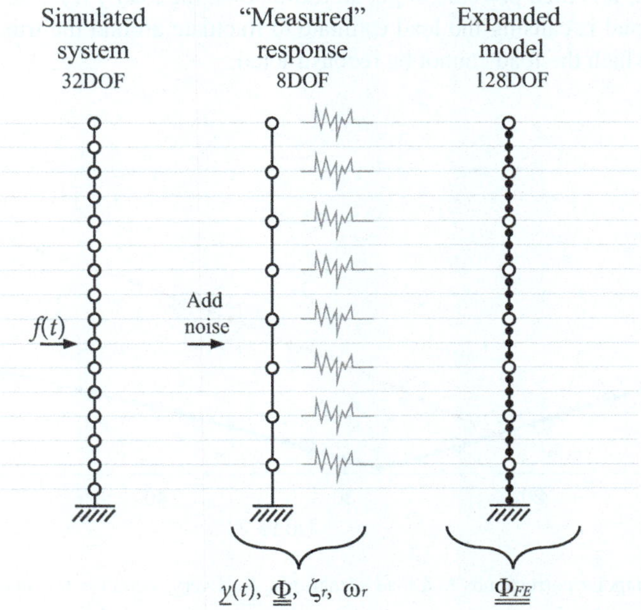

Fig. 15.1 Simulation case setup. Note: only every second node visible on the expanded model

$$\underline{y_{FE}}(t) = \underline{\underline{\Phi_{FE}}}\, \underline{\hat{q}}(t), \tag{15.9}$$

where index FE indicates expanded response using Finite Element, FE, mode shapes. Several other expansion techniques are available for this task, e.g. SEREP [11] or LC [12], but in this case the FE mode shapes are adopted presuming high correlation between the FE model and the physical model. The size of $\underline{\underline{\Phi_{FE}}}$ is determined by the desired resolution during load identification, but is limited to the same number of experimentally determined mode shapes.

A new receptance FRF matrix is formed based on the obtained information from the OMA and an updated FE model.

$$\underline{\underline{H}}(\omega) = \sum_{r=1}^{N} \left(\frac{Q_r \underline{\psi_r} \cdot \underline{\psi_r}^T}{i\omega - \lambda_r} + \frac{Q_r^* \underline{\psi_r^*} \cdot \underline{\psi_r^*}^T}{i\omega - \lambda_r^*} \right) \tag{15.10}$$

where Q_r is a modal scaling constant and $\underline{\psi_r}$ is the expanded and mass normalized mode shape r.

Letting $\underline{\underline{F_0}}$ contain a set of different predefined load distributions $\underline{f_0}$. The distributions can be arbitrary chosen but for this case say $\underline{\underline{F_0}}$ being equal to the identity matrix meaning that each column corresponds to a single localized force or moment. That is

$$\underline{\underline{F_0}} = \left[\underline{f_0}^{(1)}, \underline{f_0}^{(2)} \dots \underline{f_0}^{(N)} \right] = \begin{bmatrix} 1\ 0 \dots 0 \\ 0\ 1 \dots 0 \\ \vdots\ \vdots\ \ddots\ \vdots \\ 0\ 0 \dots 1 \end{bmatrix}_{N \times N} \tag{15.11}$$

where N is the total number of free degrees of freedom in the system. Then by evaluating the different distributions, $\underline{f_0}$, one by one it is possible to get an estimate of the approximated response corresponding to that distribution, Eqs. (15.5), (15.6) and (15.7). Letting the error on the estimate, ε, be quantified as

$$\varepsilon = \sum_{all\ DOFs} RMS\left(\underline{y}(t) - \underline{\hat{y}}(t) \right) \tag{15.12}$$

In Fig. 15.2, the error for each load scenario is shown. The curve for the rotational DOFs is included merely for academic interest as the most likely physical input will be a force rather than a moment. In this case a clear minimum is observed at a force acting at node 28. From Fig. 15.1 it can be seen that node 28 in the expanded model coincide with the position of the load. By selecting the load distribution, $\underline{f_0}$, causing the minimum error and multiplying by the scaling function, $g(t)$, associated with this distribution, it is then possible to get an estimate on the load $\underline{f}(t)$. The result is shown in Fig. 15.3. The added noise in the response signal is causing the load estimate to fluctuate around the true value. In the frequency domain this yields a noise floor under which the load cannot be reconstructed.

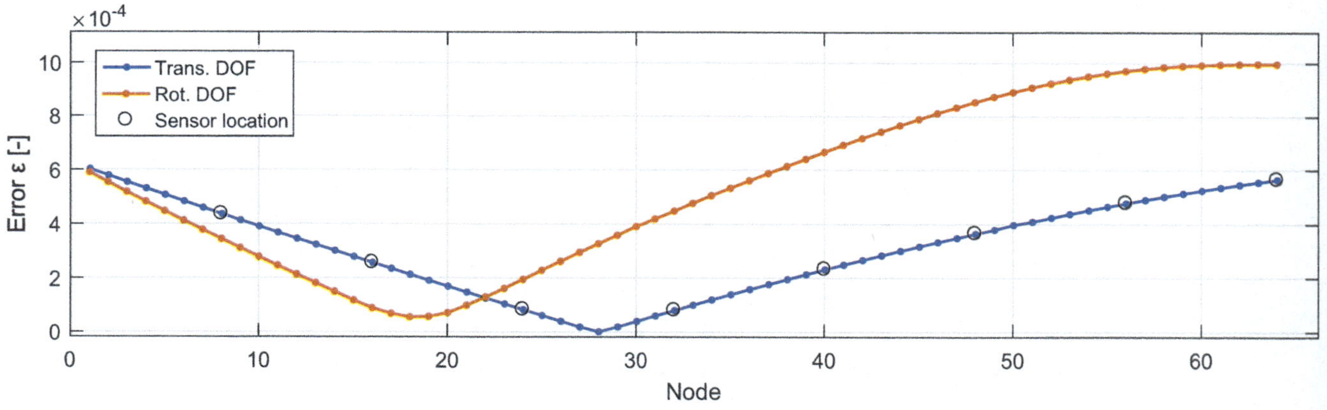

Fig. 15.2 Calculated error. Each datapoint corresponds to a load scenario in $\underline{\underline{F_0}}$. Every second is loading to a translational DOF and rotational DOF respectively, hence the separation of curves

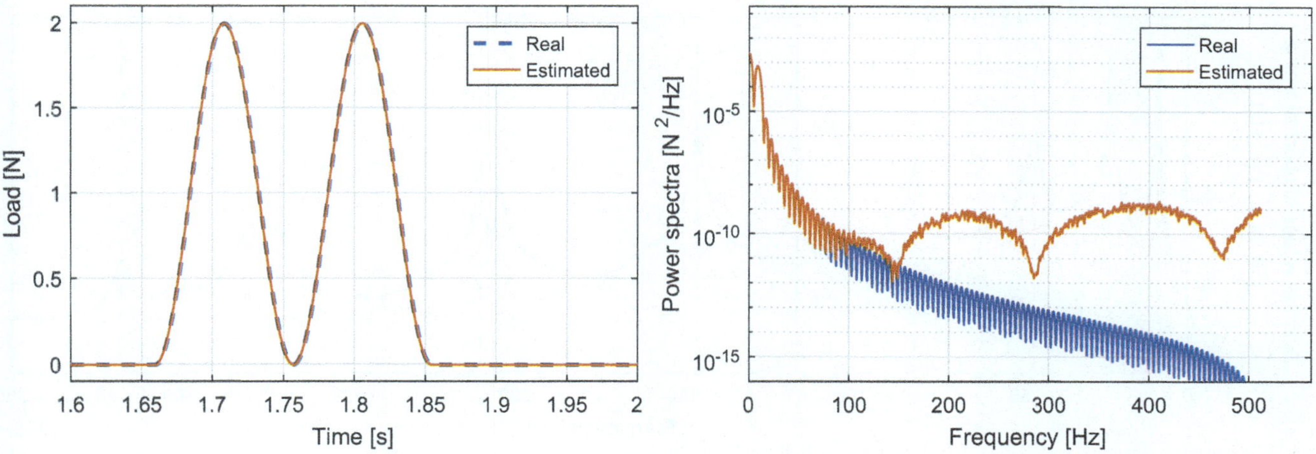

Fig. 15.3 Time- and frequency domain estimate. Five modes were included to establish the frequency response function. Power spectra computed using Welch averaging with a block size of 2^{11}

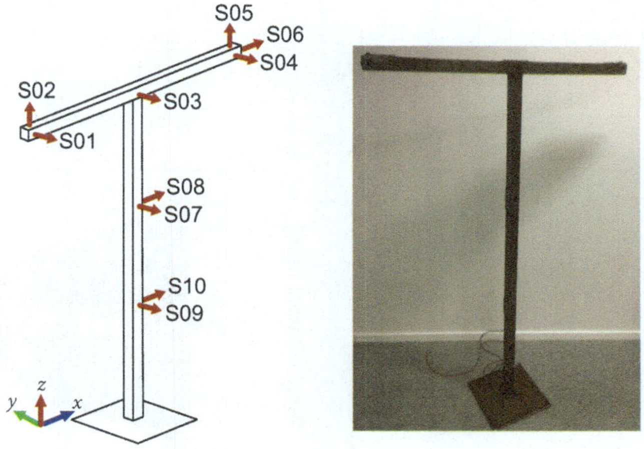

Fig. 15.4 Left: Schematic sensor layout. Right: Photo of test subject

15.4 Experimental Case

An experimental case study has been performed at the facility at Aarhus University. A cantilever T-structure made from RHS $40 \times 40 \times 2$ is used for the experiment. The T measures 1.5 m in height and is 1 m wide. The T is clamped to the floor through the base plate seen in Fig. 15.4. Ten uni-axial accelerometers are distributed as shown in Fig. 15.4. Accelerometers used are Brüel & Kjær type 4508-B 100 mV/g.

First, a test with random excitation from compressed air is recorded from which modal parameters are extracted. Identification technique used for this task is the frequency domain decomposition, FDD, [13]. The response is recorded for 300 s at a sampling frequency of 1652 Hz. The recordings from the test are represented in terms of singular values from the spectral density matrix, see Fig. 15.5. Note the harmonic peaks at 50 and 100 Hz caused by the power grid. Damping ratio is for the first mode estimated as 0.25% and is assumed to be applicable for all modes. Ten modes were identified in the frequency band of 0–350 Hz and their frequencies are listed in Table 15.2.

Secondly an impact hammer is used to generate and record loading of the structure. A Brüel & Kjær impact hammer type 8206 22.5 mV/N is used for the experiment. The hammer is equipped with a soft tip and an impact is made between sensor S03 and S07 in the y direction. The exact position is 12 cm below the sensor S03 i.e. 1/4 of the distance between the two sensors, see Fig. 15.4.

Figure 15.6 shows the raw acceleration response recorded during the impact. The accelerations are integrated twice in the frequency domain to obtain the displacements. During integration a highpass filter of 5 Hz is used. In order to improve computational efficiency only 12/60 s are included in the analysis. In addition the signal is decimated with a factor 4 pushing the Nyquist frequency down to 207 Hz [14].

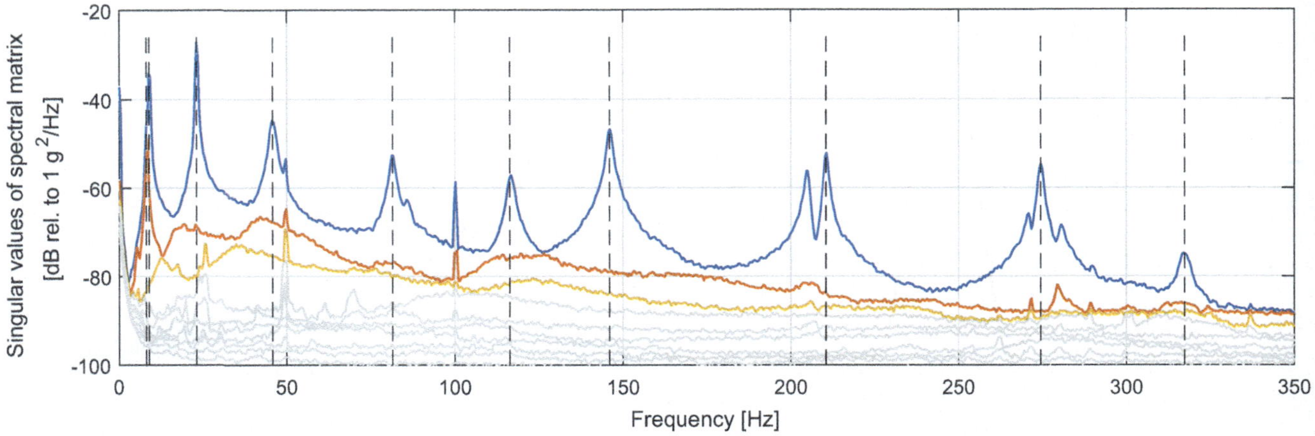

Fig. 15.5 Singular value decomposition of the response from random loading. Spectral matrix computed using Welch averaging with a block size of 2^{12}

Table 15.2 Natural frequencies

Mode	1	2	3	4	5	6	7	8	9	10
Frequency OMA [Hz]	8.17	9.08	23.1	44.8	81.4	116	146	211	274	317
Frequency FE [Hz]	8.34	8.83	24.6	45.3	81.1	116	147	207	277	315

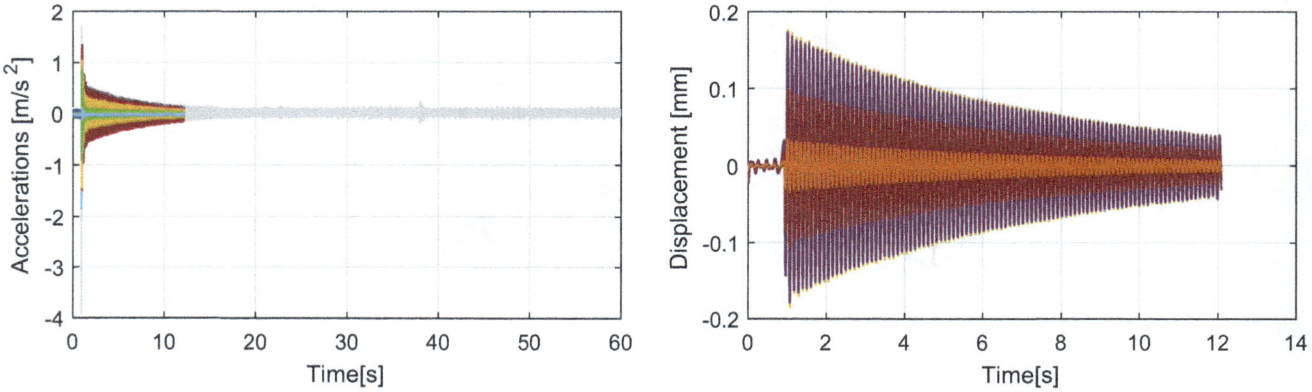

Fig. 15.6 Impact response shown for all channels

An FE model is made and updated to increase the correlation of frequencies and mode shapes. The model is made from Bernoulli beam elements with six degrees of freedom at each node and distributed mass. Meshing of the structure is done so that the sensor locations match an FE node. The number of FE nodes are chosen with respect to convergence of the modes of interest and the desired spacial resolution for the successive load identification. The mesh density may be seen in Fig. 15.9.

The model is then updated in two subsequent steps. (1): Material density increased until the total weight of the FE model corresponds to the physical model. (2): The rotational stiffness at the base of the T, Rx, Ry, Rz are updated until a best fit for the first 10 modes is achieved. The MAC matrix (after updating) can be seen in Fig. 15.7. The MAC matrix is computed using DOF pairs between the FE model and the experimental DOFs. Once an FE model is created and updated to an acceptable level, the frequency response function can yet again be formed using Eq. (15.10). The mass normalized mode shapes from the updated FE model are used along with the eigenfrequencies identified from the OMA. All ten mode shapes are included in the process.

When defining the set of possible load distributions, $\underline{\underline{F_0}}$, again the identity matrix is used as a basis. However, this time, omitting the rotational DOFs. Since the expanded model contains 41 nodes, (including the fixed node at the base), this yields in total 120 different load scenarios to be evaluated. The error associated with each load scenario is shown in Fig. 15.8. The results have been broken into three groups depending on their direction. As seen in Fig. 15.8 impact in the y direction yields much less error than for the other directions. The minimum is more indistinct compared to the simulated case study in Fig. 15.2. However a minimum is observed at the location of impact – node 23. Since the load distributions used in this

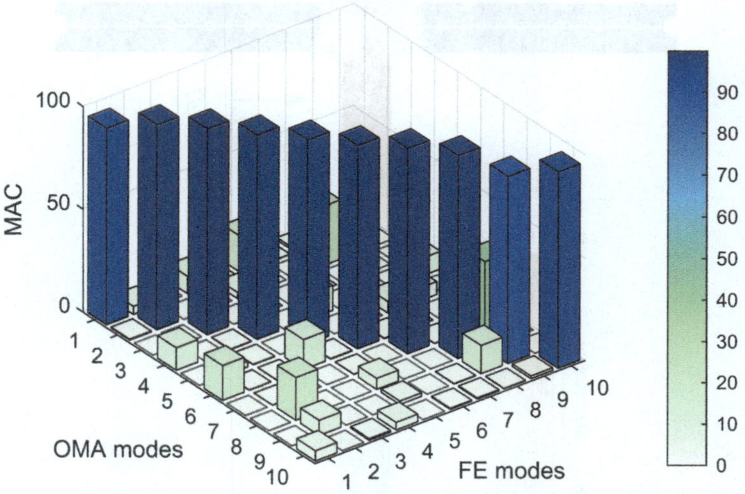

Fig. 15.7 Modal assurance after updating

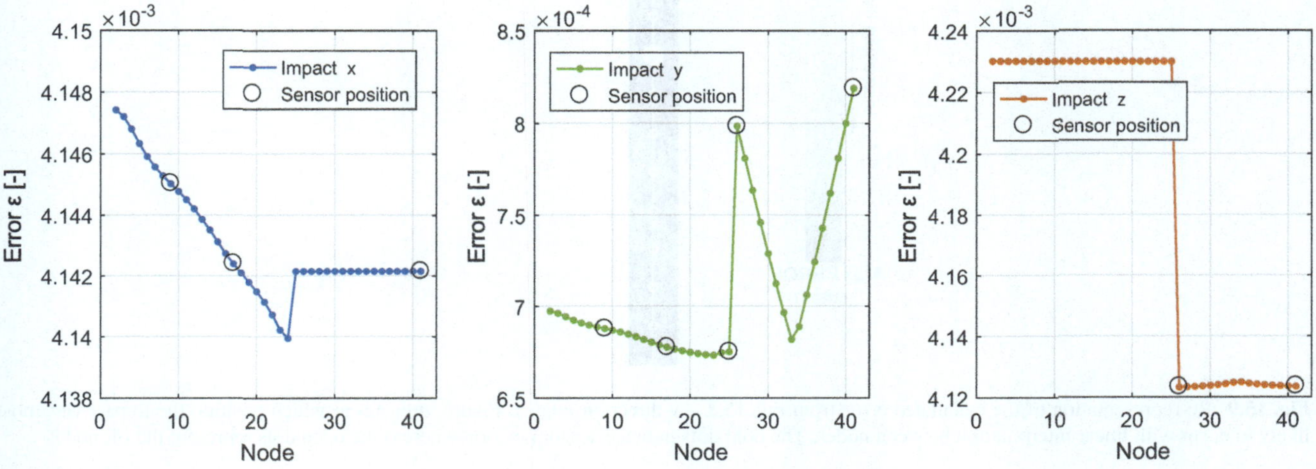

Fig. 15.8 Calculated error. Node 1 is located at the base of the T and node 25 is located at the joint

case consist of localized forces acting at one node at a time, the errors associated with each distribution can be assimilated with a spacial position and provide a visual feedback on where an impact was most likely to have happen. This is shown in Fig. 15.9.

The load distribution causing the minimum error can be multiplied by its scaling function and an estimate for the temporal variation can be obtained as it was done in the simulation case study. The result on the load estimate is shown in Fig. 15.10. In the time domain, the estimate follows the trend of the hammer with a peak deviation of 3%. It is observed that a negative load is estimated prior to the impact and some ringing afterwards. Looking at the response data in Fig. 15.6 it is observed, that the structure is not at complete rest at the time of impact. Since the load estimate is directly related to the response of the structure any ambient vibration will skew the identification of hammer input. In the frequency domain the estimate shows some discrepancy around the resonance frequencies. This may be due to the assumption that all modes bear the same damping ratio. Noise spikes are observed at 50 and 100 Hz, these are a remaining product of the harmonic noise in the response signal.

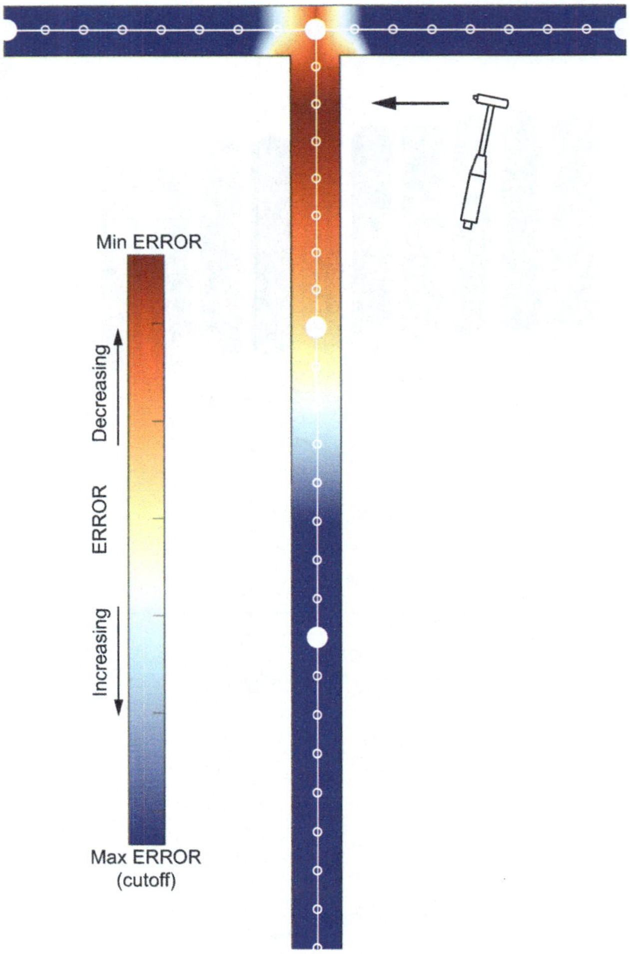

Fig. 15.9 Re-representation of the calculated error from Fig. 15.8 – *y* direction impact. Figure indicates at which position the impact was most likely to occur with linear interpolation between nodes. The bold dots indicate sensor position whereas the open dots represent the FE nodes

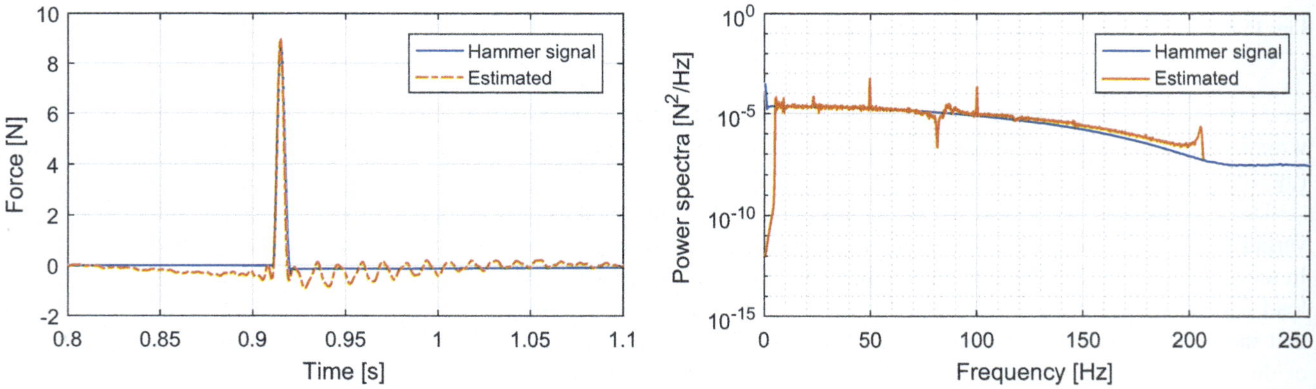

Fig. 15.10 Identified load. Power spectra computed using Welch averaging with a block size of 2^{11}

15.5 Conclusion

A method for indirect load identification has been presented. The method is titled "scenario based" as a keystone to success is having a set of good load distributions to select from. The method was demonstrated by a numerical simulation and a physical experiment. Some of the main observations are recaptured here:

By use of expanded mode shapes, the method shows potential in load identifications even at locations where the response is not recorded immediately. In the two cases the method proves not to yield any false-positive estimates for the position of the impact load, although the estimated location of impact is more outspoken for the numerical study than for the physical experiment. The rank deficiency of the FRF matrix is tamed by assuming a load distribution prior to making a least square fit. This makes the method insensitive to modal truncation, but remains reliant on a well-updated FE model. The noise and harmonic disturbance to the recorded response signal is causing the load identification to deviate but the algorithm remains stable.

Obviously one of the shortcomings to this method is the need for engineering judgement in order to develop a set of reasonable load distributions. Since this method is scenario based it is crucial that a fairly good distribution has been foreseen prior to the analysis otherwise the estimate may be erroneous.

Another challenge to this method arises if the distribution changes with time. Segmenting the response may improve this limitation but further work is needed in this area.

It is fundamental that the response is recorded at the time of loading. I.e. the structure must be at rest in the beginning of a time series otherwise this may alter the load estimates. This assumption seems to be violated to some extent in the physical experiment presented in this paper. From Fig. 15.6 it seems like the T was vibrating from excitation not caused by the hammer.

15.6 Future Work

In principle there are no limitations to the load distribution and the method could be applied on other cases, where some prior knowledge on loading pattern exists. For instance wave loading on an offshore structure.

Some limiting factors, which have not been covered in great detail, could be a basis for further work. For example, what is the sensitivity of the estimated modal parameters and how will a small change influence the load identification process or how this method will perform on highly damped systems.

Acknowledgements The authors acknowledge the funding received from Centre for Oil and Gas – DTU/Danish Hydrocarbon Research and Technology Centre (DHRTC).

References

1. Stevens, K.: Force identification problems – an overview. In: Proceedings of the SEM Spring Conference on Experimental Mechanics, pp. 838–844 (1987)
2. Brincker, R., Ventura, C.: Introduction to Operational Modal Analysis. Wiley, New York (2015)
3. Liu, Y., Shepard, W.: Dynamic force identification based on enhanced least squares and total least-squares scheme in the frequency domain. J. Sound Vib. **282**, 37–60 (2005)
4. Aenlle, M., Brincker, R., Fernández, P., Fernández, A.: Load estimation from modal parameters. In: Proceedings of the 2nd International Operational Modal Analysis Conference, vol. 1, pp. 39–50 (2007)
5. Léger, P., Ricles, J.M., Robayo, L.J.: Reducing modal truncation error in the wave response analysis of offshore structures. Commun. Appl. Numer. Methods **6**, 7–16 (1990)
6. Maes, K., Iliopoulos, A., Weijtjens, W., Devriendt, C., Lombaert, G.: Dynamic strain estimation for fatigue assessment of and offshore monopile wind turbine using filtering and modal expansion algorithms. Mech. Syst. Signal Process. **76–77**, 592–611 (2016)
7. Maes, K., Smyth, Q.W., De Roeck, G., Lombaert, G.: Joint input-state estimation in structural dynamics. Mech. Syst. Signal Process. **70–71**, 445–466 (2016)
8. Parloo, E., Verboven, P., Guillaume, P., van Overmeire, M.: Force identification by means of in-operation modal models. J. Sound Vib. **262**, 161–173 (2002)
9. Wang, B., Chiu, C.: Determination of unknown impact force acting on a simply supported beam. Mech. Syst. Signal Process. **17**(3), 683–704 (2003)
10. Lourens, E., Reynders, E., De Roeck, G., Degrande, G., Lombaert, G.: An argumented Kalman filter for force identification in structural dynamics. Mech. Syst. Signal Process. **27**, 446–460 (2012)
11. O'Callahan, J., Avitabile, P., Riemer, R.: System equivalent reduction expansion process. In: Proceedings of the 7th International Modal Analysis Conference (IMAC) (1989)
12. Brincker, R., Skafte, A., Anelle, M., Sestieri, A., D'Ambrogio, W., Canteli, A.: A local correspondence principle for mode shapes in structural dynamics. Mech. Syst. Signal Process. **45**, 91–104 (2014)
13. Brincker, R., Zhang, L., Andersen, P.: Modal identification from ambient responses using frequency domain decomposition. In: Proceedings of the International Modal Analysis Conference, IMAC, pp. 625–630 (2000)
14. Lyons, R.: Understanding Digital Signal Processing. Pearson Education, Upper Saddle River (2011)

Chapter 16
The Realisation of an Inerter-Based System Using Fluid Inerter

Predaricka Deastra, David J. Wagg, and Neil D. Sims

Abstract Many lightly damped flexible structures suffer from unwanted vibrations. Typically a tuned-mass-damper (TMD) can be used to reduce unwanted vibrations of a specific mode of vibration. The inerter is a novel passive vibration control device offering a wide range of potential applications in engineering practice. It has been analytically proven to be an effective device for controlling unwanted vibrations in structural systems. One of the most effective control strategies employing an inerter is the tuned inerter damper (TID) whose inerter element is connected in series with parallel connected spring-damper. When the inerter element is in parallel with the damper element, it is then called Parallel Viscous Damper Inerter (PVID). In this paper, we will introduce a new passive modal vibration control strategy for the PVID based on a fluid inerter combined with a linear spring connected in parallel. The fluid inerter produces inertance by the acceleration of the fluid inside a helical pipe coiled around the outside of the main fluid chamber. The fluid inerter has both inertance and damping in one device and these properties are coupled to each other. Hence, it is a particular challenge to tune both parameters to fit with optimized values resulting from a design analysis. In this paper, a new analysis will be presented for this device that demonstrates how the PVID with a fluid inerter can be modelled to achieve the targeted parameters.

Keywords Flexible structure · Passive vibration control · Parallel viscous inerter damper · Fluid inerter · Linear spring

16.1 Introduction

It is well known that inertia is the mechanical analogue of a capacitor with one grounded terminal. In contrast, inerters are the mechanical analogue of a capacitor with no restriction on the connection of the capacitor terminals. They have become an attractive research area since the late 1990s. In some fields of study, the application of inerters is strongly beneficial for vibration reductions. For example, in civil engineering, inerters have been shown to be able to reduce the structural oscillation due to earthquakes [1, 2].

In general, there are three types of inerter: mechanical geared inerter [3], fluid inerter [4, 5], and ball-screw inerter [1]. The use of an inerter as a damper has been proposed by combining the inerter element with viscous damping. In the literature recently, most contributions focus on the analytical study of the inerter-based damper. There are three layouts that have been studied extensively in the existing literature: Tuned Inerter Damper (TID) [2], Parallel Viscous Inerter Damper (PVID) [6], and Tuned Mass Damper Inerter (TMDI) [7].

The TID has the inerter element connected in series with a viscous damping element, whereas the PVID has them in parallel. In [2] the authors showed that the TID has similar performance to the Tuned Mass Damper (TMD) but with the benefit of a smaller total physical mass. The authors also discuss how the TID can be tuned to reduce all the vibration modes of a multi-degree-of-freedom (MDOF) system by placing the device underneath the structure, which is shown to be the best placement of the device. An extensive discussion about the optimum tuning of both TID and PVID devices, using algebraic solutions, can be found in [8]. Most recently, the optimum tuning of the PVID, using the demand-based method for damped structures, is presented in [6].

The realisation of the inerter-based damper has also been proposed in various other studies. In [1] for example, the authors discussed a concept of Tuned Viscous Mass Damper (TVMD) and showed how it is beneficial for reducing the seismic response of structures. The TVMD uses a ball-screw inerter combined in parallel with a viscous damper element. The spring element is provided externally via bracing. The application of this system in a real structure in Japan can be found in [9]. In a similar way, Hessabi and Mercan in [10] investigated the used of a mechanical geared inerter using a Gyro

P. Deastra (✉) · D. J. Wagg · N. D. Sims
Department of Mechanical Engineering, The University of Sheffield, Sheffield, UK
e-mail: pdeastra1@sheffield.ac.uk; david.wagg@sheffield.ac.uk; n.sims@sheffield.ac.uk

S. Pakzad (ed.), *Dynamics of Civil Structures, Volume 2*, Conference Proceedings of the Society for Experimental Mechanics Series, https://doi.org/10.1007/978-3-319-74421-6_16

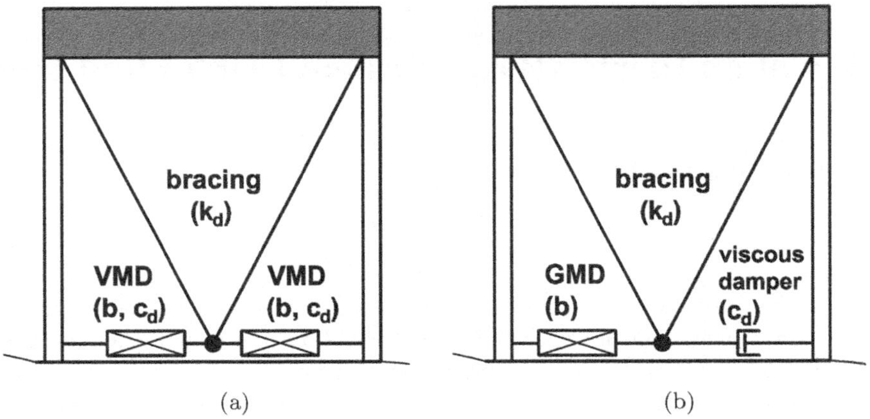

Fig. 16.1 An SDOF structure with (**a**) TVMD [9] (**b**) GVB [10]

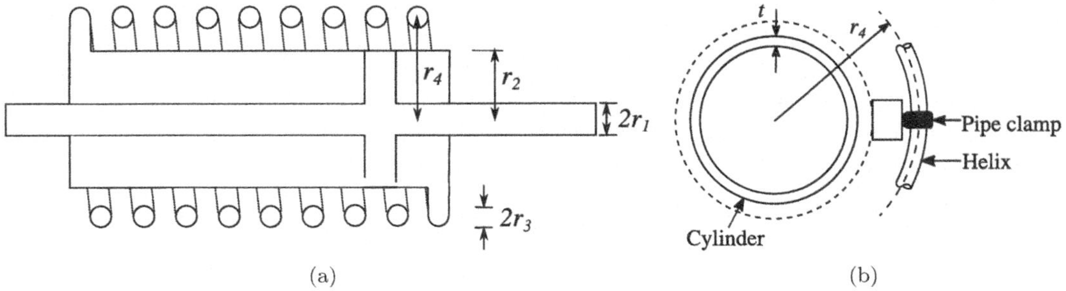

Fig. 16.2 Schematic diagram of the helical fluid inerter showing the (**a**) longitudinal cross section, and (**b**) the top view of the system. (Taken from [11])

Mass Damper (GMD) in combination with dampers via bracing. This system is called GMD-Viscous damper-Brace (GVB) system. The distinction between GVB and TVMD is the location of the viscous damping element. This is separated from the inerter element of the GVB, whereas in the TVMD both are in one device. Both the GVB and TVMD have the same layout which is the PVID layout, where the inerter and viscous damping element are connected in parallel and together these two elements are connected in series with the spring element given by the bracing. This is illustrated in Fig. 16.1.

The existing literature mostly uses the mechanical inerter for the realization of the inerter-based damper layout, either using a mechanical geared inerter or a ball-screw inerter. To the authors' knowledge, there is no study on the application of fluid inerters for the realisation of the inerter-based damper layouts. In this paper, we investigate the use of a helical fluid inerter for a PVID layout. In Sect. 16.2, we discuss the concept of helical fluid inerter. Sections 16.3 and 16.4 are dedicated to the application of the helical fluid inerter via bracing in an SDOF structure. In Sect. 16.5, we discuss the design of the fluid inerter parameters. Finally, the conclusions are given in Sect. 16.6.

16.2 Helical Fluid Inerter

The configuration of a helical fluid inerter is shown in Fig. 16.2. With reference to Fig. 16.2, the damping force of the fluid inerter with terminal velocity v is given in [4] as follows

$$F_{damper} = 0.03426\frac{2\rho l A_1}{\sqrt{D_h R}}\left(\frac{A_1}{A_2}\right)^2 v^2 + 17.54\frac{2\mu l A_1}{D_h^2}\left(\frac{A_1}{A_2}\right)v, \tag{16.1}$$

where μ and ρ are the fluid viscosity and density, l is the length of the channel, D_h is the channel hydraulic diameter, R is bend radius, and A_1 and A_2 are the area of the piston and the channel respectively, given by $A_1 = 2\pi(r_2 - r_1)^2$ and

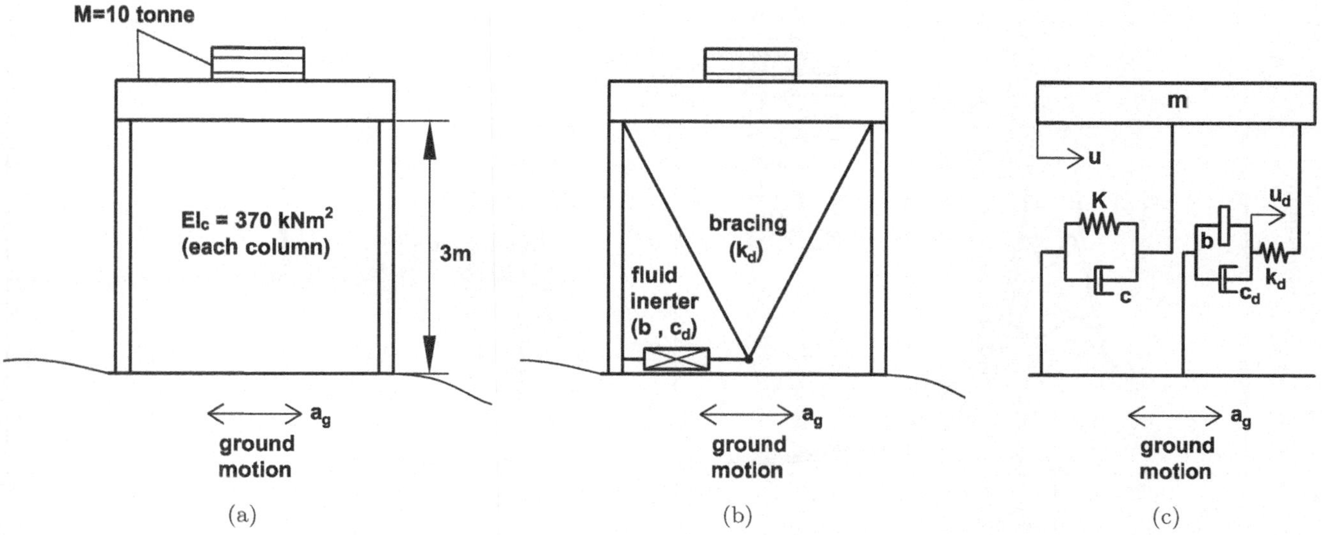

Fig. 16.3 (**a**) Considered structure [12] (**b**) Considered structure with fluid inerter and bracing (**c**) Analytical model

$A_2 = 2\pi r_3^2$. The constant inertance with the physical parameters described above is calculated using proposed formula in [11],

$$b = \frac{m_{hel}}{1 + \left(\frac{h}{2\pi r_4}\right)^2} \left(\frac{A_1}{A_2}\right)^2, \tag{16.2}$$

where m_{hel} is the mass of the fluid inside the helic, r_4 is the distance from the center of the helical channel to the center of the longitudinal axis of the cylinder, and h is the pitch of the helix.

16.3 Description of the SDOF Structure with a PVID

One existing structure adopted from [12] was used in this analysis. The structure is illustrated in Fig. 16.3a. The displacement transmissibility of the structure with a PVID as shown in Fig. 16.3b, c subjected to base excitations, can be expressed as follows

$$\left|\left(\frac{X}{R}\right)^2\right| = \frac{[1 - \mu q^2(\lambda + 1)]^2 + [2\zeta(\lambda + 1)q]^2}{[1 - q^2(\lambda\mu + 1 + \mu) + \lambda\mu q^4]^2 + [2\zeta(1 - \lambda q^2 + \lambda)q]^2} \tag{16.3}$$

where X and R are the Laplace transform of the main mass displacement and the base displacement, $\mu = \frac{b}{m}$, is the inertance-to-mass ratio, $q = \frac{\omega}{\omega_n}$, is the frequency ratio, $\omega_n = \sqrt{\frac{K}{m}}$, is the natural frequency of the system $\lambda = \frac{K}{k_d}$, is the stiffness ratio, and $\zeta = \frac{c_d}{2m\omega_n}$, is the damping ratio of the device. It should be noted, that the damping of the structure is given as 2%, which is relatively small and has therefore been neglected in the frequency domain analysis.

In order to optimize the parameters of the PVID, the equations proposed in [8] were used. The optimum damping ratio (ζ_{opt}) is given as

$$\zeta_{opt} = \sqrt{\frac{\zeta_P^2 + \zeta_Q^2}{2}}, \tag{16.4}$$

where

$$\zeta_{P.Q}^2 = \frac{\left(1 - \mu(1 + \lambda)q_{P.Q}^2\right)\left(1 + 2\mu + 2\mu\lambda - 3\mu\lambda q_{P.Q}^2\right)}{4(\lambda + 1)\lambda q_{P.Q}^2} \tag{16.5}$$

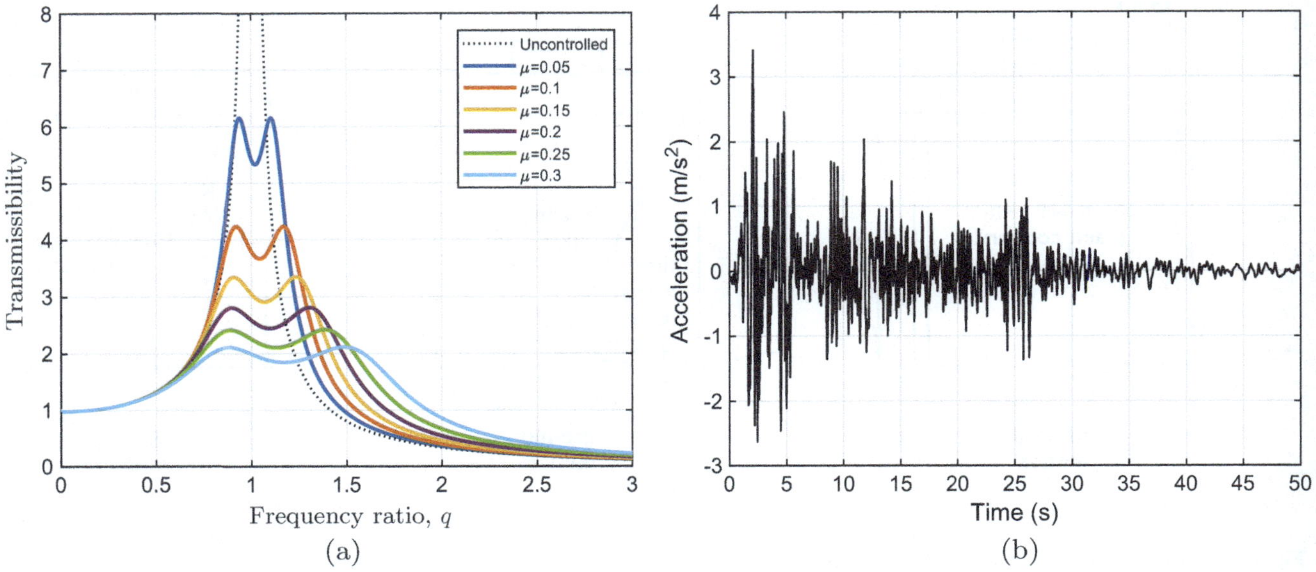

Fig. 16.4 (**a**) Optimization of the PVID (**b**) El Centro NS 1940

and

$$\lambda_{opt} = \frac{1}{2\mu}(1 - 2\mu + \sqrt{1 - 2\mu}). \tag{16.6}$$

Here q_P and q_Q are the frequency ratio at which the invariant points P and Q are located, given by:

$$q_{P,Q}^2 = \frac{1}{4\mu\lambda(\lambda + 1)}\left(1 + 2\lambda + 2\mu(1 + \lambda)^2 \pm \sqrt{(2\mu(1 + \lambda)^2 + 1 - 2\lambda)^2 + 8\lambda}\right) \tag{16.7}$$

16.4 Response of the SDOF Structure with a PVID

The frequency response of the structure with optimized PVID parameters obtained by using Eqs. 16.4, 16.5, 16.6, and 16.7 for various inertance-to-mass ratio is shown in Fig. 16.4a. As would be expected, increasing the inertance-to-mass ratio will decrease the frequency response of the structure as well as increasing the width of the peaks. As a result, the cut-off frequency is also moving to the right.

The time history response of the structure with and without the PVID was also investigated. The equation of motion of the system with a PVID as shown in Fig. 16.3c can be expressed as

$$\begin{cases} \ddot{u}(t) + 2\zeta\omega\dot{u}(t) + \omega^2 u(t) + \dfrac{1}{\lambda}\omega^2[u(t) - u_d(t)] = -a_g(t) \\ \mu\ddot{u}_d(t) + 2\bar{\zeta}\omega\dot{u}_d(t) - \dfrac{1}{\lambda}\omega^2[u(t) - u_d(t)] = 0 \end{cases} \tag{16.8}$$

where $\bar{\zeta} = \frac{c_d}{2m\omega_n}$ and $u(t)$ and $u_d(t)$ are the main mass and damper displacements as functions of time.

In order to evaluate the performance of the structure subjected to external vibration, an El Centro ground motion [13] as illustrated in Fig. 16.4b, was used in the analysis.

As predicted, using $\mu = 0.05$, the response of the structure is significantly reduced as presented in Fig. 16.5a. The frequency response analysis, as shown in Fig. 16.5b, also demonstrates that the PVID gives a significant reduction of the response of the main mass of the structure around the natural frequency at $f = 0.91$ Hz.

Increasing the inertance-to-mass ratio also results on the reduction of both displacements of the main mass and the damper as shown in Fig. 16.6. It is obvious that the damper displacement is more sensitive to increasing μ in comparison with the main mass displacement.

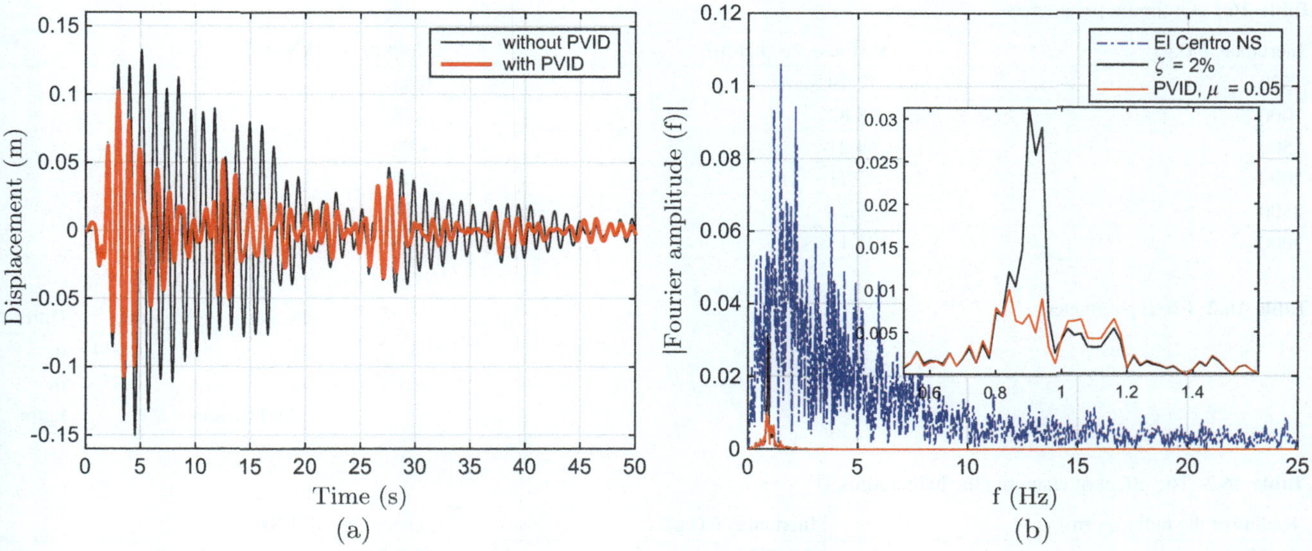

Fig. 16.5 (**a**) Displacement response of the main mass, $\mu = 0.05$ (**b**) Fourier spectra of the main mass displacement

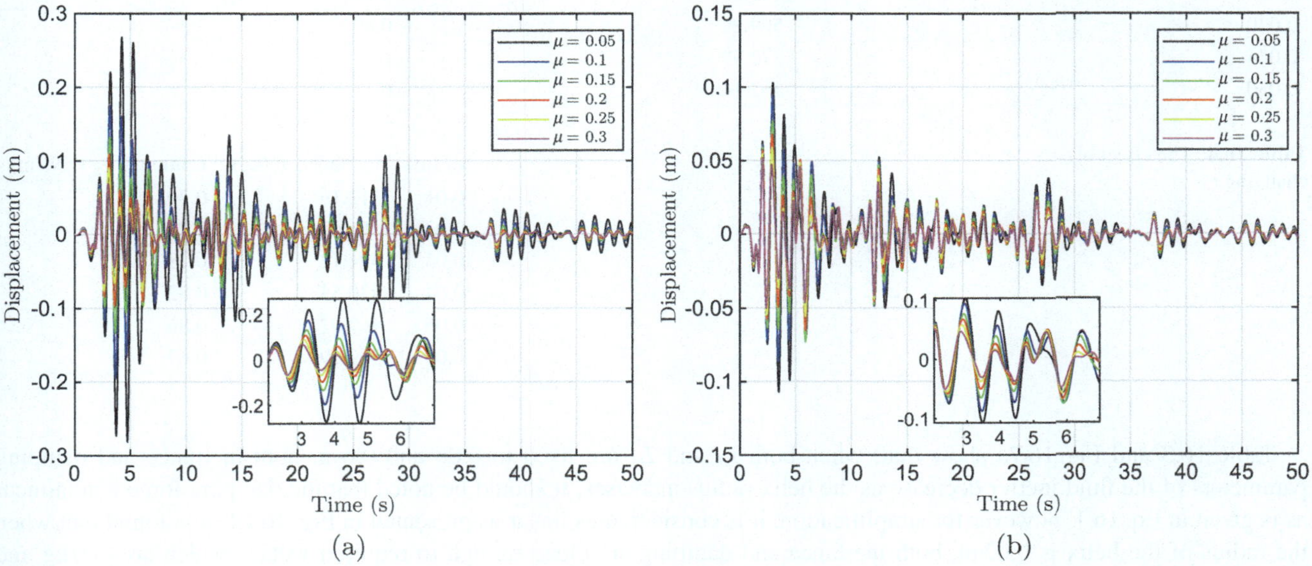

Fig. 16.6 Displacement history of the (**a**) damper (**b**) main mass

16.5 Parametric Study of the Fluid Inerter

This parametric study is aiming to design the physical parameters of the fluid inerter that give inertance b and damping c_d close enough to the required optimal values when using $\mu = 0.05$ as shown in the Table 16.1. Similar to the GVB and TVMD systems, the stiffness element of the system can be easily achieved by the design of the bracing system. However, the design of the fluid inerter is quite challenging due to the inertance and the damping parameters being coupled to each other.

In the design process, some parameters are fixed as given in the Table 16.2. The helical channel is assumed to turn around along the length of the tube, with no space in between ($h = 2r_3$). Hence, the total length of the channel is calculated using $l = 2n\pi r_4$, where n is the number of turns in the helix, given by $n = L_t/2r_3$, where L_t is the length of the tube, and r_3 is the radius of the helix. The distance between outer surface of the tube and the outer surface of helical channel,r_d, is given by $r_d = r_4 - (r_3 + r_2)$, where r_2 is the radius of the tube.

We note that in Eqs. 16.1 and 16.2, that both inertance and damping are strongly influenced by the area of the helical channel A_1 and the channel length l. Thus in order to try to achieve the design requirements, three parameters are studied: r_3, r_d, and L_t.

Table 16.1 Optimum parameters

Inertance, b (kg)	Stiffness, k_d (kN/m)	Damping, c_d (kNs/m)
500	17.79	0.83
1000	38.82	2.48
1500	64.21	4.86
2000	95.71	8.05
2500	136.23	12.23
3000	191.13	17.78

Table 16.2 Fixed parameters

Property	Value	Units
r_1	0.014	m
r_2	0.05	m
Fluid density	999	kg/m^3

Table 16.3 The effect of changing the helix radius

Radius of the helix, r_3 (m)	Inertance, b (kg)	Damping, c_d (kNs/m)
0.005	28,897.5	2000
0.010	3767.32	38.92
0.015	1161.88	4.28
0.020	509.29	0.89
0.025	270.49	0.27
0.030	162.14	0.1

Table 16.4 The effect of changing r_d

r_d (m)	Inertance, b (kg)	Damping, c_d (kNs/m)
0.01	312.19	0.70
0.02	351.67	0.74
0.03	391.12	0.78
0.04	430.53	0.82
0.05	469.92	0.86
0.06	509.29	0.89

Table 16.3 and Fig. 16.7a show that, when both r_d and L_t are fixed to 0.06 and 0.6 m, both inertance and damping parameters of the fluid inerter decrease as the helix radius increases. It should be noted that the damping force is nonlinear as is given in Eq. 16.1, however for simplification, it is considered as linear as presented in Fig. 16.7d. It is found that when the radius of the helix is 0.02 m, both inertance and damping are close enough to required values, which are 500 kg and 0.83 kNs/m given in Table 16.1.

Tables 16.4 and 16.5 show the influence of changing r_d when r_3 and L_t are fixed to 0.02 and 0.6 m, and the influence of changing L_t when r_3 and r_d are fixed to 0.02 and 0.06 m. It can be seen that both r_d and L_t are less sensitive to the changing of inertance and damping compared to r_3, as also can be seen in Fig. 16.7b, c.

Finally, it is concluded that the parameters obtained in order to achieve the targeted parameters of b and c_d are: $r_3 = 0.02$ m, $r_d = 0.06$ m, and $L_t = 0.6$ m. Detailed dimension of the device are presented in Fig. 16.8.

16.6 Conclusion

In this paper we have shown how a helical fluid inerter can be used to achieve a PVID concept. As a result, in a range of applications, the PVID can be realised using a fluid inerter to reduce the peak response of the host structure due to base excitation. Furthermore, it has been demonstrated that, as would be expected, increasing the inertance-to-mass ratio will decrease displacements of both the main mass and the PVID.

The issue of simultaneously designing for the required damping and inertance values has also been addressed. It was shown, for a selected example, that the inertance and damping parameters of the fluid inerter can be effectively tuned to fit

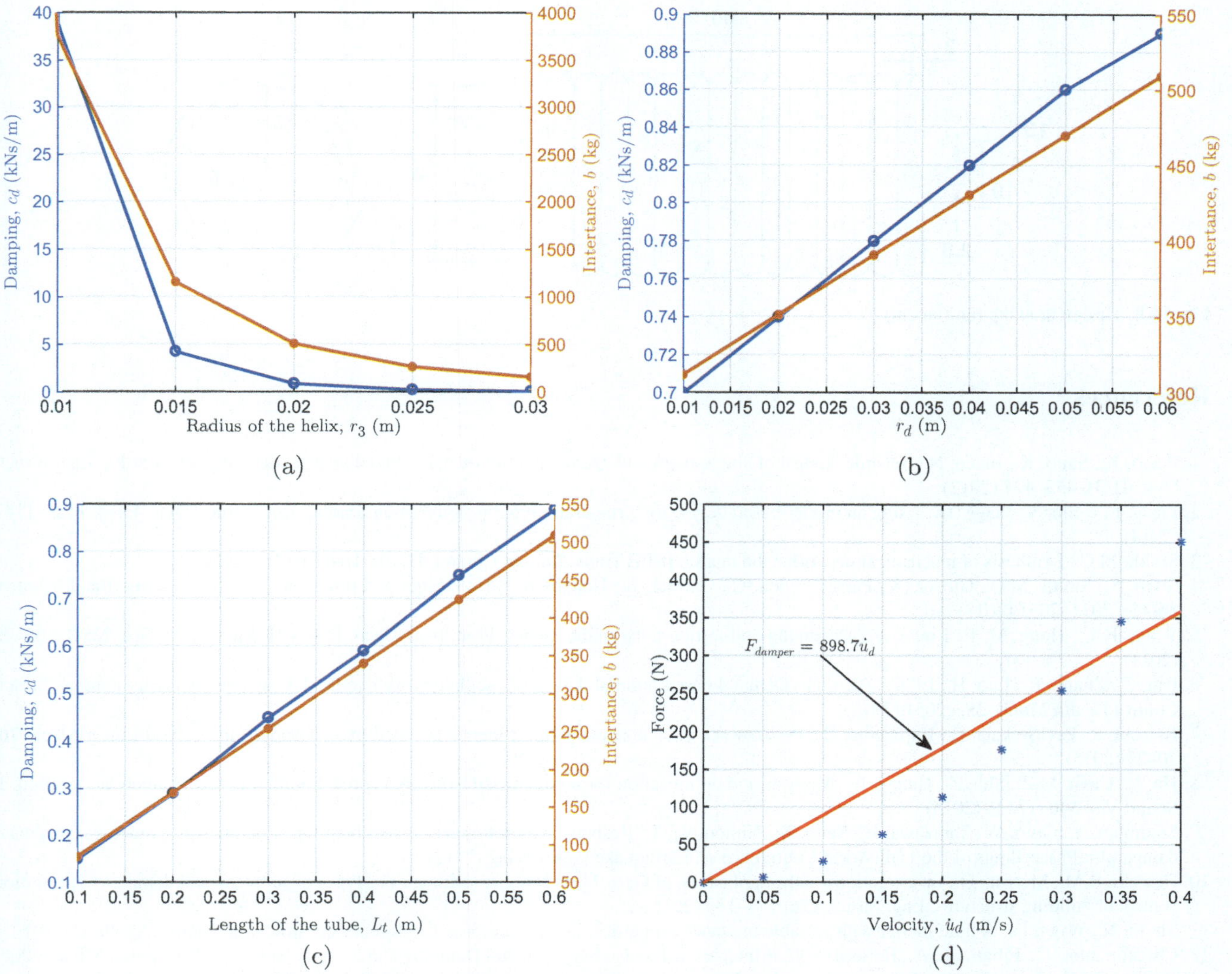

Fig. 16.7 The influence of changing (**a**) r_3 (**b**) r_d and (**c**) L_t to damping and inertance values, and (**d**) Force versus velocity of the damper when $r_3 = 0.02$ m, $r_d = 0.06$ m, and $L_t = 0.6$ m

Table 16.5 The effect of changing the tube length

Length of the tube, L_t (m)	Inertance, b (kg)	Damping, c_d (kNs/m)
0.1	84.88	0.15
0.2	169.76	0.29
0.3	254.65	0.45
0.4	339.53	0.59
0.5	424.41	0.75
0.6	509.29	0.89

the required values resulting from a fixed-point theory of optimization. This was achieved by varying three parameters: r_3, r_d, and l (functions of L_t). The radius of the helix, r_3, is found to be the most sensitive to changing the inertance and damping in a useful way, whereas the length of the tube is found to be the least sensitive. This demonstrates that the helical inerter can be used efffectively to achieve the PVID concept in practice.

Acknowledgements Predaricka Deastra is funded by Indonesian Endowment Fund For Education (LPDP). The authors gratefully acknowledge this funding.

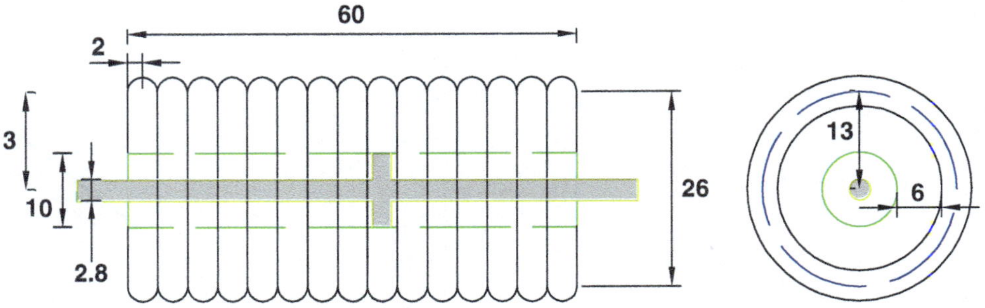

Fig. 16.8 Design drawing (unit in cm)

References

1. Ikago, K., Saito, K., Inoue, N.: Seismic control of single-degree-of-freedom structure using tuned viscous mass damper. Earthq. Eng. Struct. Dyn. **41**(3), 453–474 (2012)
2. Lazar, I., Neild, S., Wagg, D.: Using an inerter-based device for structural vibration suppression. Earthq. Eng. Struct. Dyn. **43**(8), 1129–1147 (2014)
3. Smith, M.C.: Synthesis of mechanical networks: the inerter. IEEE Trans. Autom. Control **47**(10), 1648–1662 (2002)
4. Swift, S., Smith, M.C., Glover, A., Papageorgiou, C., Gartner, B., Houghton, N.E.: Design and modelling of a fluid inerter. Int. J. Control **86**(11), 2035–2051 (2013)
5. Wang, F.-C., Hong, M.-F., Lin, T.-C.: Designing and testing a hydraulic inerter. Proc. Inst. Mech. Eng. C J. Mech. Eng. Sci. **225**(1), 66–72 (2011)
6. Pan, C., Zhang, R., Luo, H., Li, C., Shen, H.: Demand-based optimal design of oscillator with parallel-layout viscous inerter damper. Struct. Control Health Monit. **25**, e2051 (2018)
7. Brzeski, P., Pavlovskaia, E., Kapitaniak, T., Perlikowski, P.: The application of inerter in tuned mass absorber. Int. J. Non Linear Mech. **70**, 20–29 (2015)
8. Hu, Y., Chen, M.Z., Shu, Z., Huang, L.: Analysis and optimisation for inerter-based isolators via fixed-point theory and algebraic solution. J. Sound Vib. **346**, 17–36 (2015)
9. Sugimura, Y., Goto, W., Tanizawa, H., Saito, K., Nimomiya, T.: Response control effect of steel building structure using tuned viscous mass damper. In: Proceedings of the 15th World Conference on Earthquake Engineering (2012)
10. Hessabi, R.M., Mercan, O.: Investigations of the application of Gyro-Mass Dampers with various types of supplemental dampers for vibration control of building structures. Eng. Struct. **126**, 174–186 (2016)
11. Smith, N., Wagg, D.: A fluid inerter with variable inertance properties. In: 6th European Conference on Structural Control, Sheffield (2016)
12. Christopoulos, C., Filiatrault, A., Bertero, V.V.: Principles of Passive Supplemental Damping and Seismic Isolation. IUSS press, Milan (2006)
13. Building Performance Standardization Association. Retrieved from https://www.seinokyo.jp/jsh/top/ (2017)

Chapter 17
Experiences from the Five-Year Monitoring of a Long-Span Pontoon Bridge: What Went Right, What Went Wrong and What's Next?

Knut Andreas Kvåle, Ole Øiseth, and Anders Rønnquist

Abstract The Bergsøysund Bridge is a 930-m-long end-supported pontoon bridge located in Norway, and has been the target of a 5-year-long, extensive monitoring program. Herein, we will describe the unique structural characteristics of the bridge. The monitoring system has been under continuous expansion and revision, and consists of sensors monitoring both the excitation and the response of the bridge. Quantification of the uncertainties of the modelling methodology for structures of this nature has been the main goal, for which purpose modal analysis has been an indispensable tool. Modal analysis has also been used to study the effects the environment has on the structure's dynamic behaviour. We discuss the limitations of the results from modal analyses. Furthermore, we rise the question of how long monitoring campaigns may continue to provide useful information of this bridge and similar civil structures.

Keywords Floating bridge · Monitoring campaign · Modal analysis · Numerical prediction · Structural health monitoring

17.1 Background

The western coast of Norway is the part of Norway that is put on postcards. Immense fjords surrounded by dramatic mountains dominates the region. The same region is also housing industry that is producing and refining more than half of the Norwegian export, such as oil, gas and fish. Ferries are currently a necessity for both private and industrial transport in the region. The combination of these two factors is what made the government initiate the funding of the massive road infrastructure project *Ferry-free Coastal Highway E39*, organized and led by the Norwegian Public Roads Administration (NPRA), with the intention of substituting all the ferries along the highway with permanent road links. However, the geography does pose a big challenge: the deep and wide fjords imply that several world records have to be broken. The research community has therefore played an important role in finding and assessing technological solutions for the fjord crossings. Floating bridges are possible options for several of the crossings.

Floating bridges take advantage of the buoyancy of the water they are resting on, making them very stiff vertically. This makes them circumvent the problem that constrains the lengths of suspension bridges – gravity. However, a floating structure is very sensitive to forces induced by water waves. For some of the crossings required in the NPRA project, side-anchoring is not feasible. With spans of 4–5 km this will result in very flexible structures. To better understand how floating bridges behave, and how to model them, experience and observations of existing subjects are highly valuable. The Bergsøysund Bridge is an existing end-supported floating bridge with a floating section of 830 m, exposed to harsh marine environment. The bridge is comprised of a horizontally arc-shaped steel truss superstructure, that is supported on seven concrete pontoons. A photograph of the bridge is shown in Fig. 17.1. This bridge has been the target of a 5-year-long, extensive monitoring program.

K. A. Kvåle (✉) · O. Øiseth · A. Rønnquist
Department of Structural Engineering, Faculty of Engineering Science, NTNU, Norwegian University of Science and Technology, Trondheim, Norway
e-mail: knut.a.kvale@ntnu.no

S. Pakzad (ed.), *Dynamics of Civil Structures, Volume 2*, Conference Proceedings of the Society for Experimental Mechanics Series, https://doi.org/10.1007/978-3-319-74421-6_17

Fig. 17.1 The Bergsøysund Bridge. (Photograph by NTNU/K.A. Kvåle)

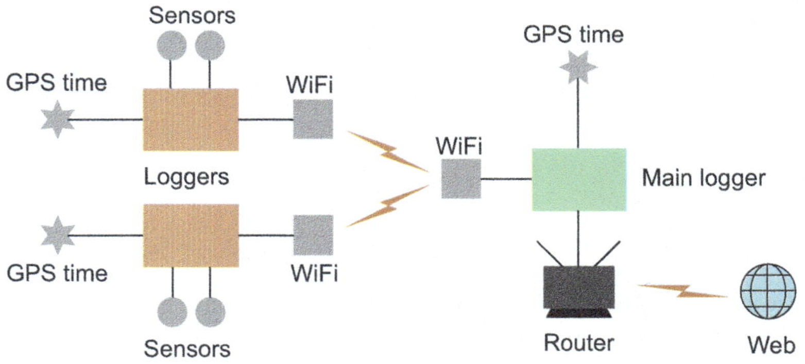

Fig. 17.2 Data acquisition structure. (Reproduced from [1], with permission from Elsevier)

17.2 Structural Monitoring

The bridge has been, and is still, under extensive monitoring. Currently, six wave radars and five anemometers are measuring the environmental excitation, whilst 14 tri-axial accelerometers and one Global Navigation Satellite System (GNSS) sensor are measuring the dynamic response of the bridge. The sensor data is sent digitally to logger units that communicate with a main logger unit via Wi-Fi (Fig. 17.2). The system ensures synchronous data by frequently obtaining common time stamps with global positioning system (GPS) units. The monitoring system is described in detail in [1]. Since the first version of the monitoring system was installed early 2013 [2], the system has been under continuous expansion and modification. The different stages of the monitoring system are indicated in Table 17.1.

The monitoring system was installed to address the main objective of studying and quantifying the accuracy of the numerical methodology for response predictions of floating bridges. This encompasses both the structural part of the model and the excitation model. This may furthermore be subdivided in the following sub-tasks: (i) evaluation of accuracy of structural model and excitation model, with reference to measurements; (ii) model updating [3]; (iii) system identification, both as a part of the first objective and by itself to study the behaviour of the structure [4]; and (iv) force identification [5].

17.3 What Went Right?

The system has provided data with quality sufficient for its purpose. State-of-the-art wind and wave sensors have been used to ensure the highest quality possible, with sampling rates high enough to provide useful information from a dynamic perspective. Digital sensors ensure the no measurement noise accumulate in the wiring. By transferring data wirelessly from the logger units to the main logger, problems concerning voltage drops for sensor power and extensive cable lengths have been avoided. The GPS time stamping have provided data with no noticeable time lags. Down-time of the network or

Table 17.1 Stages of system set-up

Set-up no	Start-up time	Description	Purpose with revision
1	Early 2013	Eight accelerometers [2]	–
2	December 2014	Ten accelerometers	Better spatial description of vibrations
3	March 2015	Fourteen accelerometers, one temporary anemometer and six wave radars	Better spatial description of vibrations, all pontoon accelerations are measured (good mode shapes). Measurement of excitation
4	September 2015	Fourteen accelerometers, six wave radars [1], five anemometers and one GNSS sensor	Check accuracy of GNSS sensor and verify displacements from integrated accelerations. Better representation of excitation situation
5	May 2017	As above, but with rearranged wave radars [6]	Enable a better characterization of the wave field from measurements

electricity grid did not affect the data collection, as the system was logging data autonomously upon triggering and all units were equipped with backup batteries for robustness. This has ensured that the system has been operable when the power is down and the weather is potentially at its harshest.

17.4 What Went Wrong?

As mentioned above, the different stages of the monitoring system are shown in Table 17.1. Mid-campaign changes are not ideal regarding long-term consistency, as you end up with recordings of different data structures to handle, but nonetheless it has been an unavoidable consequence of economic and practical concerns. Some of the changes have been made because different sensor networks have served different purposes, scientifically. The most obvious example concerns the two goals of the wave radars: (i) characterize the homogeneity of the wave field across the strait and (ii) characterize both the frequency- and direction-dependency of the local wave field. These two applications required entirely different setups, and it was not economically feasible or reasonable to acquire more sensors to cover both needs simultaneously. To handle changing sensor networks, generic metadata files for all time periods of the different stages was used. These files defined all necessary properties to enable global analyses of the data, for instance Cartesian coordinates and transformation matrices describing the local coordinate system of all the sensors. The close-to inevitable down-time of sensors must also be handled with care when analysing data bulk-wise. Stability issues with consumer-level network modems resulted in the need of digital electric timers to reboot the modem regularly.

17.5 What's Next?

For the applications mentioned above, the monitoring system has provided high-quality data that combined with state-of-the-art numerical methods has yielded good results. After five years of monitoring, we are, however, left with the question: what's next? How can we take advantage of new data from the monitoring system? What will more recordings provide, that the data already available does not? Structural health monitoring (SHM) is a topic that deservedly has received a lot of research attention. Vibration-based SHM is normally based on the modal parameters. During the first years of the operation of a structure, it is expected to experience some settling of the structural properties, leading to changes in its dynamic behaviour and thus modal parameters. After some time, however, the structure will probably attain some sort of converged state. It is only at the end of life that significant changes are expected to arise again. Observable changes in modal parameters might require drastic changes of the structure during its mid-life period.

Modal decomposition has historically been needed to reduce the number of degrees of freedom in a system, to make it manageable. In recent years, following the revolution of computational power, the time-saving factor is not as crucial. Rather, modal decomposition serves as a tool to synthesize and analyse complicated mechanical systems. Measurement-based modal analysis has served as an indispensable tool this far in the monitoring campaign, to assess the dynamic behaviour and to compare it with numerical predictions. Still, we feel that modal analysis, or more specifically, the interpretation of the behaviour by its fundamental vibration modes, is not yet fully utilized in this case.

Acknowledgements The research was funded by the Norwegian Public Roads Administration. The authors gratefully acknowledge this support.

References

1. Kvåle, K.A., Øiseth, O.: Structural monitoring of an end-supported pontoon bridge. Mar. Struct. **52**, 188–207 (2017)
2. Dahlen, A., Lystad, T.M.: Instrumentering av Bergsøysundbrua og Gjemnesundbrua. Master thesis, Norwegian University of Science and Technology (2013)
3. Petersen, Ø., Øiseth, O.: Sensitivity-based finite element model updating of a pontoon bridge. Eng. Struct. **150**, 573–584 (2017)
4. Kvåle, K.A., Øiseth, O., Rønnquist, A.: Operational modal analysis of an end-supported pontoon bridge. Eng. Struct. **148**, 410–423 (2017)
5. Petersen, Ø.W., Øiseth, O., Nord, T.S., Lourens, E.-M.: Model-based estimation of hydrodynamic forces on the bergsoysund bridge BT – dynamics of civil structures. In: Proceedings of the 34th IMAC, a Conference and Exposition on Structural Dynamics 2016, vol. 2, pp. 217–228. Springer International Publishing, Cham (2016)
6. Kvåle, K.A., Øiseth, O.: Characterization of the wave field around an existing end-supported pontoon bridge from simulated data. In: Proceedings of the International Conference on Earthquake Engineering and Structural Dynamics, Reykjavik (2017)

Chapter 18
Development of a 3-DOF Structural Displacement Sensor Based on a Two-Stage Kalman Filter

Jun Yeon Chung, Kiyoung Kim, Jaemook Choi, and Hoon Sohn

Abstract Structural displacement is one of the important indicator for monitoring and assessing the safety of civil infrastructures. GPS-RTK has been one of the widely used sensor for displacement measurement, but the GPS-RTK has low sampling rate and its precision and accuracy are easily affected by stability of satellite and environmental conditions. To overcome the limitations of GPS-RTK, a novel 3-DOF structural displacement sensor is developed in this study. The developed sensor measures 3-DOF displacement, velocity and acceleration of large-scale civil structures based on data fusion of acceleration measured from a force feedback accelerometer, and velocity and displacement obtained from a low cost GPS-RTK using two-stage Kalman filtering. The developed 3-DOF structural displacement sensor offers the following advantages over the existing GPS-RTK sensors: (1) The proposed sensor can measure 3-DOF displacement, velocity and acceleration simultaneously, (2) A better accuracy (around 2 mm) and a better sampling rate (up to 100 Hz) can be achieved, compared to the conventional GPS-RTK sensors, and (3) The performance is less affected by weather conditions and multi path problems, which deteriorate the performance of conventional GPS-RTK sensors. The performance of the proposed sensor was validated through a series of lab scale tests and a field test conducted on Yeongjong Grand Bridge.

Keywords Displacement estimation · Sensor data fusion · Two-stage Kalman filter · Civil engineering structures monitoring · GPS-RTK

18.1 Introduction

For structural health monitoring of civil structure, the measurement of displacement is important, since it is utilized for deciding whether the target civil structure is safe or dangerous. LVDT (Linear Variable Differential Transformer), LDV (Laser Doppler Vibrometer) and GPS-RTK are widely used for measuring displacement of civil structures. However they have some limitations described as followed. LVDT is a typical displacement measurement sensor, which has accuracy of 1 μm. In spite of its performance, the installation requirement of the LVDT restricts the application in civil structures like offshore bridges since it needs a fixed ground support [1]. LDV could be an alternative of LVDT. However, the high cost of LDV obstructs the multiple node measurement to large civil structures. Additionally the installation requirement of LDV, that the LDV should be supported by the rigid ground where any vibration could not happen, hinders the sensor application to water way crossing bridges [2]. For these reasons, GPS-RTK has been widely used in displacement measurement in large-scale civil structures. However, GPS-RTK has limitations: (1) The measurement reliability of GPS-RTK is highly dependent on environmental and weather conditions by easily changing from fixed to float mode. (2) The sampling rate of GPS-RTK is limited to 20 Hz, and the accuracy is around 2 cm. (3) The low frequency noise components may be observed in the GPS-RTK displacement signal due to the multipath problem.

J. Y. Chung · K. Kim · J. Choi · H. Sohn (✉)
Department of Civil and Environmental Engineering, Korea Advanced Institute of Science and Technology, Daejeon, South Korea
e-mail: jjysh014@kaist.ac.kr; kiyoungkim@kaist.ac.kr; cjmook@kaist.ac.kr; sohnhoon@kaist.ac.kr

S. Pakzad (ed.), *Dynamics of Civil Structures, Volume 2*, Conference Proceedings of the Society for Experimental Mechanics Series, https://doi.org/10.1007/978-3-319-74421-6_18

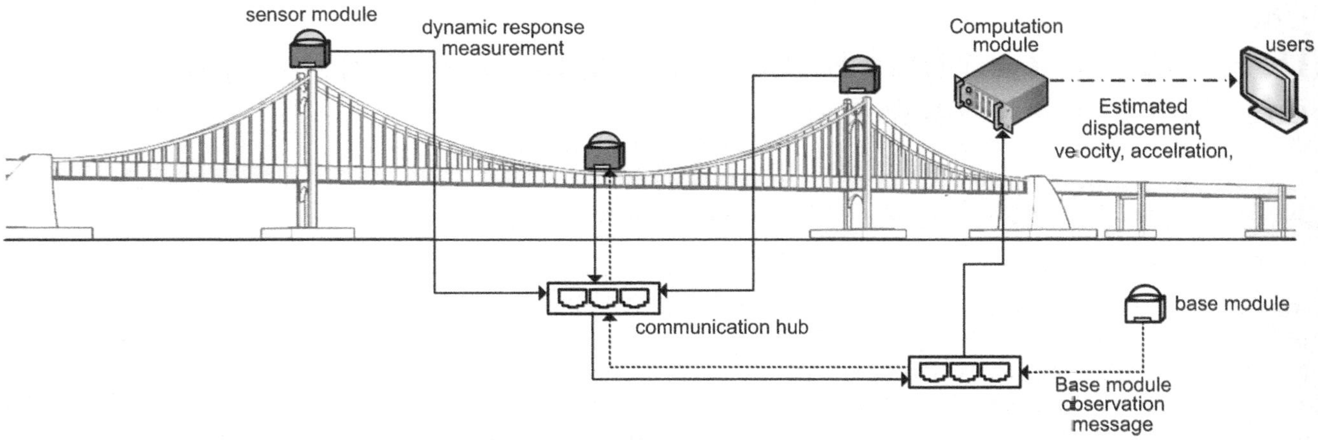

Fig. 18.1 Overview of the proposed measurement system including sensor, base and computational modules

18.2 The Proposed Displacement Sensor

To overcome the limitations of GPS-RTK, a novel structural displacement sensor is developed in this study. The developed sensor measures 3-DOF displacement and acceleration of large-scale civil structures based on data fusion of acceleration measured from a force feedback accelerometer, velocity measured from a single GPS sensor and displacement obtained from a low cost GPS-RTK using two-stage Kalman filtering [3].

The proposed displacement sensor system consists of a sensor module, a base module, and a computation module shown in Fig. 18.1. The sensor module and the base module receive satellite signals independently. The base module transmits its satellite signals to the sensor module through communication hubs. Then, the sensor module calculate RTK position and velocity value with two satellite signals acquired by the sensor module and the GPS base module, measures acceleration with its accelerometer, and time-synchronize all the measurements. The dynamic response measurement data are transmitted to computation module online. The computation module estimates the precise displacement, velocity, acceleration by applying a series of data processing techniques including a reliability assessment algorithm and a two-stage Kalman filter to the received raw data from sensor module.

The proposed 3-DOF structural displacement sensor offers the following advantages over the existing GPS-RTK sensors commonly used for civil engineering structures monitoring: (1) The proposed sensor can achieve a better accuracy(around 2 mm) and a better sampling rate (up to 100 Hz) compared to the conventional GPS-RTK sensors, and (2) The performance is less affected by weather conditions, multipath problems and signal blockage, which typically deteriorate the performance of conventional GPS-RTK sensors.

18.3 Experimental Verification

For verification of the proposed displacement measurement system, a field experiment at Yeongjong Grand Bridge in South Korea has been conducted. The displacement estimated by the proposed sensor was compared with the LDV reference displacement in Fig. 18.2. As a result, the performance of the proposed displacement sensor is demonstrated with high accuracy with the RMSE of 1.55 mm.

18.4 Conclusion

In this paper, a dynamic displacement measurement system is proposed for monitoring large-scale civil infrastructures. The system is composed of a sensor module, a base module, and a computation module. Compared to other existing sensor or data fusion techniques, it has an advantage in estimating both acceleration, velocity and displacement with better accuracy at once with real time calculation. The measurement performance of the system is validated by various lab-scale tests and

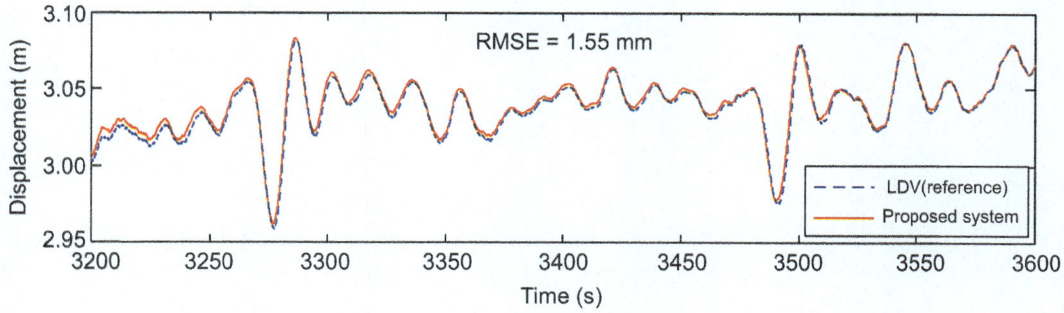

Fig. 18.2 Sensor measurements obtained from the experiment

several field applications. In the field applicability test, the measurement performance of an ambient vibration was verified to have the RMSE of 1.55 mm. As a future work, the proposed system would adopt GPS chipsets which support not only L1 but also L2 GPS signal for enhancing the quality of estimated dynamic responses.

Acknowledgment This research was supported by the Field-oriented Support of Fire Fighting Technology Research and Development Program funded by the Ministry of Public Safety and Security (MPSS-소방안전-2015-72)

References

1. Nassif, H.H., Gindy, M., Davis, J.: Comparison of laser Doppler vibrometer with contact sensors for monitoring bridge deflection and vibration. NDT&E Int. **38**(3), 213–218 (2005)
2. Castellini, P., Martarelli, M., Tomasini, E.: Laser Doppler Vibrometry: development of advanced solutions answering to technology's needs. Mech. Syst. Signal Process. **20**(6), 1265–1285 (2006)
3. Kim, K., Choi, J., Koo, G., Sohn, H.: Dynamic displacement estimation by fusing biased high-sampling rate acceleration and low-sampling rate displacement measurements using two-stage Kalman estimator. Smart Struct. Syst. **17**(4), 647–667 (2016)

Chapter 19
Effects of Pedestrian Excitation on Two Short-Span FRP Footbridges in Delft

Stana Živanović, Justin Russell, Marko Pavlović, Xiaojun Wei, and J. Toby Mottram

Abstract Reported in this paper is an evaluation of the vibration behaviour of two footbridges in The Netherlands having main spans of about 15 m. Short-span footbridges over canals and rivers are embedded in the landscape of Delft and elsewhere. Increasingly, these bridges are made of Fibre-Reinforced Polymer (FRP) composites, utilising the high-strength and light-weight nature of the material, and taking advantage of fast installation and low maintenance costs. Due to low mass, these FRP bridges might be sensitive to dynamic excitation by human actions. The two footbridges investigated in this paper have the main bearing structure which consists of either two or four longitudinal beams made of vacuum infused FRPs with foam cores connected by an FRP deck. Modal testing revealed that fundamental vertical natural frequency of the two structures at 4.8 and 6.1 Hz is in the range typical of similar structures made of concrete and steel, whilst the corresponding damping ratio for the wider, slightly cambered, bridge was exceptionally high at 7.9%. The vibration response to dynamic force by people walking was typically up to 1 m/s^2. While both of these light-weight structures performed satisfactorily under the regular pedestrian loading, the higher frequency – higher damping structure represents an example of successful control of pedestrian-induced vibrations by means of longitudinal restraints at the end supports and slightly curved shape of the main structure.

Keywords FRP composites · Short-span footbridges · Modal properties · Vibration · Walking

19.1 Introduction

Short-span footbridges made of high-strength, light-weight Fibre-Reinforced Polymer (FRP) composites are frequently employed in crossing canals and rivers in the Netherlands. These structures are fast to install and require little maintenance even in aggressive environments. The light-weight nature of the FRP material, with density that is about one-quarter that of steel, means that a 15 m footbridge can weigh as little as two tonnes (2000 kg), is easy to manipulate on site, and offers savings in energy use and cost of transport, lifting requirements and design of substructure foundations [1]. Whilst the low mass of FRP structures makes them increasingly popular choice in bridge engineering, it is the key factor that might lead to sensitivity to dynamic excitation by pedestrians.

There is little information in the public domain on dynamic performance of as-built FRP footbridges. There exists awareness that these structures might be lively under dynamic excitation. At the same time, the authors' experience with a number of short (e.g. 15–20 m) span FRP structures is that they mostly exhibit acceptable vibration performance during typical loading conditions. Testing full-scale structures is a necessary step towards enhancing understanding of their dynamic behaviour and ensuring informed and confident structural engineering designs in the future.

To contribute to the above goal, the University of Warwick in collaboration with TU Delft conducted a programme of testing on two footbridges in Delft. This paper describes the two structures first, followed by description of experiments to determine the modal properties. Then the vibration performance of the two structures under single pedestrian excitation as well as groups of pedestrians is analysed. Finally, discussion and conclusions from the studies are presented.

The testing was conducted in cold weather conditions (air temperature 3–10 degrees Celsius, humidity 74–86%, light-to-gentle breeze) from the 6th to the 8th December 2016.

S. Živanović (✉) · J. Russell · X. Wei · J. T. Mottram
School of Engineering, University of Warwick, Coventry, UK
e-mail: s.zivanovic@warwick.ac.uk

M. Pavlović
TU Delft, Department of Structural Engineering, Delft, The Netherlands

S. Pakzad (ed.), *Dynamics of Civil Structures, Volume 2*, Conference Proceedings of the Society for Experimental Mechanics Series, https://doi.org/10.1007/978-3-319-74421-6_19

19.2 Description of Two Footbridges

Bridge 1 is a pedestrian bridge at the edge of TU Delft's campus. It consists of two (relatively) independent spans of 15 m and 10 m with deck width of 2 m. Each span has two longitudinal beams made of vacuum infused FRPs with foam cores connected by an FRP deck, moulded as one piece, and resting on neoprene pads at the ends of each span. Surface is an epoxy layer with embedded gravel. Handrails are made of steel and they are continuous along the spans. Total mass of superstructure for the 15 m span is estimated at about 4500 kg. View of the two spans, as well as of the data acquisition station, is shown in Fig. 19.1.

Bridge 2 is a 14.7 m long by 4.5 m wide pedestrian and cyclist bridge, with allowance for 12 tonne service vehicles. Structure is made of four vacuum infused FRP foam core beams connected by a composite cover. It has a slight camber and sits on concrete supports, with neoprene pads which also restrain it longitudinally. Hand railing is provided by a series of 1 m composite posts bolted to side plates. Total mass is 6600 kg. Fig. 19.2. shows the bridge and the data acquisition station.

Fig. 19.1 Bridge 1 – left: side view and right: data acquisition station

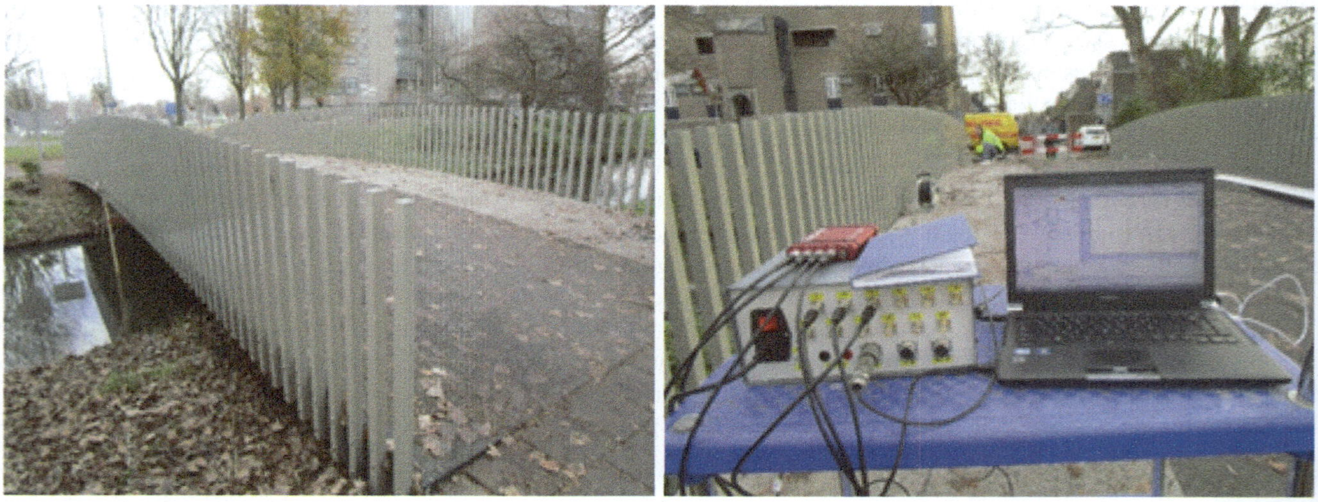

Fig. 19.2 Bridge 2 – left: side view and right: data acquisition station

19.3 Modal Properties

To determine the dynamic properties (mode shapes, natural frequencies and damping ratios) of the two structures, the vertical vibration response of the deck was measured using three force balance accelerometers (model QA750 by Honeywell) having nominal sensitivity of 1300 mV/g, where g is the acceleration of gravity. Two types of excitation were considered: natural (ambient) excitation due to wind and impact excitation applied manually using an instrumented impact hammer (Model 5803A by Dytran, nominal sensitivity 0.23 mV/N). Since deck response due to natural excitation was very low, the decision was made to perform the modal testing using hammer excitation. This led to measuring Frequency Response Functions (FRFs). During the measurement periods the bridges were closed for pedestrian/bicycle traffic.

The force and response signals (in the vertical direction only) were recorded using four-channel data acquisition card (SignalCalc Quattro by Data Physics). The results of modal testing are presented in the reminder of Sect. 19.3.

19.3.1 Bridge 1

To test Bridge 1, a grid of 24 test points (TPs) on the deck, over the two spans, was utilised (Fig. 19.3). The 15 m span accommodates TPs 1-7 and 51-57, whilst the remaining TPs belong to the 10 m span. The longer span is of more interest for this study since it has the lower fundamental natural frequency. For this reason, the results from the 15 m span only will be reported.

TP2 was chosen as a reference point to which the impact force was employed. The impact was applied five times in each measurement setup and each time the data acquisition window lasted 8 s. The force and response signals were sampled at 1280 Hz. The 8 s data acquisition window resulted in sufficiently fine frequency resolution of 0.125 Hz. An exponential window was applied to all signals to minimise the leakage and noise effects [2]. Three roving accelerometers were moved across the test grid until all 14 TPs were covered.

The measured accelerance FRFs were analysed using ME'scope software [3] to identify modal properties of vibration modes in the frequency range up to 20 Hz. Damping ratio for each vibration mode was corrected to eliminate the artificial damping introduced by applying an exponential window to the raw data records [4]. The identified vibration modes are shown in Fig. 19.4. The isometric and side views of mode shapes are presented, with the corresponding values for natural frequencies and damping ratios. Two vertical flexural modes resemble the shapes typical of a simple beam structure. The three torsional modes look similar; this is a consequence of apparent flexibility in the boundary conditions and potential vibration transfer between the two spans via the handrails. Figure 19.4 shows that the supports engage in motion of either Mode 2 or Mode 3, whilst they are relatively stationary in Mode 4.

According to the Setra guidelines [5], only the first mode has potential to compromise vibration serviceability of the structure, due to its natural frequency being less than 5 Hz. On the other hand, this mode, and all the others, have the damping ratio above 1.2%. This amount of damping is a positive feature of the bridge given that structures that exhibit excessive vibrations often have damping ratios well below 1%.

The actual damping ratio in the first mode is likely to be lower than the estimated 1.2% due to presence of the hammer operator on the bridge throughout the testing. To eliminate the operator's interaction with the structure and his influence on the modal properties of the first mode, the natural frequency and damping were calculated from the free decay response measured in the test in which a person walked at 2.4 Hz to excite resonance in Mode 1. Using the logarithmic decrement method, the natural frequency was determined to be unchanged at 4.8 Hz, whilst the damping ratio was lower at 0.7%, i.e. 58% of the value determined in the bridge – hammer operator dynamic system.

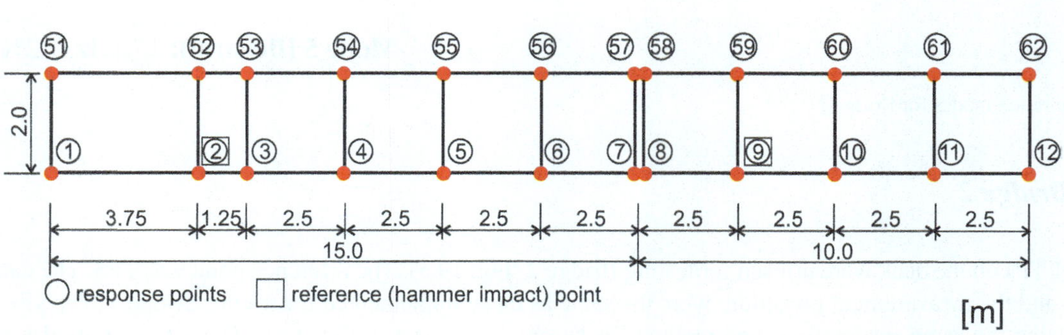

Fig. 19.3 Bridge 1 – measurement grid

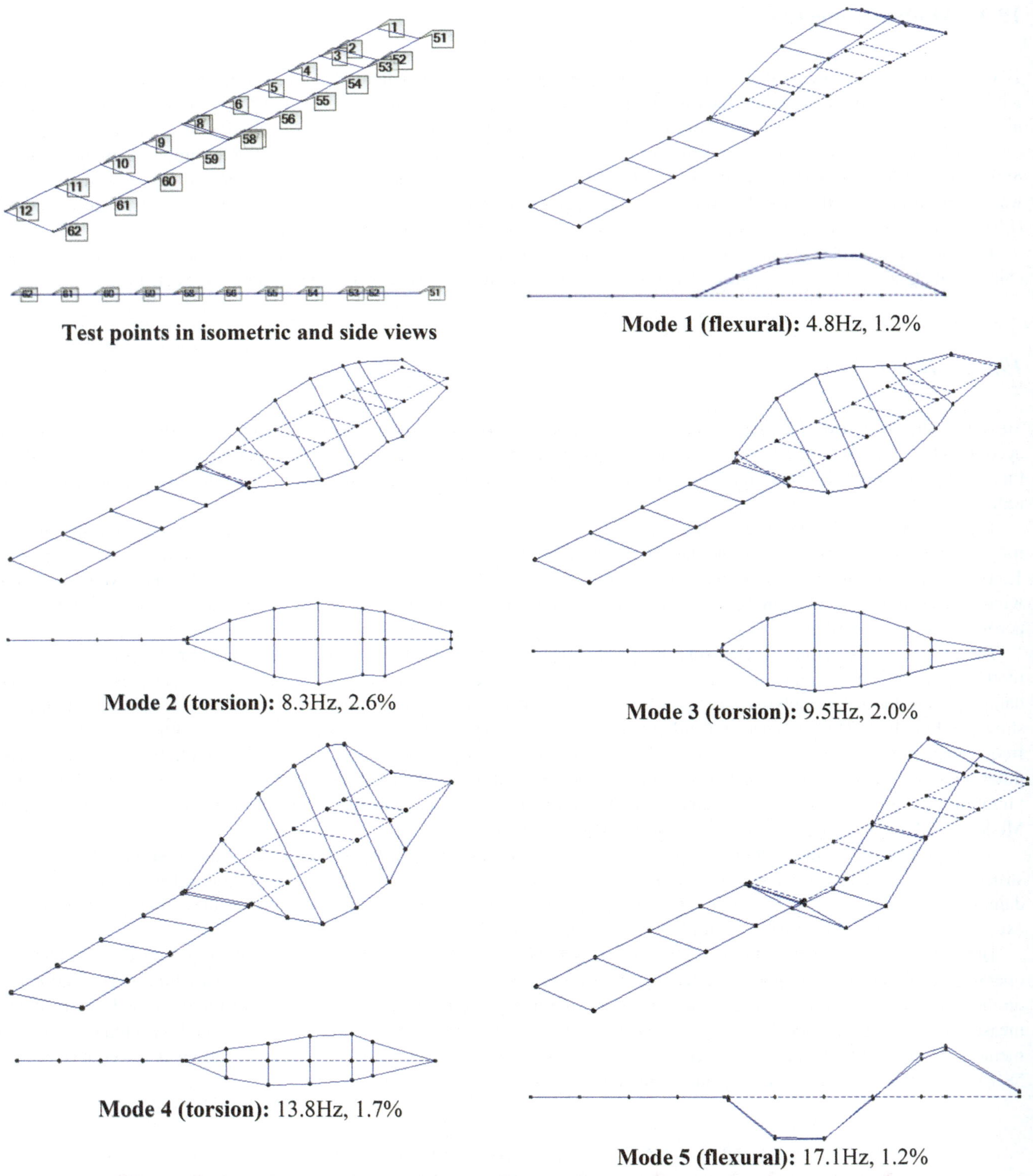

Fig. 19.4 Vibration modes for Bridge 1

19.3.2 Bridge 2

A grid of 21 TPs on the deck were utilised for testing Bridge 2 (Fig. 19.5). The reference point was TP3. The data acquisition parameters and the measurement procedure were the same as those implemented for testing Bridge 1. The vibration modes, identified using the same assumptions and methods as before, are shown in Fig. 19.6. The vibration modes of this bridge

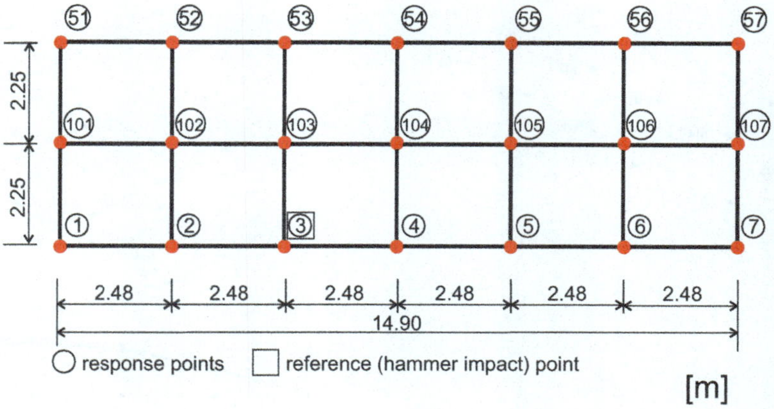

Fig. 19.5 Bridge 2 – measurement grid

are all above 5 Hz, and according to Setra guidelines [5], the risk of exciting resonance on this structure is negligible. In addition, the first two modes that, given their frequencies, might be excitable by higher harmonics of the human induced dynamic loading both have significant damping, reinforcing the conclusion that the vibration performance of this bridge is unlikely to be an issue in normal use. Note that interaction of bridge dynamics and that of the hammer operator is not of interest, due to inherent high frequency of vibration.

The two bridges have fundamental natural frequencies at 4.8 Hz and 6.1 Hz. This belongs to the natural frequency range of 3-7 Hz typical of concrete and steel bridges having similar span length of 15-20 m [6]. The damping ratio of Mode 1 of Bridge 1 is similar of data typical of reinforced concrete structures (that normally have higher damping than steel, composite steel-concrete or pre-stressed concrete structures), whilst the damping ratio of Bridge 2 is exceptionally high. This is believed to be due to this bridge having the structural engineering features of being slightly cambered and having longitudinal restraints.

19.4 Vibration Performance

Two loading scenarios were considered for both bridges. First single pedestrian loading was applied. Tests were performed at controlled pacing rates by two members of the testing team. Second loading scenario involved groups of pedestrians walking. The test configurations and the measured vibration responses are listed in Table 19.1. The response was measured at TPs 3 and 4 on both bridges.

The response tests on Bridge 1 show that Mode 1 at 4.8 Hz responded most to walking excitation by both single walker and groups of people. The second harmonic of the force generated by Walker 1 and Walker 2, when they walked at 2.4 Hz, excited the resonance in Mode 1, which is the reason that walking at this frequency generated the highest vibration response. In free walking tests by the groups, strong vibrations were intermittent, and did not exceed 1 m/s^2. According to Setra, this bridge provides mean comfort for its users, which is an acceptable comfort classification given the location and the usage of Bridge 1.

In walking tests on Bridge 2, the response is Mode 1 was very low owing to significant amount of damping in this mode. As a result, Mode 2 at 10.1 Hz responded most; but this response was accompanied with response in higher modes and at multiple frequencies of the higher forcing harmonics. For such a high frequency content, the recorded peak response of 1.5 m/s^2 to a group of walkers is not presenting the comfort challenge to pedestrians using the bridge. In the single walker tests, impulse response to each heel impact was easily identifiable in the data recorded, suggesting that this bridge exhibits dynamic behaviour similar to high frequency floors [7, 8]. Humans walking on this bridge cannot excite the resonance of the structure in any of the vibration modes, and the bridge is likely to provide very good comfort for its users.

It can be concluded that both bridges exhibit satisfactory vibration serviceability performance in normal use. If exposed to larger (so called vandal) dynamic loads, such as due to jumping, Bridge 1 has potential to respond strongly due to high likelihood of exciting resonance at 4.8 Hz. On the other hand, the resonance response of Bridge 2 is extremely unlikely, and therefore this bridge is unlikely to be seen as uncomfortable even when exposed to extreme dynamic loads generated by human actions.

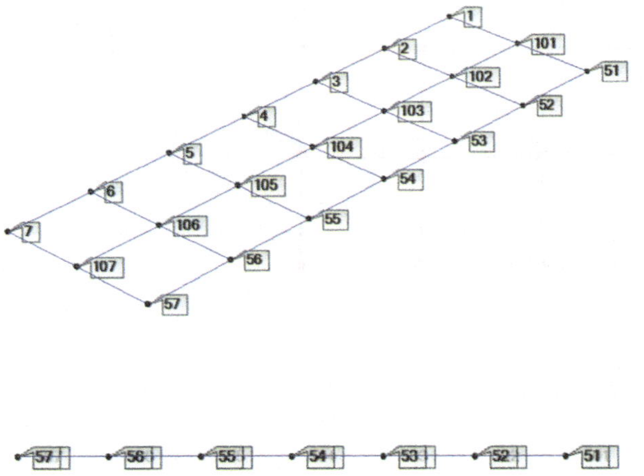

Test points in isometric and side views

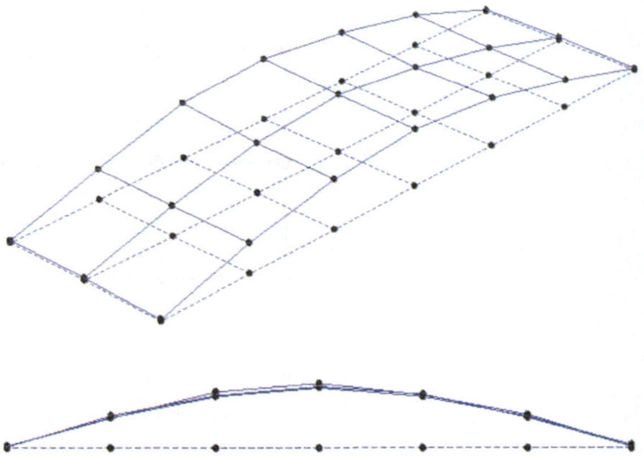

Mode 1 (flexural-longitudinal): 6.1Hz, 7.9%

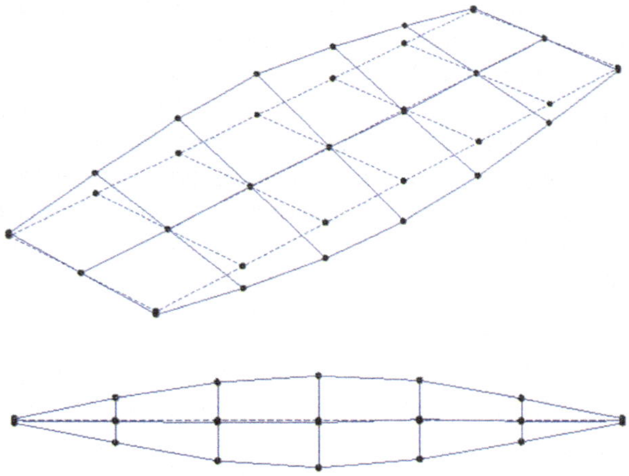

Mode 2 (torsion): 10.1Hz, 4.4%

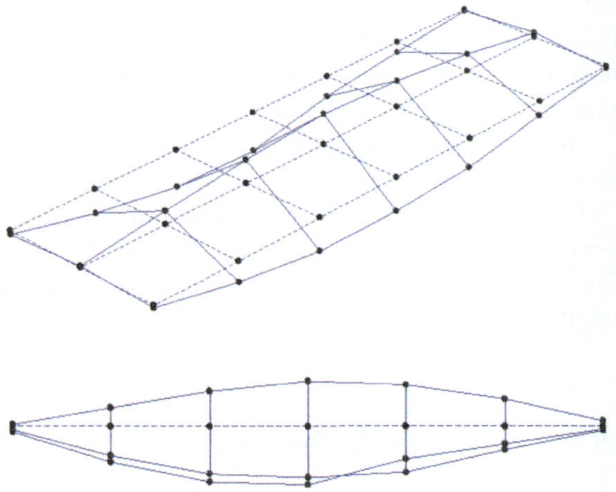

Mode 3 (flexural-transverse): 17.1Hz, 1.0%

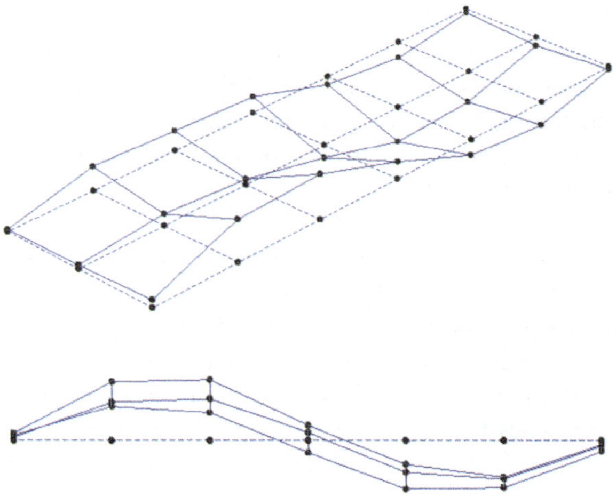

Mode 4 (flexural-longitudinal & transverse): 18.9Hz, 2.1%

Fig. 19.6 Vibration modes for Bridge 2

Table 19.1 Information on loading arrangements and corresponding vibration response

Test #	Loading	Pacing frequency [Hz]	Max. acceleration [m/s^2] (TP number)	Stongest response frequency [Hz]	Comment
Bridge 1					
1	Walker 1	1.4–2.5 (in 0.1 Hz steps)	1.2 (TP4)	4.8 (Mode 1)	Walking at 2.4 Hz
2	Walker 2	1.4–2.5 (in 0.1 Hz steps)	0.9 (TP4)	4.8 (Mode 1)	Walking at 2.4 Hz
3	8 walkers	Free	0.6 (TP4)	4.8 (Mode 1)	Walking for 10 min
4	15 walkers	Free	1.0 (TP4)	4.8 (Mode 1)	Walking for 10 min
Bridge 2					
5	Walker 1	1.4–2.5 (in 0.1 Hz steps)	0.45 (TP3/4)	10.1 (Mode 2)	Walking at 2.5 Hz
6	Walker 2	1.4–2.5 (in 0.1 Hz steps)	0.35 (TP3/4)	10.1 (Mode 2)	Walking at 2.5 Hz
7	8 walkers	Free	1.5 (TP3/4)	10.1 (Mode 2)	

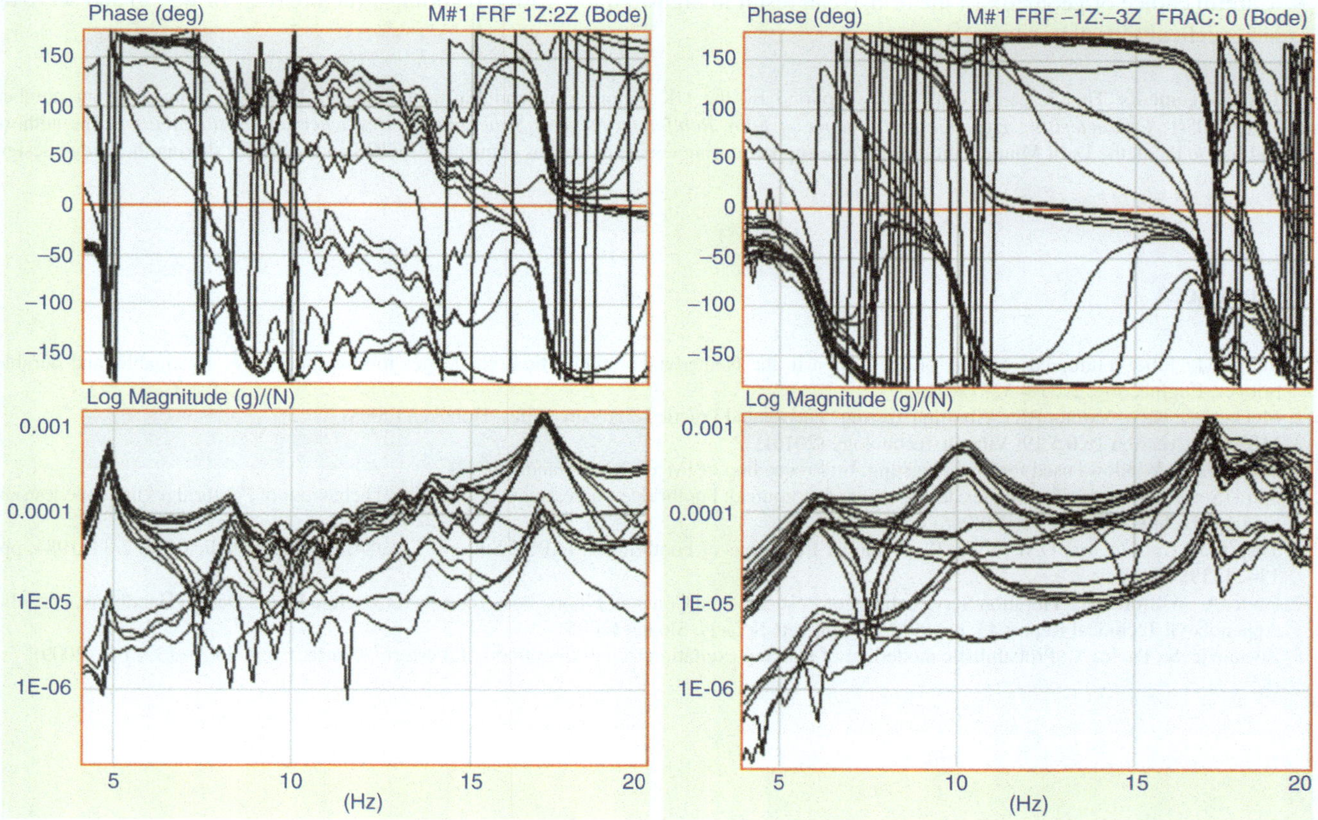

Fig. 19.7 Phase and magnitude of measured FRFs on – left: Bridge 1 and right: Bridge 2

Some of the key features of the two bridges are observable in the measured FRFs, presented in the four plots in Fig. 19.7. For example, the first two modes of Bridge 2 (first two peaks in the magnitude graph in Fig. 19.7: right) are unusually broad and for highly damped situation. Bridge 1, on the other hand, exhibits complex behaviour in 9–15 Hz frequency range (Fig. 19.7: left). This finding might be due to a strong interaction between the railings and the two deck spans. Finally, both bridges have a strong mode at around 17 Hz. This relatively high frequency is unlikely to be of interest for evaluation of pedestrian comfort.

19.5 Conclusions

The following conclusions can be drawn from the experimental analysis of the two short-span FRP bridges located in Delft:

- The fundamental natural frequencies of the two structures at 4.8 and 6.1 Hz belong to the frequency range of 3-7 Hz typical of the 15–20 m span footbridges built of traditional construction materials, such as steel and concrete.
- Damping ratio of the two bridges is either similar to that seen in most damped traditional structures (0.7% for Bridge 1) or much higher (7.9% for Bridge 2).
- Both bridges exhibit satisfactory vibration behaviour in normal use, suggesting that the FRP composites are viable alternatives to traditional construction materials for bridging what is classified as short spans.
- The two bridges exhibit fundamentally different vibration behaviour, with Bridge 1 able to be excited in resonance, whilst Bridge 2 exhibits vibration performance similar to high-frequency floors in buildings.
- Careful choice of geometry of the bridge can result in high-frequency high-damping structure (e.g. Bridge 2) and serve as an excellent vibration control measure.

Acknowledgements This research work was supported by the UK Engineering and Physical Sciences Research Council [grant number EP/M021505/1: *Characterising Dynamic Performance of Fibre Reinforced Polymer Structures for Resilience and Sustainability*]. The authors would like to thank the Delft Municipality for logistic support during testing as well as companies MOCS and FWD for sharing the bridge design data.

References

1. Smiths, J.: Fiber-reinforced polymer bridge design in the Netherlands: architectural challenges toward innovative, sustainable, and durable bridges. Engineering. **2**, 518–527 (2016)
2. McConnell, K.G., Varoto, P.S.: Vibration Testing: Theory and Practice, 2nd edn. Wiley, Hoboken (2008)
3. ME'scope, version 16.0.5.19: Vibrant Technology (2016)
4. Fladung, B.: Windows used for impact testing. In: Proceeding of IMAC-XV. Orlando (1997)
5. Sétra (Service d'Etudes Techniques des Routes et Autoroutes): Footbridges, assessment of vibrational behaviour of Footbridges under pedestrian loading. Technical Guide, Paris (2006)
6. Tilly, G.P., Cullinngton, D.W., Eyre, R.: Dynamic Behaviour of Footbridges, IABSE Surveys S-26/84, IABSE Periodica, No. 2/84, 1984, pp. 13–24 (1984)
7. Pavic, A., Willford, M.: Vibration Serviceability of Posttensioned Concrete Floors. Post-Tensioned Concrete Floors Design Handbook, 2nd edn. Appendix G, Technical Report 43, pp. 99–107. Concrete Society, Slough (2005)
8. Živanović, S., Pavic, A.: Probabilistic modelling of walking excitation for building floors. J. Perform. Constr. Facil. **23**(3), 132–143 (2009)

Chapter 20
Vibrational Response of Structures Exposed to Human-Induced Loads

Jonas Syders Knudsen, Nikolaj Grathwol, and Svend Ole Hansen

Abstract Structures like pedestrian bridges, staircases and floors become lighter and more slender. As a consequence human-induced vibrations in resonance with the structure become increasingly important when considering the serviceability of a structure. The present paper describes the vibrational response using models for pedestrian loads and the accompanying assessment of the serviceability of a structure exposed to human-induced vibrations. The presented approach uses natural frequencies, modal masses and structural damping to determine the structural vibrations. This allows for more flexible and elegant structures when considering human comfort, which often is dimensioning for light structures with large spans. A criteria based on frequency limits is not sufficient to ensure satisfactory vibration serviceability, especially for light structures. The approach is compared with vibration measurements on a structure before and after the installation of tuned mass dampers.

Keywords Human-induced vibrations · Vibration measurements · Tuned mass dampers · Vibration criteria · Structural dynamics

20.1 Introduction

Human-induced vibrations become increasingly important as structures are getting lighter and longer. A spectral load model to predict human-induced vibrations was presented in [1]. The load model can be reduced to a simple analytical expression, which has been applied for practical applications for more than 15 years [2]. The simplified method may be applied to model rhythmic loading from people jumping, which is relevant for the design of grand stands and fitness centers [1]. The focus of this paper is to assess vertical human-induced vibrations of floors, pedestrian bridges and staircases with the spectral model and the criteria of human comfort on general office floors.

20.2 Serviceability Criteria for Structural Vibrations

The level of comfortable structural vibrations is influenced by several factors such as: individual perception, environment, serviceability, situation, duration and body position. Some of the more strict criteria are found for laboratories, while structures like grand stands and fitness centers have much more moderate criteria for satisfactory vibration levels. Continuous and intermittent vibrations are considered to be less tolerable than sudden impact vibrations in the same order of magnitude. Human-induced vibrations caused by walking in a general office are considered to be continuous or intermittent, as the events are expected to occur several times per day.

It is recommended in the Eurocode EN 1990 [3] to apply the International Standard ISO 10137 [4] for satisfactory vibrations in buildings. The ISO 10137 criteria for continuous vertical vibrations in general offices is applied in this paper, see Fig. 20.1, which shows the standard deviation of the accelerations criteria as a function of the response frequency. The vibration limit of $0.02\ \mathrm{m/s^2}$ between 4 and 8 Hz for general offices is established by the ISO base curve [4] with a multiplying

J. S. Knudsen (✉) · N. Grathwol
Svend Ole Hansen ApS, Copenhagen, Denmark
e-mail: jsk@sohansen.dk

S. O. Hansen
Svend Ole Hansen ApS, Copenhagen, Denmark

SOH Wind Engineering LLC, Burlington, VT, USA

S. Pakzad (ed.), *Dynamics of Civil Structures, Volume 2*, Conference Proceedings of the Society for Experimental Mechanics Series, https://doi.org/10.1007/978-3-319-74421-6_20

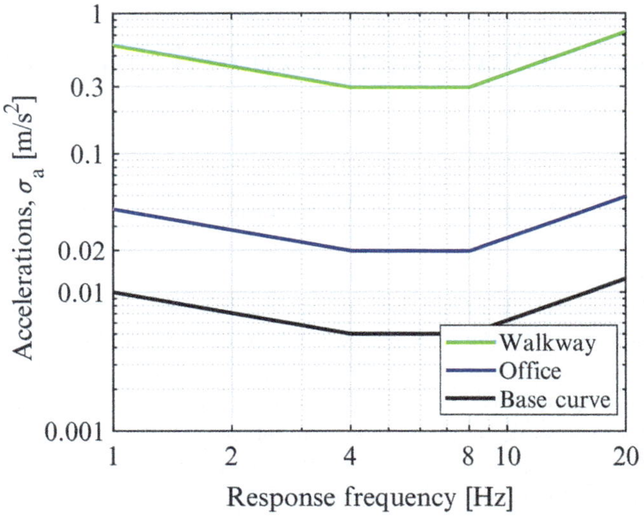

Fig. 20.1 ISO criteria [4] for the standard deviation of the vertical accelerations for continuous floor vibrations in general offices and walkways as a function of the structural response frequency. The criteria are based on the ISO base curve with a multiplying factor of 4 and 60. The response frequencies are divided into 1/3 octave band [4]

factor of 4. The level of acceptable vibrations in general offices is similar to the recommended limit incorporated in the Danish National Annex [5], where the levels are based on measurements and tests on general office floors with employees working at their desks. The tests indicated that the amount of complains from employees sitting at their desks increase significantly if the level of accelerations become higher than the recommended limit. This recommended level of vibrations for general offices has been applied for more than 10 years in Danish building designs, and the specifications have turned out to be in line with the user expectations.

The curves in Fig. 20.1 show the limits for the standard deviation of the accelerations for different frequencies, implying that people are mostly sensitive to vertical vibrations with a frequency of 4–8 Hz. This may be due to natural frequencies of the human body being in this frequency range. For a human-induced load in resonance with the structure, the response frequency in Fig. 20.1 is equal to the relevant natural frequency of the structure.

20.3 Mathematical Model of Human-Induced Vibrations

This section is divided into a spectral load model [1] and a response model presented in Sects. 20.3.1 and 20.3.2, respectively. The response model is based on a typical damped one degree-of-freedom system.

20.3.1 Human-Induced Loads

A spectral approach may be used to describe the load and associated load effects of humans walking or jumping. Neglecting the static load, the human-induced vertical load from a group of people may be determined as a sum of k harmonic components [1]

$$F(t) = N m_{\mathrm{p}} g \sum_{j=1}^{k} \alpha_j C_j \cos(2\pi j n_{\mathrm{p}} t + \phi_j), \tag{20.1}$$

where t is time, N is the number of persons, n_{p} is the movement frequency, m_{p} is the human mass, g is the gravitational acceleration and α_j, C_j and ϕ_j are the amplitude factor, crowd reduction factor and phase lag for the j'th harmonic load component, respectively. Assuming no phase lag and that the resonant term is dominant, the magnitude of the single harmonic load component may be estimated as

$$\max|F(t)| \approx N m_{\mathrm{p}} g \alpha_j C_j, \tag{20.2}$$

Table 20.1 Vertical load coefficients for human-induced vibrations

	Walking on horizontal surface [2]	Walking on staircase [6]	Rhythmic jumping [2], free possibility of movement	Rhythmic jumping [2], reduced possibility of movement
n_p	1.6–2.4 Hz	1.6–2.4 Hz	0–3 Hz	0–3 Hz
α_1	0.4	1.00	1.6	0.4
α_2	0.1	0.21	1.0	0.25
α_3	0.06	0.08	0.2	0.05
ρ_1	0	0	1.0	1.0
ρ_2	0	0	0.3	0.1
ρ_3	0	0	0.03	0.01

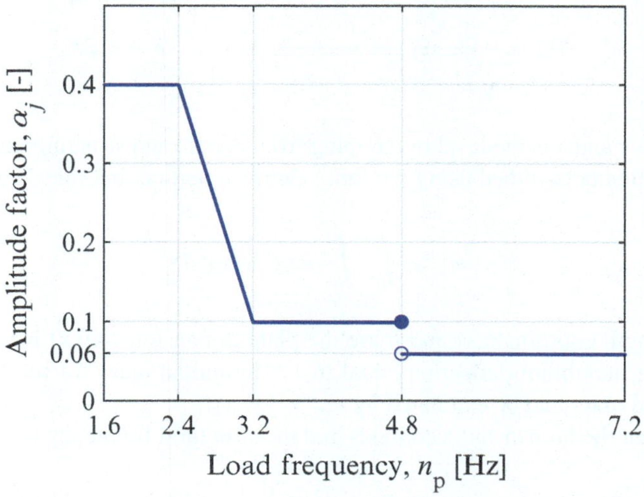

Fig. 20.2 Load amplitude factor α_j for people walking on a horizontal surface as a function of the load frequency. The gab of the amplitude factor near 4.8 Hz is caused by model uncertainties. A linear dependency is assumed for 2.4–3.2 Hz

where C_j is the crowd reduction factor describing the correlation between loads from various persons at different locations on the structure. The crowd reduction factor is defined as [1]

$$C_j = \sqrt{\rho_j + \frac{1 - \rho_j}{N}\lambda}, \tag{20.3}$$

where ρ is the correlation coefficient and λ is a scaling of the equivalent number of persons given as

$$\lambda = \frac{1/N \sum_{i=1}^{N} \gamma_i^2}{(1/N \sum_{i=1}^{N} \gamma_i)^2}, \tag{20.4}$$

where the mode shape value γ_i is assumed to have the same sign for all N people. Constant mode shapes with $\lambda = 1$, are conservative when assessing the comfort of a vibrating structure. For a simply supported beam with uniform load distribution, λ is equal to 4/3. For a load from a single person, the crowd reduction factor is $C_j = 1$ and for a group of people walking, the crowd reduction factor is $C_j = \sqrt{\lambda/N}$. The relevant amplitude factors and correlation coefficients for people walking and jumping may be found in Table 20.1.

The amplitude factor for people walking on a horizontal surface is shown as a function of the load frequency in Fig. 20.2. Load frequencies up to 7.2 Hz, corresponding to $k = 3$, will often result in accurate estimates of the human-induced vibrations. However, higher harmonics may be relevant for light structures with natural frequencies above 7.2 Hz. Values for jumping people should be applied for grand stands and fitness centers [1].

A correlation coefficient of $\rho = 0$, as shown in Table 20.1 for people walking, may be applied when the load from different persons is uncorrelated. As a consequence, the correlation coefficients in Table 20.1 are not applicable for people marching. The correlation coefficient is especially important for rhythmic loads such as people jumping, which is relevant for the design of grand stands and fitness centers.

20.3.2 Structural Response from Human-Induced Loads

For a harmonic forced response of a damped one degree-of-freedom system, the standard deviation of the response acceleration may be calculated as a function of the load amplitude given by Eq. (20.1) and the modal mass m_g of structure

$$\sigma_a = \frac{N m_p g}{m_g} \sqrt{\frac{1}{2} \sum_{j=1}^{k} (\alpha_j C_j H_j)^2}, \tag{20.5}$$

where H_j is the dynamic amplification factor defined as

$$H_j = \frac{1}{\sqrt{(1 - (\frac{j n_p}{n_1})^2)^2 + (\frac{\delta}{\pi} \frac{j n_p}{n_1})^2}}, \quad \delta << 2\pi, \tag{20.6}$$

where n_1 is the natural frequency and δ is the level of damping from people and structure given as the logarithmic decrement (LD). The modal mass m_g is often determined using the finite element method, but may be calculated for a simple plate as

$$m_g = \int_0^a \int_0^b m(x) \gamma(x)^2 dx, \tag{20.7}$$

where x is the independent length coordinate, a and b are the plate dimensions, $m(x)$ is the mass per area and $\gamma(x)$ is the mode shape normalized with a maximum deflection equal to 1. The modal mass for the first mode of a simple supported plate with uniformly distributed mass maybe estimated by $m_g = \frac{1}{4} m_{total}$.

Assuming resonance between the human-induced loads and the structural frequency implies

$$j n_p = n_1, \quad j = 1, 2, \ldots, k, \tag{20.8}$$

where n_p is the moving frequency of the people. For a structure with a natural frequency of 4 Hz, the moving frequency of the people is assumed to be 2 Hz resulting in the second harmonic load component to be in resonance with the structure.

Assuming that the resonance term of the loading is dominant, similar to Eq. (20.2), results in a simple analytical expression to estimate the standard deviation of the structural accelerations

$$\sigma_{a,j} \approx \frac{1}{\sqrt{2}} \frac{N m_p g \alpha_j C_j}{m_g} \frac{\pi}{\delta}, \tag{20.9}$$

A typical design load for general office floors is a single person walking in resonance [4] with the natural frequency of the floor.

20.4 Model Results

A comparison between the full model Eq. (20.5) and the single harmonic approach Eq. (20.9) is illustrated in Fig. 20.3 based on the ISO criteria described in Sect. 20.2 for general offices. The simple expression in Eq. (20.9) is un-conservative as it underestimates the response. It is fairly accurate when considering heavy-weighing and lightly damped structures. For light-weighing and highly damped structures, the accuracy decreases, but will often give good indications to whether the comfort of a structure is satisfactory. The deviations become higher than 10% for natural frequencies of 6 Hz with a damping of 0.12 LD and the deviations decrease for natural frequencies of 4 or 2 Hz, see Fig. 20.3.

The results in Fig. 20.3 are found using the load from a single person with mass $m_p = 75$ kg. Points above the relevant curve imply that the comfort of the structure is satisfactory. For example, a floor with a natural frequency of 4 Hz, modal mass of 100 ton and a damping of 0.2 LD have satisfactory comfort. However, if the natural frequency is 2 Hz and the other parameters remain unchanged, the comfort will be unsatisfactory.

The model shows that the important parameters to determine the magnitude of structural vibrations are the natural frequency, modal mass and damping. Applying the model and the ISO criteria shown in Fig. 20.1 result in specific parameters

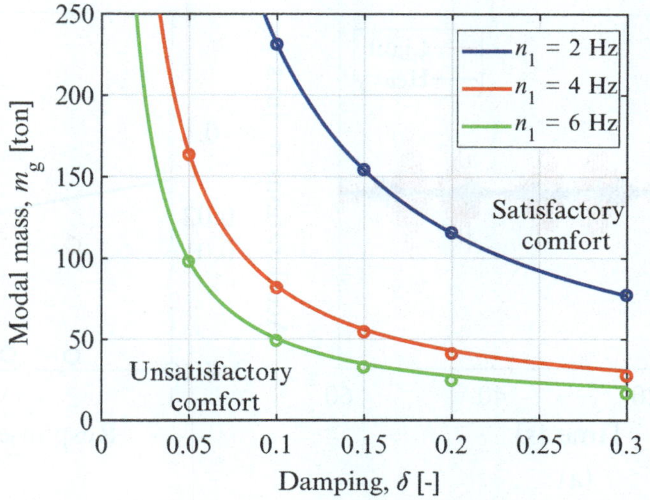

Fig. 20.3 Comfort of general office floors exposed to a load from a single person walking. Comparison of full model Eq. (20.5) marked by lines and the simple model Eq. (20.9) marked by circles. Points above the relevant curve imply satisfactory comfort

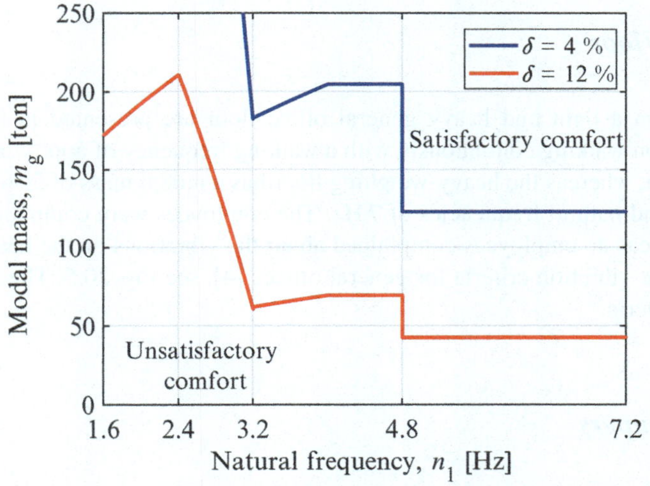

Fig. 20.4 Comfort of general office floors exposed to a load from a single person walking. The comfort is calculated by the full model Eq. (20.5). Points above the relevant curve imply that the comfort of the structure is satisfactory

to determine the vibration comfort in general offices, see Fig. 20.3 and 20.4. Figure 20.4 shows that relatively large modal mass and damping is criteria for natural frequencies up to 4.8 Hz where the relevant load harmonic shifts from $j = 2$ to $j = 3$. The necessary modal masses are approximately 125 and 45 ton for damping levels of 0.04 and 0.12 LD respectively, as the natural frequencies become higher than 4.8 Hz.

20.5 Measurements

Calculated structural vibrations are associated with uncertainties from constrain-types, connections, masses, structural damping levels etc. Some examples of the measured response on structures such as a pedestrian bridge, a staircase and general office floors are presented in this section.

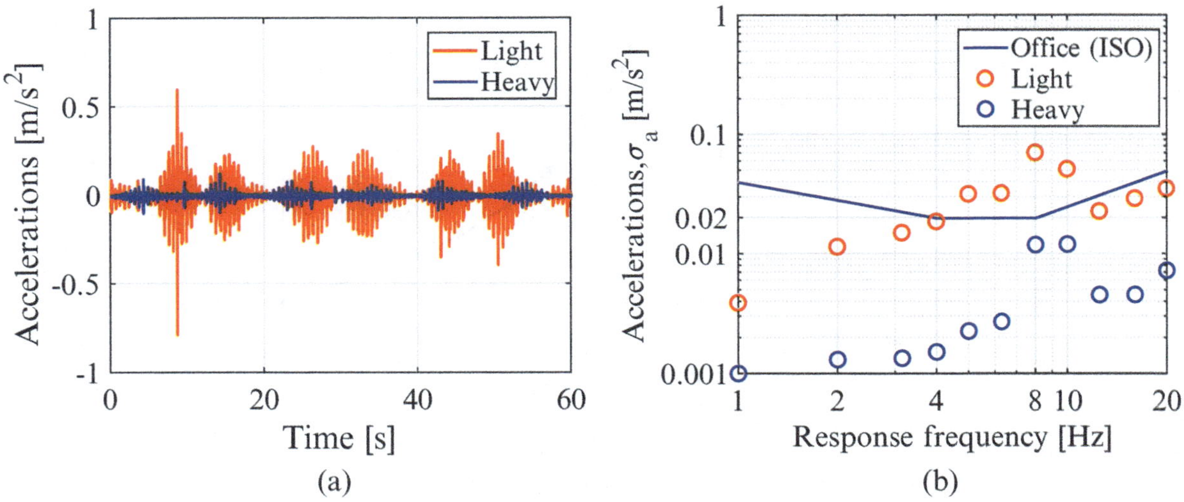

Fig. 20.5 Measured accelerations during a single person walking continuously with a walking frequency of approximately 2 Hz. The accelerations are measured on a light and heavy general office floor. The response frequencies are divided into 1/3 octave band [4]

20.5.1 General Office Floor

The measured accelerations on a light and heavy general office floor are presented in Fig. 20.5. The accelerations are measured during a single person walking continuously with a walking frequency of approximately 2 Hz. The light-weighing floor has a modal mass of 5 ton, whereas the heavy-weighing floor has a modal mass of 50 ton. The two general office floors have similar damping levels and natural frequencies of 7 Hz. The employees were complaining about the vibrations on the light general office floor, whereas no employees complained about the vibrations on the heavier floor. The vibrations of the light-weighing floor exceed the vibration criteria for general offices [4], see Fig. 20.5. The level of vibrations on the heavy floor is below the vibration criteria.

20.5.2 Tuned Mass Dampers

Structural damping is influenced by the type of material, connections, amplitude etc. Typically levels of structural damping are 0.02 and 0.12 LD for welded steel and reinforced concrete structures, respectively. In some design situations it might be advantageously to add damping to reduce vibrations. Tuned mass dampers (TMD) are very usable to increase damping levels as they are cost effective and the mechanical components are relatively simple. Tuned mass dampers have been used during the last 70 years on structures such as: bridges, chimneys, tall buildings, floors, cables etc. A tuned mass damper weighs typically 1–10% of the structural modal mass and contains simple elements such as: a mass, springs, dashpots. The design of a typical TMD is shown in Fig. 20.8.

The theory about tuned mass dampers is well known. It was determined and improved by Den Hartog [7] during the 1950s, and the theory was based on harmonic excitation loading. Tuned mass dampers are often installed on structures exposed to stochastic loads such as wind gust. The load type is not without consequence as the effectiveness of TMDs decreases for structures exposed to stochastic loads compared to harmonic loads. The behavior of TMDs exposed to stochastic loads is described by Krenk [8]. The load caused by several persons walking on a horizontal surface is similar to stochastic loads. The influence of reduced effectiveness should be considered when designing TMDs for human-induced vibrations.

Measured accelerations are shown in Fig. 20.6 during a decay test on a Danish pedestrian bridge with a modal mass of approximately 15 ton and a vertical natural frequency of 3 Hz. The bridge is exited by an identical harmonic load with and without a tuned mass damper. The consequence of the TMD is significant, as the damping level is increased from 0.025 to approximately 0.25 LD.

The level of vibrations from a single person walking on a staircase with installed TMDs are shown in Fig. 20.7. The staircase is located inside the public science center *Experimentarium* in Copenhagen, see Fig. 20.8. The staircase has a modal mass of approximately 10 ton and a natural frequency of 5 Hz. The accelerations are measured during a single person walking continuously with a walking frequency of approximately 2 Hz, see Fig. 20.7. The measured level of vibration is clearly lower than the criteria [4].

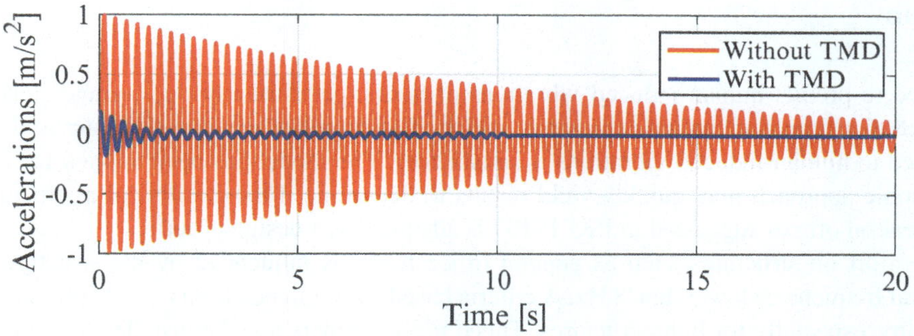

Fig. 20.6 Measured accelerations during decay tests on a pedestrian bridge with and without a TMD. Both decay tests are carried out with an identical excitation force

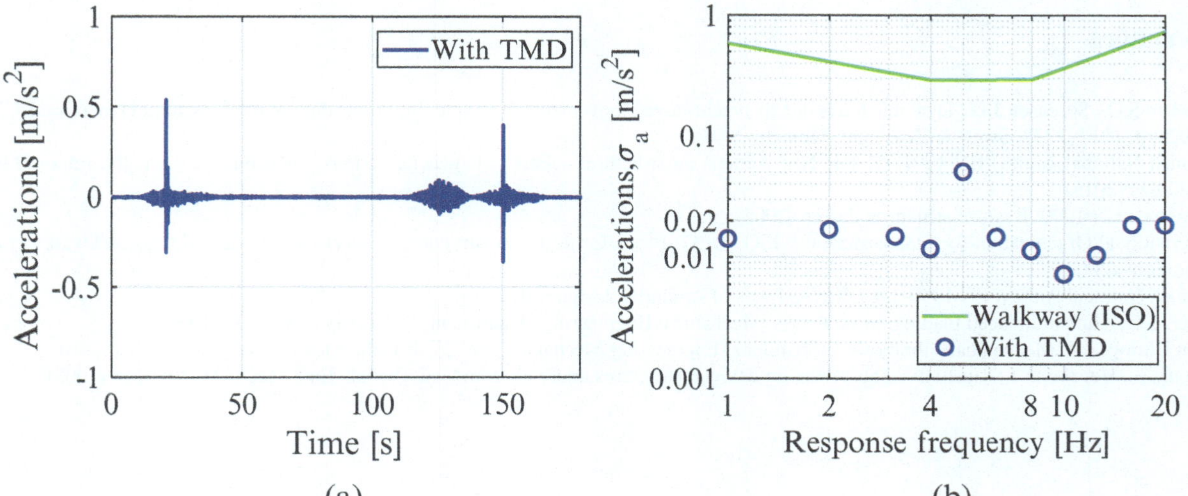

Fig. 20.7 Accelerations measured on a staircase with installed TMDs during a single person walking continuously with a walking frequency of approximately 2 Hz. The response frequencies are divided into 1/3 octave band [4]

(a) (b)

Fig. 20.8 Left: Copper plated helical staircase located inside the public science center *Experimentarium* in Copenhagen, Denmark. Right: Typical design of a TMD. (**a**) Photo by: CSK Stålindustri A/S, Tom Jersøe. (**b**) Tuned mass damper

20.6 Conclusion

A model is presented to predict human-induced vibrations of floors, pedestrian bridges, grand stands and staircases. The model can be reduced to a simple analytical expression, which often provides accurate estimates to determine if the comfort of a structure exposed to human-induced vibrations is satisfactory. The model has been applied for more than 15 years [2] and the single harmonic approach may quickly yield results to determine if the comfort of a structure is satisfactory. The vibration level for general offices suggested in ISO 10137 is adequate as a design criteria.

The vibration comfort on structures such as general office floors is influenced by the structural mass, especially for structures with natural frequencies lower than 8 Hz. A criteria based on frequency limits is not sufficient to ensure satisfactory vibration serviceability, especially for light structures. Tuned mass dampers may be installed to increase the damping levels to obtain a better comfort. Figures 20.3 and 20.4 may provide a simple design approach to determine whether a general office floor will have satisfactory comfort when exposed to people walking.

References

1. Hansen S.O., Sørensen J.D.: Dynamic loads due to synchronized movements of people. In: Proceeding of the Structural Dynamics, EURODYN 2002, pp. 1217–1222. Swets & Zeitlinger, Munich (2002)
2. Danish National Annex: EN 1991-1-1 DK NA. Actions on structures – Part 1–1: general actions – densities, self-weight, imposed loads for buildings (2015)
3. Eurocode EN 1990: Basis of structural design (2002)
4. International Organization for Standardization: ISO 10137. Bases for design of structures – serviceability of buildings and walkways against vibrations (2007)
5. Danish National Annex: EN 1990-1-1 DK NA. Basis of structural design (2013)
6. Keer, S.C.: Human induced loading on staircases. Mechanical Engineering Department, University of London (1998)
7. Den Hartog, J.P.: Mechanical Vibrations. Probabilistic Engineering Mechanics, vol. 23, 4th edn. Mcgraw-Hill, New York (1956)
8. Krenk, S., Høgsberg, J.: Tuned mass absorbers on damped structures under random load. Probab. Eng. Mech. **23**, 408–415 (2008)

Chapter 21
Protection of Critical Assets from the Effects of Ground Vibrations

Brad Pridham and Nick Walters

Abstract The approaches to control of vibrations in critical science, technology, and healthcare facilities focus on three key elements: (1) the vibration source; (2) the vibration path; and, (3) the vibration receiver. Techniques for modelling, assessment, and control vary depending on the nature of the vibration disturbance. Among the most challenging sources to control are ground motions generated by construction activities and/or heavy industrial processes that contaminate the site environment near vibration-sensitive facilities.

In this paper we discuss recommended approaches to control and management of the effects of ground vibration effects on ultra-sensitive research facilities. Approaches to design assessments and concepts for control are presented, with focus placed on management of the source, characterization of the path, and treatment of the vibration receiver. Illustrative examples and key findings from field testing, modelling, and monitoring are presented from a project involving a research facility located next to a construction site. A discussion on current challenges is provided together with suggestions for future research.

Keywords Construction vibration · Hospital · Pile driving · Vivarium · Microscopy

21.1 Introduction

The impacts of site vibrations on research spaces vary according to the source characteristics, site geology, building construction, and sensitivity of the receptor. Sources may be impulsive (e.g., pile driving, stamping, demolition), narrow-band quasi-stationary (e.g., vibratory compaction), or random (e.g., road and rail traffic). Geological conditions affect the efficiency of transmission to the building. Foundation size and depth also plays a role, although at very low levels of ground motion our data suggests that these factors may be less important. Vibration sensitivities of modern research spaces typically fall at or well below the threshold of human perception depending on use.

Assets critical to the operations of research facilities, such as equipment/toolsets, and specimens (e.g., animals, cell cultures), can be adversely affected by vibrations as follows:

- failure of tools to perform at the necessary resolution for the research being conducted;
- disruption of experiments that forces re-start, or produces false/misleading results;
- an inability to perform procedures correctly (e.g., micro-surgery, precise radiation treatments); and,
- undue stress on research animals that can cause problems with mortality, reproduction, and experiment outcomes.

These effects can produce significant losses to a facility due to operational downtime, irreproducible experimental outcomes, poor tool performance, and in extreme cases, physical damage. For vivarium spaces the psychological impacts of vibrations on animal populations are an added concern. As a result, facility owners and user groups are usually interested in mitigating risk of ground vibration impacts as much as is practical.

In this paper we describe a framework for protection of critical assets from ground vibration disturbances. The framework incorporates control concepts related to the source, path and receiver in approaches to identifying, qualifying, and mitigating risk. Application of the framework is illustrated in a case study on addressing construction vibration impacts on a research facility housing a vivarium and sensitive equipment.

B. Pridham (✉) · N. Walters
Novus Environmental Inc., Guelph, ON, Canada
e-mail: bradp@novusenv.com

S. Pakzad (ed.), *Dynamics of Civil Structures, Volume 2*, Conference Proceedings of the Society for Experimental Mechanics Series, https://doi.org/10.1007/978-3-319-74421-6_21

21.2 A Framework for Protection of Critical Assets

A general framework that can be applied for the protection of critical assets includes three phases: (1) identification of vibration exposure, (2) qualification of risk, and (3) mitigation of risk. Each of these phases is discussed below.

21.2.1 Identifying Exposure

The first phase is identifying the anticipated exposure of critical assets to ground vibrations. The degree of exposure can be determined qualitatively based on the sensitivity of the receiver, the type(s) of source(s), and proximity of receivers to the source(s). In many cases exposure is identified at the onset of the project based on knowledge of the site and program. Experienced designers often understand when a Vibration Consultant should be engaged on a project, and look to them to provide clarity on vibration sensitivity and other concerns related to ground vibrations. The consultant can provide clarity on vibration sensitivity of critical receptors, and experiential guidance regarding the degree of exposure of critical assets.

21.2.1.1 Criteria Specification

Specification of vibration criteria is essential to understanding exposure and qualifying risk. For research facilities, the Vibration Class curves are usually used for specification of criteria, and can be applied to a wide range of equipment and spaces (see Fig. 21.1 and [1]). Criteria for most critical assets fall at or below the Operating Theatre curve (also known as the ISO Base Curve, or the threshold of human perception).

In general, the vibration class criteria represent the most sensitive equipment, space, or specimen within a given class. Some examples of critical assets and associated criteria are listed in Table 21.1.

Vendor criteria, when available, can be helpful to understanding spectral sensitivities and to identify ground vibration sources of concern. For example, heightened sensitivity in the 12–15 Hz band could mean increased exposure to disturbances from roadway traffic, whereas heightened sensitivity in the 20–40 Hz band could mean increased exposure to disturbances from railway traffic. It is always helpful to overlay vendor criteria on the generic curves as a means of ranking exposures and ruling out concerns with specific items.

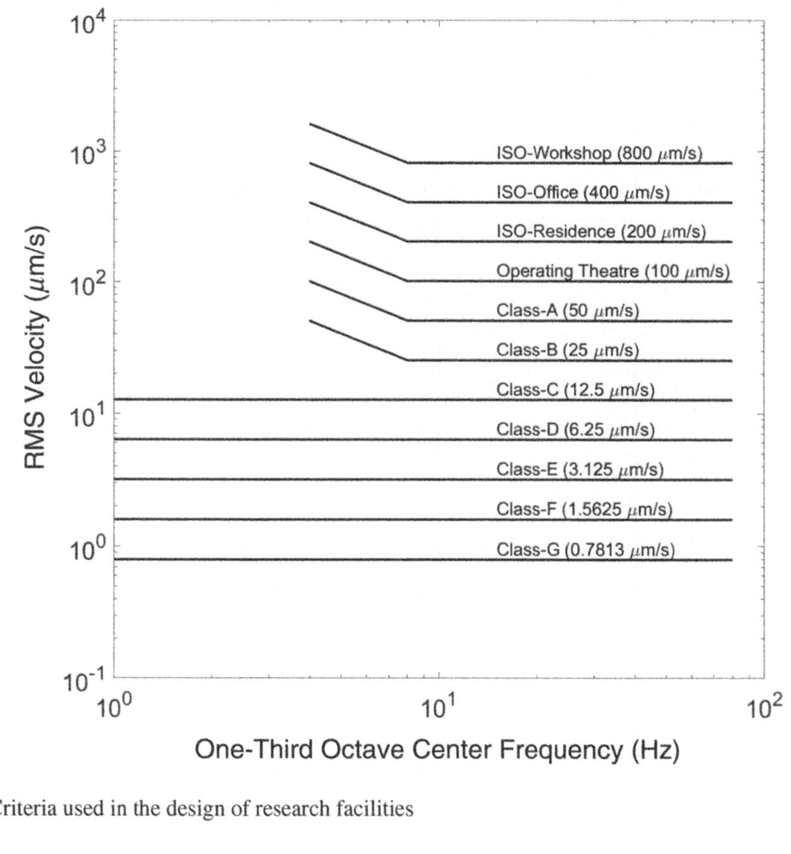

Fig. 21.1 Vibration Class Criteria used in the design of research facilities

Table 21.1 Examples of vibration criteria for critical assets in research facilities

Critical assets	Vibration class criteria
General surgical suites (human and animal), animal holding rooms	ISO Operating Theatre (100 μm/s)
Microscopes (400× magnification), micro-balances	Class A (50 μm/s)
Tissue and cell cultures, CT scanners	Class B (25 μm/s)
Microscopes (1000× magnification), long-term cultures	Class C (12.5 μm/s)
Confocal microscopes, NMR, MRI	Class D (6.25 μm/s)
SEM/TEM/STEM, FIB, lasers	Class E or lower (3.125 μm/s)

Table 21.2 Setback distances associated with different vibration criteria levels for various sources

	Critical setback distance (m)			
Source	ISO-Op	Class B	Class D	Class E
Bus and truck traffic, heavy	5	20	40	60
Heavy rail (freight, locomotive)	40	120	210	250
Light rail	10	50	120	160
Compaction	20	70	90	100
Piling	100	200	300	400
Excavation and demolition	50	180	200	250

21.2.1.2 Types and Proximity of Sources

The types and setbacks of ground vibration sources at the site are reviewed together with criteria to confirm exposure. Table 21.2 summarizes minimum preferred setbacks for typical ground vibration sources for various receptor sensitivities. These conditions are based on measurements at numerous sites and predictions using published desktop models used in impact assessments [2].

Note that the values in Table 21.2 are guidelines and that ground vibrations due to the activities indicated can be highly variable. For example, vibrations from pile driving depend on the hammer weight and drop height (energy rating); roadway vibrations depend on roughness of the pavement surface, speed of vehicles etc. A detailed qualification of risk of impacts is recommended when the setback of critical assets is equal to or less than the values indicated in Table 21.2.

21.2.2 Qualification of Risk

Once exposure to ground vibrations has been established it is necessary to qualify the risk of impacts to the facility. This process typically includes baseline testing at the site and/or modelling of ground vibration impacts to confirm the extent of impacts and the need for mitigation measures.

21.2.2.1 Baseline Testing

Site measurements are used to quantify the existing vibration environment and spectral transmission characteristics of geological conditions. For qualification of risk on existing facilities, measurements of building attenuation can be helpful for use in the assessment. The worst-case vibration levels from all relevant sources should be measured, if possible. Long duration measurements may be necessary to quantify temporal variations in levels at the site. While some sources may produce vibrations that exceed recommended criteria, they may be intermittent or of such low recurrence that they are of no consequence to operations. Long records of continuous vibrations are processed to produce centile spectral levels for comparison with criteria and identification of temporal variability of specific sources. An example of centile spectra measured at a site are shown in Fig. 21.2. 'Bulges' in the spectra at 5, 31.5 and 63 Hz are indicative of intermittent or transient events, while the 'pinch' at 12.5 Hz indicates a steady/continuou source.

Qualifying risk of vibration impacts from future sources or activities at the site requires more effort and usually includes testing involving staged or coordinated activities. Examples include ground transmissibility testing for use in rail vibration models; staged vehicle movements at locations of future parking and roads; and staged construction activities to quantify

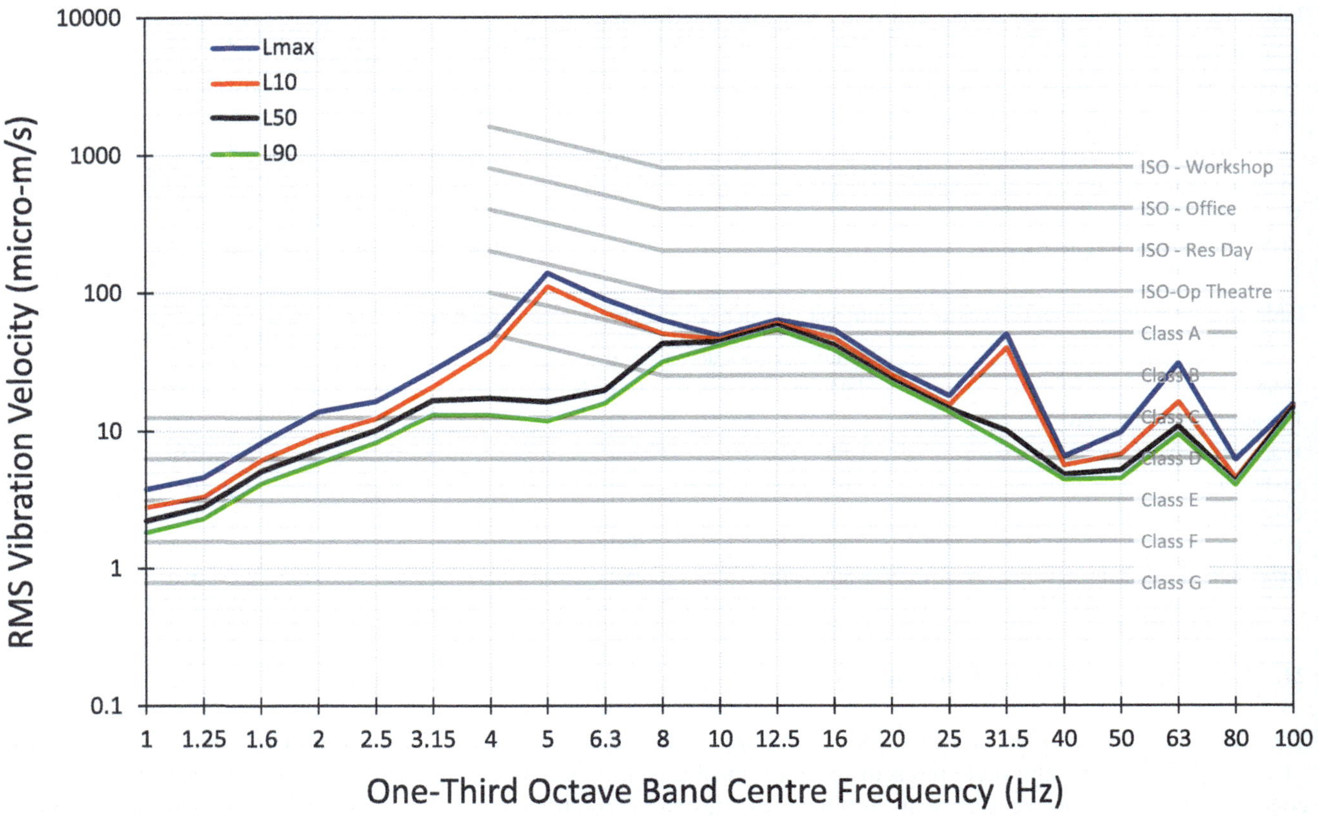

Fig. 21.2 Example of centile spectra associated with a long-term measurement

ground and building transmission for vibratory and impulsive sources (e.g., compaction, pile driving). These types of tests can be costly and may require significant coordination between vibration consultant, owner and contractor. However, they can be of great benefit to qualifying risk of ground vibration impacts and often the costs of testing are negligible compared with the costs associated with vibration disturbances to critical assets or the cost of implementing control strategies (i.e. changes to foundation type, equipment energy ratings, barriers).

21.2.2.2 Ground Vibration Impact Modelling

Qualifying impacts from future sources involves development of a prediction model using both published and empirical parameters extracted from site testing. The general impact model has the following form (see for example [2, 3]):

$$Receiver\ Level = Source\ Level + Ground\ Transmission\ Effects + Building\ Transmission\ Effects \tag{21.1}$$

The reliability of this model depends heavily on the availability of site-specific information. Published source levels, or those measured at other sites, can be highly variable, but are typically all that is available. Thus, a reasonable amount of conservatism is necessary when specifying these values. Ground transmission effects can usually be measured at the site, and it may also be possible to measure building effects if the facility exists. A comprehensive model includes parameters at each spectral line in the relevant frequency range, so it is possible to compute spectral estimates of the receiver level for comparison with the criteria. When it is not possible to measure parameter values, published data in references [2, 3] can be used. However, one should be aware that the published values are often conservative and are not site-specific, and as such can be grossly in error.

All pertinent sources and receptors are reviewed using the impact model and it is thus possible to qualify which source/receptor combinations are at greatest risk and require protection from ground vibration disturbances. If risk has been identified, there is potential for impacts of ground vibrations on operations and it is necessary to proceed to the next phase in the framework.

21.2.3 Mitigation of Risk

Effective mitigation of risk includes controls that target one or a combination of the source, path and receiver. The best strategies usually involve a combination of controls, but this is not always possible for some sites and sources. Some general approaches to mitigation are discussed here.

21.2.3.1 Source Control

The ability to control source emissions is typically limited, but can involve some of the following strategies:

- scheduling operation/use of source during periods of minimum impact (e.g., testing a generator after hours);
- vibration isolation of the source (e.g., a stamping press);
- relocation of the source; and,
- specification of low-vibration equipment, means and methods (e.g., construction).

Some of these approaches may be impossible, particularly if the source is owned by others or impractical to control. In our experience this strategy is most effective for mitigation of construction vibrations where it is possible to control the types of equipment used and the source emission levels.

21.2.3.2 Path Control

Control of the vibration path can include:

- strategic layouts to increase setback and screening between source and receiver;
- changes to the transmission medium including trenches and barriers; and,
- specification of allowable setback distances associated with 'moving' sources (e.g., construction equipment, new roadways etc.).

21.2.3.3 Receiver Control

Receiver controls are typically more practicable than source and path controls. Particularly when addressing vibrations from road and rail or adjacent buildings owned by others. Common strategies include:

- relocation to further setbacks or to other buildings;
- vibration isolation;
- scheduling operations around 'quiet' periods; and,
- vibration monitoring.

Monitoring can be used a means of both source and receiver control, and is usually implemented as part of an overall risk management strategy. While real-time measurements of vibrations do not explicitly protect assets from disturbances, they can be very useful for scheduling, and in scenarios involving construction are helpful to managing construction equipment selection, setback distances and real-time complaint/alarm management.

21.3 Case Study: Managing Ground Vibrations During Construction

Application of the framework is illustrated with a case study involving management of construction vibration impacts on a research facility. As Vibration Consultants on the project, the authors were responsible for developing a risk mitigation strategy to ensure appropriate protection of the research facility and surrounding land uses during construction and demolition.

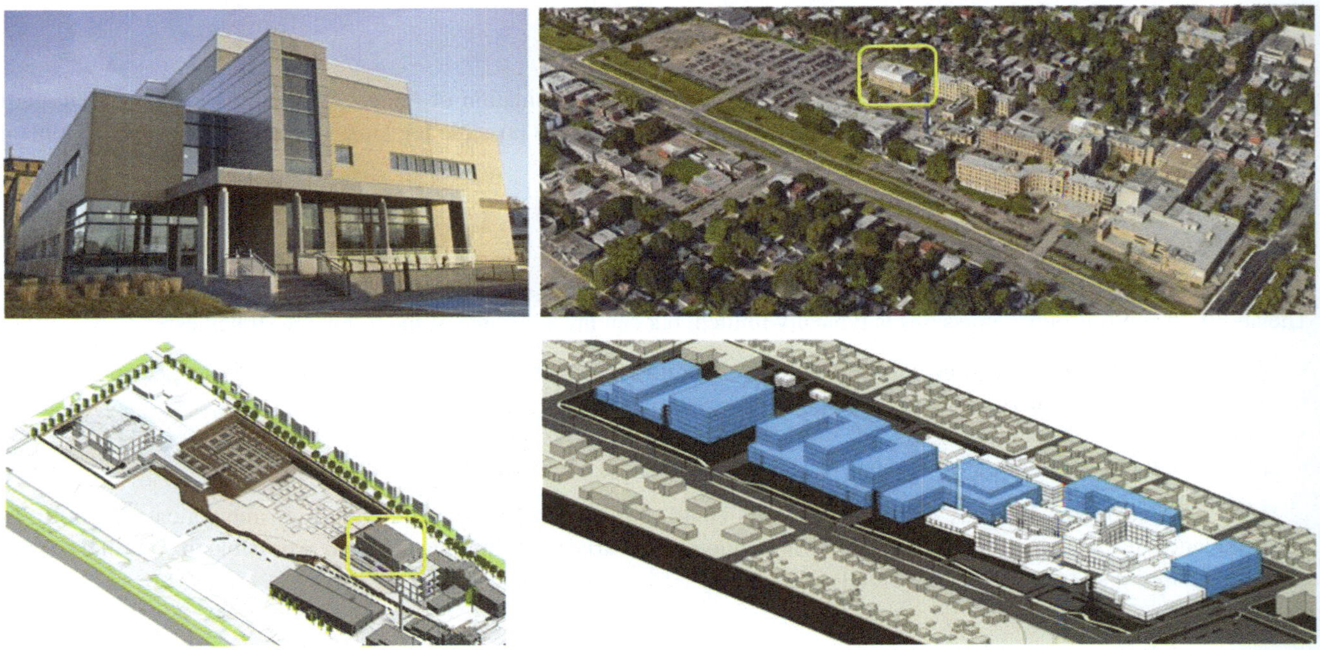

Fig. 21.3 Photos of the research facility and renderings of campus expansion and shoring works

21.3.1 Facility Description and Identification of Exposure

The case study building is a research facility on a hospital campus adjacent to the planned location of a major expansion to the campus. The expansion will include construction of several multi-storey buildings and construction works that will include shoring, construction of raft slabs, installation of Franki piles, and wide-spread compaction throughout the site. The research facility houses numerous pieces of research equipment as well as a vivarium at basement level. Protection of research assets was identified as a major concern during construction due to potential effects of ground vibrations on performance of sensitive equipment, long-term experiments (e.g., cell culture development), and on the health of research animals located at basement level. Photos of the existing site and research facility, and renderings of the future expansion are shown in Fig. 21.3.

Vibration criteria were established based on experience with the design of similar facilities and studies by others [1, 4]. A summary of applicable criteria are as follows:

- Protection of research animals from undue disturbances: ISO Operating Theatre (100 μm/s)
- Confocal microscopy, cell cultures: Class B (25 μm/s)

The facility houses approximately 700 mice, each costing thousands of dollars. As such, any vibration disturbances causing psychological or physical damage to the animals could result in significant losses to the facility. Additionally, long duration cell cultures could be disturbed leading to erroneous or spoiled test results. Given the criticality of these assets to operations, and the proximity of the planned construction works, a high risk of exposure was identified by the owner and stakeholders, and thus the authors were retained to complete a detailed qualification of risk and assist with developing a vibration mitigation strategy.

21.3.2 Qualification of Risk

To qualify specific risks to the facility, a site vibration model was developed that included information on source levels, ground transmission, and building effects. Site tests were completed in two phases: phase one included baseline measurements of typical site levels and measurements of simulated construction activity, and phase two included measurements of pile driving used to update the site model.

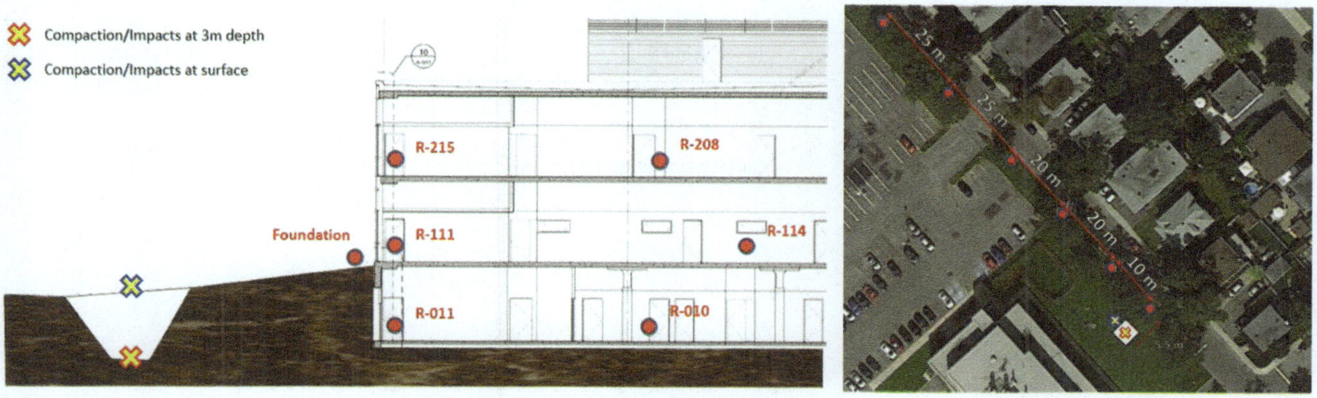

Fig. 21.4 Sketches of the phase one testing setup showing source and measurement locations

21.3.2.1 Phase One Testing

The first phase of testing involved quantification of ground and building attenuation characteristics. An excavator armed with a plate compactor was used to apply impacts and vibratory compaction adjacent to the research facility. Vibration monitors were installed at six locations within the facility and outside next to the foundation. Compaction and impacts (bucket hits) were applied at surface and in a pit at a depth corresponding approximately to the base of piles on the new buildings. Measurements were also completed at various setbacks on the ground surface to quantify ground vibration propagation characteristics. Sketches of the test setup are shown in Fig. 21.4.

21.3.2.2 Phase Two Testing

Preliminary modelling was conducted following the first phase of tests to establish potential impacts of Franki Piling. Model parameters were extracted from the excavator testing as well as historical measurements of Franki pile installations at other sites. The results indicated that the installation of Franki piles had the potential to significantly disrupt experiments and animals in the research building, in addition to potential cosmetic damage to residential buildings close to the pile locations.

To accurately evaluate the risk of the pile driving activity it was determined that test piles on site would be required. Two test piles were installed using a 7000 lb. hammer at varying drop heights increasing from 5 ft. to 20 ft. Vibration monitors were installed on the ground surface near the piling rig, within the research facility and hospital, and at the foundations of surrounding residential buildings. Results from the Franki pile tests were then used to refine the construction vibration model developed with updated parameters related to large impact events.

21.3.2.3 Site Vibration Model

Results from the site tests were used to develop a model to evaluate vibration impacts from a range of construction equipment and activities. The model was defined as follows:

$$V_r = PPV_s\, D\alpha + CL + \beta d \tag{21.2}$$

Where:

V_r = predicted vibration level;
$PPV_s\, D\alpha$ = transmission of vibrations in the ground;
CL = attenuation by the building foundation; and,
βd = transmission of vibration within the building.
PPV_s = construction equipment source vibration level;
D = distance between equipment and building foundation;
α = ground vibration decay factor;
CL = foundation attenuation factor;
β = structural floor decay factor; and,
d = distance to interior space from foundation wall.

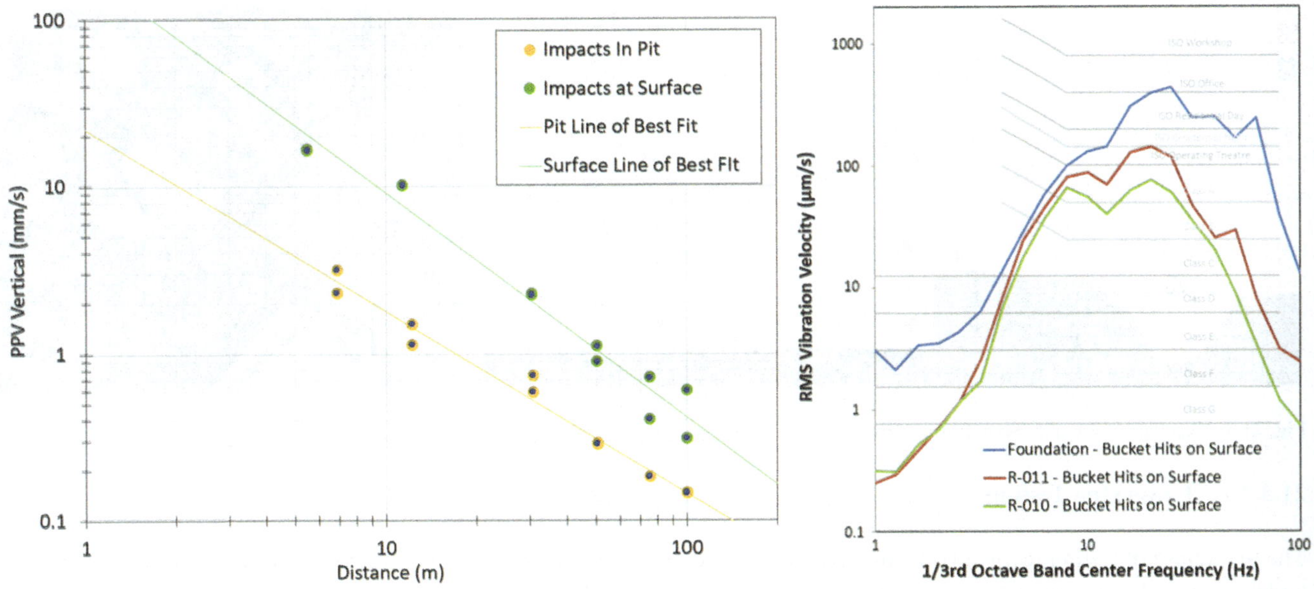

Fig. 21.5 Data measured during ground (left) and building (right) transmission testing

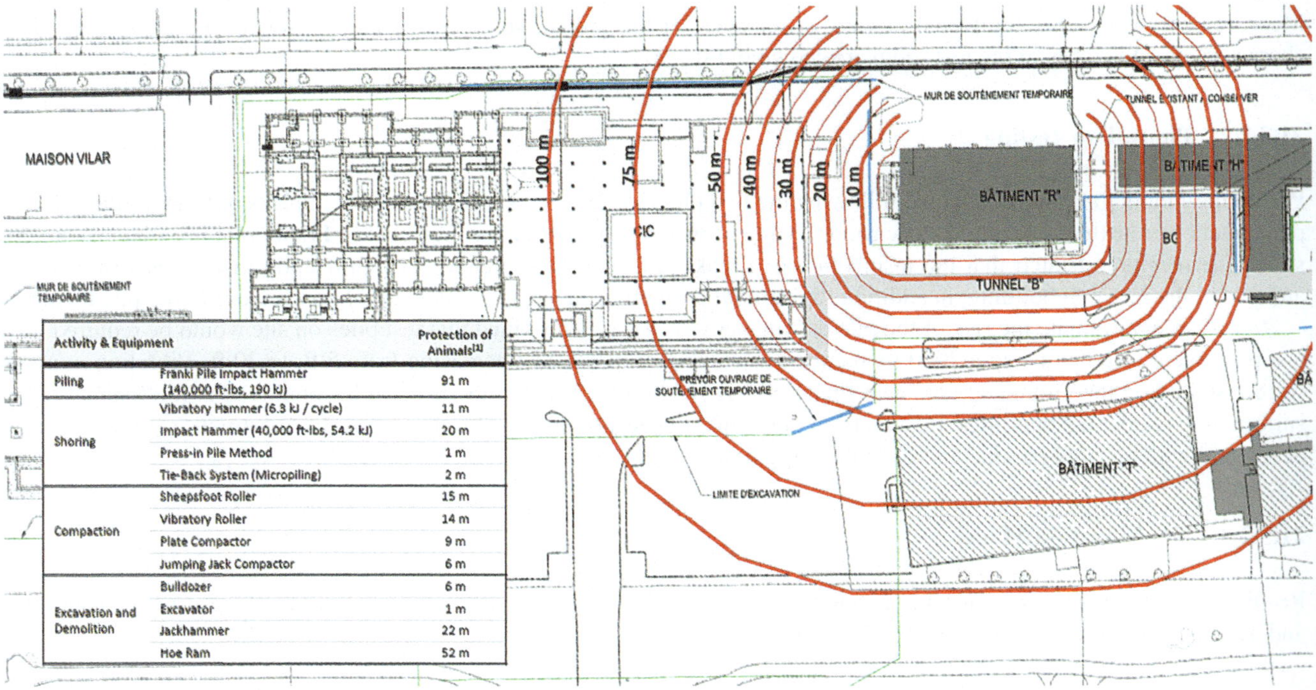

Activity & Equipment		Protection of Animals[1]
Piling	Franki Pile Impact Hammer (140,000 ft-lbs, 190 kJ)	91 m
Shoring	Vibratory Hammer (6.3 kJ / cycle)	11 m
	Impact Hammer (40,000 ft-lbs, 54.2 kJ)	20 m
	Press-in Pile Method	1 m
	Tie-Back System (Micropiling)	2 m
Compaction	Sheepsfoot Roller	15 m
	Vibratory Roller	14 m
	Plate Compactor	9 m
	Jumping Jack Compactor	6 m
Excavation and Demolition	Bulldozer	6 m
	Excavator	1 m
	Jackhammer	22 m
	Hoe Ram	52 m

Fig. 21.6 Allowable setbacks computed from the site vibration model

Inputs to the model were specified using the site testing data. Examples of data used to specify the ground and building transmission parameters are shown in Fig. 21.5.

The source vibration levels (PPV_S) were specified using measurements from equipment on other sites together with published values. The model in Eq. (21.2) produces an estimate of the peak particle velocity at the receiver, which is then converted to an equivalent RMS one-third octave velocity level using conversion factors established from phase one and two test data. The model was then used to predict safe setbacks associated with specific construction activity to prevent undue vibration disturbances on critical assets. An example of the results from this process is shown in Fig. 21.6.

21.3.3 Mitigation of Risk

While the construction vibration model did include site specific parameters and was verified by on site testing and measurements, a vibration monitoring protocol was required to guide the construction team with management of vibration impacts. Safe operating set-backs were provided as a general recommendation; however, these can be difficult to enforce on site in some cases, and do not include all possible types of equipment.

To mitigate risk a vibration monitoring protocol was developed. This included monitor locations, alarm and trigger levels to change work, and stop-work conditions. Over the course of the demolition and construction activities monitoring data were continually reviewed for refinement of control requirements where appropriate.

21.4 Concluding Remarks

A framework is presented for protection of critical assets in research facilities from the adverse effects of ground vibrations. The framework includes three phases: (1) identification of exposure, (2) qualification of risk of vibration impacts, and (3) mitigation of risk of vibration impacts. Application of the framework is illustrated in a case study involving the protection of a research facility from ground vibrations during construction. The framework can be applied to any facility where exposure to ground vibrations is a concern and risk of vibration disturbances need to be managed.

References

1. Amick, H., Gendreau, M., Busch, T., Gordon, C.: Evolving criteria for research facilities: I – Vibration. In: Proceedings of SPIE Conference 5933: Buildings for Nanoscale Research and Beyond, San Diego (2005)
2. Transit Noise and Vibration Impact Assessment, Federal Transit Administration (FTA), Technical Report No. FTA-VA-90-1003-06, May 2006
3. Dowding, C.: Construction Vibrations, 2nd edn. Prentice Hall, Upper Saddle River (2000)
4. Carman, R., et al.: Vibration effects on laboratory mice during building construction. J. Acoust. Soc. Am. (2008). https://doi.org/10.1121/1.2935010

Chapter 22
Experimental Characterisation of Dynamic Properties of an All-FRP Truss Bridge

Xiaojun Wei, Giosue Boscato, Justin Russell, Alessandro Adilardi, Salvatore Russo, and Stana Živanović

Abstract Fibre Reinforced Polymers (FRPs) have increasingly been utilised for construction of pedestrian bridges due to high strength- and stiffness-to-weight ratios, low maintenance costs and quick installation. Their relatively low mass and stiffness make these bridges potentially susceptible to vibration serviceability problems, which increasingly govern the design. Currently, the wider application of FRPs in civil engineering is hindered by the lack of experimental insight in dynamic performance of as-built structures. This paper presents an experimental investigation on a 25 m long glass-FRP truss footbridge in Italy. Ambient vibration tests were conducted to identify the dynamic properties. The peak-picking method and stochastic subspace identification approach were employed for modal parameter identification. The two methods produced very consistent results. Eight vibration modes were identified in the frequency range up to 10 Hz. Two lateral flexural vibration modes having natural frequencies of 5.8 and 9.6 Hz were identified, as well as two vertical flexural modes (at 7.5 and 8.1 Hz) and four torsional modes (at 2.1, 2.7, 4.8 and 9.3 Hz). Damping ratios for all modes up to 10 Hz except the eighth mode were above 1.2%.

Keywords FRP truss footbridge · Dynamic properties · Ambient vibration testing · Peak picking method · Stochastic subspace identification method

22.1 Introduction

The Fibre Reinforced Polymer (FRP) composites have become a popular choice of structural material for footbridges and road bridges [1–5] thanks to their favourable properties, including high strength- and stiffness-to-weight ratios, good durability and short installation time. Owing to relatively low stiffness and proneness to large deformation, the FRP composites are usually limited to short- to medium- span beam bridges. Longer bridge spans could be achieved if the FRP composites were used in truss bridges, suspension bridges or cable-stayed bridges, in which materials can be utilised to their highest potential.

Vibration serviceability is increasingly found to govern the design of FRP structures. A sound knowledge and understanding of their dynamic characteristics is important for developing robust serviceability design procedures. Currently, there is lack of experimental data on the dynamic characteristics of FRP structures. Existing design guidance used in conventional material designs is usually employed in the design of FRP structures, which may lead to a conservative solution and compromise the benefits of using FRP composites. In addition, insufficient knowledge about the dynamic properties of FRP structures lowers the confidence of structural engineers, which hinders the wide application of this new material. Extensive and comprehensive experimental testing of dynamic performance of as-built FRP structures is required to inform design of FRP structures against dynamic loading. This paper provides a rare insight into the dynamic features of a truss footbridge at the outskirts of Prato, Italy.

X. Wei (✉) · J. Russell · S. Živanović
School of Engineering, University of Warwick, Coventry, UK
e-mail: x.wei.3@warwick.ac.uk

G. Boscato · S. Russo
Iuav University of Venice, Venice, Italy

A. Adilardi
Comune di Prato, Prato, Italy

S. Pakzad (ed.), *Dynamics of Civil Structures, Volume 2*, Conference Proceedings of the Society for Experimental Mechanics Series, https://doi.org/10.1007/978-3-319-74421-6_22

Fig. 22.1 The Prato Bridge: (a) side view; (b) deck view

The paper starts with a description of the FRP bridge in Sect. 22.2, followed by a description of the ambient-based modal testing in Sect. 22.3. Identified modal parameters are reported in Sect. 22.4 with conclusions from their evaluation made in Sect. 22.5.

22.2 Bridge Description

The Prato Bridge is a simply supported 25 m truss bridge for pedestrians and cyclists. It was open in 2008 and it crosses a busy dual carriageway at the outskirts of Prato, Italy (Fig. 22.1). The bridge is made of pultruded glass-FRP channel sections, with bolted connections consisting of stainless steel plates and bolts. FRP plates provide additional lateral bracing. The deck is 2.5 m wide at the middle and 3.6 m at the ends and is made of a number of pultruded FRP planks, each 5 m long and 500 mm wide and 40 mm deep, bolted at the ends as well as at the midspan to transverse members below. Additional beam elements provide contact support at the planks quarter points. A high metal mesh provides a barrier at the slides of the bridge. The bridge weighs a total of around 8 tonnes [6] and rests on two concrete piers, each 5.7 m high from the ground (see Fig. 22.1a).

22.3 Ambient Vibration Testing

The dynamic characteristics of the Prato footbridge were identified using ambient vibration testing, which requires measuring the acceleration responses of the deck under the natural excitation (in this case the wind and road traffic passing underneath the bridge). The testing programme was conducted by the University of Warwick (UK) and Iuav University of Venice (Italy) on 29 November 2016. The bridge was closed to pedestrian traffic during data recording.

To identify the vibration modes of the bridge, the measurement grid shown in Fig. 22.2 was employed, which includes 22 measurement stations on the deck at the location of each truss joint. Lateral and vertical acceleration measurement campaigns were carried out separately using two sets of different data acquisition systems and accelerometers. Vertical accelerations at all the MSs except the MS 54 were recorded using eleven PCB 393C accelerometers having nominal sensitivity of 1000 mV/g (Fig. 22.3a). The vertical acceleration measurement campaign was divided into two set-ups. Lateral accelerations were recorded at the MSs 1, 3, 4, 6, 8, 9 and 11 using four Dytran accelerometers having nominal sensitivity of 500 mV/g (Fig. 22.3b). Lateral and vertical accelerometers at MS 4 were used as reference vibration sensors, whilst the other accelerometers are the roving transducers. The duration for each measurement set-up was 10 min. The sampling frequencies for vertical and lateral acceleration measurement were 600 Hz and 256 Hz.

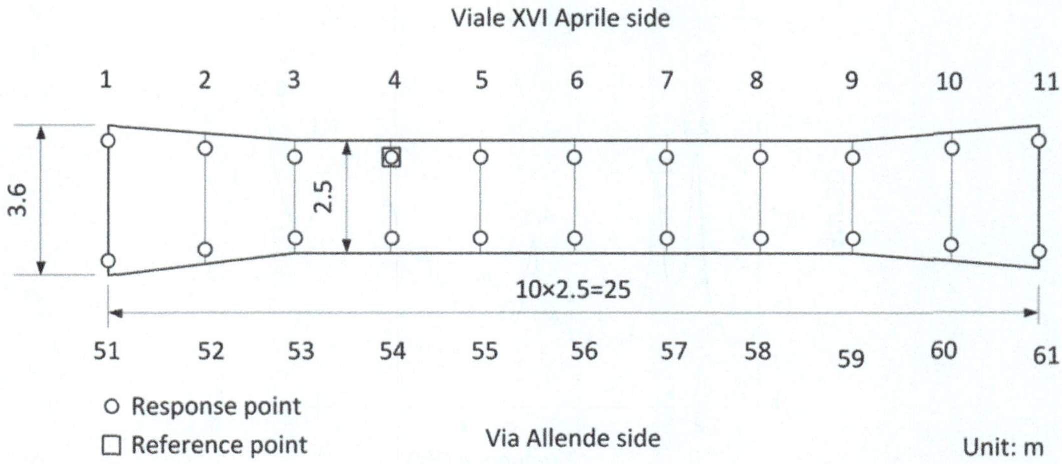

Fig. 22.2 Measurement grid on the deck

Fig. 22.3 Accelerometers on the deck: (**a**) Vertical: PCB 393C accelerometer; (**b**) Lateral: Dytran accelerator

22.4 Modal Parameter Identification

The modal parameters of the Prato footbridge were identified using the frequency-domain peak-picking method and the time-domain stochastic Sub-Space Identification (SSI) method. Whilst the peak-picking method was chosen to identify natural frequencies only, the SSI method was employed to identify natural frequencies, damping ratios and mode shapes.

22.4.1 Peak-Picking Method

Since certain modes may not be observable in the power spectrum density at a particular measurement station, the Averaged Normalised Power Spectral Densities (ANPSD) [7, 8] of all the measurement stations were used to identify the natural frequencies. Figures 22.4 and 22.5 display the ANPSDs of the vertical and lateral accelerations, respectively. The peaks indicate the natural frequencies of the bridge. The analysis identified six peaks at frequencies of 2.1, 2.7, 4.8, 7.4, 8.1 and 9.3 Hz that correspond to either vertical bending or torsional modes. By contrast, Fig. 22.5 shows that there are two lateral bending modes at 5.8 and 9.6 Hz.

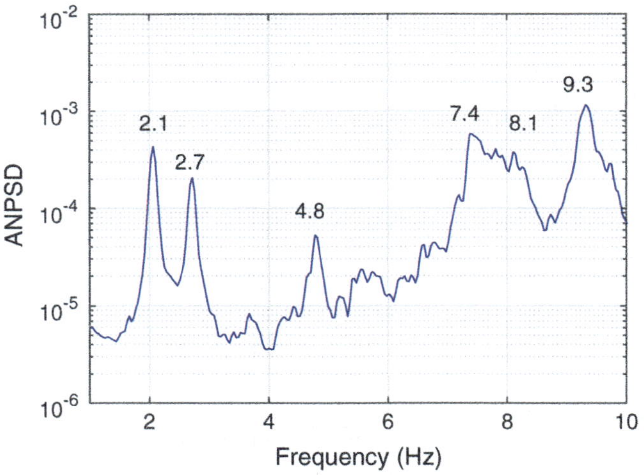

Fig. 22.4 ANPSD for vertical acceleration records

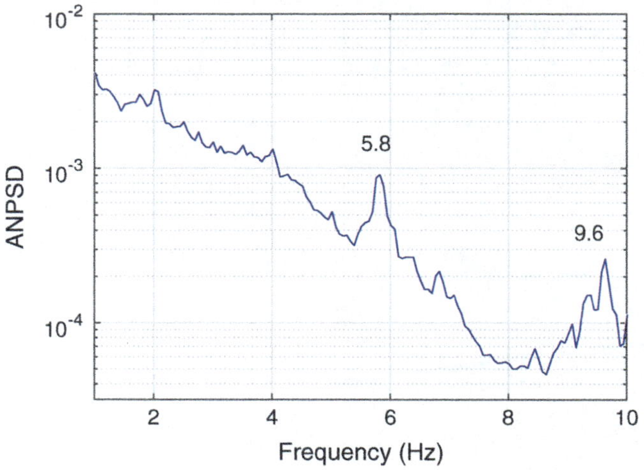

Fig. 22.5 ANPSD for lateral acceleration records

22.4.2 Stochastic Subspace Identification

A reference-based data-driven stochastic SSI algorithm, built in the Matlab toolbox MACEC 3.2 [9–12], was used for data pre-processing and modal parameter identification. The measured data were first visually inspected. It was found that the vertical accelerations corresponding to MS 10 and MS 59 were corrupted by intermittent spikes probably due to connection problems. These two abnormal signals were excluded from further signal processing. All the other measured data were de-trended before further analysis.

The vertical accelerations at the MSs presented in Fig. 22.6 were filtered using a low-pass filter with a cut-off frequency of 48 Hz and resampled at 60 Hz. The state-space model order parameter was set to 160. The stabilisation criteria were set to 1% for frequency, 5% for damping, 1% for modal assurance criterion and 0.8 for the low bound of the modal phase collinearity. Figure 22.7 shows the stabilisation diagram from the vertical acceleration measurement at MSs 1 – 9 and 10, with the power spectral density of all the signals superimposed. Modes below 10 Hz were inferred by the stable poles depicted by large red circles. It can be seen that the stable poles for natural frequencies are clearly identified, except for the one around 8.1 Hz. In total six modes were identified using the vertical accelerations, including four torsional and two vertical bending modes, as illustrated in Figs. 22.8, 22.9, 22.10, 22.11, 22.12 and 22.13. All the vertical and torsional modes found by the peak-picking method are identified by the SSI method. The vertical bending modes shown in Figs. 22.11 and 22.12 are very similar. One of the possible reasons for the two similar vertical bending modes is that the whole bridge is a truss bridge but the test was only conducted on the deck. That is, the vertical bending modes might be parts of the global modes of the whole structure, including top chords.

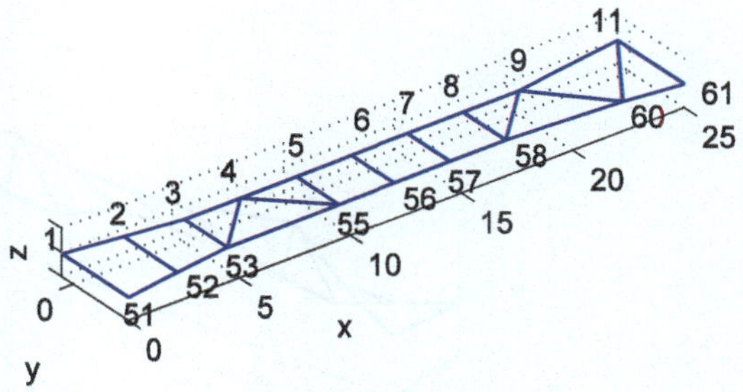

Fig. 22.6 MSs with vertical acceleration used for modal parameter identification

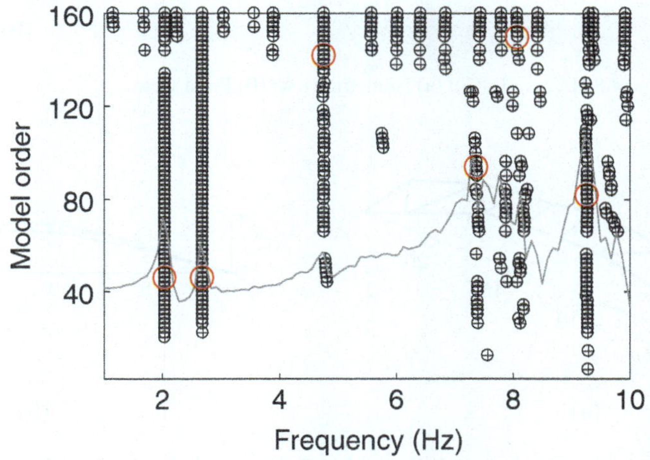

Fig. 22.7 Stabilisation diagram from the vertical acceleration measurement at MSs 1 – 9 and 10

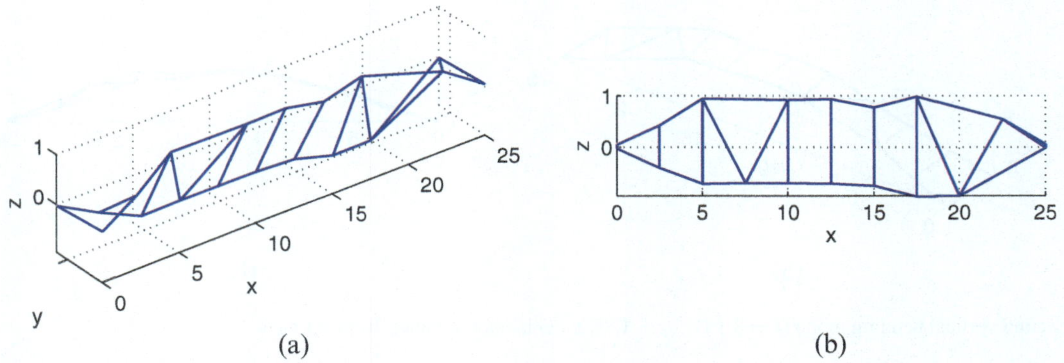

Fig. 22.8 First torsional mode ($f = 2.1$Hz, $\zeta = 1.6\%$): (**a**) Isometric view; (**b**) Front view

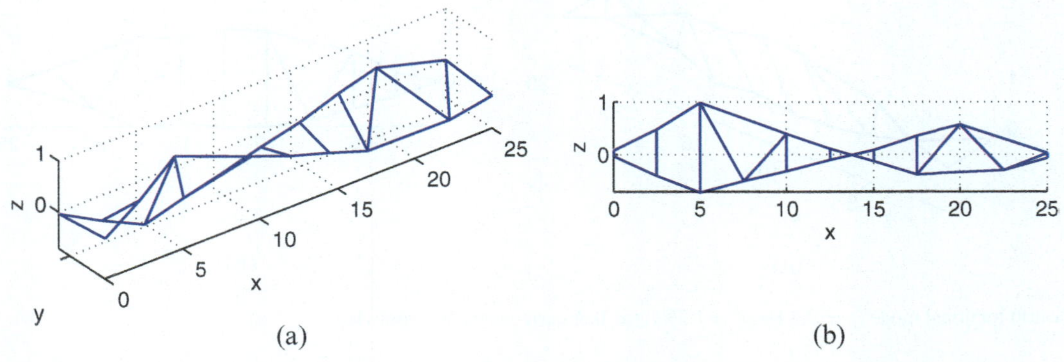

Fig. 22.9 Second torsional mode ($f = 2.7$ Hz, $\zeta = 1.3\%$): (**a**) Isometric view; (**b**) Front view

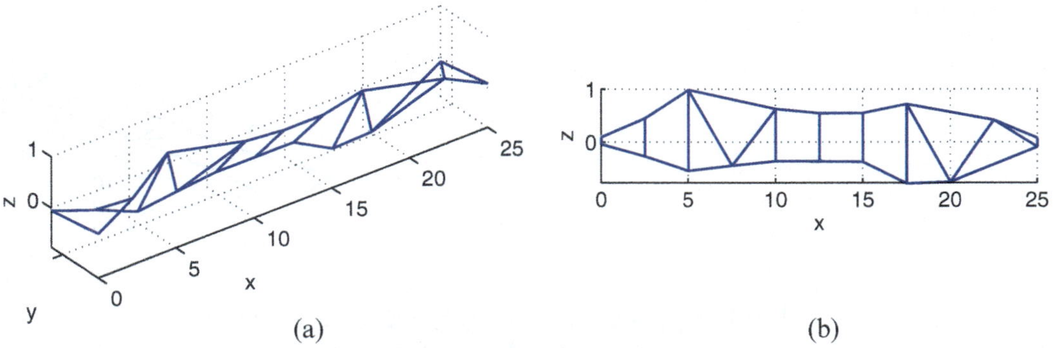

Fig. 22.10 Third torsional mode ($f = 4.8$ Hz, $\zeta = 1.4\%$): (**a**) Isometric view; (**b**) Front view

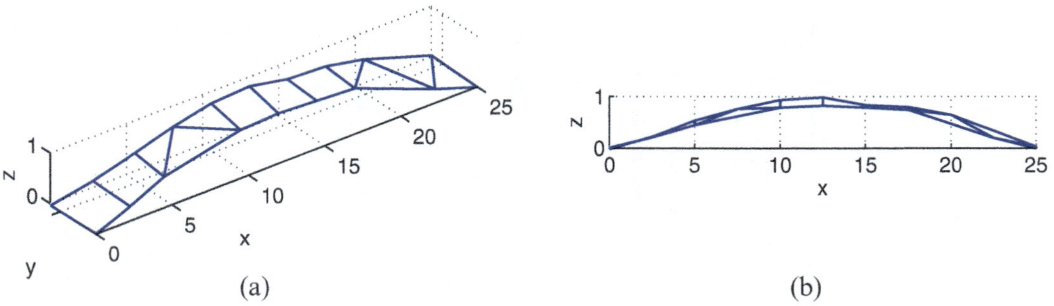

Fig. 22.11 First vertical bending mode ($f = 7.5$ Hz, $\zeta = 2.6\%$): (**a**) Isometric view; (**b**) Front view

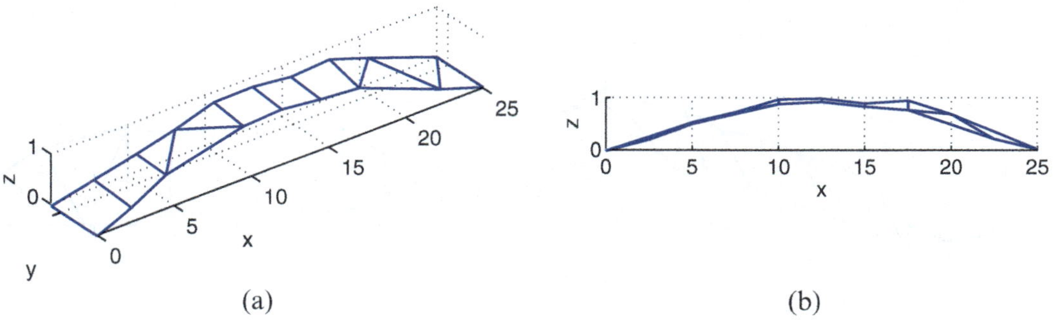

Fig. 22.12 Second vertical bending mode ($f = 8.1$ Hz, $\zeta = 1.7\%$): (**a**) Isometric view; (**b**) Front view

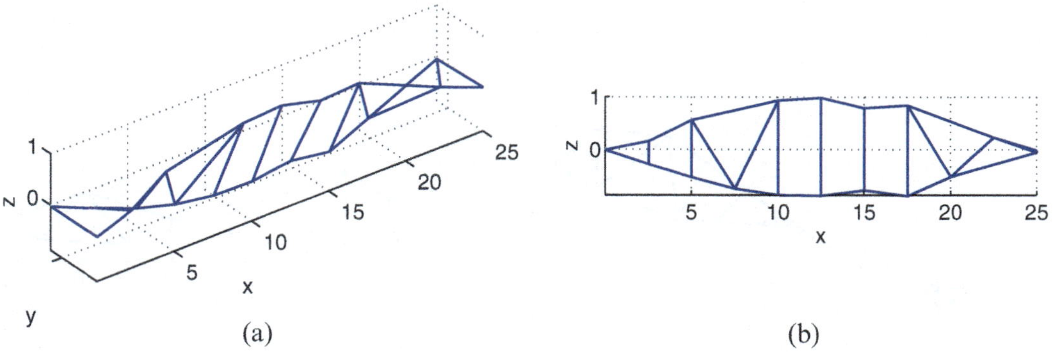

Fig. 22.13 Fourth torsional mode ($f = 9.3$ Hz, $\zeta = 1.2\%$): (**a**) Isometric view; (**b**) Front view

Similarly, the lateral accelerations at the MSs 1, 3, 4, 6, 8, 9 and 11 (Fig. 22.14) were filtered using a low-pass filter with a cut-off frequency of 34.1 Hz (80% of the resampling frequency) and resampled at 42.7 Hz (a fifth of the sampling frequency). All other parameters for system identification were the same as those used in processing vertical accelerations. The lateral vibrations at the MSs 51, 53, 54, 56, 58, 59 and 61 are slave to those at the MSs 1, 3, 4, 6, 8, 9 and 11. Two lateral bending modes were identified, as shown in Figs. 22.15 and 22.16, which were also inferred by the peak-picking method.

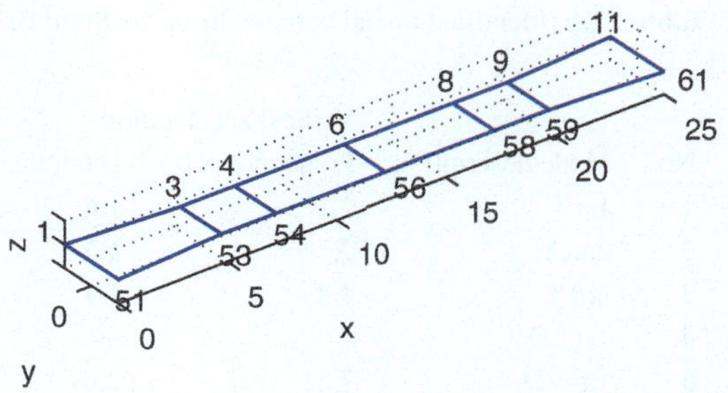

Fig. 22.14 Grid for lateral mode shape demonstration

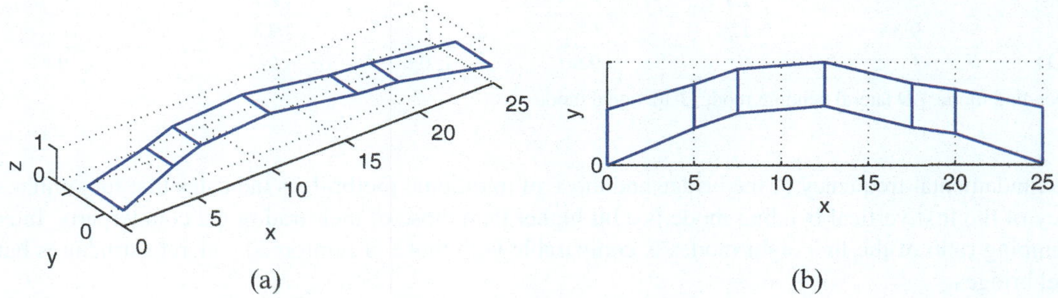

Fig. 22.15 First lateral mode ($f = 5.8$ Hz, $\zeta = 1.8\%$): (**a**) Isometric view; (**b**) Front view

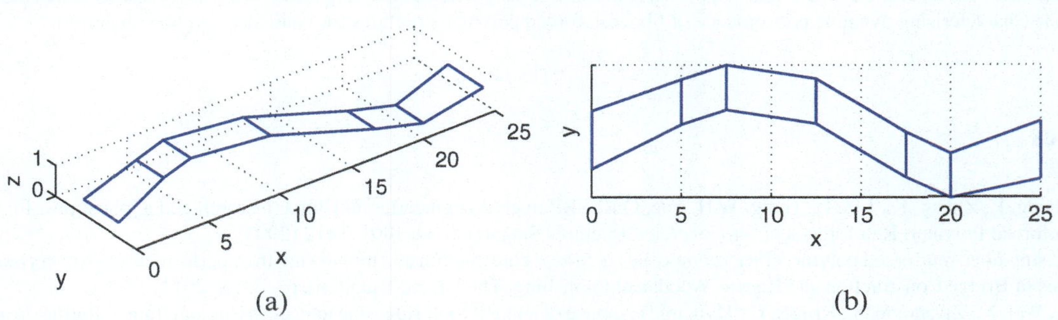

Fig. 22.16 Second lateral mode ($f = 9.6$ Hz, $\zeta = 0.6\%$): (**a**) Isometric view; (**b**) Front view

The identified modes are summarised in Table 22.1. All the modes, including four torsional, two vertical bending and two lateral bending modes, inferred by the peak-picking method, were identified by the SSI method. The natural frequencies of the modes identified by both methods are in excellent agreement. The first torsional, lateral and vertical modes are of frequencies 2.1 Hz, 5.8 Hz and 7.5 Hz, respectively. The damping ratios of all the modes except the second lateral mode are no smaller than 1.2%. The frequency of the first vertical bending mode is a bit higher than those of traditional structures of the similar spans [13]. The averaged damping ratio for the eight modes is about 1.5%, indicating that the damping ratio of FRP structures might be comparable with those for structures made of reinforced concrete and higher than those for steel structures [14]. In addition, there is a relatively high density of torsional modes.

22.5 Conclusions

In this paper, the dynamic properties of a truss pedestrian bridge made of FRP components were measured using ambient vibration testing. The method of peak picking of the averaged normalised power spectral densities in the frequency-domain was employed to identify the natural frequencies, whilst the stochastic subspace identification method, in the time-domain, was employed to the identify natural frequencies, damping ratios and mode shapes. The two methods generated very consistent results in terms of the number of modes and natural frequencies. In total, eight modes were found in the frequency range up to 10 Hz. These include four torsional, two vertical bending and two lateral bending modes. The first torsional, lateral and vertical modes are of frequencies 2.1 Hz, 5.8 Hz and 7.5 Hz, respectively. The damping ratios of all the modes except the second lateral mode are no smaller than 1.2%. There is a relatively high density of torsional modes. Comparison

Table 22.1 Identified modal parameters of the Prato Bridge

No.	Mode description	SSI				Peak-picking	
		Vertical acceleration		Lateral acceleration		Vertical acceleration	Lateral acceleration
		Frequency (Hz)	Damping (%)	Frequency (Hz)	Damping (%)	Frequency (Hz)	Frequency (Hz)
1	1st T	2.1	1.6	/	/	2.1	/
2	2nd T	2.7	1.3	/	/	2.7	/
3	3rd T	4.8	1.4	/	/	4.8	/
4	1st LD	/	/	5.8	1.8	/	5.8
5	1st VD	7.5	2.6	/	/	7.4	/
6	2nd VD	8.1	1.7	/	/	8.1	/
7	4th T	9.3	1.2	/	/	9.3	/
8	2nd LD	/	/	9.6	0.6	/	9.6

VD vertical bending mode, *LD* lateral bending mode, *T* torsional mode

between the fundamental frequency of the bridge and those of traditional footbridges indicates that for similar span lengths, the frequency of the first vertical bending mode is a bit higher than those of their traditional counterparts. Interestingly, the averaged damping ratio of the first eight modes is comparable with those of reinforced concrete structures but higher than those of steel bridges.

Acknowledgements This research work was supported by the UK Engineering and Physical Sciences Research Council [grant number EP/M021505/1: Characterising dynamic performance of fibre reinforced polymer structures for resilience and sustainability].

References

1. Ye, L.P., Feng, P., Zhang, K., Lin, L., Hong, W.H, Yue, Q.R: FRP in civil engineering in China: research and applications. In: Proceedings of Fiber Reinforced Polymer: Reinforcement for Concrete Structures Singapore, pp. 1401–1412 (2003)
2. Wan, B.: Using fiber-reinforced polymer (FRP) composites in bridge construction and monitoring their performance: an overview. In: Advanced Composites in Bridge Construction and Repair, Woodhead Publishing, The United Kingdom, pp. 3–28 (2014)
3. Russell, J., Wei, X., Živanović, S., Kruger, C.: Dynamic response of an FRP footbridge due to pedestrians and train buffeting. Procedia Eng.**199**, 3059–3064 (2017)
4. Wei, X., Russell, J., Živanović, S., Mottram, J.T.: Experimental investigation of the dynamic characteristics of a glass-FRP suspension footbridge. In: Caicedo J, Pakzad S. (eds.) Dynamics of Civil Structures, Proceedings of the 35th IMAC, A Conference and Exposition on Structural Dynamics 2017, vol. 2, pp. 37–47. Springer, Cham (2017)
5. Burgoyne, C., Head, P.: Aberfeldy Bridge–an advanced textile reinforced footbridge. Techtextil Syposium, pp. 1–9 (1993)
6. Adilardi, A., Frasconi, L: Design of a pedestrian bridge made with pultruded profiles of fibreglass-reinforced plastics in Prato. In: Footbridge 2008 – Footbridges for Urban Renewal, Porto, pp. 345–346 (2008)
7. Felber, A.J.: Development of a hybrid bridge evaluation system. PhD Thesis, The University of British Columbia (1994)
8. Ren, W.-X., Peng, X.-L., Lin, Y.-Q.: Experimental and analytical studies on dynamic characteristics of a large span cable-stayed bridge. Eng. Struct. **27**, 535–548 (2005)
9. Reynders, E., Schevenels, M., Roeck, G.D.: MACEC 3.2: a matlab toolbox for experimental and operational modal analysis. Department of Civil Engineering, KU Leuven (2014)
10. Peeters, B., De Roeck, G.: Reference-based stochastic subspace identification for output-only modal analysis. Mech. Syst. Signal Process. **13**, 855–878 (1999)
11. Peeters, B., De Roeck, G.: Stochastic system identification for operational modal analysis: a review. J. Dyn. Syst. Meas. Control. **123**, 659–667 (2001)
12. Reynders, E., Roeck, G.D.: Reference-based combined deterministic–stochastic subspace identification for experimental and operational modal analysis. Mech. Syst. Signal Process. **22**, 617–637 (2008)
13. Tilly, G.P., Cullington, D.W., Eyre, R.: Dynamic behaviour of footbridges. IABSE Surveys S-26/84, IABSE Periodica, pp. 13–24 (1984)
14. Bachmann, H., Pretlove, A.J., Rainer, J.H.: Vibrations induced by people. In: Bachmann, H., Ammann, W.J., Deischl, F., Eisenmann, J., Floegl, I., Hirsch, G.H., et al. (eds.) Vibration Problems in Structures: Practical Guidelines, pp. 1–28. Birkhäuser Basel, Basel (1995)

Chapter 23
Modal Parameter Uncertainty Estimates as a Tool for Automated Operational Modal Analysis: Applications to a Smart Building

Rodrigo Sarlo and Pablo A. Tarazaga

Abstract The knowledge of modal parameter uncertainties derived from operational modal analysis (OMA) can greatly improve automated decisions by providing information about the quality of the modal identification. Yet so far, this information has been largely ignored in continuous monitoring studies on civil infrastructure, especially with respect to buildings. In this paper, we implement an automated version of Covariance Based Stochastic Subspace Identification on a highly instrumented smart building. An expansion of the technique estimates uncertainty bounds for all modal parameters. Through a series of full scale experiments, we demonstrate how uncertainties are valuable tools in various contexts of automation. These include the identification and removal of badly-fitted modes, the identification of periods of high signal-to-noise ratio, and the validation of reference sensors selection.

Keywords Modal analysis · Buildings · Instrumentation · Uncertainty · Automation

23.1 Introduction

Operational Modal Analysis (OMA) takes advantage of ambient vibrations which occur during regular operation, thus making it a convenient method for continuous structural analysis. Since all OMA methods require some user judgment, a non-trivial degree of automation is necessary to bridge the gap between traditional and continuous OMA. Recently, there has been a growing body of research in automated OMA for civil structures. Methods generally consist of automated peak-picking for frequency domain methods [1, 2] or clustering methods for stabilization diagram interpretation [3, 4]. Applications of these methods in the literature focus heavily on bridges and the effects of environmental and operational conditions on their modal parameters, primarily natural frequency. Environmental studies deal with factors such as temperature [5, 6] and wind [5, 7], while for the operational studiers consider factors such as vehicular traffic [8, 9] and pedestrians [10]. Understanding these effects is useful in Structural Health Monitoring (SHM) in order to reduce the likelihood of false positives in damage detection. Compared to bridges, automated OMA of buildings has seen relatively little development, due in part to the more complex nature of these structures. Not only is building construction much more elaborate than bridges, but buildings also experience highly variable levels of excitation, including very low amplitudes. This makes it more challenging to extract reliable modal information consistently.

Another missing piece to the automation puzzle is the inclusion of modal parameter uncertainties, particularly frequency and mode shape, in automated decision-making. These are an essential tool in determining the reliability of the data and significance of observed modal parameter variations. Even further, they may help to determine whether modal-based damage detection methods are capable of achieving the level of sensitivity required for commercial applications. Several methods exists for estimating uncertainty bounds on modal parameters obtained from OMA [11–14]. But despite an understanding that uncertainties are useful for automated choices, there exist very few automation approaches which incorporate them.

In this paper, we demonstrate some of the challenges to automated OMA when applied to a five-story smart building and how uncertainty estimates can be used to address them. The on-campus building, named Goodwin Hall, is permanently instrumented with 225 high-sensitivity accelerometers, making it an excellent platform for studying continuous monitoring. Herein, we address three challenges: (1) Cleaning up the results of current automation strategies, (2) distinguishing high from

R. Sarlo (✉) · P. A. Tarazaga
Department of Mechanical Engineering, Virginia Tech Smart Infrastructure Lab (VTSIL), Blacksburg, VA, USA
e-mail: sarlo@vt.edu

S. Pakzad (ed.), *Dynamics of Civil Structures, Volume 2*, Conference Proceedings of the Society for Experimental Mechanics Series, https://doi.org/10.1007/978-3-319-74421-6_23

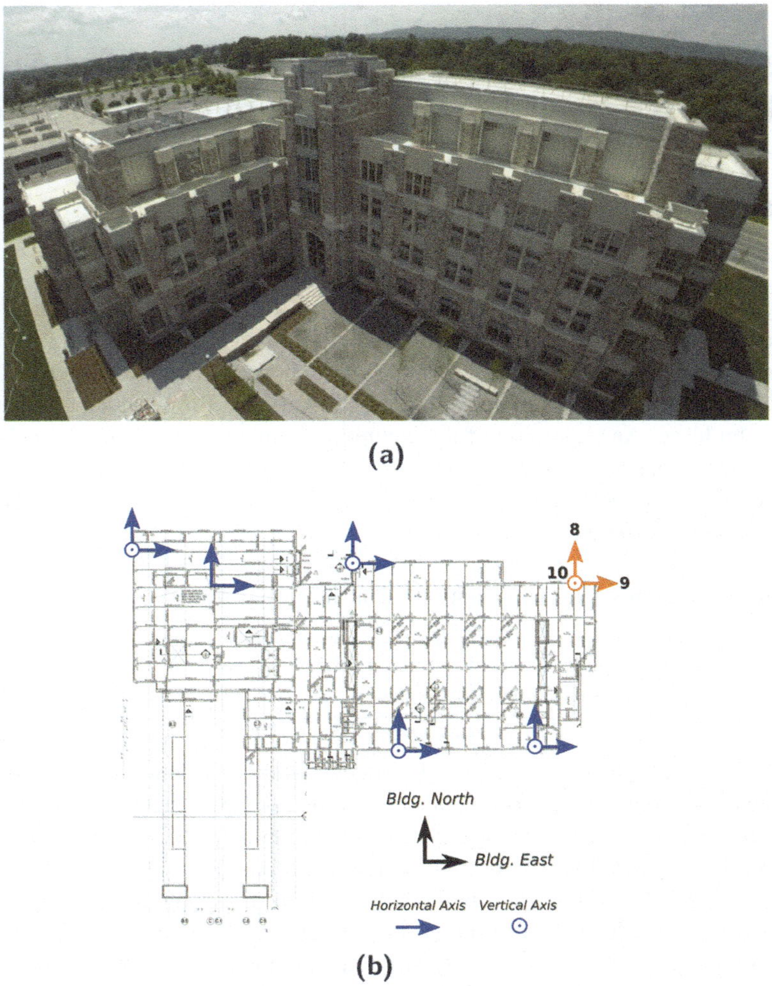

Fig. 23.1 (**a**) Aerial view of Goodwin Hall. (**b**) Sensor layout for the fifth floor, provided for reference

low quality data sets, and (3) ensuring a good choice of reference sensors. Then through a series of preliminary experiments and results, we demonstrate how uncertainty bounds can be incorporated to address these challenges. The general goal is to present a few key ideas about the properties of modal parameter uncertainties and how they are useful. Thorough analysis of such properties is reserved for future work.

23.2 Approach

Goodwin Hall is a 160,000 sq-ft five-story classroom and laboratory building on the Virginia Tech Blacksburg campus. It has a steel, concentrically-braced frame with a limestone facade. An aerial view of the building can be seen in Fig. 23.1a. Currently, the building houses 225 total accelerometers (PCB 393B04). The accelerometers are mounted to the building frame through welded stainless steel mounts. For this study we used a subset of 98 sensors to conduct OMA. The sensor layout for the fifth floor of the building is shown in Fig. 23.1b as an example. For more information on the sensor configuration as well a full modal analysis on the building, readers are referred to Sarlo et al. [15].

The primary sources of ambient excitation for the building were wind and human occupants. Covariance-based Stochastic Subspace Identification (SSI-Cov) [16, 17] was chosen as the OMA method mainly due to its suitability for uncertainty estimation. In addition to this, the method is known to be robust to noise and has successfully been implemented in many studies. Its main drawback is its large computational requirements. The modal parameter uncertainty estimates were obtained from the variance propagation method outlined by Döhler et al. [11].

23.3 Results

To automate the SSI-Cov analysis, we have implemented a variation of the stabilization diagram clustering method described by Reynders et al. [3]. The method uses so-called *validation criteria*, which represent some "distance" between modes, to group similar modes together into a *cluster*. To get a final list of physical modes, one representative mode would be taken from each cluster. In our case, stabilization diagrams were generated for model orders from 5 to 30 modes. Seven validation criteria were used: (1) Frequency Difference, (2) Damping Difference, (3) Pole distance, (4) Modal Assurance Criterion (MAC), (5) Participation Factor, (6) Modal Transfer Norm (MTN), and (7) Modal Phase Collinearity (MPC). Descriptions of these criteria can be found in Reynders et al. [3]. Figure 23.2a shows an example cluster (contained within the dashed line) taken from a stabilization diagram. Figure 23.2b shows all the selected clusters in frequency-damping space, with the associated 95% uncertainty bounds for each mode. Ideally, the cluster would represent a vertical line of modes on the stabilization diagram and a tight grouping in the frequency-damping plot. This is almost the case except for the mode indicated by the arrow. This mode can be observed as an outlier and is likely due to underfitting at a low model order. It could potentially be discriminated based on its frequency and damping values alone, however, the extremely high uncertainty bounds demonstrate a clear sensitivity to this type of error, making them a very good tool for discarding badly-fitted modes.

In order to "clean up" the clusters using this additional uncertainty information, three techniques were tested: (1) adding uncertainties to the list of validation criteria, (2) applying a hard threshold prior to clustering using the original validation criteria, and (3) applying robust outlier detection (Matlab `robustcov`) on the final clusters. The first technique failed to

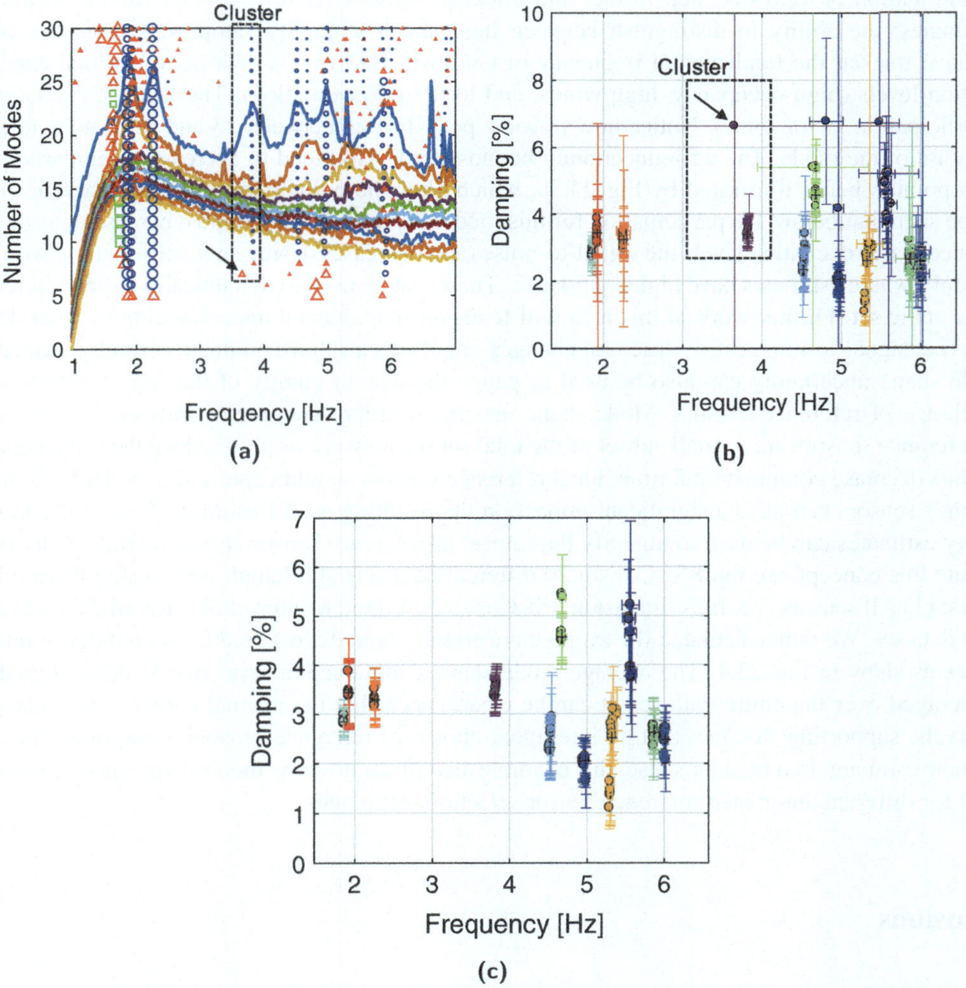

Fig. 23.2 (**a**) Stabilization diagram for a hour of wind excitation data. Blue ○'s = stable modes; red △'s = unstable modes; green □'s = non-structural stable modes. The dashed line denotes a selected cluster of modes. (**b**) Automatically identified mode clusters (stable lines) from the stabilization diagram with damping and frequency 95% confidence intervals included. The cluster enclosed by the box corresponds to the cluster denoted in (**a**). The arrow points to the same mode from both cases. (**c**) Clusters after robust outlier removal

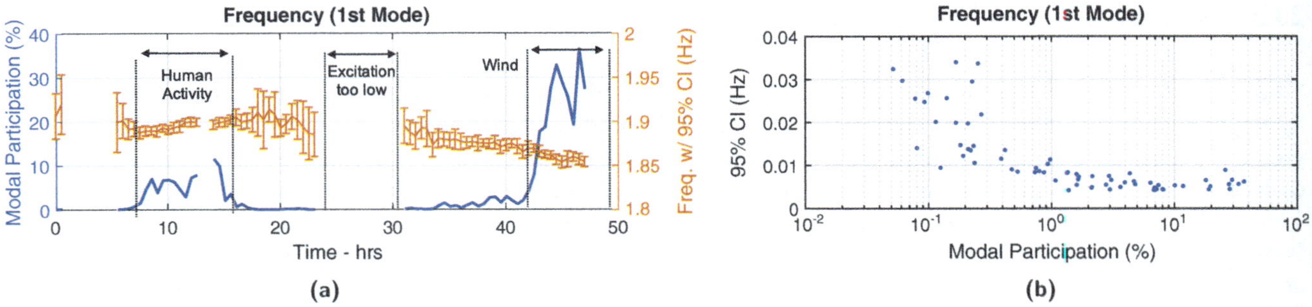

Fig. 23.3 (**a**) Frequency and modal participation of the first fundamental building mode over a 48 h period, estimated every 30 min. The error bars represent the 95% confidence intervals on the estimates. (**b**) Confidence intervals from (**a**) plotted against the modal participation

identify a large number of stable lines and so was deemed unsuitable. The second and third techniques both showed good results, however, the third technique (robust outlier detection) has the added advantage that it does not require a user-defined threshold. The results from this technique are shown in Fig. 23.2c. Comparing this to Fig. 23.2b shows tighter clusters and tighter uncertainty bounds for all modes. This methodology demonstrates one of the ways in which uncertainty estimates can be used to improve the performance of existing automation strategies. Future work in this area will seek to quantify these improvements over a large sets of data and for a variety of excitation conditions.

Taking the automation process one step further and tracking modes over time reveals another useful characteristic of uncertainty estimates: the ability to distinguish between high and low quality samples. In the case of the example in Fig. 23.3a, we have tracked the fundamental frequency of Goodwin Hall over a 48 h period, which experienced a diverse range of excitation levels (high occupancy, high winds, and little to no excitation). The level of excitation of the mode is related to its participation factor (blue). Notice how in some periods (e.g. between 23 and 31 h) no data exists. This is due to lack of excitation of the mode. The 95% uncertainty bounds (error bars) tend to decrease when participation is high and vice-versa. This point is better illustrated by Fig. 23.3b, which plots the participation factor versus the confidence bounds directly. It can be seen that below 1% participation for this mode, the uncertainty bounds increase considerably. This makes logical sense, since at low excitation levels the signal-to-noise ratio decreases. Using such information, we can identify points in the modal history where estimates have high confidence. These can be used to dynamically define reference modes, rather that relying on a static set. Further work in this area will focus on fundamental understanding of what degree of influence physical factors (i.e. signal-to-noise ratio, structural changes, input spectra) have on the uncertainty estimates.

Finally, mode shape uncertainty can also be used to gauge the overall quality of the sensor selection, especially with respect to the choice of reference sensors. Mode shape uncertainty represents the bounds on each element of the mode shape vector. Reference sensors are a small subset of the total sensor network used to reduce the dimensions of the SSI-Cov algorithm and thus decrease computational time. Ideal reference sensors should capture the modal behavior of interest. The choice of reference sensors can have a significant impact on the resulting modal estimates, especially mode shapes. Mode shape uncertainty estimates can be used to quantify the impact of reference sensor choice on the quality of the mode shape itself. To illustrate this concept, we ran SSI-Cov on 20 different data sets (1 h length each) using three different choices of reference sensors: (1) All sensors (i.e. full-dimension SSI-Cov), (2) A fixed manual choice for all data sets and (3) Randomly chosen for each data set. We then calculated the average uncertainty bounds (referred to as confidence interval) for the first five mode shapes as show in Fig. 23.4. The average mode shape confidence interval (CI) is defined as the average of the element CI's averaged over the entire data set. It can be clearly seen that the manual choice consistently results in lower confidence intervals, supporting that fact that an intelligent choice of reference sensors is important to obtain the highest quality mode shape estimate. In a broader sense, this example also illustrates how mode shape uncertainties could be used as a validation tool for different automated reference sensor selection techniques.

23.4 Conclusions

In this paper we implemented an automated operational modal analysis algorithm on a highly instrumented smart building. The advantage of the chosen SSI-Cov technique is that it can be expanded to provide uncertainty bounds for all modal parameters. These uncertainty estimates provide important information about the quality of the modal identification which complements automated decisions based on modal parameters alone. One possible application is the identification and

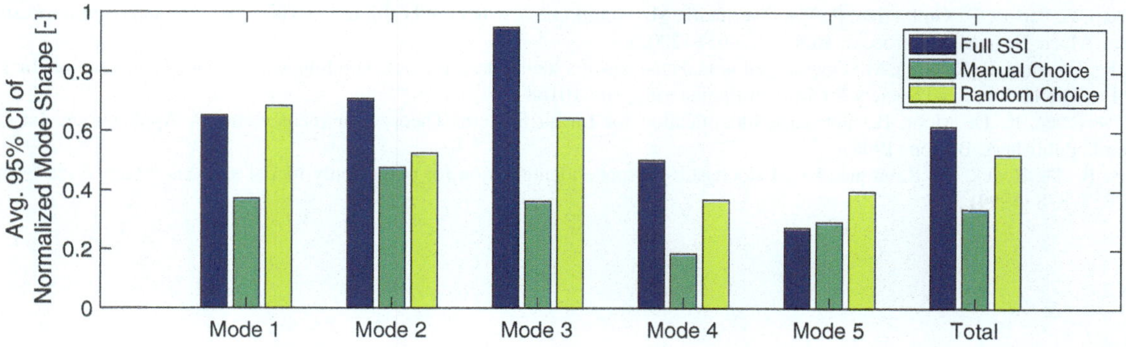

Fig. 23.4 Average estimated 95% confidence interval (CI) for 20 data sets of 1 h, using three different choices of reference sensors: all sensors (full SSI), manually chosen, randomly chosen. The CI is taken for a normalized mode shape with a maximum magnitude of 1

removal of badly-fitted modes. A second application is the identification of time periods of high signal-to-noise ratios when used in conjunction with participation factor estimates. This can be useful for selecting reliable data from a series of data sets, for example, a representative set of modes for the month of January. Finally, uncertainties can also validate the choice of reference sensors, which are essential for both mode shape resolution and computational efficiency. Future work in these areas will focus on simulations and laboratory experiments aimed at a fundamental characterization of the effects of certain sources of noise on SSI-Cov uncertainty estimates. The noise sources will include variable excitation spectra, low signal-to-noise ratios, and noise at the individual sensor level.

Acknowledgements The authors acknowledge the support as well as the collaborative efforts provided by our sponsors, VTI Instruments, PCB Piezotronics, Inc.; Dytran Instruments, Inc.; and Oregano Systems. The authors are particularly appreciative for the support provided by the College of Engineering at Virginia Tech through Dean Richard Benson and Associate Dean Ed Nelson as well as VT Capital Project Manager, Todd Shelton, and VT University Building Official, William Hinson. The authors would also like to acknowledge Gilbane, Inc. and in particular, David Childress and Eric Hotek. We are especially thankful to the Student Engineering Council (SEC) at Virginia Tech and their financial commitment to this project. Dr. Tarazaga is also thankful for the support provided by the John R. Jones III Fellowship. The work was conducted under the patronage of the Virginia Tech Smart Infrastructure Laboratory and its members.

References

1. Brincker, R., Andersen, P., Jacobsen, N.-J.: Automated frequency domain decomposition for operational modal analysis. In: Proceedings of the 13th SEM International Modal Analysis Conference, Nashville, pp. 661–667 (1995)
2. Rainieri, C., Fabbrocino, G.: Automated output-only dynamic identification of civil engineering structures. Mech. Syst. Signal Process. **24**(3), 678–695 (2010)
3. Reynders, E., Houbrechts, J., De Roeck, G.: Fully automated (operational) modal analysis. Mech. Syst. Signal Process. **29**, 228–250 (2012)
4. Cabboi, A., Magalhães, F., Gentile, C., Cunha, Á.: Automated modal identification and tracking: application to an iron arch bridge. Struct. Control Health Monit. **24**(1), e1854 (2017)
5. Cross, E.J., Koo, K.Y., Brownjohn, J.M.W., Worden, K.: Long-term monitoring and data analysis of the Tamar Bridge. Mech. Syst. Signal Process. **35**, 16–34 (2012)
6. Peeters, B., De Roeck, G.: One-year monitoring of the Z 24-Bridge: environmental effects versus damage events. Earthq. Eng. Struct. Dyn. **30**(2), 149–171 (2001)
7. Siringoringo, D., Fujino, Y.: System identification of suspension bridge from ambient vibration response. Eng. Struct. **30**(2), 462–477 (2008)
8. Magalhães, F., Cunha, Á., Caetano, E.: Online automatic identification of the modal parameters of a long span arch bridge. Mech. Syst. Signal Process. **23**(2), 316–329 (2009)
9. Zhang, Q.W., Fan, L.C., Yuan, W.C.: Traffic-induced variability in dynamic properties of cable-stayed bridge. Earthq. Eng. Struct. Dyn. **31**, 2015–2021 (2002)
10. Hu, W.-H., Moutinho, C., Caetano, E., Magalhães, F., Cunha, Á.: Continuous dynamic monitoring of a lively footbridge for serviceability assessment and damage detection. Mech. Syst. Signal Process. **33**, 38–55 (2012)
11. Döhler, M., Mevel, L., Andersen, P.: Efficient uncertainty computation for modal parameters in stochastic subspace identification. In: Proceedings of the International Cnference on Uncertainty in Structural Dynamics, pp. 4793–4805 (2012)
12. Reynders, E., Pintelon, R., De Roeck, G.: Uncertainty bounds on modal parameters obtained from stochastic subspace identification. Mech. Syst. Signal Process. **22**(4), 948–969 (2008)
13. Reynders, E., Maes, K., Lombaert, G., De Roeck, G.: Uncertainty quantification in operational modal analysis with stochastic subspace identification: validation and applications. Mech. Syst. Signal Process. **66–67**, 13–30 (2016)

14. Verboven, P., Parloo, E., Guillaume, P., Van Overmeire, M.: Autonomous structural health monitoring–part I: modal parameter estimation and tracking. Mech. Syst. Signal Process. **16**(4), 637–657 (2002)
15. Sarlo, R., Tarazaga, P., Kasarda, M.: Operational modal analysis of a steel-frame, low-rise building with L-shaped construction. In: Proceedings of SPIE – The International Society for Optical Engineering, vol. 10168 (2017)
16. Van Overschee, P., De Moor, B.: Subspace Identification for Linear System: Theory – Implementation – Applications, 1st edn. Kluwer Academic Publishers, Boston (1996)
17. Peeters, B., De Roeck, G.: Reference-based stochastic subspace identification for output-only modal analysis. Mech. Syst. Signal Process. **13**(6), 855–878 (1999)

Chapter 24
Modeling Human-Structure Interaction Using Control Models: External Excitation

Ahmed T. Alzubaidi and Juan M. Caicedo

Abstract Modeling the human-structure interaction *(HSI)* phenomenon on flexible structure when a person is performing an activity such as dancing remains a challenge. Human activity on a flexible structure can generate excessive vibrations that can cause serviceability problems. Traditional models for human-structure interaction considered the human as a mass-spring-damper system that do not allow the use of external excitation to the human (e.g. music). Recently, researchers have proposed the use of control systems to model the human that will allow the use of external excitation. However, the use of control laws for modeling human-structure interaction has only been performed when the human is standing. This paper focuses on expanding these models to model people in motion. The paper presents an experimental study describing testing on a cantilever structure. The uncertainty of the models is estimated using Bayesian inference.

Keywords Human-structure interaction · Control theory · Human activity · Structural dynamics · Bayes inference

24.1 Introduction

The advancements of new materials allows structural engineers to utilize lightweight and slender members in the designing of structures such as gyms, theaters, dance floor, and stadium. These new material provide appropriate strength for the structure but are susceptible to human-induced vibration problems due to a higher live to dead load ratio [1]. Examples of human-induced vibration problems include The Millennium Bridge [2], a railway suspension bridge at Niagara Falls [3], and a footbridge failure happened in 1831 in Broughton, UK [4]. In all of these cases, not only the excitation was produced by people, but the characteristics of the human-structure system might have changed because of the presence of the crowd. For this reason, the need to model the combined human-structure system when people are standing, sitting, and/or moving. The dynamic effect of the human body on structures has traditionally been modeled as single or multiple degree of freedom system (MDS) constructed from lumped masses, dampers, and springs [5–8]. In addition to MDS, models based on control theory have been proposed by Ortiz et al.[1]. In their work, Ortiz-Lasprilla and Caicedo use as Proportional, Integrative and Derivative (PID) controller to model a standing person or group of people. PIDs have also been utilized to model human balance [9] for medical applications. In addition to proposing the use of control theory to model the human body, Ortiz-Lasprilla and Caicedo utilized Bayes inference to estimate the uncertainty in the parameters of the model.

The phenomenon of human-structure interaction is caused by the interdependence of structural and human dynamic systems when people perform activities such as standing, jumping, and/or dancing. This interdependence can be expressed as new properties of the overall dynamic system [10]. In general, two essential issues are studied when the structure is occupied by a person or crowd. First, the dynamic characteristics of the combined human-structure system such as the natural frequency and the damping ratio. This combined characteristics can make the structure prone to excessive dynamic behavior due to relatively small loads. The second issue is the synchronization of the occupants themselves either because of crowd dynamics (e.g. people walking in sync in a large crowd) or because of interactions with the structural system (e.g. people walking in sync because the floor is moving) [10, 11]. This can lead to higher loads that do not increased linearly with the number of people [6] such as the case of the London Millennium Bridge. During the opening day on June 10 2000, pedestrians induced a high level of lateral forces on the structure and the bridge was immediately closed [2]. While particular examples have been studied at length, there is a lack of fundamental knowledge about modeling human-structure interaction in general [12]. As a result, most civil engineers do not have the appropriate tools to consider the influence of crowds during the design of structures. The current state of the practice is to consider live loads on the structure as additional mass only [6].

A. T. Alzubaidi (✉) · J. M. Caicedo
Department of Civil and Environmental Engineering, University of South Carolina, Columbia, SC, USA
e-mail: alzubaa@email.sc.edu; caicedo@cec.sc.edu

S. Pakzad (ed.), *Dynamics of Civil Structures, Volume 2*, Conference Proceedings of the Society for Experimental Mechanics Series, https://doi.org/10.1007/978-3-319-74421-6_24

In this study, the work presented in [1] is expanded to include other external excitation to a person such as music. Experimental data of the empty and occupied structure are collected and Bayesian model updating is used to estimate the distributions of the model parameters. This paper is organized as follows, the models for the human and structure are proposed in Sect. 24.2. Section 24.3 discussed the experimental setup and briefly describes how the probability density function of the parameters is obtained. Finally, Sect. 24.4 discusses the posterior probability densities found through the analysis.

24.2 Models

Control systems have been utilized for a large variety of applications such as automobile steering control, industrial robotics, oven, furnace, radar antenna, etc. These systems which are not only applied in the mechanical and electrical engineering fields but also in civil, chemical, and environmental engineering [13]. The fundamental concepts of control system were developed for linear systems using feedback concept. Control system are based on input and output relationship and can be classified as open and closed loop control system. In this research, we will focus on closed loop control system to model human-structure interaction. The idea of using feedback is not new. The Greek used feedback back in 300 to B.C. in developing the float regulator mechanisms [13]. Closed loop control systems use an output of the plant (system to be controlled) and a feedback loop to achieve the desired behavior [13]. The essential aim of the feedback concept is to reduce the error between the input and output in the system. However, depending on the values adopted for the controller the system can be unstable. In the last decade, control systems have been used to model the human body. In the area of human-structure interaction, researchers have used control theory to represent a human standing in a flexible structure and model human-structure interaction of a human with bent knees [1].

24.2.1 Human-Structure Model

Generally, closed loop control with negative feedback has been used to reduce the error in a system [13, 14]. Here, a closed loop is used to model the interaction between a person standing in a flexible structure and the structure itself. The block diagram of a typical single input single output closed loop system consists of a plant $G(s)$ and controller $H(s)$. While in other applications the plant is the system that will be controller, in this paper the plan represents the structural system. The controller $H(s)$ represents the dynamics of the person on the structure. The excitation force on the structure is represented by the term $R(s)$ whereas the output acceleration is represented by the term $C(s)$. This variable is used as the controller (or human model) input [1]. The transfer function of the combined structure and human is denoted by equation the equation

$$TF(s) = \frac{G(s)}{1 + G(s)H(s)} \tag{24.1}$$

In this paper we expand the model used by Ortiz and Caicedo [1] to include other external excitations to the person. In particular, we are modeling how people respond to sound. Therefore, the human model allows two inputs, accelerations from the structure $C(s)$ and excitation in the form of sound (e.g. music) $M(s)$ as shown in Fig. 24.1.

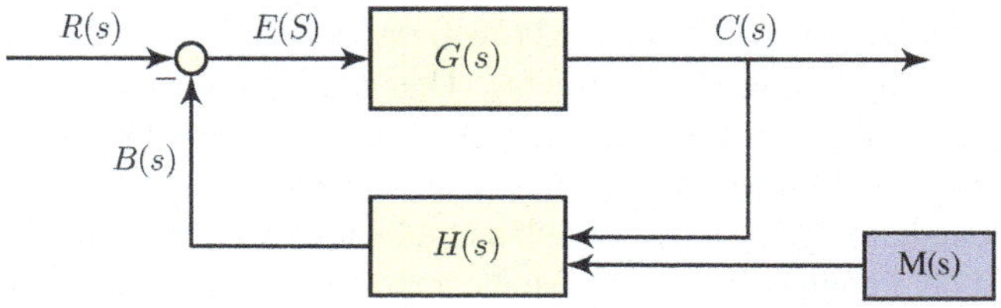

Fig. 24.1 Block diagram of system

Fig. 24.2 Experimental structure

24.2.2 *Human Model*

The proposed model for the human is a proportional, integrative and derivative (PID) controller similar to [1]. In this particular application two PID controllers are used. One that takes the floor acceleration as input and a second one that takes sound as input. The output of both PIDs are added to model the force that the person. The three constants of each PID (k_p, t_i, and t_d) are estimated using Bayesian inference as explained in Sect. 24.3.1.

24.2.3 *Structural Model*

The test structure is a cantilever consisting of a steel frame and concrete blocks used to change the dynamic characteristics of the system as shown in Fig. 24.2. An internal support can also be used to change the dynamic properties of the system. The length of the cantilever used in this paper is 124in. The structure is modeled using poles and zeros. In this particular application the model consists of one pole and no zeros (single degree of freedom system) by the equation:

$$G(s) = \frac{K}{(s - p_1)(s - p_2)} \tag{24.2}$$

where K is the gain of system. The pole are expressed in terms of the natural frequency (ω_n) and damping ratio ζ of the structure as shown in the following equation [15]

$$p_{1,2} = -\zeta\omega_n \pm \sqrt{(\zeta\omega_n)^2 - \omega_n^2} \tag{24.3}$$

24.3 Experimental Testing and Model Updating

One PCB 3701D1FA20G accelerometer with a sensitivity of 100.4 mV/g and a PCB 096D50 impact hammer with a sensitivity of 0.2305 mV/N were used to measure accelerations of the structure and applied an impulsive force respectively. The accelerometer measured accelerations in the vertical direction and the impact hammer was utilized to excite the structure in the middle of the concrete block in the vertical direction as well (See Fig. 24.2). An NI CompactDAQ data acquisition system with an N19234 module was used to collect the records of force and acceleration. A sampling frequency of 1707 Hz was utilized to collect records of 20 s. The experimental frequency response function (FRF) was estimated using the equation [16].

$$\mathrm{TF}(f) = \sqrt{H_1(f) * H_2(f)} \tag{24.4}$$

where

$$H_1(f) = \frac{S_{yx}(f)}{S_{yy}(f)}, \qquad H_2(f) = \frac{S_{xx}(f)}{S_{xy}(f)} \tag{24.5}$$

where S_{xf} is the cross spectral density between the output acceleration and the input force, S_{fx} is the cross spetral density between the input force and the acceleration. S_{ff} and S_{xx} represent the auto spectral densities for the force and acceleration respectively.

Three different tests were performed. First, the structure was tested empty and a model of the structure alone was updated. Then, the structure was updated with the person with bent knees and without moving. The data for the second test was used to update the first PID controller. Finally, the person was asked to excite the structure at a specific beat provided by a metronome. This final dataset was used to update the complete model.

24.3.1 Bayes Inference

The parameters of the structure and human models are update using the Bayesian inference. Bayes theorem is expressed using the equation:

$$P(\Theta|D) = \frac{P(D|\Theta)P(\Theta)}{P(D)} \tag{24.6}$$

where $P(\Theta|D)$ is the posterior probability of the parameters Θ given the observed data D. $P(D|\Theta)$ is the likelihood, and $P(\Theta)$ is the prior probability density function of the parameters. The prior PDf expresses the analysts belief about the parameters values before updating. The posterior is sampled using Markov Chain Monte Carlo methods (MCMC) [17–19].

24.3.2 Empty Structure

The experimental transfer function of the empty structure is shown in Fig. 24.3 where the natural frequency (ω_n) is approximately 3 Hz.

The prior probability density function PDF distribution of natural frequency of empty structure is assumed to be the normal distribution $\mathrm{P}(\omega_n) = N(3.5, 0.35)$. The prior for the damping ratio was assumed as $\mathrm{P}(\zeta) = N(0.03, 0.006)$. The likelihood function was assummed Normal with a standard deviation modeled by an Inverse Gamma with shape $\alpha = 10$ and scale $\beta = 0.2$. A MCMC algorithm was utilized to sample $P(\Theta = \{\omega_n, \zeta\}|D)$. The posterior distributions for ω_n and ζ were used as prior distributions for the model of the human-structure system of the person with bent knees.

24.3.3 Person with Bent Knees

A test similar to that of the empty structure was performed but now with a person standing on the structure. The person was not moving and with bent knees. The transfer function between the hammer excitation and the acceleration of the structure was calculated and used to update the model shown in Fig. 24.1 without the sound excitation $M(s)$. The model updated contained the parameters $\Theta = \{\omega_n, \zeta, K_p, T_d, T_i\}$ corresponding to the structural system and the first PID controller modeling the person. The posterior $P(\Theta|D)$ was used to inform the prior for the model of the model where the person is exciting the structure.

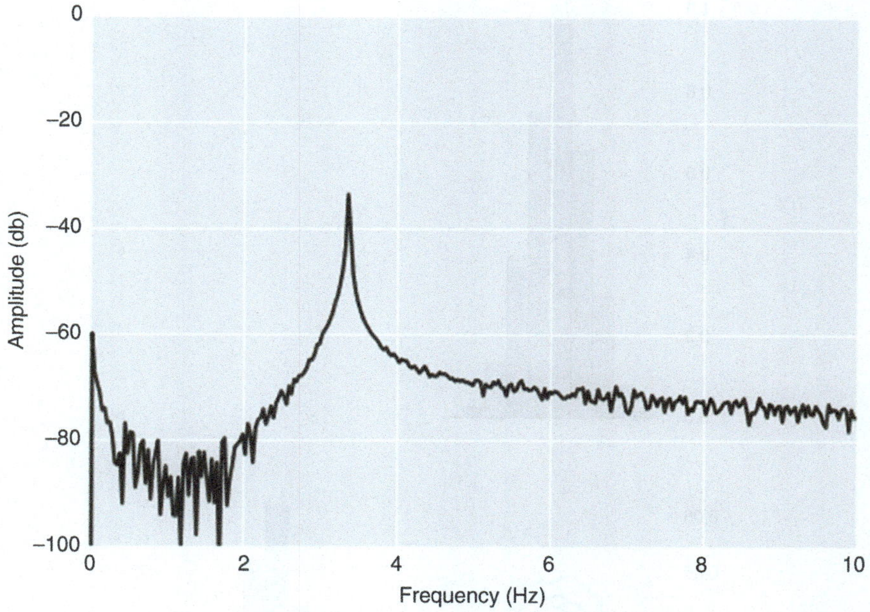

Fig. 24.3 Experimental transfer function of empty structure

Table 24.1 Prior PDF for model of structure excited by the person

Parameter	PDF	Mean, Standard deviation
ω_n	Normal	$\mu = 3.0$, $\sigma = 0.6$
ζ	Normal	$\mu = 0.03$, $\sigma = 0.006$
gain	Normal	$\mu = 21.0$, $\sigma = 4.2$
K_p	Normal	$\mu = 45$, $\sigma = 9.0$
T_d	Normal	$\mu = 0.035$, $\sigma = 0.007$
T_i	Normal	$\mu = 0.045$, $\sigma = 0.009$
K_{p1}	Normal	$\mu = 40$, $\sigma = 14$
T_{d1}	Normal	$\mu = 0.05$, $\sigma = 0.015$
T_{i1}	Normal	$\mu = 0.05$, $\sigma = 0.015$

24.3.4 Structure Excited by the Person

Generally, one can classify human activity in a floor, such as dancing, in terms of the contact between the person and the floor. The first type, which is the simplest to model and used in this paper, is when the person is in contact with the floor 100% of the time. This is, the feet of the person do not leave the floor at any time. The case when the person jumps or partially loose contact with the floor is more complex and not considered in this paper [20, 21]. A metronome was set to 180 bpm and the person standing in the structure was asked to excite the structure with their feet by bending their knees at the frequency of the sound. Acceleration records and the times of the ticks of the metronome were acquired with the data acquisition system. The data was used to update a model that now contains both PID controllers and the model of the structure.

24.4 Results and Discussion

The complete model has parameters $\Theta = \{\omega_n, \zeta, K_p, T_d, T_i, K_{p1}, T_{d1}, T_{i1}\}$. Here the parameters K_p, T_d, and T_i are for the PID that simulates the response of the person to the floor vibrations and the parameters K_{p1}, T_{d1}, and T_{i1} belong to the second PID that uses $M(s)$ as an input in Fig. 24.1. The prior values to update the model were obtained from the posterior PDFs of the prior tests (empty structure and person with bent knees) and are shown in Table 24.1.

The experimental transfer function between the beat of the metronome, assumed as an impulse at 180 bpm, and the floor accelerations are calculated to update the last model. The posterior distributions are shown in Figs. 24.4 and 24.5. Figure 24.4 shows the posterior for the natural frequency and the damping ratio of the structure. The natural frequency of the posterior is

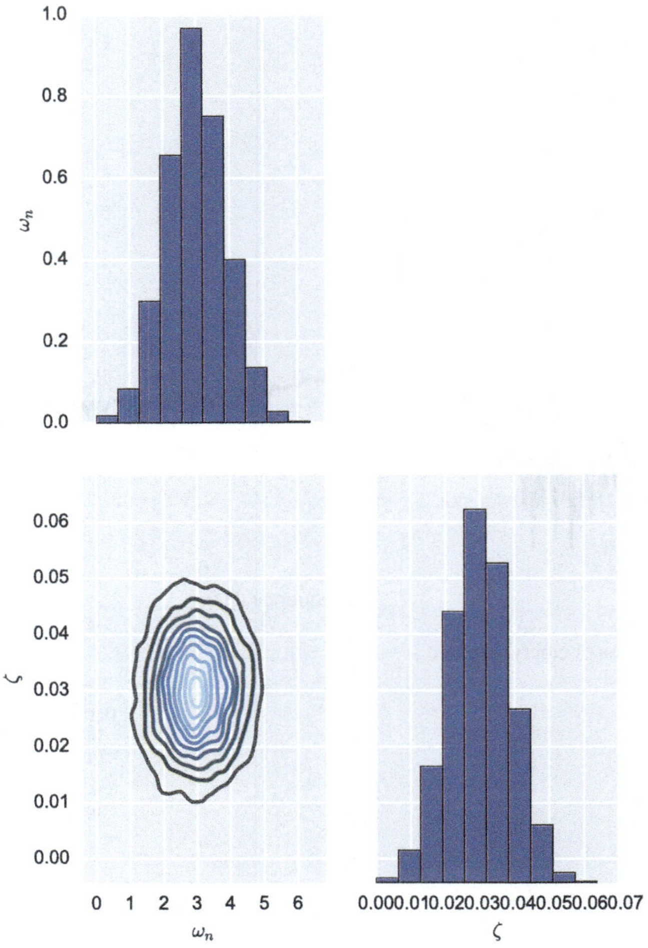

Fig. 24.4 Posterior for structural parameters

centered at 3 Hz with a damping ratio of approximately 3%. This matches well the peak observed in Fig. 24.3. The posterior PDFs of the PIDs in Fig. 24.5 shows range of parameters similar those found by Ortiz and Caicedo [5]. An interesting correlation is observed between the k_p and t_{i1} parameters (Table 24.2).

24.5 Conclusions

This paper proposes a human-structure interaction model excited by a person bending their knees. The model is based on control theory and includes not only the feedback between the structure and the person but an additional input to the human model. the parameters of the model are updated in a probabilistic fashion using Bayes inference. Results show that the mean of the posterior density functions of the structural system correlate well with the peaks of an experimental transfer function. The values of the controller parameters are similar to those found on previous studies that did not consider the person exciting the structure. In the future this model could be expanded to include other type human activity.

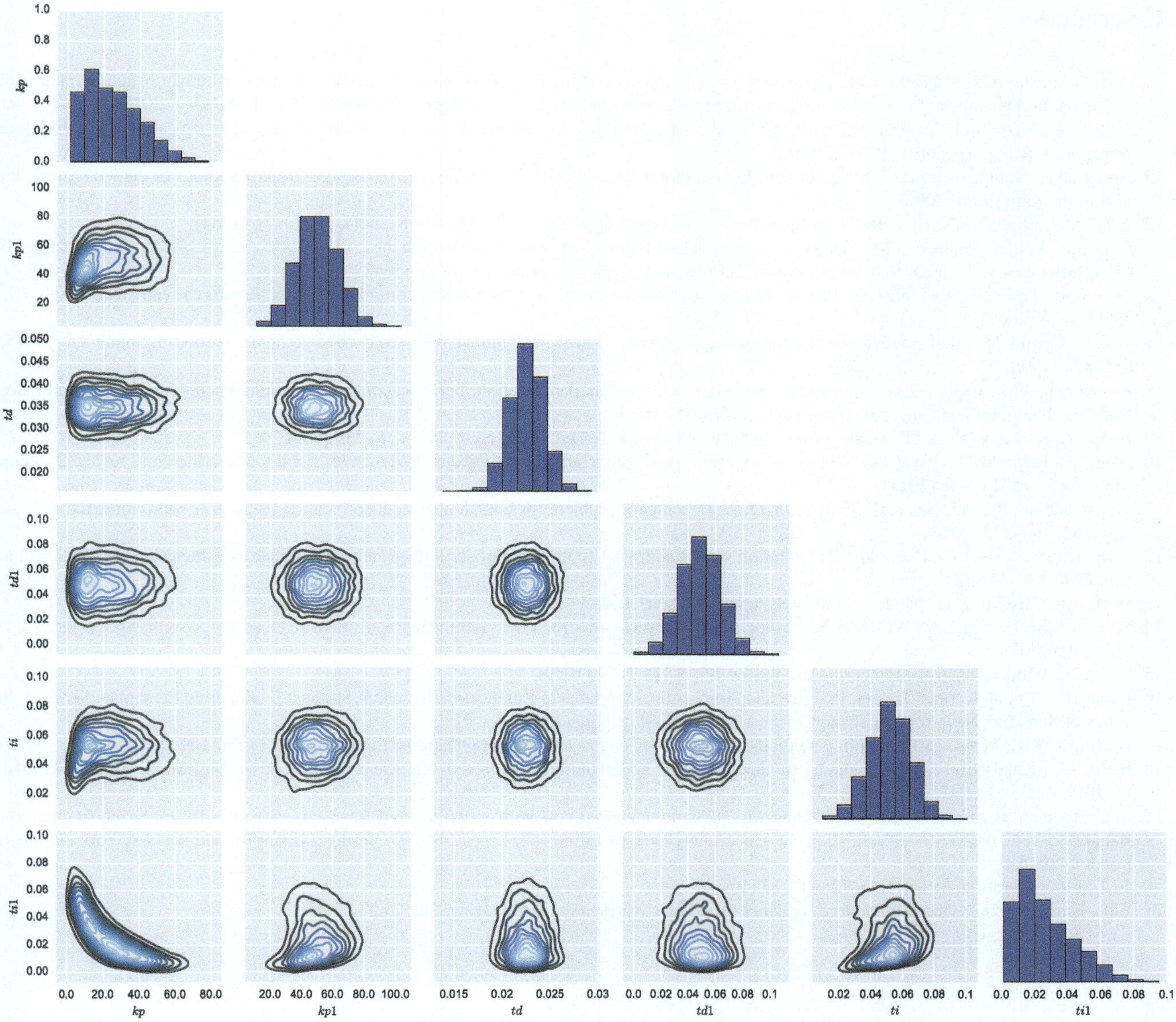

Fig. 24.5 Posterior for parameters of PIDs

Table 24.2 Moments of random variables describing the parameters of the human, music and structure

Parameter	ω_n $[Hz]$	ζ	$gain$	K_p	T_d	T_i	K_p1	T_d1	T_i1
Mean	2.988	0.03	20.909	24.936	3.5E-2	5.2E-2	49.712	4.9E-2	2.6E-2
St. Deviation	1.1E-2	0.009	6.33	14.972	3E-3	1.3E-2	13.242	1.5E-2	17E-3
95% HPD	(1.28, 4.78)	(1.2E-2, 4.8E-2)	(8.314, 33.087)	(1.822, 52.56)	(2.8E-2, 4.2E-2)	(2.4E-2, 7.6E-2)	(23.961, 75.299)	(2.1E-2, 8E-2)	(1E-3, 5.9E-2)

References

1. Ortiz-Lasprilla, A.R., Caicedo, J.M.: Comparing closed loop control models and mass-spring-damper models for human structure interaction problems. In: Dynamics of Civil Structures, vol. 2, pp. 67–74. Springer Cham Heibelderg New York Dordrecht London (2015)
2. Dallard, P., Fitzpatrick, T., Flint, A., Low, A., Smith, R.R., Willford, M., Roche, M.: London millennium bridge: pedestrian-induced lateral vibration. J. Bridg. Eng. **6**(6), 412–417 (2001)
3. Taylor, D.P.: Damper retrofit of the London millennium footbridge–a case study in biodynamic design. In: Proceedings of the 73D Shock and Vibration Symposium (2002)
4. Tilly, G., Cullington, D., Eyre, R.: Dynamic behaviour of footbridges. In: IABSE Surveys S-26/84, pp. 13–24 (1984)
5. Lasprilla, A.R.O., Caicedo, J.M., Ospina, G.A.: Modeling human–structure interaction using a close loop control system. In: Dynamics of Civil Structures, vol. 4, pp. 101–108. Springer Cham Heidelberg New York Dordrecht London (2014)
6. Sachse, R., Pavic, A., Reynolds, P.: Human-structure dynamic interaction in civil engineering dynamics: a literature review. Shock Vib. Dig. **35**(1), 3–18 (2003)
7. Wei, L., Griffin, M.: Mathematical models for the apparent mass of the seated human body exposed to vertical vibration. J. Sound Vib. **212**(5), 855–874 (1998)
8. Brownjohn, J.M.: Energy dissipation in one-way slabs with human participation. In: Proceedings of the Asia-Pacific Vibration Conference, Nanyang Technological University, Singapore, pp. 13–15 (1999)
9. Hidenori, K., Jiang, Y.: A PID model of human balance keeping. IEEE Control Syst. **26**(6), 18–23 (2006)
10. Jones, C., Reynolds, P., Pavic, A.: Vibration serviceability of stadia structures subjected to dynamic crowd loads: a literature review. J. Sound Vib. **330**(8), 1531–1566 (2011)
11. Madarshahian, R., Caicedo, J.M., Zambrana, D.A.: Benchmark problem for human activity identification using floor vibrations. Expert Syst. Appl. **62**, 263–272 (2016)
12. Živanović, S., Pavic, A., Reynolds, P.: Vibration serviceability of footbridges under human-induced excitation: a literature review. J. Sound Vib. **279**(1), 1–74 (2005)
13. Dorf, R.C., Bishop, R.H., Modern Control Systems. Pearson, Boston (2011)
14. Sim, J., Blakeborough, A., Williams, M.: Modelling of joint crowd-structure system using equivalent reduced-dof system. Shock Vib. **14**(4), 261–270 (2007)
15. Ogata, K.: Modern control engineering. Book Rev. **35**(1181), 1184 (1999)
16. Ewins, D.: Modal Testing: Theory, Practice and Application. Mechanical Engineering Research Studies: Engineering Dynamics Series, John Wiley & Sons Inc. New York Brisbane Chichester Toronto Singapore (2003)
17. Hastings, W.K.: Monte carlo sampling methods using Markov chains and their applications. Biometrika **57**(1), 97–109 (1970)
18. Beck, J.L., Katafygiotis, L.S.: Updating models and their uncertainties. I: Bayesian statistical framework. J. Eng. Mech. **124**(4), 455–461 (1998)
19. Madarshahian, R., Caicedo, J.M.: Reducing MCMC computational cost with a two layered Bayesian approach. In: H. Sezer Atamturktur, Babak Moaveni, (eds.) Model Validation and Uncertainty Quantification, vol. 3, pp. 291–297. Springer Cham Heidelberg New York Dordrecht London (2015)
20. Ji, T.: Floor vibration. Struct. Eng. **72**(3/1), 37 (1994)
21. Ellis, B., Ji, T.: Floor vibration induced by dance-type loads: verification. Struct. Eng. **72**, 37–37 (1994)

Chapter 25
Measurement of Human Loads Using Computer Vision

Ozan Celik, Chuan-Zhi Dong, and F. Necati Catbas

Abstract The applications of computer vision techniques in civil engineering are becoming increasingly popular with their promising capabilities such as easy and low-cost deployment, contactless measurement solutions and accurate reconstruction of structural finite element models. This paper aims to explore the possibilities of using several computer vision techniques on the reconstruction of load time histories. The motivation is to find a suitable method that could be applied both in the laboratory environment and in the field for the prediction of excessive loads that are exerted upon assembly type structures and specifically on stadiums. Virtual feature extraction and tracking methods are applied on the video recordings of test subjects while jumping and bobbing on structures instrumented with load cells and accelerometers. The results are compared with the sensor based measurements to assess the accuracy levels and feasibility of the methods.

Keywords Computer vision · Loading · Digital image correlation · Human structure interaction · Vibration serviceability

25.1 Introduction

A widely accepted vibration serviceability criterion is to evaluate the system as a three-step framework namely excitation, path and response [1–4]. A great deal of research that has been done on human induced load measurement and modeling in civil engineering indicates that the estimation of jumping type loads (jumping/bouncing/bobbing/etc.) of both individuals and crowds are critical since these may create extreme vibration responses [5–7]. In case these loads are acted upon flexible structures such as stadiums, concert venues, dance floors, which have low frequency behavior by design requirements, vibration and displacement levels could be exacerbated to the point where catastrophic damages may occur (Jones et al. 2011a). On the other hand, when the motion of the crowd is in question, the severity of loading may vary significantly and may not be as strong depending on the level of synchronization between individuals.

The objective of the study is to introduce alternative load time history measurement technique for singular individuals and crowds using an advanced computer vision algorithm along with its application in both laboratory and real life. First, the algorithm is briefly explained and its application is presented on a modular grandstand simulator that is instrumented with different types of conventional sensors.

25.2 Force Measurement Using Computer Vision

Utilization of computer vision allows the body motion of individuals or crowds to be measured with a camera from a long distance, without any contact and on any structure regardless of their flexibility. In case each body part is tracked individually, according to Newton's second law of motion the ground reaction force is estimated as follows:

$$F_{GR} = \sum_{i=1}^{n} m_i \left(a_i - g\right) \tag{25.1}$$

O. Celik · C.-Z. Dong · F. N. Catbas (✉)
Department of Civil, Environmental and Construction Engineering, University of Central Florida, Orlando, FL, USA
e-mail: ozancelik@knights.ucf.edu; ceczdong@knights.ucf.edu; catbas@ucf.edu

S. Pakzad (ed.), *Dynamics of Civil Structures, Volume 2*, Conference Proceedings of the Society for Experimental Mechanics Series, https://doi.org/10.1007/978-3-319-74421-6_25

Where m_i and a_i are mass and acceleration of the center of the mass of the i-th body segment,g is the standard gravity and n is the total number of body segments. The first step for force estimation is to find the displacement of the body. Once the displacement time history is known, acceleration record can be determined by taking either the second numerical derivative or the numerical gradient of the record. To capture the motion an algorithm based on the well-known optical flow estimation called Lucas-Kanade tracking is given.

25.2.1 Lucas-Kanade Tracking

The optical flow measurement along with its improved different versions have been accepted as a reasonable and fairly accurate estimation of displacement and velocity and has had numerous applications such as recognition, tracking, motion modeling, segmentation, etc. in different engineering fields. In the scope of this paper, a well-established variations proposed by [8–10] for sparse motion estimation is applied.

Determining optical flow requires the assumption of constant brightness in spatial and temporal space of consecutive frames by

$$f(x, y, t) = f(x + dx, y + dy, t + dt) \tag{25.2}$$

Where f represents the image frame with the parameters x, y and t indicating pixel locations and time respectively. Expanding Eq. 25.2 into Taylor series gives

$$f(x, y, t) = f(x, y, t) + \frac{\partial f}{\partial x}(x + dx - x) + \frac{\partial f}{\partial y}(y + dy - y) + \frac{\partial f}{\partial t}(t + dt - t) \tag{25.3}$$

Canceling the same terms on both sides and simplifying Eq. 25.3 result in

$$f_x u + f_y v = -f_t \tag{25.4}$$

Equation 25.4 is the so called optical flow equation indicating a line having a slope of $-f_x/f_y$ with the two unknowns u and v corresponding to motion in x and y. The solution of Eq. 25.4 necessitates one more assumption which eventually gives name to different variations on the computation of optical flow. Lucas-Kanade method assumes that the motion within the small neighborhood of the pixel of interest is the same and small thereby generating more equations than unknowns:

$$\begin{bmatrix} f_{x_1} & f_{y_1} \\ f_{x_2} & f_{y_2} \\ \vdots & \vdots \\ f_{x_n} & f_{y_n} \end{bmatrix} \begin{Bmatrix} u \\ v \end{Bmatrix} = \begin{Bmatrix} -f_{t_1} \\ -f_{t_2} \\ \vdots \\ -f_{t_n} \end{Bmatrix} \tag{25.5}$$

It is now possible to solve this system of two equations with the two unknowns. By rearranging Eqs. 25.3 and 25.5 the solution is found:

$$\begin{Bmatrix} u \\ v \end{Bmatrix} = \frac{1}{\sum f_{x_i}^2 \sum f_{y_i}^2 - \left(\sum f_{x_i} f_{y_i}\right)^2} \begin{bmatrix} \sum f_{y_i}^2 & -\sum f_{x_i} f_{y_i} \\ -\sum f_{x_i} f_{y_i} & \sum f_{x_i}^2 \end{bmatrix} \begin{Bmatrix} -\sum f_{x_i} f_{t_i} \\ -\sum f_{y_i} f_{t_i} \end{Bmatrix} \tag{25.6}$$

25.3 Laboratory Verification

A grandstand simulator is constructed and used for further series of tests Fig. 25.1. The grandstand Fig. 25.1 can be adjusted to have one or two clear spans with continuous beams across the middle supports. It has two 18 ft. girders (S3x5.7 steel section) in the longitudinal direction. The 3 ft transverse beam members are used for lateral stability Fig. 25.1. The simulator can be modified to reveal different structural stiffness and damage cases by altering specially designed intermediate and boundary conditions (e.g., pin supports, rollers, fixed support, and semi-fixed support).

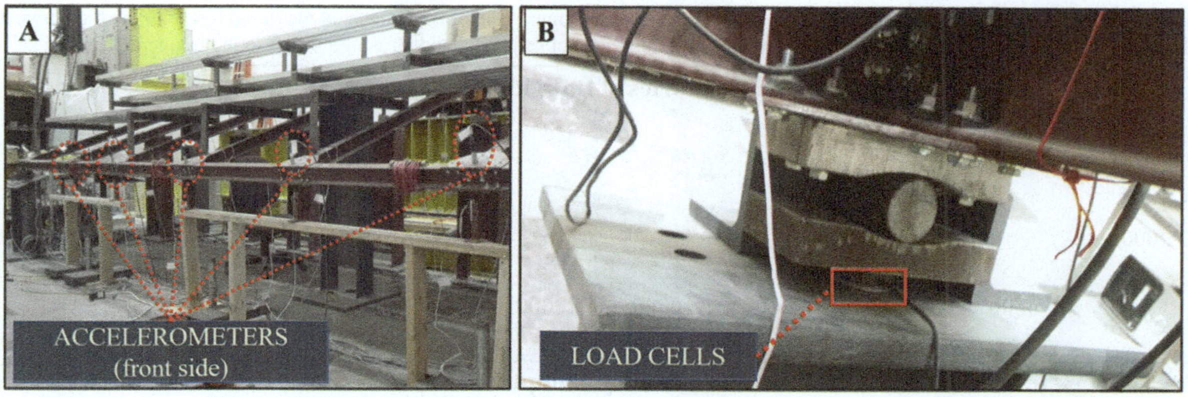

Fig. 25.1 The grandstand simulator along with the details of its elements

Fig. 25.2 Jumping test in progress

The motivation behind the construction of the grandstand simulator is to build a model that can serve for the solution of various vibration serviceability problems alongside with the scope of this work. A test is designed to show the applicability and efficiency of the Lucas-Kanade method in tracking the subjects and on an almost rigid platform as an alternative to force plates. The second group of tests is an expansion of the first group with more subjects and on the flexible grandstand Fig. 25.2. For the test, subjects are asked to accompany the metronome beats of 1, 2 and 3 Hz which is the common way of controlled jumping tests in the literature. In addition to that, subjects also jump to a commonly played song that they are familiar with especially from the football games that are hosted in local games. This study is focused mainly on this last specific beat.

Table 25.1a shows the force time history of each individual breaking down the time of arrival differences of each pulse along with their sum. The second harmonic of the force pattern is close to the first fundamental frequency of the simulator which creates extreme deflections that sometimes cause the subject to lose balance and even stop. The time frames that the jumpers are in and out of phase because of this imbalance are seen in Table 25.1b and are captured by the proposed method. The general morphology is almost an exact match with the exception of the amplitudes. Total force history captured by the sensors are higher in amplitudes than the vision based methods (Table 25.1c). This observation is only made when the structure is in flexible setup with visible deflections. As the subjects jump off of the structure, an uplift force is generated thereby causing the inertial forces to be involved. The frequency domain representation in Table 25.1d smears over the other bands indicating the imbalance and out-of-phase jumping trend.

Table 25.1 (a) Force time histories of each individual and their sum (b) Comparison of estimated and measured forces of the four subjects on the flexible grandstand; (c) Close-up view of the comparison; (d) Comparison of estimated and measured forces in the frequency domain

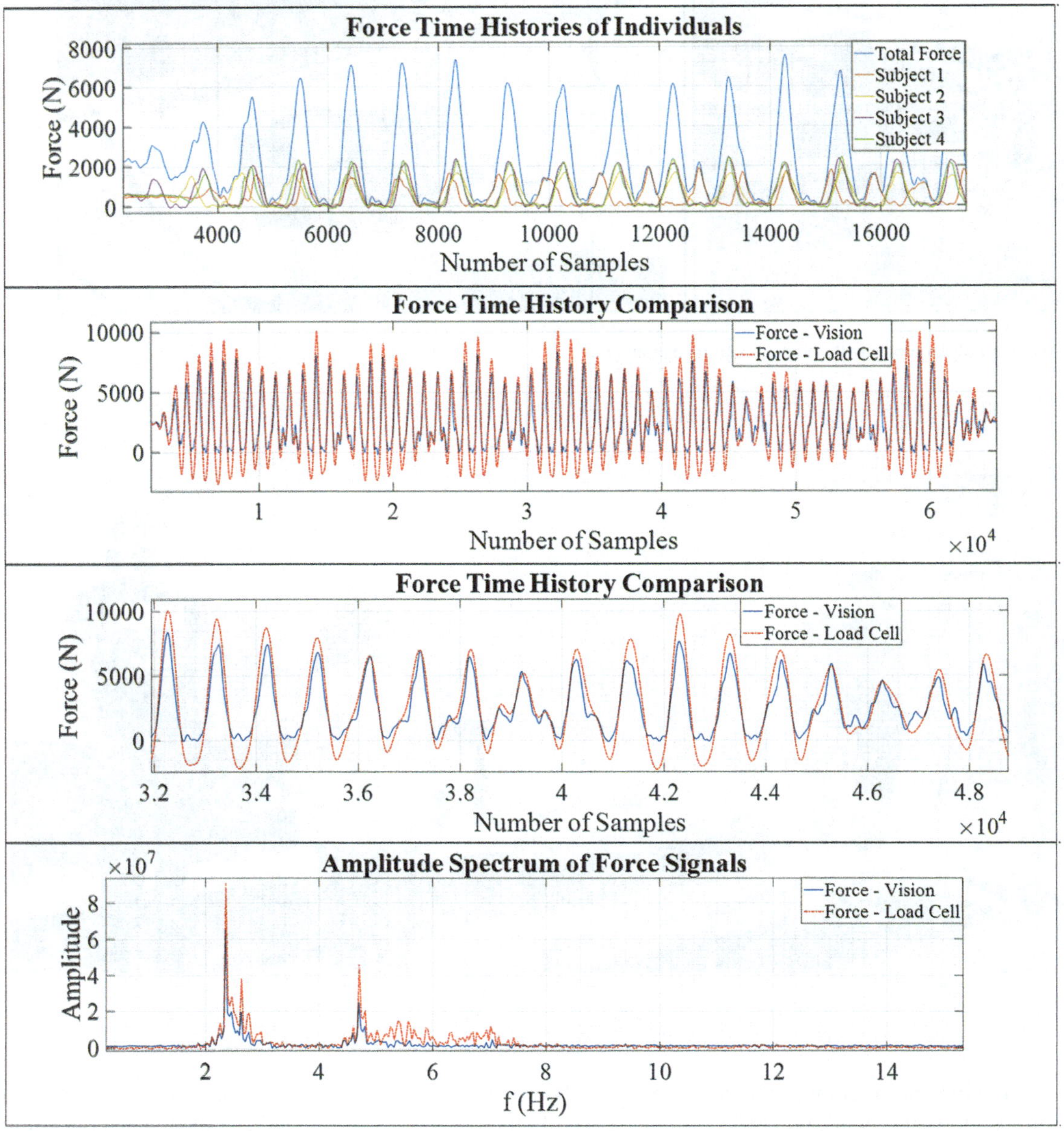

25.4 Conclusions

The paper presents alternative load time history measurement techniques for singular individuals and crowds and proves the feasibility of such based on the results acquired from both laboratory and real life. Although there are now successful synthetic load time history creation methods for individuals, crowd load modeling is still a missing piece. The use of computer vision methods can free researchers from limited laboratory conditions and pave the way for creating accurate databases of real, large crowds to be used for robust serviceability design. The future work of the authors inludes making improvements in improving the results and conducting field experiments.

25.5 Funding Statement

The financial support for this research was provided by Qatar National Research Fund [QNRF (a member of Qatar Foundation)] via the National Priorities Research Program (NPRP), Project Number: NPRP 6-526-2-218 and NSF Division of Civil, Mechanical and Manufacturing Innovation [grant number 1463493]. The statements made herein are solely the responsibility of the authors.

References

1. Catbas, F.N., Celik, O., Avci, O., Abdeljaber, O., Gul, M., Do, N.T.: Sensing and monitoring for stadium structures: a review of recent advances and a forward look. Front. Built Environ. **3**, 38 (2017)
2. Celik, O., Do, N.T., Abdeljaber, O., Gul, M., Avci, O., Catbas, F.N.: Recent issues on stadium monitoring and serviceability: A review. In: Allen, M., Mayes, R., Rixen, D. (eds.) Dynamics of Coupled Structures, Volume 4. Conference Proceedings of the Society for Experimental Mechanics Series. Springer, Cham (2016)
3. ISO/TC/98/SC 2: ISO 10137:2007, Bases for design of structures – Serviceability of buildings and walkways against vibrations (2007)
4. Jones, C.A., Reynolds, P., Pavic, A.: Vibration serviceability of stadia structures subjected to dynamic crowd loads: a literature review. J. Sound Vib. **330**(8), 1531–1566 (2011)
5. Georgiou, L., Racic, V., Brownjohn, J.M.W., Elliot, M.T.: Coordination of groups jumping to popular music beats. In: Caicedo, J., Pakzad, S. (eds.) Dynamics of Civil Structures, Volume 2. Conference Proceedings of the Society for Experimental Mechanics Series. Springer, Cham (2015)
6. Sim, J., Blakeborough, A., Williams, M.S., Parkhouse, G.: Statistical model of crowd jumping loads. J. Struct. Eng. **134**(12), 1852–1861 (2008)
7. Yao, S., Wright, J.R., Pavic, A., Reynolds, P.: Experimental study of human-induced dynamic forces due to jumping on a perceptibly moving structure. J. Sound Vib. **296**(1–2), 150–165 (2006)
8. Lucas, B.D., Kanade, T: An iterative image registration technique with an application to stereo vision: In Proceedings of the 7th International Joint Conference on Artificial Intelligence, vol. 2, pp. 674–679 (1981)
9. Farneback, G.: Very high accuracy velocity estimation using orientation tensors, parametric motion, and simultaneous segmentation of the motion field: In Proceedings Eighth IEEE International Conference on Computer Vision. ICCV 2001, vol. 1, pp. 171–177 (2001)
10. Farnebäck, G.: Two-frame motion estimation based on polynomial expansion. In: Bigun, J., Gustavsson, T. (eds.) Image Analysis. SCIA 2003. Lecture Notes in Computer Science, vol. 2749. Springer, Berlin, Heidelberg (2003)

Chapter 26
Automatic Detection of Structural Deficiencies Using 4D Hue-Assisted Analysis of Color Point Clouds

Ali Khaloo and David Lattanzi

Abstract Recent developments in the fields of robotics and remote sensing technologies such as 3D laser scanners and photogrammetric approaches have provided an unprecedented opportunity to collect a massive amount of data from infrastructure systems in a contactless and nondestructive manner, which can potentially improve the structural health monitoring process. However, the complex nature of these geometrically accurate and high-resolution 3D models makes it inefficient and time-consuming to manually analyze and manipulate them and automating this process continue to pose a challenge. Thus, procedures that automate the data processing in order to detect a variety of damages are desired to make full use of these modern inspection technologies as a tool for infrastructure integrity assessment and asset management. The aim of this paper is to present a new algorithm to automatically identify and evaluate structural deficiencies in massive 3D point clouds of complex infrastructure systems. This approach takes advantage of both local geometry and color data properties associated with each point to improve the damage detection capabilities in a variety of scenarios. Linear and non-linear transformations from the RGB color space to non-RGB spaces were performed to increase separability between the damage and the structure and to achieve robustness to changes in illumination. Recently, a complex and large-scale gravity dam in Maryland, USA has served as a test bed for the developed methodology. In this experiment, a multi-scale photogrammetric computer vision approach was utilized to generate accurate and highly detailed 3D models of the targeted dam. In order to maximize the accessibility and to overcome geometric constraints, different multi-rotor Unmanned Aerial Vehicle (UAV) platforms with varied payload and maneuverability capabilities, each equipped with different optical sensors were used in this study. Experimental results demonstrate that the presented 4D point cloud analysis method can accurately detect and quantify a variety of anomalies from spalling to moisture infiltration in exposed concrete structures.

Keywords 3D point cloud · Unmanned Aerial Vehicle (UAV) · Computer vision · Structural health monitoring · Damage detection

26.1 Introduction

Currently, infrastructure assessment protocols mandate evaluating a given structure for a variety of defects and aging phenomena with a primarily visual methodology. Such visual inspection methods are the predominant method for evaluating the structural integrity of almost all infrastructure systems, and often result in subjective and highly variable inspection assessments [1]. The costs of these inspections are a financial burden for many operators and owners, and comprise an outsized portion of annual operating budgets. Furthermore, shortcomings in inspection and maintenance can have a severely negative impact on local economies [2]. As such, there is a need to reduce the costs and improve the results of inspection processes.

Due to the availability of low cost, high quality and easy-to-use visual sensing technologies, the rate of creation and deployment of computer vision methods for civil engineering applications has been rapidly increasing over the last decade [3]. Computer vision, referred to in this context as the processes of automatic extraction, analysis and understanding of useful information from a single digital image or sequence of images, can augment the visual inspection process by enabling quantitative assessments of visually observable damage [4, 5]. In particular, enabling the automatic location, assessment, quantification, and tracking of visible (or otherwise detectable) damage in visual datasets would substantially enhance the inspection practices in use currently.

A. Khaloo (✉) · D. Lattanzi
Department of Civil, Environmental, and Infrastructure Engineering, George Mason University, Fairfax, VA, USA
e-mail: akhaloo@gmu.edu; dlattanz@gmu.edu

S. Pakzad (ed.), *Dynamics of Civil Structures, Volume 2*, Conference Proceedings of the Society for Experimental Mechanics Series, https://doi.org/10.1007/978-3-319-74421-6_26

One emerging computer vision approach is to generate and analyze photorealistic three-dimensional (3D) models of a targeted structure, with that model serving as the basis of a long-term structural record. As recently investigated [6, 7], utilizing modern techniques such as building information modelling (BIM) and 3D virtual reconstruction can facilitate more efficient and cost-effective management of infrastructure systems. While in the past the acquisition of such 3D data (also known as *point clouds*) was restricted to specialized and expensive cameras and laser scanners, this has changed drastically during the last few years [8]. Although these 3D models provide geometrically accurate and densely sampled representations of the targeted structure, they do not inherently provide information as to the changes in a structure over time, and they do not provide explicit information regarding structural integrity. Thus, automated data processing procedures to detect a variety of damages (e.g., deformation, spalling, and moisture infiltration) are desirable.

Prior work in 3D point cloud change analysis for structural monitoring includes deformation analysis [9, 10], post-disaster damage estimation [11, 12], bridge clearance measurement [13, 14], and detection of flatness defects of concrete surfaces [15]. However, these studies are mostly focused on 3D change analysis based on geometric information and cannot accurately detect textural deterioration and surficial damages such as corrosion, rust and spall off. In order to overcome this limitation, researchers [16–18] have investigated using the reflection intensity and color information acquired by 3D laser scanners to identify textural damages such as moisture affectation [19]. However, the dependency of these algorithms on laser response intensity data makes them unsuitable for photogrammetric-based 3D point cloud models.

This research presents a new method to automatically identify and evaluate structural deficiencies in massive 3D point clouds of complex infrastructure systems. This approach takes advantage of both local geometry and color data properties associate with each point to improve the damage detection capabilities in a variety of scenarios independent from the source of the 3D data. By incorporating photometric invariant color representations, it was possible to identify textural manifestations of damages under severe illumination changes. To evaluate the proposed automated condition assessment methodology on a large-scale infrastructure system, a complex gravity dam in Maryland, USA (the Brighton Dam) has served as a test bed. A series of experiments involving the analysis of both simulated and real structural damage were designed and performed.

26.2 Methodology

The automated damage detection and quantification method is designed to identify small deteriorations (e.g., minor cracks, section loss, or scour) where all the primary structural elements are sound. To achieve this, point clouds of both the damaged and undamaged system are compared analytically. This computer vision-based method has four key phases: (i) fine point cloud registration, (ii) determination of point correspondences between clouds, (iii) color space transformation and (iv) color distance measurement. By combining both 3D spatial data and color features of points, it is possible to create a unique spatio-temporal model of the targeted structures through four-dimensional (4D) visual data analytics.

26.2.1 Fine Point Cloud Registration

Prior to estimating and localizing any changes in 3D models, the reference and compared (undamaged and damaged) point clouds, P_r and P_c, are automatically aligned and globally registered against each other using a robust variant of the Iterative Closest Point (ICP) algorithm [20]. The overall aim of the ICP algorithm is to estimate a rigid transformation (usually composed of scaling factor s, rotation matrix R, and translation vector t) between points $p_i \in P_r$, for a 3D point cloud P_r, and point-correspondences $q_i \in P_c$, from the 3D cloud to be registered against through minimizing least squares of the residuals:

$$E(R,t) = \min_{R,t} \sum_i \| q_i - (Rp_i + t) \|^2 \tag{26.1}$$

Where the summation is taken over all 3D data points.

Since ICP is an approximate algorithm essentially doing local convergence, its results depends on the initial pose of one 3D model with respect to the other. Another factor driving the difficulty of point cloud registration is noise/outliers in the point clouds. In the presence of excessive points that do not correspond to the same features in both models, ICP runs the risk of converging to a local optimum driven by false matches.

In order to minimize the accumulative error in solving this least squares problem, a well-known technique called Generalized (isotropic) Procrustes Analysis (GPA) was utilized [21] to determine rigid transformation parameters with the following objective function:

$$e = tr\left[\sum_{i=1}^{n}\sum_{j=i+1}^{n}\left((s_iX_iR_i + u\otimes t_i) - (s_jX_jR_j + u\otimes t_j)\right)^T \times \left((s_iX_iR_i + u\otimes t_i) - (s_jX_jR_j + u\otimes t_j)\right)\right] \quad (26.2)$$

Where $X_1, X_2, \cdots, X_n$ are n sets of point clouds, each having p points in a q dimensional space with u representing a $p \times 1$ unit vector and $\otimes$ denoting the Kronecker product. As discussed in [5], implementing GPA reduces the average error of the final alignment by a factor of 2 with a computational complexity of $\mathcal{O}\left[w^2 \times n\log(n)\right]$ where n stands for the number of points and w number of point clouds. Prior to fine registration, an initial alignment of point clouds were performed through pair-wise matching of 10 manually picked point correspondences in order to reduce the inherent sensitivity of the ICP algorithm regarding the initial positions of the 3D models.

26.2.2 Finding Point Correspondences Between Clouds

Upon registration of the P_r and P_c point clouds, matching point pairs are formed by searching for correspondences between point clouds prior to change analysis. In this work, the Fast Library for Approximate Nearest Neighbors (*FLANN*) [22] was implemented to reduce the computational complexity of the nearest neighbor search by using the randomized *k-d* trees approach.

26.2.3 Color-Space Transformation

Among color spaces, *RGB* (*Red*, *Blue*, and *Green*) is the most widely used representation for color display, processing and storing of digital visual datasets. However, high correlation between its components, significant perceptual non-uniformity, and mixing of chrominance and luminance data make the *RGB* color space unsuitable for most color analysis and color-based recognition algorithms [17, 23]. In addition, since *RGB* color difference measurements are not in a uniform scale, it is impossible to reliably evaluate color similarity of two points from their distance in *RGB* space [24]. For effective color-based defect detection, it is important to select an appropriate color space to minimize the effects of both illumination and viewpoint changes, and to maximize the separability between the deficiency and the actual structure. It is possible to derive a variety of color spaces from *RGB* representations through linear or nonlinear transformations. Although there are many possible color spaces such as *YIQ*, *HSI*, and normalized *RGB* [25, 26], selecting the most suitable color representation for defect detection purposes is still a challenging task [24]. In this study, five color spaces have been evaluated in order to determine the most reliable color representation for a robust damage detection algorithm: *YIQ*, *XYZ*, normalized *RGB*, *HSI*, and $L^*a^*b^*$ [27–29].

The *YIQ* color space is widely used in video compression and transmission standards, and consists of three components: luminance (*Y*), hue (*I*), and saturation (*Q*) [27]. The first component, luminance, represents grayscale information, while the last two components make up chrominance (color information). The *YIQ* space can partly lessen the correlation of the red, green and blue components through a linear transformation, which makes it well suited for defect detection in visual datasets [26].

Another primary color space used to describe pure spectral content is *XYZ*. This representation also maps the *Y* axis to the luminance (perceived relative brightness) and pure white to a (equal-valued) vector. One of the characteristics of this system is that the tristimulus values *X*, *Y*, and *Z*, are always positive for all color stimuli [28]. Since *XYZ* color representation is just a linear transformation of the *RGB* space, it tends to be sensitive to the viewing direction, illumination change, and object surface orientation [29].

Normalized *RGB* (*rgb*) is a color representation that can be easily obtained from the *RGB* values using a simple normalization procedure [28]:

$$r = \frac{R}{R+G+B},\quad g = \frac{G}{R+G+B},\quad b = \frac{B}{R+G+B} \quad (26.3)$$

This nonlinear color transformation make colors independent from changes in lighting intensities. This normalization is an efficient method to get the variations of intensities uniformly across the spectral distribution. A remarkable property of this representation is that for matte surfaces, *rgb* is invariant to changes in surface orientation relative to the light source [25]. However, due to the nonlinear transformation from the *RGB* space, the normalized colors (*rgb*) are very noisy under low intensities [24].

The *HSI* (Hue, Saturation and Intensity) color space is one of the most suitable models for color analysis since it separates color from brightness characteristics [27]. In this model, hue represents a pure color (pure yellow, orange, or red), and is determined by the dominant wavelength in the spectral distribution of light wavelengths. Saturation refers to purity of color or the amount of the relative white content of a color, and intensity represent the overall brightness or the amount of light. The following equations show how the *RGB* components can be converted to hue (H), saturation (S) and Intensity/Brightness (I):

$$H = \cos^{-1}\left\{\frac{\frac{1}{2}[(R-G)+(R-B)]}{\sqrt{(R-G)^2+(R-B)(G-B)}}\right\} \tag{26.4}$$

$$S = 1 - \left(\frac{3}{R+G+B}\right)[\min(R, G, B)] \tag{26.5}$$

$$I = \frac{1}{3}(R+G+B) \tag{26.6}$$

These components are the perceptual features of color that most closely represent how humans perceive color. Humans can distinguish different hues easily, which cannot be described directly by the *RGB* color representation. Since hue value is independent from intensity, it has been utilized in many color-based visual data analysis algorithms such as color assisted point cloud registration [30], image-based corrosion detection [31], and concrete detection for construction monitoring using image datasets [25]. In this study, the normalized hue value (range from 0 to 1) associated with each point was used for the presented defect detection method.

The CIE (Commission International de l'Eclairage) color system was developed to represent perceptual uniformity, where color differences among various hues are perceived uniformly. The nonlinear transformation from *RGB* to $L^*a^*b^*$ color space is designed to lessen the shortcomings in color appearance and difference by making $L^*a^*b^*$ as a *colorimetric* (i.e., colors perceived as matching are encoded identically) color representation [29].

$$L^* = 116f(Y/Y_n) - 16 \tag{26.7}$$

$$a^* = 500[f(X/X_n) - f(Y/Y_n)] \tag{26.8}$$

$$b^* = 200[f(Y/Y_n) - f(Z/Z_n)] \tag{26.9}$$

Where $f(I) = I^{1/3}$ if $I > \left(6/29\right)^3$ and $f(I) = \left(841/108\right)(I) + 4/29$ if $I \le \left(6/29\right)^3$. In addition, X_n, Y_n, and Z_n represent a reference white (perfectly reflecting diffuser) as defined by a CIE standard *D65* illumination. Like the *HSI* color space, the $L^*a^*b^*$ system is an excellent decoupler of intensity (represented by lightness L^*) and color (represented by a^* for red minus green and b^* for green minus blue), which makes it suitable for color model-based object detection tasks [32].

In this study, in order to make the algorithm robust to illumination changes, due to the lighting changes during the data acquisition, the illuminance component of the color spaces such as, L^* for $L^*a^*b^*$, Y for *XYZ*, I for *HSI*, and Y for *YIQ*, was not considered in the color difference analysis.

26.2.4 Color Distance Measurement

The *color distance* between corresponding 3D points $p_i \in P_r$ and $q_i \in P_c$ can be measured by defining a distance metric between the color features of each point. By measuring color similarity between colored 3D point clouds at different times, it is possible to identity and track visual textural damages in the targeted structure.

The color distance Δ between color information of two corresponding points can be measure using the Euclidean distance:

$$\Delta E = \sqrt{\left(v_{1,1} - v_{2,1}\right)^2 + \left(v_{1,2} - v_{2,2}\right)^2 + \left(v_{1,3} - v_{2,3}\right)^2} \tag{26.10}$$

Where $\overrightarrow{v_i} = \left[v_{i,1}, v_{i,2}, v_{i,3}\right]^T$ is a color triplet. This 3D Euclidean distance can be modified based on the dimensionality of the utilized color space. Although using Euclidean distance as a metric to calculate color differences has been implemented in several studies, its performance on the *RGB* color space and its linear derivations (e.g., *XYZ*) is not robust due to their fundamental shortcoming in separating hue values from other color components [24, 33].

Among color representations, the CIE $L^*a^*b^*$ space is the only one that is associated with an optimized and well-defined color difference formulation [29]. This measurement is a quantitative description of the perceived color difference between a pair of colors which also considers perceptual non-uniformities.

$$\Delta E_{94}^* = \sqrt{\left(\Delta L^* / \left(K_L S_L\right)\right)^2 + \left(\Delta C_{ab}^* / \left(K_C S_C\right)\right)^2 + \left(\Delta H_{ab}^* / \left(K_C S_H\right)\right)^2} \tag{26.11}$$

Where $K_L = K_C = K_H = 1$, $C_{ab}^* = \sqrt{a^{*2} + b^{*2}}$, $S_L = 1$, $S_C = 1 + 0.045C_{ab}^*$ and $S_H = 1 + 0.015C_{ab}^*$. ΔL^*, ΔC_{ab}^*, and ΔH_{ab}^* represent the difference between the lightness, chroma and hue values in the above formulation.

26.3 Experimental Analysis

Images of the Brighton dam (Fig. 26.1), in-service since 1944 and located in Brookeville, MD, were collected on two separate occasions (January 19th and April 13th, 2017) in order to generate both reference and compared, P_r and P_c, 3D point clouds.

To generate geometrically complete and high-resolution models, two multi-rotor unmanned aerial vehicles (UAV) and a fixed-wing UAV were used to acquire both oblique and nadir image sets. A DJI Inspire 1 aircraft mounted with a 12 megapixel camera was used to image both the downstream and upstream faces of the dam, while a smaller quadcopter aircraft (DJI Phantom 4 with a 16 megapixel sensor) was utilized inside each structural bay due to its improved maneuverability.

Fig. 26.1 (**a**) The Brighton Dam; (**b**) Dense 3D point cloud (January'17 dataset)

Fig. 26.2 Dense 3D point cloud of the Bay#5: (**a**) Reference model P_r; (**b**) Compared model P_c

Fig. 26.3 Color-assisted linear dimensional change analysis: (**a**) 3D point cloud rendering and (**b**) automatic mapping of the simulated damages

In this study, a multi-scale photogrammetric 3D reconstruction methodology previously developed by the research team and referred to as Hierarchical Point Cloud Generation (HPCG) [5], was used to generate two highly detailed 3D models of the dam.

In total 1618 oblique images captured using the quadcopter platforms, along with 558 nadir images (from an area of approximately 174,502 m^2) were used to generate the P_r point cloud of 1,997,005,831 points. For the compared point cloud, P_c, a total of 2567 images were collected, resulting in a model with 1,244,648,688 points.

Fig 26.2 illustrates Bay #5 of the dam for both inspection intervals. In this bay, a series of simulated damage patterns with known dimensions were applied prior to generating the compared point cloud, to serve as quantifiable benchmarks for evaluation of the color-assisted measurement algorithm. As shown in Fig. 26.2, due to the lighting changes during the image data acquisition, there was an apparent need to utilize illumination invariant color spaces in the developed change analysis algorithm.

To evaluate the measurement accuracy and repeatability of the presented algorithm in identifying linear crack propagation, based on different color spaces, lines with varying length (range from 76.2 mm - 152.4 mm) and width (range from 1 to 3 mm) were drawn using white chalks on the sides of the Bay#5 (Fig. 26.3).

Table 26.1 Linear dimensional evaluation of the change analysis algorithm for different damage scenarios

Linear dimensional analysis			
Damage actual width and length (mm)	Color Model	Estimated length (mm)	Length deviation (%)
W: 1 – L: 152.4	*YIQ*	140.0	7.9
	XYZ	125.2	17.8
	rgb	141.8	7.0
	Hue	148.9	**2.3**
	$L^*a^*b^*(\Delta E)$	148.7	2.4
	$L^*a^*b^*(\Delta E^*_{94})$	–	–
W: 3 – L: 76.2	*YIQ*	60.8	20.2
	XYZ	55.3	27.4
	rgb	63.0	17.3
	Hue	66.2	13.1
	$L^*a^*b^*(\Delta E)$	68.5	**10.1**
	$L^*a^*b^*(\Delta E^*_{94})$	30.7	59.7
W: 3 – L: 152.4	*YIQ*	145.5	4.5
	XYZ	132.7	12.9
	rgb	133.7	12.3
	Hue	147.3	3.4
	$L^*a^*b^*(\Delta E)$	148.3	**2.7**
	$L^*a^*b^*(\Delta E^*_{94})$	120.5	20.9

Table 26.1 summarized the comparisons between the ground-truth dimensions of the simulated crack damages and the estimated values utilizing the presented change analysis algorithm. Upon automatic detection of the simulated damages using the presented algorithm, the dimensions of the extracted flaws were manually measured by computing the distance between two furthest points on each of the identified regions (Fig. 26.3).

The experimental results show that the ΔE^*_{94} color distance metric associated with the $L^*a^*b^*$ color space performed poorly due to its consideration of lightness (L^*) in measuring color similarity. However, utilizing the 2D Euclidean distance (based on a^* and b^* chrominance components) was shown to overcome this issue.

As it was expected, the linear transformation from the *RGB* tristimulus coordinates to the *XYZ* and its highly correlated nature resulted in limited robustness to illumination changes in comparison with other color spaces in identifying simulated cracks even after omitting the illuminance component. Furthermore, the non-linear re-mapping of the *XYZ* space to the $L^*a^*b^*$ provided a more perceptually uniform differences in luminance or chrominance, which resulted in a relatively robust performance in detecting cracks with varying dimensions (mean error of %5).

In case of the *YIQ* color space, due to less correlation between its color information and by only incorporating the *I* and *Q* components (which jointly describe the hue and saturation) in the Euclidean color distance metric, it was possible to achieve better results in compare with other linear transformations of the *RGB* space. The major problem of linear color spaces (e.g., *RGB*, *YIQ* and *XYZ*) is the high correlation of the three components, which makes the three components dependent upon each other and associate strongly with intensity. Hence, nonlinear color spaces performed relatively better in detecting the simulated damages on the wall of the Bay#5.

To assess the area measurement accuracy of the presented algorithm, a series of white square patches (Fig. 26.4) with dimensions varying from 645 mm^2 to 16,129 mm^2 (1 $inch^2$ to 25 $inch^2$) were marked on the wall of the targeted bay.

As presented in Table 26.2, by utilizing non-*RGB* color spaces and dropping their illuminance component, the color-assisted change detection algorithm provided a reliable quantification of changes even under severe changes in lighting conditions, as illustrated in Fig. 26.2.

Similar to the linear dimensional analysis, the presented algorithm performs well in detecting the pre-defined and controlled simulated damages. Since the $L^*a^*b^*$ color space controls intensity and color information more independently and better than other color models, it provided a better and more consistent measurements of even small color differences between the P_r and P_c point clouds.

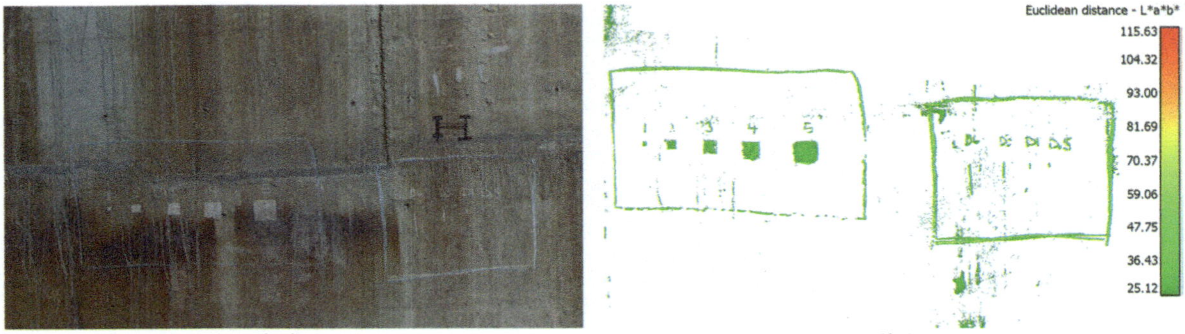

(a) (b)

Fig. 26.4 Color-assisted surface area change analysis: (**a**) 3D point cloud rendering and (**b**) automatic mapping of the simulated damages

Table 26.2 Surface area evaluation of the change analysis algorithm for different damage scenarios

Surface area analysis			
Damage actual area (mm^2)	Color Model	Estimated area (mm^2)	Area deviation (%)
645	*YIQ*	440.8	31.7
	XYZ	379.9	41.1
	rgb	393.4	39.0
	Hue	454.9	29.5
	$L^*a^*b^*(\Delta E)$	508.3	**21.2**
	$L^*a^*b^*(\Delta E^*_{94})$	320.3	50.3
5806	*YIQ*	5313.8	8.5
	XYZ	4120.0	29.0
	rgb	5142.3	11.4
	Hue	5401.0	**7.0**
	$L^*a^*b^*(\Delta E)$	5306.0	9.1
	$L^*a^*b^*(\Delta E^*_{94})$	4210.8	27.5
16,129	*YIQ*	15839.2	2.9
	XYZ	13775.1	14.6
	rgb	15913.0	**1.3**
	Hue	15949.9	1.8
	$L^*a^*b^*(\Delta E)$	15999.0	**1.3**
	$L^*a^*b^*(\Delta E^*_{94})$	14477.6	10.2

26.4 Conclusions

In this study, a new point cloud change analysis algorithm was developed that combines geometric and color information to identity the textural damages in infrastructure systems without dependency to the source of the spatial data. This makes the presented method applicable for both laser scanned and image-based colored point clouds. Both linear and non-linear transformations from the *RGB* space to other color representations were implemented to increase the separability between the defect and the targeted structure and to achieve robustness to changes in illumination. To find the optimal combination of color space and color distance measurement in identifying deficiencies, the performance of five color representations in tandem with two color distance metrics (e.g., Euclidian and ΔE^*_{94}) was comparatively analyzed.

The presented algorithm has been tested on massive 3D models of a large and complex gravity dam generated by utilizing the acquired image dataset from multiple UAV platforms. The comparative study showed that by implementing illumination invariant non-linear color spaces, the algorithm was capable of providing millimeter-level accuracy in the measurements. Among different color representation, the $L^*a^*b^*$ and *HSI* models consistently performed more robustly in automatically identifying deficiencies with varying dimensions using the Euclidean color distance metric. Future work seeks to improve the accuracy of the developed damage detection technique, by utilizing hybrid color spaces which are composed of different non-linear color representations.

Acknowledgements This material is based upon the work supported by the National Science Foundation (NSF) under Grant No. CMMI-1433765. The authors would also like to acknowledge the support made by NVIDIA Corporation with the donation of a Tesla K40 GPU used in this research. Any opinions, findings, and conclusions or recommendations expressed in this publication are those of the authors and do not necessarily reflect the views of the NSF or NVIDIA.

References

1. Phares, B.M., Washer, G.A., Rolander, D.D., Graybeal, B.A., Moore, M.: Routine highway bridge inspection condition documentation accuracy and reliability. J. Bridg. Eng. **9**, 403–413 (2004)
2. Chang, P.C., Flatau, A., Liu, S.C.: Review paper: health monitoring of civil infrastructure. Struct. Health Monit. **2**, 257–267 (2003)
3. Koch, C., Georgieva, K., Kasireddy, V., Akinci, B., Fieguth, P.: A review on computer vision based defect detection and condition assessment of concrete and asphalt civil infrastructure. Adv. Eng. Inform. **29**, 196–210 (2015)
4. Lattanzi, D., Miller, G.: Review of robotic infrastructure inspection systems. J. Infrastruct. Syst. **23**, 04017004 (2017)
5. Khaloo, A., Lattanzi, D.: Hierarchical dense structure-from-motion reconstructions for infrastructure condition assessment. J. Comput. Civ. Eng. **31**, 04016047 (2016)
6. McGuire, B., Atadero, R., Clevenger, C., Ozbek, M.: Bridge information modeling for inspection and evaluation. J. Bridg. Eng. **21**, 04015076 (2016)
7. Khaloo, A., Lattanzi, D., Cunningham, K., Dell'Andrea, R., Riley, M.: Unmanned aerial vehicle inspection of the Placer River Trail Bridge through image-based 3D modelling. Struct. Infrastruct. Eng. **14**(1), 124–136 (2018)
8. Fathi, H., Dai, F., Lourakis, M.: Automated as-built 3D reconstruction of civil infrastructure using computer vision: achievements, opportunities, and challenges. Adv. Eng. Inform. **29**, 149–167 (2015)
9. Jafari, B., Khaloo, A., Lattanzi, D.: Deformation tracking in 3D point clouds via statistical sampling of direct cloud-to-cloud distances. J. Nondestruct. Eval. **36**, 65 (2017)
10. Pesci, A., Casula, G., Boschi, E.: Laser scanning the Garisenda and Asinelli towers in Bologna (Italy): detailed deformation patterns of two ancient leaning buildings. J. Cult. Herit. **12**, 117–127 (2011)
11. Kashani, A.G., Graettinger, A.J.: Cluster-based roof covering damage detection in ground-based lidar data. Autom. Constr. **58**, 19–27 (2015)
12. Gong, J., Maher, A.: Use of mobile lidar data to assess hurricane damage and visualize community vulnerability. Transp. Res. Rec. J. Transp. Res. Board. **2459**, 119–126 (2014)
13. Riveiro, B., González-Jorge, H., Varela, M., Jauregui, D.V.: Validation of terrestrial laser scanning and photogrammetry techniques for the measurement of vertical underclearance and beam geometry in structural inspection of bridges. Measurement. **46**, 784–794 (2013)
14. Liu, W., Chen, S., Hasuer, E.: Bridge clearance evaluation based on terrestrial LIDAR scan. J. Perform. Constr. Facil. **26**, 469–477 (2012)
15. Tang, P., Huber, D., Akinci, B.: Characterization of laser scanners and algorithms for detecting flatness defects on concrete surfaces. J. Comput. Civ. Eng. **25**, 31–42 (2010)
16. Armesto-González, J., Riveiro-Rodríguez, B., González-Aguilera, D., Rivas-Brea, M.T.: Terrestrial laser scanning intensity data applied to damage detection for historical buildings. J. Archaeol. Sci. **37**, 3037–3047 (2010)
17. Hou, T.-C., Liu, J.-W., Liu, Y.-W.: Algorithmic clustering of LiDAR point cloud data for textural damage identifications of structural elements. Measurement. **108**, 77–90 (2017)
18. Jiang, C., Tsai, Y.J.: Enhanced crack segmentation algorithm using 3D pavement data. J. Comput. Civ. Eng. **30**, 04015050 (2016)
19. Conde, B., Del Pozo, S., Riveiro, B., González-Aguilera, D.: Automatic mapping of moisture affectation in exposed concrete structures by fusing different wavelength remote sensors. Struct. Control Health Monit. **23**, 923–937 (2016)
20. Besl, P.J., McKay, N.D.: Method for registration of 3-D shapes. Presented at the Robotics-DL tentative (1992)
21. Toldo, R., Beinat, A., Crosilla, F.: Global registration of multiple point clouds embedding the Generalized Procrustes Analysis into an ICP framework. In: 3DPVT 2010 Conference (2010)
22. Muja, M., Lowe, D.G.: Fast approximate nearest neighbors with automatic algorithm configuration. VISAPP 1. **2**, 2 (2009)
23. Vezhnevets, V., Sazonov, V., Andreeva, A.: A survey on pixel-based skin color detection techniques. In: Proc. Graphicon, pp. 85–92. Moscow (2003)
24. Cheng, H.-D., Jiang, X.H., Sun, Y., Wang, J.: Color image segmentation: advances and prospects. Pattern Recogn. **34**, 2259–2281 (2001)
25. Son, H., Kim, C., Kim, C.: Automated color model–based concrete detection in construction-site images by using machine learning algorithms. J. Comput. Civ. Eng. **26**, 421–433 (2012)
26. Siegel, M., Gunatilake, P.: Remote enhanced visual inspection of aircraft by a mobile robot. In: Proc. of the 1998 IEEE Workshop on Emerging Technologies, Intelligent Measurement and Virtual Systems for Instrumentation and Measurement (ETIMVIS'98), pp. 49–58 (1998)
27. Gonzalez, R.C., Woods, R.E.: Digital Image Processing. Prentice Hall, New Jersey (2008)
28. Szeliski, R.: Computer Vision. Springer London, London (2011)
29. Westland, S., Ripamonti, C., Cheung, V.: Computational Colour Science Using MATLAB. Wiley, West Sussex (2012)
30. Men, H., Pochiraju, K.: Hue-assisted automatic registration of color point clouds. J. Comput. Des. Eng. **1**, 223–232 (2014)
31. Choi, K.Y., Kim, S.S.: Morphological analysis and classification of types of surface corrosion damage by digital image processing. Corros. Sci. **47**, 1–15 (2005)
32. Aligholi, S., Lashkaripour, G.R., Khajavi, R., Razmara, M.: Automatic mineral identification using color tracking. Pattern Recogn. **65**, 164–174 (2017)
33. Wesolkowski, S.: Color image edge detection and segmentation: a comparison of the vector angle and the euclidean distance color similarity measures. https://uwspace.uwaterloo.ca/handle/10012/937 (1999)

Chapter 27
Human Activity Benchmark Classification Using Multilayer Artificial Neural Network

Ramin Madarshahian, Juan M. Caicedo, and Nicholas Haerens

Abstract Human induced floor vibrations have recently been proposed to track human activity for a number of applications such as health care and surveillance. For example, floor vibrations can be used to identify human falls, or the existence of an intruder in a room. In these applications, the acceleration signals should be classified accurately to eliminate false positives. In this paper, a multi-layer artificial neural network is used to classify floor vibrations. Data from a previously published benchmark problem, which consists of seven types of human activities, is used to train and test the algorithm. Results show the capabilities of a multilayer artificial neural network in human activity classification.

Keywords Deep learning · Floor vibration · Human activity · Classification

27.1 Introduction

Human induced vibrations in a structure are easily measured using piezoelectric accelerometers [1–5]. Davis et al. [1] used wireless sensors to collect human activity induced vibrations. They simulated different types of human activity such as people walking, ball drop, Vacuuming, Fall - Butt to Back, and Fall - Knee on a floor, of a two story wood building. Focusing on fall activities, they illustrated that fall acceleration records are noticeably different than other types of activities. Madarshahian et al. designed an experiment to study the effects of different factors, such as source distance to the sensor, as well as type and severity of human induced activity on a classification algorithm [6]. In extension of their work, they gathered acceleration records of seven types of activity at five different locations of a lab room to provide a benchmark for developing and testing new proposed human activity classification algorithms [7]. The benchmark also suggested seven cases of different difficulties to let researchers examine their proposed classification algorithms in different scenarios. For example, Madarshahian and Caicedo utilized the collected data to introduce an application of logistic regression algorithm on human activity recognition [2]. The classification algorithm used by them was able to assign a probability to different predefined human activity candidates. These probabilities are helpful in selecting the most competing candidates amongst all candidates in the pool of human activities. In this work, applications of a neural network on classifying the human activity signals are studied. First a one-layer neural network is trained and its capability to distinguish the signals is investigated, then the results are compared with the results of a two-layer neural network. It is shown that a multilayer artificial neural network (MANN) with only two layers can classify the testing data efficiently. The results suggest that MANN with more layers can potentially be used when bigger size data with more different human activities is used.

27.2 Data Structure

The benchmark developed by Mardarshahian et al. is used as source of the data for this paper [7]. The benchmark has two sets of data, one for training and one for testing. Each set contains seven types of human activities which are repeated in five specified locations of a lab room and are recorded by four sensors. Readers are encouraged to read the details of the experiments in [7]. In this paper, only the data related to three types of activity and one location of excitation are considered. Acceleration signals are related to dropping a ball and a bag of K'nex pieces on a floor, and a human jump, each repeated 100 times for training data and 15 times for the testing data. Figure 27.1 shows an example of recorded signal for ball and

R. Madarshahian · J. M. Caicedo (✉) · N. Haerens
University of South Carolina, Columbia, SC, USA
e-mail: mdrshhn@email.sc.edu; caicedo@cec.sc.edu

S. Pakzad (ed.), *Dynamics of Civil Structures, Volume 2*, Conference Proceedings of the Society for Experimental Mechanics Series, https://doi.org/10.1007/978-3-319-74421-6_27

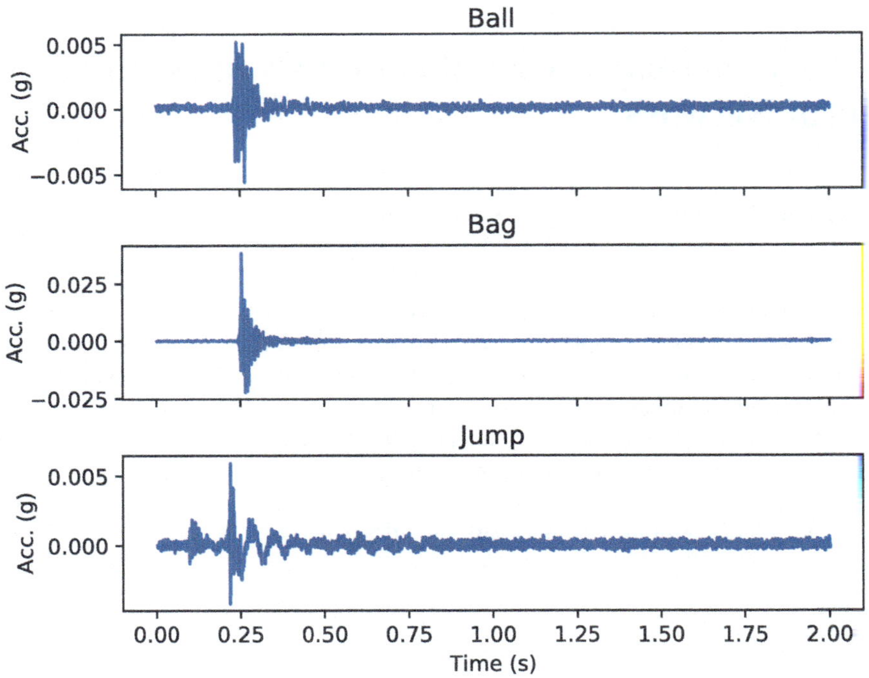

Fig. 27.1 Typical signals in the dataset

bag drop and human jump. As seeing in this figure the characteristics of each type of activity results in a different floor acceleration profile. For example, the acceleration signals of the jump appear to have a lower frequency content than the other two signals. The classification algorithm is designed to identify the type of each human activity when a recorded signal is fed to the algorithm as an input.

27.3 Neural Network

Two Neural networks are designed and compared using the Tensorflow package of Python [8]. In the first one, only one layer is used with 3305 neurons (input samples) that are connected to 3 neurons (three cases i.e. Ball, Bag, Jump). The Softmax function, defined as a generalized logistic function with values between zero and one, is used as an activation function. The model in the Neural network is defined by Eq. 27.1:

$$C = \text{softmax}(S.W + B) \tag{27.1}$$

where S denotes input samples and C denotes related cases for an input signal. W and B are weights and biases, respectively. Figure 27.2 shows the diagram of the first neural network. In the second Neural network, a hidden layer is added to the network. The new layer contains 200 Neurons. Here, a Relu activation function [9] is used as an activation function between inputs and the hidden layer, while softmax function is kept for the last layer. Softmax output, a number between zero and one, can be interpreted as a probability for case prediction. Figure 27.3 shows a diagram representing the second Neural network. For each network, 300 training signals are used in 10,000 batches of 100. Signals in each batch are selected without replacement from the training set of signals. Gradient Descent Optimizer [10] with a decayed learning rate is used to train the data.

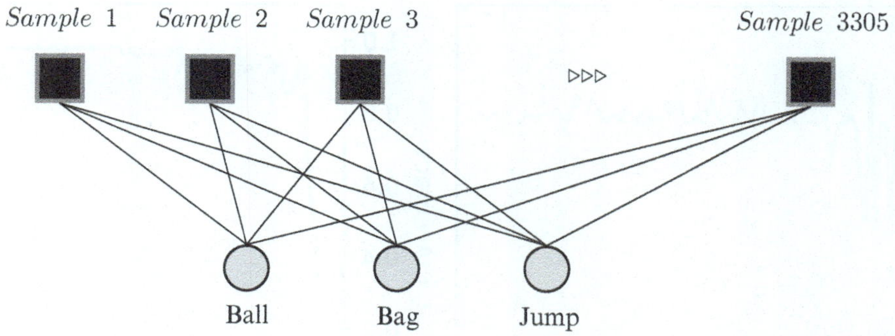

Fig. 27.2 The first neural network

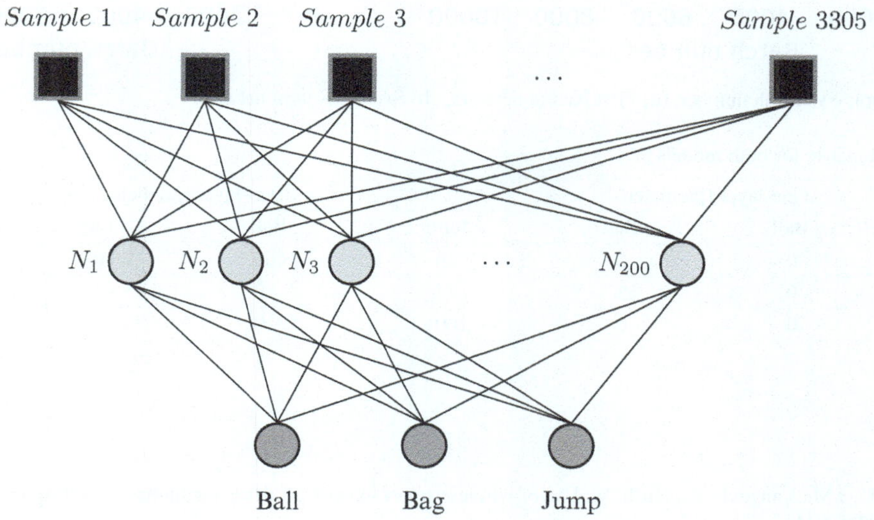

Fig. 27.3 The second neural network

27.4 Results

When the first Neural network is used, an accuracy of 0.78 is obtained after 10,000 batches are used for training. This accuracy has a considerable amount of false positives predictions which is not favorable but it is better that a random choice (0.33). Adding one more layer increases the accuracy to 0.98. Figure 27.4 shows how the accuracy in testing data is changed by the number of batches in each Neural network. Table 27.1 shows a confusion matrix detailing the results for each network. In the first one 20% of "Ball" signals are wrongly classified as "Jump", and 47% of "Bag" signals are classified as "Jump". By adding one layer to the network, false positives are eliminated for the "Ball" signals and reduce to 7% (1 out of 15 testing signals) for the "Bag" signals. All "Jump" signals are correctly classified as well.

27.5 Concluding Remarks

In this paper, two neural networks have been used to classify human induced activities. In the data set, three types of human activities were considered including dropping a bag of K'nex legos and a ball on the floor and the jump of a person. The first neural network contains one layer, and resulted in an accuracy of 0.78. Many "Ball" and "Bag" related signals are wrongly classified as signals created by human jumps. Using a two-layer network improved the results considerably, such that all of the "Ball" signals are classified correctly and only one signal of the "Bag" case is wrongly classified as "Jump". This paper shows that multilayer neural networks have the potential to be used as a strong classification tool for classifying human induced activities.

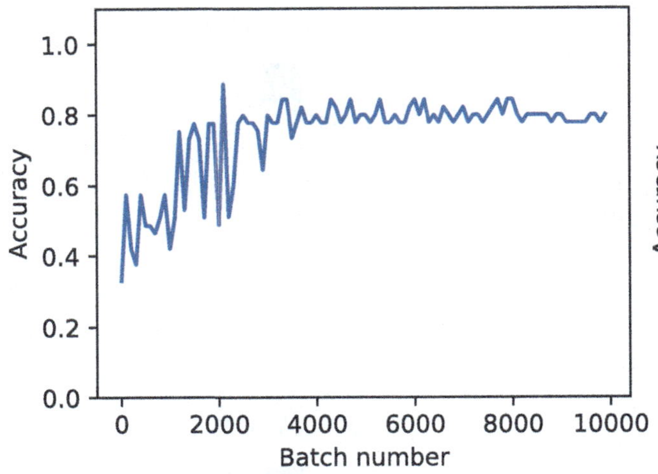

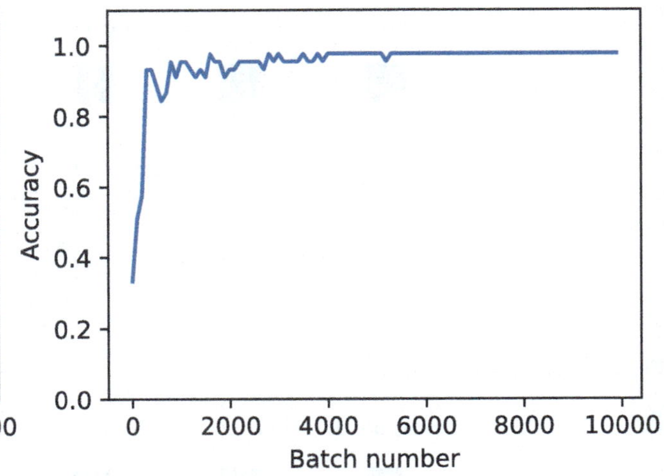

Fig. 27.4 Testing accuracy vs batch number. (**a**) First Neural network. (**b**) Second Neural network

Table 27.1 Confusion matrix for both models (%)

	One-layer (predicted)			Two-layer (predicted)		
	Ball	Bag	Jump	Ball	Bag	Jump
Ball (Actual)	80	0	20	100	0	0
Bag (Actual)	0	53	47	0	93	7
Jump (Actual)	0	0	100	0	0	100

References

1. Davis, B.T., Caicedo, J.M., Langevin, S., Hirth, V.: Use of wireless smart sensors for detecting human falls through structural vibrations. Civ. Eng. Top. **4**, 383–389 (2011)
2. Madarshahian, R., Caicedo, J.M.: Human Activity Recognition Using Multinomial Logistic Regression, pp. 363–372. Springer, Cham (2015)
3. Pan, S., Bonde, A., Jing, J., Zhang, L., Zhang, P., Noh, H.Y.: Boes: building occupancy estimation system using sparse ambient vibration monitoring. In: SPIE Smart Structures and Materials+ Nondestructive Evaluation and Health Monitoring, pp. 90611O–90611O (2014)
4. Pan, S., Wang, N., Qian, Y., Velibeyoglu, I., Noh, H.Y., Zhang, P.: Indoor person identification through footstep induced structural vibration. In: Proceedings of the 16th International Workshop on Mobile Computing Systems and Applications, pp. 81–86. ACM (2015)
5. Mirshekari, M., Pan, S., Zhang, P., Noh, H.Y.: Characterizing wave propagation to improve indoor step-level person localization using floor vibration. In: SPIE Smart Structures and Materials+ Nondestructive Evaluation and Health Monitoring, pp. 980305–980305 (2016)
6. Madarshahian, R., Caicedo, J.M., Zambrana, D.A.: Evaluation of a time reversal method with dynamic time warping matching function for human fall detection using structural vibrations. In: Model Validation and Uncertainty Quantification, vol. 3, pp. 171–176. Springer, Cham (2014)
7. Madarshahian, R., Caicedo, J.M., Zambrana, D.A.: Benchmark problem for human activity identification using floor vibrations. Expert Syst. Appl. **62**, 263–272 (2016)
8. Abadi, M., Agarwal, A., Barham, P., Brevdo, E., Chen, Z., Citro, C., Corrado, G.S., Davis, A., Dean, J., Devin, M., et al.: Tensorflow: large-scale machine learning on heterogeneous distributed systems. arXiv preprint arXiv:1603.04467 (2016)
9. Nair, V., Hinton, G.E.: Rectified linear units improve restricted Boltzmann machines. In: Proceedings of the 27th International Conference on Machine Learning (ICML-10), pp. 807–814 (2010)
10. Bottou, L.: Stochastic gradient learning in neural networks. In: Proceedings of the Neuro-Nimes'91 (1991). http://leon.bottou.org/publications/pdf/nimes-1991.pdf

Chapter 28
Cracking Influence on Dynamic Parameters of Reinforced Concrete Floors

William Ferreira Miranda, Suzana Moreira Avila, and Graciela Nora Doz

Abstract In view of the need to satisfactorily provide good performance of structures in their service situations, it is desired to understand the influence of some reinforced concrete particularities on the dynamic parameters of the structure. These particularities are fundamental in static analysis of pavements design. Starting from spans, geometry, and structural typologies common in real buildings, this paper studies the influence of cracking and non-constant reinforcements on the vibration frequency of concrete floors. It is shown that the reinforced concrete properties significantly impact the floor's dynamic parameters; this impact is stronger when the floor is slender and deformable. Some of the common structural typologies, as the ribbed/waffle slabs, are even more susceptible to those influence. Knowing the main variables that may be responsible for excessive vibration phenomenon, it was possible to prevent issues during the design process.

Keywords Dynamics · Floor vibrations · Reinforced Concrete · Cracking · Serviceability

28.1 Introduction

Structural engineering have provided long-span concrete floors with low dimensions, that make them even more slender and susceptible to excessive vibration.

Pavic [4] says these excessive vibrations problems are becoming more common due to the lack of research and scientific interest in the performance of this type of pavements, since concrete floors are usually known for their good performance against vibrations in service conditions.

Most of the literature's research focus on controlling vibrations of existent buildings or on understanding the general phenomena. To the best of our knowledge, little is said about real concrete structures particularities and their additional variables, such as loads, steel reinforcements, cracking and long-term deformations.

Carmona [2] emphasizes his concern with the cracking occurred in a test platform during its construction when the shoring was removed. Great effort was made to repair these cracks and try to bring the structure back to the expected natural frequency, since it was found (both experimentally and in numerical simulations) that the natural frequency of the platform reduced significantly during the cracking.

In order to fill this gap, this paper analyzes the influence of concrete floors proprieties such as cracking and non-constant reinforcements on the dynamic parameters of the pavements. Previously acquired numerical data of intact concrete floors are compared with syntactic data generated in new numerical simulations that consider those proprieties.

28.2 Methodology

The data used on the present study were collected from modeling and processing several concrete floors on the TQS (18.18.14 version). TQS is a commercial software, focused on stress analysis, deflections, dimensioning and verification of concrete structures [5].

The software was set to use a grillage analysis process to determine the natural frequencies of the pavement, which was composed by beams and slabs. Both were considered simultaneously [4].

W. F. Miranda (✉) · S. M. Avila · G. N. Doz
Universidade de Brasília – UnB, Brasília, DF, Brazil
e-mail: avilas@unb.br; graciela@unb.br

S. Pakzad (ed.), *Dynamics of Civil Structures, Volume 2*, Conference Proceedings of the Society for Experimental Mechanics Series, https://doi.org/10.1007/978-3-319-74421-6_28

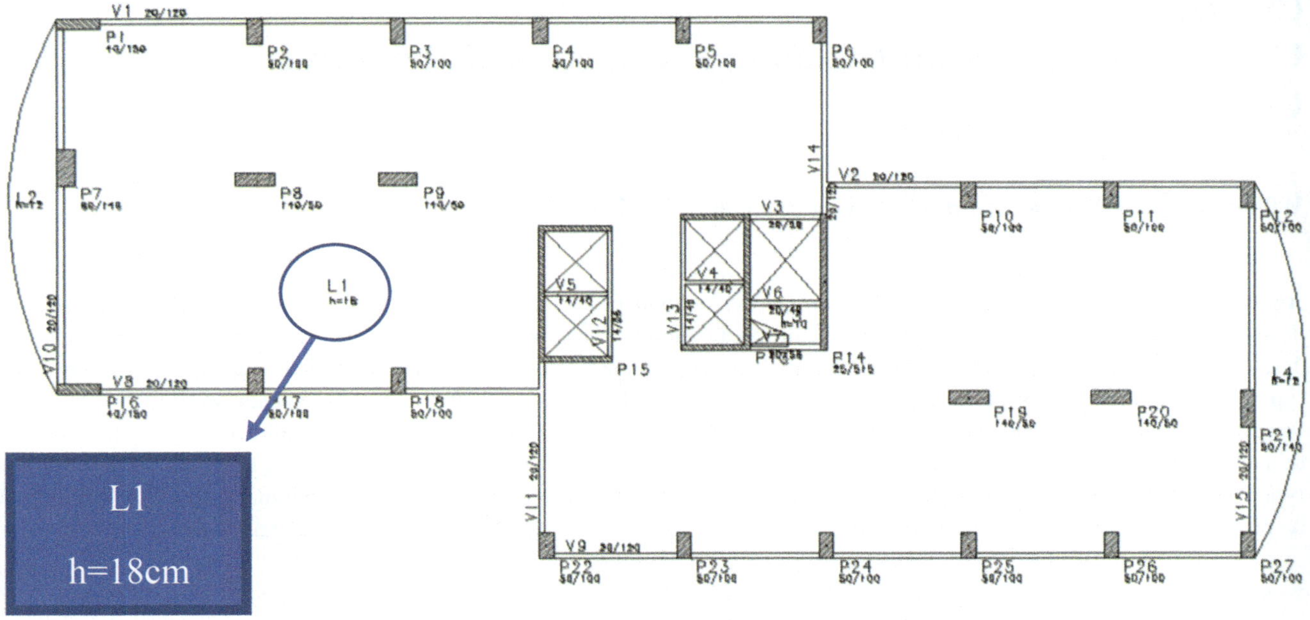

Fig. 28.1 Structural typology of main analyzed concrete floor (CILAMCE 2017)

Starting from previous parametric analysis, where the bending inertia properties were only defined by the linear geometric and material properties of the structural elements, this work studies the effects of concrete cracking and non-constant reinforcements in the floor's sections upon the first modal frequencies.

The algorithm calculates the stress caused on the thousands of the discrete sections and compares with the cracking limit of the concrete. When this limit is exceeded, the software overwrites the inertia of the corresponding section by a lower value calculated using the Branson's technique [1]. It also calculates the inertia increase due to the steel reinforcements that should be added on each section to resist the static loads.

The steel reinforcements where determined considering all the static loads suggested by the Brazilian regulation Association – ABNT [1] to an office building, while the cracking took only 60% of the non-permanent load (in addition to all the permanent ones), in order to simulate an average common situation.

Figure 28.1 shows one of the analyzed pavements. The pillars and spans were kept constants in all models, while varying the structural typology and/or slab thickness.

28.3 Results and Discussion

In order to access the influence of the cracking effect and non-constant steel reinforcements on concrete floors where pillars support the slabs directly (without beams, besides the ones on the contour), the model presented in Fig. 28.1 was calculated varying the thickness of the main slab. The frequency of the first vibration modes for each thickness are shown on Table 28.1.

Note that the vibration modes are not the same during the intact slab processing and when the slab was considered cracked. Figure 28.2 shows these differences.

Since the vibration modes are not the same between the presented situations, the frequency changes need to be carefully analyzed. On the other hand, the natural frequency, who is related to the first vibration mode, is very similar in both cases.

As expected, if the thickness of the main slab increases, the influence of the cracking analysis reduces. However, in the thinner slabs, where greater deformability and cracking are expected, the natural frequency reduction was significant. According to the Brazilian regulations [1], the main serviceability vibration check is based on the comparison between the natural frequency of the floor (without any cracking or reinforcements considerations) and limits that depends on the building's occupation. These limits could be overpassed when cracking is considered.

Figure 28.3 shows the grid of elements before and after the cracking analysis. The red colored bars represent sections which reached concrete's cracking limit. The white bars were also tested, but they did not reach the cracking criteria. Mesh density can slightly change bar colors by superposition (white to grey).

Table 28.1 Influence of cracking and reinforcements in floor frequency

Thickness of the main slab	Vibration mode	Frequency of intact concrete situation (Hz)	Frequency of cracked and reinforced situation (Hz)	Frequency change (%)
h = 18 cm	1°	4821	4367	−9,4
	2°	5228	4767	−8,8
	3°	5385	4915	−8,7
	4°	5529	5110	−7,6
	5°	5816	5629	−3,2
h = 20 cm	1°	5430	5150	−5,2
	2°	5885	5595	−4,9
	3°	5983	5820	−2,7
	4°	6195	6023	−2,8
	5°	6524	6515	−0,1
h = 22 cm	1°	6036	5947	−1,5
	2°	6531	6472	−0,9
	3°	6598	6655	0,9
	4°	6822	6837	0,2
	5°	7223	7317	1,3
h = 24 cm	1°	6648	6693	0,7
	2°	7165	7239	1,0
	3°	7227	7432	2,8
	4°	7471	7556	1,1
	5°	7925	8096	2,2

Taking a close look at one of the midspan bars, it is possible to study the variation in the bending moment inertia. In one of the bars, for example, the intact concrete inertia was 9,72E−5 cm^4, increasing up to 10,54E−5 cm^4 when considering the steel reinforcements, and reducing to 4,29E−5 cm^4 (−56%) during the cracked floor analysis. This little inertia increasing due to steel reinforcements explains why some frequency were higher in models where the cracking effect was nearly irrelevant.

Similar procedures were applied to other structural typologies. The ribbed/waffle slab system – Figs. 28.4 and 28.5 – shows even more susceptibility the inertia reducing effect due to cracking analysis.

Using previous pavements general geometry, a ribbed slab was modeled in order to fully attend the static requirements. Then, the dynamic parameters were studied. Massive concrete parts, where the pillars meet the main slab, had the same slab thickness (24 cm).

The first vibration mode is quite similar to the Fig. 28.1 slab, but its natural frequency during the intact concrete analysis was higher: 5.441 Hz.

When the steel reinforcements and the cracking effect were taken into account, the natural frequency reduced to 4.614 Hz (−15,2%).

Other vibration modes have shown even more variation between the intact and cracked analysis, but the modes correlations are not so direct due to their shape differences.

28.4 Conclusion

In this work, it is shown that the reinforced concrete proprieties, known to affect static analysis such as long-term deformations, also impact on the dynamic properties of concrete floors. When the studied floor is slender, thin and deformable, these effects are more evident. Some of the common structural typologies are even more susceptible to those influences.

Although this study focusses on real dimensions of concrete floors and structural elements, many variables (concrete properties, spans, pillar arranges, gaps, ...) can affect other results and simulations. Even loads, that normally do not influence natural frequencies, need to be considered as a variable since it will increase cracking and reinforcements, and compose the floor mass (self-weight is one of the main loads in concrete floors).

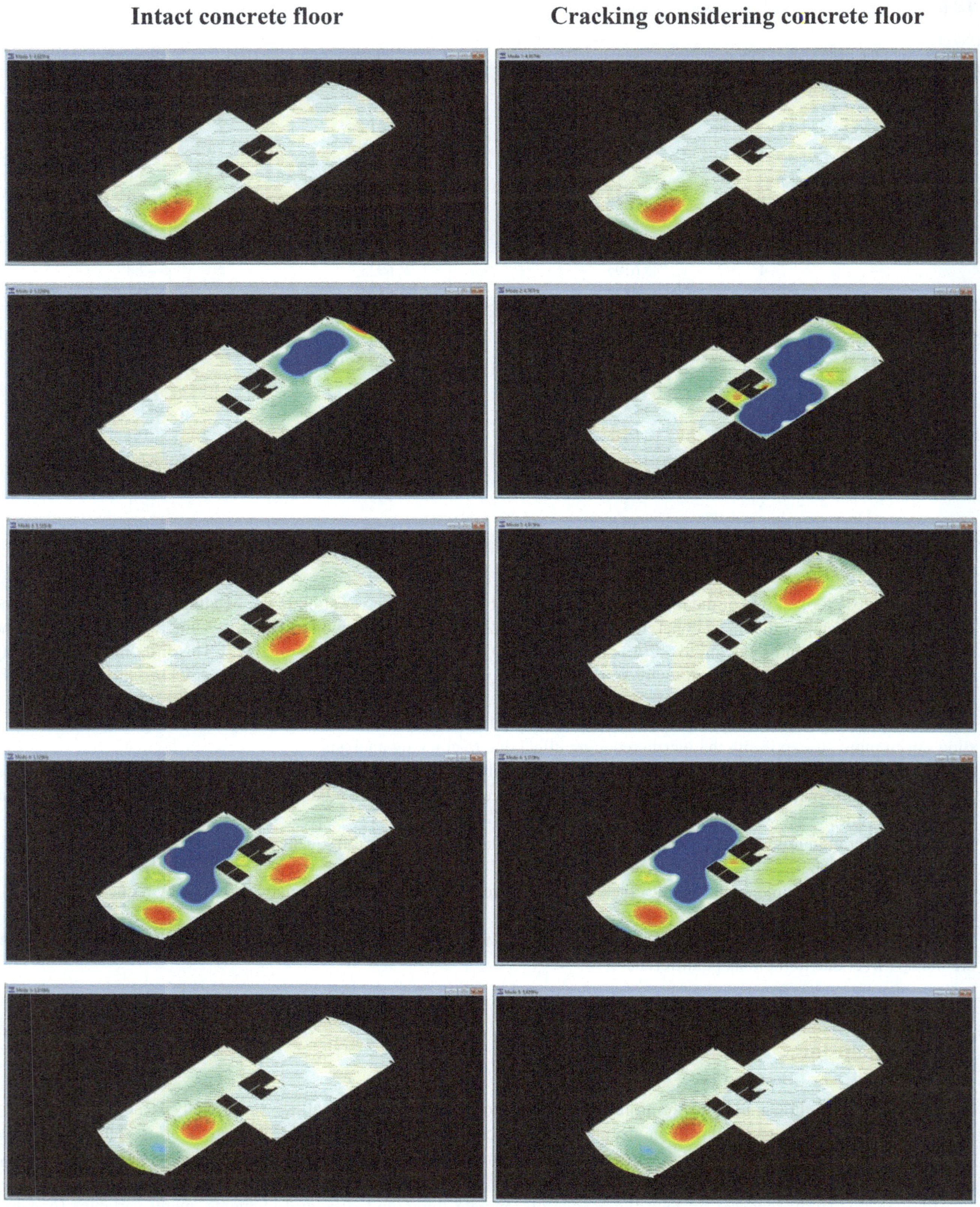

Fig. 28.2 First vibration modes for both intact and cracked cases (slab thickness 18 cm)

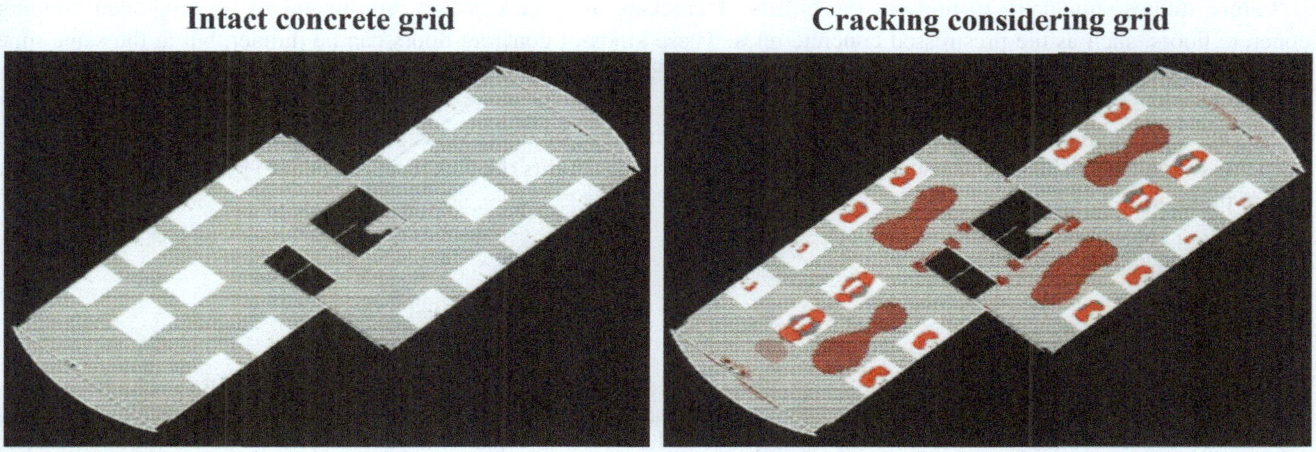

Fig. 28.3 Mesh grid for both intact and cracked cases (slab thickness 18 cm)

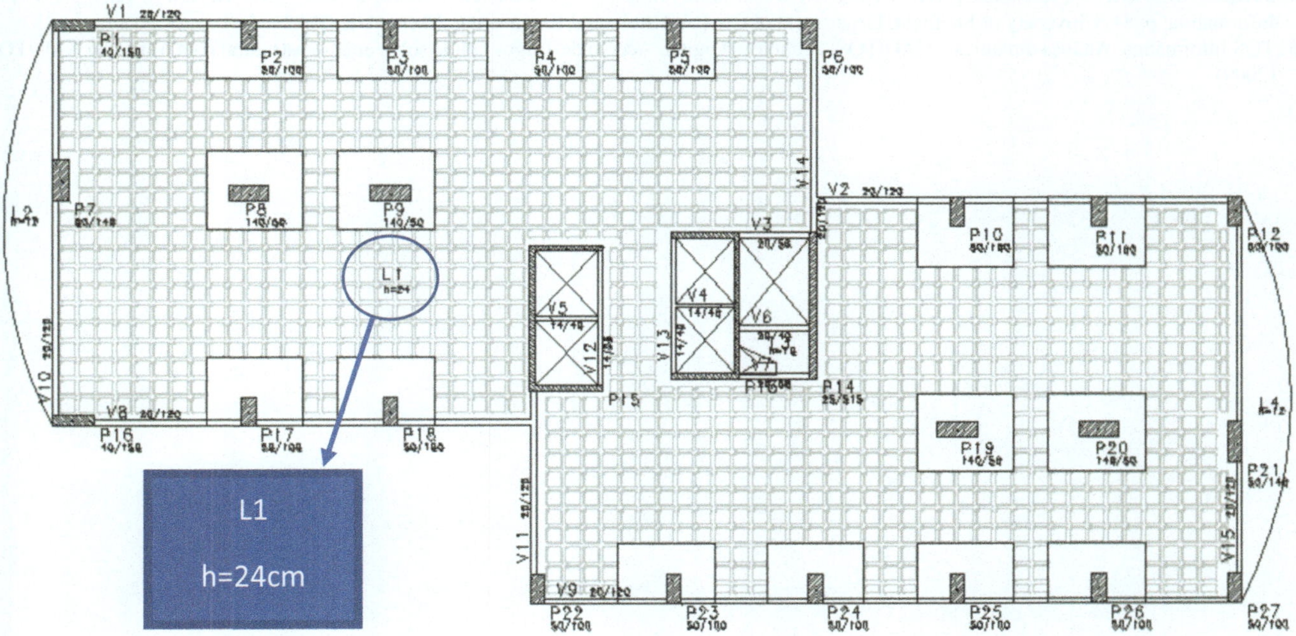

Fig. 28.4 Ribbed/Waffle main slab (CILAMCE 2017)

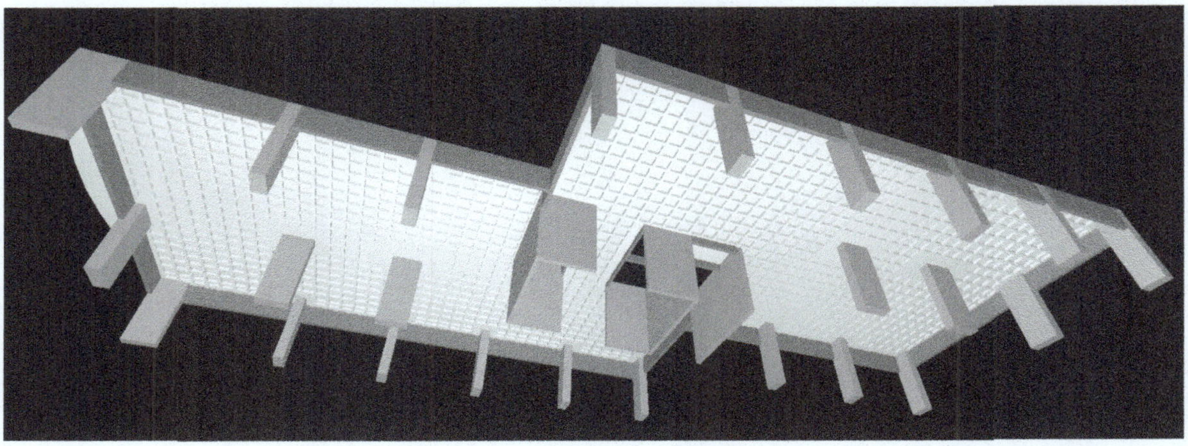

Fig. 28.5 Ribbed/Waffle main slab lower perspective view (CILAMCE 2017)

Future studies should try to find out the effects of cracking and non-constant reinforcements on long-span building concrete floors, such as the prestressed concrete ones. These kinds of concrete floors can be thinner, but at the same time the prestressed effect helps to control the concrete cracking. Experimental analysis would also help the understanding of the phenomena.

Vibration problems on concrete floors have not been a great issue on the past decades, however, studies show that it has been growing more frequent, and understanding the main variables of this phenomenon can help to prevent them during the design process.

References

1. Associação Brasileira de Normas Técnicas: NBR 6118: Projeto e Execução de estruturas de Concreto, Rio de Janeiro, Brasil (2014)
2. Carmona, J.E.: Proposta de um amortecedor para atenuação de vibrações em pisos de concreto: análise numérica e experimental, 287p. Tese de Doutorado em Estruturas e Construção Civil, Publicação E.TD-1A/16, Departamento de Engenharia Civil e Ambiental, Universidade de Brasília, Brasília, DF (2016)
3. Carvalho, R.C.: Análise não-linear de pavimentos de edifícios de concreto através da analogia de grelha, 208p. Tese de Doutorado em Engenharia, Departamento de Estruturas, Escola de Engenharia de São Carlos, Universidade de São Paulo, SP (1994)
4. Pavic, A.: In: Pavic, A., Reynolds, P. (eds.) Vibration Serviceability of Long-Span Concrete Building Floors. Part 1: Review of Background 112 Information, p. 81. University of Sheffield, Department of Civil and Structural Engineering, Exeter, UK (2002)
5. TQS Informática: Análise dinâmica – CAD/TQS. Biblioteca Digital. Acervo de artigos online sobre análise estrutural com o sistema CAD/TQS (2009)

Chapter 29
Paradigm Shift in Structural Vibration Serviceability: New Assessment Framework Based on Human's Experience of Vibration

Erfan Shahabpoor, Aleksandar Pavic, Vitomir Racic, and Hootan Rezaei

Abstract Reliable assessment of structural vibration serviceability during the design process is still a great challenge for the designers of pedestrian structures, such as footbridges and floors. Witness to this is the report of the UK Institution of Structural Engineers that approximately half of its 27,000 members, worldwide, have dealt with vibration serviceability complains related to the code-compliant designs. Although structures are meant to be designed to provide function/comfort for human users, evaluation of the 'experience' of the human users is conspicuously absent from structural design guidelines. This paper highlights the distinctive features of the Interaction-based Vibration Serviceability Assessment (I-VSA) method, proposed by the authors, and compares the results of the I-VSA with those of the current guidelines for two full-scale structures. It further proposes that: (1) the level of vibrations received by human users is a significantly more informative design parameter than maximum response levels at a certain locating on the structure, which may or may not be experienced; and (2) a deep understanding of the 'perception' of vibration by humans is needed to link the level of vibrations received by the occupants with their 'experience' from this vibration.

Keywords Human-structure interaction · Vertical vibrations · Walking forces · Perception · Experience

29.1 Introduction

Models of pedestrian dynamic loading used in contemporary vibration serviceability assessment (VSA) typically describe the vertical walking excitation as a vertical force that does not depend on structural vibrations [1]. The simplest models, such as those presented by FIB [2], ISO 10137 [3], French design guideline [4] and UK National Annex to Eurocode 1 [5], also assume that an individual walking force is periodic and presentable by a Fourier series. The frequency content of such a simple force model typically contains up to the first four dominant harmonics [1]. The design procedures usually require that one of the harmonics matches the frequency of a target vibration mode of the structure to create resonance, i.e. the worst-case scenario yielding the maximum vibration response. To account for the imperfect synchronisation between individuals in a group or crowd, the walking force of a multi-pedestrian traffic is calculated by multiplying a sum of the individual forces with factor(s) which commonly depend only on the number of pedestrians on the structure [1].

The UK recommendations for the design of permanent grandstands [6] are the only guidelines that explicitly require taking into account the interaction of both passive and active people with the grandstand they occupy and excite by jumping or bouncing in the vertical direction. Despite the satisfactory performance of this explicit modelling approach [7, 8], no other vibration serviceability design guideline has yet adopted a similar modelling concept to account for the HSI due to people walking.

E. Shahabpoor (✉)
Department of Architecture and Civil Engineering, University of Bath, Bath, UK
e-mail: e.shahabpoor@bath.ac.uk

A. Pavic
College of Engineering, Mathematics and Physical Sciences, University of Exeter, Exeter, UK

V. Racic
Department of Civil and Environmental Engineering, Politecnico di Milano, Milan, Italy

H. Rezaei
Department of Civil and Structural Engineering, Baha'i Institute for Higher Education (BIHE), Tehran, Iran

S. Pakzad (ed.), *Dynamics of Civil Structures, Volume 2*, Conference Proceedings of the Society for Experimental Mechanics Series, https://doi.org/10.1007/978-3-319-74421-6_29

Recently, the authors suggested a paradigm shift in VSA by proposing the interaction-based VSA (I-VSA) method [9]. This method explicitly addressed five main challenges of VSA when assessing the effects of walking people on structures:

1. The human-structure dynamic interaction;
2. Variability of the mass, stiffness and damping of the moving human body and the walking force due to inter- and intra-subject variability;
3. Variability of pedestrian traffic characteristics, such as traffic volume and regime (spatially unconstraint/constraint, group, etc.)
4. Varying location of each walking pedestrian on the structure, and
5. The actual level of vibration received by each pedestrian at their continuously moving location on the structure rather than the vibration response of the structure at a fixed point.

This paper reports the results of applying this methodology on two full-scale structures (Sect. 29.2) and compares these results with those of a selection of current design guidelines (Sect. 29.3). In Sect. 29.4 the 'experience' of human occupants from structural vibrations is proposed as the criteria for the next generation of VSA methods and the distinctive capabilities of the I-VSA method in realistic estimation of this experience is highlighted. Finally, the conclusions are made in Sect. 29.5.

29.2 VSA of Full-Scale Structures

A set of tests was carried out on two full-scale footbridges: a post-tensioned concrete footbridge at the University of Sheffield and a steel box girder footbridge located in Podgorica, capital of Montenegro [10]. The modal frequency, damping ratio and modal mass of the first vertical mode of the Sheffield footbridge are: 4.44 Hz, 0.6% and 7128 kg, respectively [11]. For the Podgorica footbridge these parameters are: 2.04 Hz, 0.26% and 58,000 kg, respectively [10]. Both structures are very lightly damped, and have natural frequencies in the range excitable by walking forces. In this study, only the first vertical bending mode of vibration was considered for both footbridges.

Three tests were carried out on the Sheffield footbridge with three (Test 1), six (Test 2) and ten (Test 3) pedestrians walking in a closed-loop path along the full length of the footbridge [11]. The participants were asked to walk at their normal speed and they were free to pass each other. In each test, the body mass, and time histories of structural response, walking force, direction and speed and the real-time location of each pedestrian were measured. Detailed descriptions of the tests and statistical analyses of traffic parameters are presented elsewhere [12]. Similarly, three monitoring tests, referred to as Tests 4, 5 and 6, each lasting 44 min, were carried out on the Podgorica footbridge, Montenegro, under normal pedestrian traffic. The flow of traffic was recorded using two video cameras located at both ends of the footbridge and synchronised with the recorded acceleration response. Pedestrians' crossing time, average speed and pacing frequency and the number of people on the structure at any particular moment were found using the time-stamped video footage [13]. Detailed descriptions of the tests and statistical analyses of traffic parameters are presented in Ref. [13].

29.3 Comparison with Current Design Guidelines

The performance of the interaction-based VSA method was compared with a number of the relevant design guidelines: ISO 10137 standard [3], French road authorities standard [4], UK National Annex to Eurocode 1 [5] and a method proposed by Butz [14]. For each test, the input parameters of the design guidelines were selected in a way to simulate as best as possible (within the provision of the guidelines) the corresponding walking traffic. The extensive discussion of the selected guidelines and their shortcomings were presented by Shahabpoor and Pavic [15] and Zivanovic, et al. [16] and are not repeated here.

Setra and Butz methods use response amplitude, with 95% probability of non-exceedance ($a_{95\%}$) for assessment. ISO uses peak response and UK NA suggests a mean response plus 2.5 times standard deviation ($a_{2.5\sigma}$) for a serviceability assessment. The results of the interaction-based VSA method were calculated at the anti-node response to be able to compare them with the results of the selected guidelines. The interaction-based VSA method results were also compared with their non-interactive counterparts for all tests. As can be seen in Fig. 29.1, the accuracy of the interaction-based VSA method in predicting structural response is considerably higher than all other methods in all six tests. Comparing like with like, Setra, ISO, UK NA and Butz methods have a 300–700%, 200–500%, 100–400% and 50–100% error in estimating structural response, respectively. This error range is 100–200% for the non-interactive method. In comparison, the interaction-based VSA method results show a maximum 10% error in estimating $a_{95\%}$, $a_{2.5\sigma}$ and a_{rms} and a maximum 30% error in estimating peak acceleration a_{peak}.

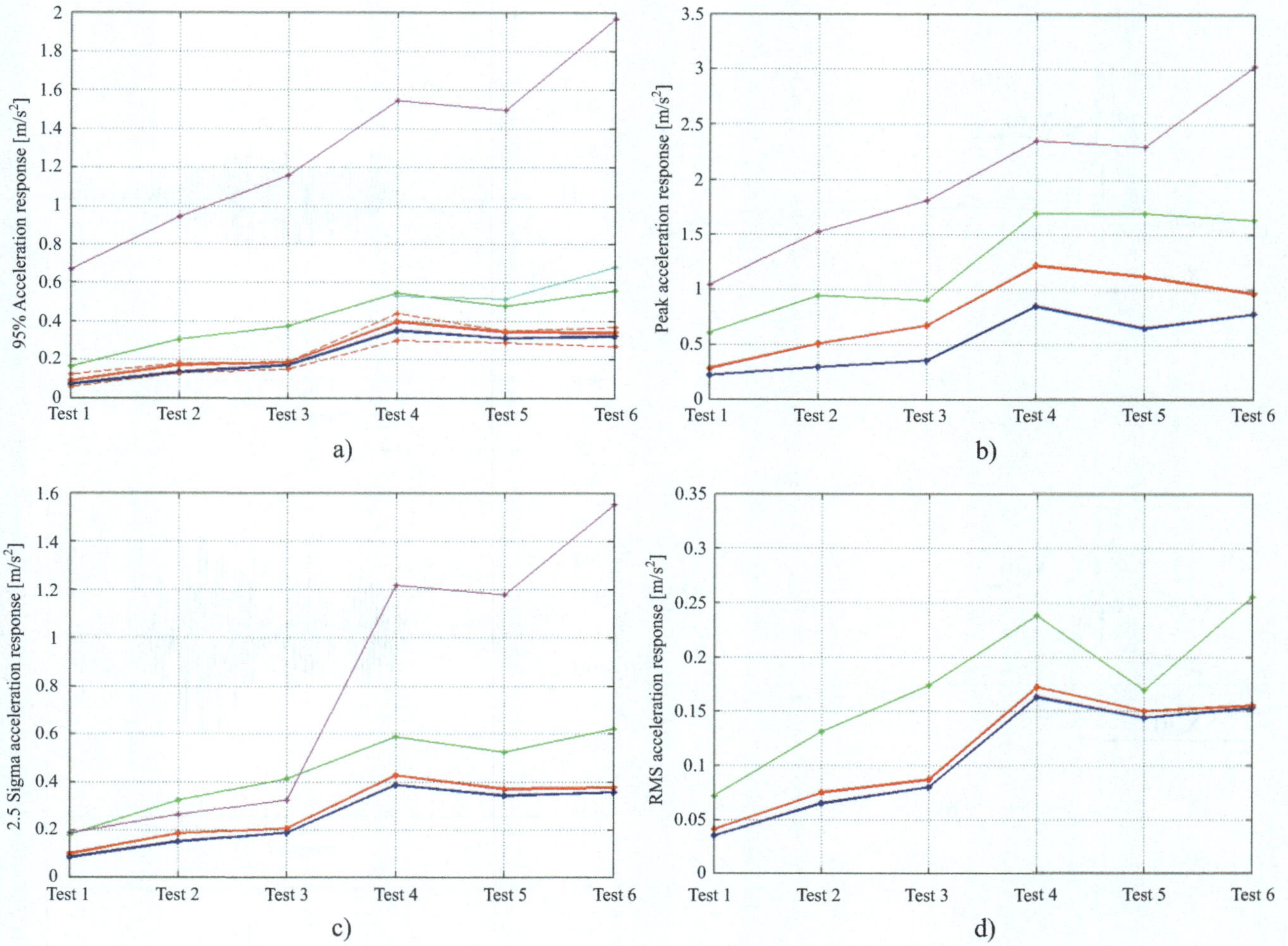

Fig. 29.1 Comparison of performance of the interaction-based VSA method with non-interactive, ISO, UK National Annex, Setra and Butz assessment methods. (**a**) Comparison of acceleration response with 95% probability of non-exceedance – experimental (blue), interactive-stabilized (solid red), interactive- max and minimum on envelope (dashed red), non-interactive (green), Setra (magenta) and Butz (cyan), (**b**) comparison of peak acceleration response – experimental (blue), interactive (red), non-interactive (green) and ISO (magenta), (**c**) comparison of acceleration response with $\mu + 2.5\sigma$ probability of non-exceedance – experimental (blue), interactive (red), non-interactive (green) and UK NA (magenta), (**d**) comparison of acceleration response RMS – experimental (blue), interactive (red), and non-interactive (green)

29.4 Human *Experience* of Structural Vibration

Going back to the basics, the ultimate goal of designing a structure is to maximize its serviceability to the intended users. Human occupants are the main users of most of the civil structures such as buildings, footbridges and stadia. However, the chronic lack of understanding of how human interacts with its surrounding environment and structures and how they perceive vibration, prevents reliable VSA of structures, and is deemed responsible for the failure of current design guidelines in providing a reliable VSA of structures. To address this problem, it is proposed that, rather than assessing the maximum structural response at a specific location on the structure, the new generation of the VSA methods must focus on assessment of a more relevant criteria, such as 'experience' of human users from structural vibrations. This entails, among other factors, taking into account (1) the way human occupants interact with their surrounding environment (usage pattern) and structure (HSI), (2) the actual level of vibration received by each pedestrian at their continuously moving location on the structure, and (3) the relation between the received level of vibrations and human 'experience' of vibration which in turn entails knowing the human perception of vibration for different activities and postures, age cohorts, health status, etc.

The I-VSA method has unique capabilities to address these challenges as it takes into account the individualised interaction of each pedestrian with the vibrating structure and surrounding environment. It further includes a realistic and versatile methodology, based on the concept of modal force, to estimate the level of vibration each pedestrian receives while in/on the structure (Fig. 29.2). The key missing component is a model of human 'perception' of vibration to link the time-

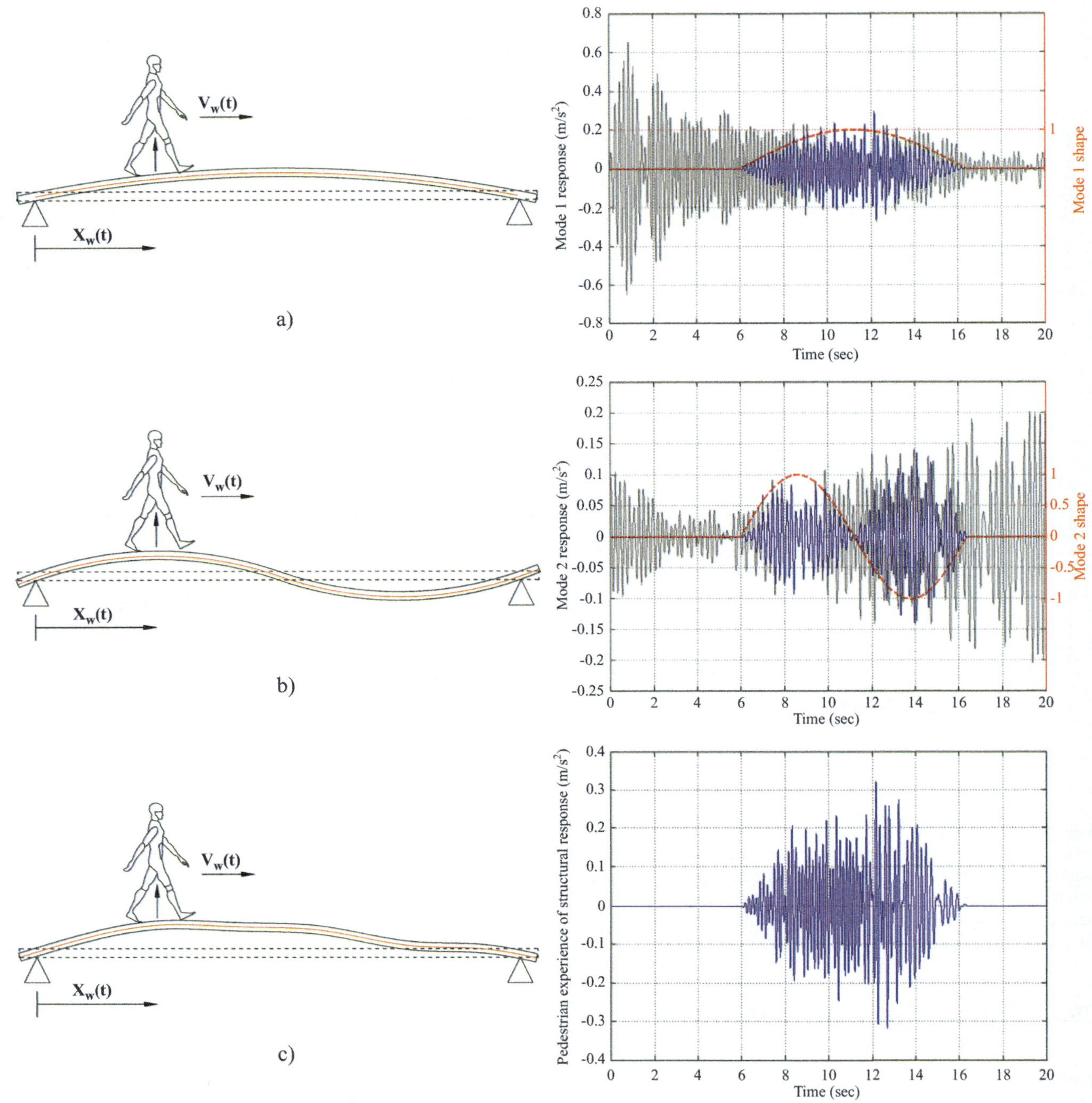

Fig. 29.2 Time-history of each pedestrian's experience as they walk along a beam structure. (**a**) Vibration levels experienced by a human user from Mode 1 response, (**b**) vibration levels experienced by a human user from Mode 2 response, (**c**) time-history of the pedestrian experience from the structural response

history of the vibration received by each human user to that user's actual experience of vibration i.e. comfort, productivity and health. Having this missing component, during the design stage, methods such as I-VSA are able to estimate the overall satisfaction of the intended users of the structure and can use this measure to assess the success of the design.

29.5 Conclusions

The interaction-based VSA method proposed by the authors is shown to addresses the most important shortcoming of the current vibration serviceability assessment guidelines for pedestrian structures: neglecting the HSI, inter- and intra-subject variability of the walking load and human body parameters, and the human 'experience' as the assessment criteria. The application of the proposed I-VSA method to two full-scale footbridge structures under different walking traffic, demonstrated that it regularly overestimates the responses by only 5–10%. This is significantly more accurate than 200–500% overestimation obtained from current international design guidelines. Human 'experience' of structural vibration is proposed as the key VSA criteria for the next generation of VSA methods. The I-VSA method have unique capabilities to estimate realistically the level of vibrations experiences by each pedestrian. However, there is still a missing component, human perception of vibration, to link the estimated level of vibration by each human to the users' satisfaction, health and productivity.

References

1. Racic, V., Pavic, A., Brownjohn, J.M.W.: Experimental identification and analytical modelling of human walking forces: literature review. J. Sound Vib. **326**(1–2), 1–49 (2009)
2. The International Federation for Structural Concrete (FIB): Guidelines for the design of footbridges (2005)
3. International Organization for Standardization (ISO): Bases for design of structures: serviceability of buildings and walkways against vibrations. ISO 10137:2007, Geneva (2007)
4. Technical Department for Transport, Roads and Bridges Engineering and Road Safety/French Association of Civil Engineering (SETRA/AFGC): Footbridges: assessment of vibrational behavior of footbridges under pedestrian loading. Technical Guide 0611, Paris (2006)
5. British Standards Institution (BSI): UK national annex to Eurocode 1: actions on structures. Part 2: Traffic loads on bridges. NA to BS EN 1991-2:2003, London (2008)
6. IStructE/DCLG/DCMS Joint Working Group: Dynamic performance requirements for permanent grandstands: recommendations for management design and assessment. Institution of Structural Engineers, London (2008)
7. Pavic, A., Reynolds, P.: Experimental verification of novel 3DOF model for grandstand crowd-structure dynamic interaction. In: The Proceeding of 26th International Modal Analysis Conference (IMAC XXVI) (2008)
8. Jones, C.A., Pavic, A., Reynolds, P., Harrison, R.E.: Verification of equivalent mass–spring–damper models for crowd–structure vibration response prediction. Can. J. Civ. Eng. **38**, 1122–1135 (2011)
9. Shahabpoor, E., Pavic, A., Racic, V.: Structural vibration serviceability: new statistical framework featuring human-structure interaction. Eng. Struct. **136**, 295–311 (2017.) https://doi.org/10.1016/j.engstruct.2017.01.030
10. Zivanovic, S., Pavic, A., Reynolds, P.: Modal testing and FE model tuning of a lively footbridge structure. Eng. Struct. **28**(6), 857–868 (2006.) https://doi.org/10.1016/j.engstruct.2005.10.012
11. Shahabpoor, E., Pavic, A., Racic, V.: Identification of mass–spring–damper model of walking humans. Structures. **5**, 233–246 (2016)
12. Shahabpoor, E., Pavic, A., Racic, V., Zivanovic, S.: Effect of group walking traffic on dynamic properties of pedestrian structures. J. Sound Vib. **387**, 207–225 (2017.) https://doi.org/10.1016/j.jsv.2016.10.017
13. Zivanovic, S.: Benchmark footbridge for vibration serviceability assessment under vertical component of pedestrian load. J. Struct. Eng. **138**(10), 1193–1202 (2012)
14. Butz, C.: Beitrag zur Berechnung fußgangerinduzierte Bruckenschwingungen. PhD Thesis (in German), Shaker Verlag Aachen (2006); ISBN-10 3-8322-5699-7
15. Shahabpoor, E., Pavic, A.: Comparative evaluation of current pedestrian traffic models on structures. In: Conference Proceedings of the Society for Experimental Mechanics Series, vol. 26, pp. 41–52 (2012)
16. Zivanovic, S., Pavic, A., Ingolfsson, E.T.: Modelling spatially unrestricted pedestrian traffic on footbridges. J. Struct. Eng. **136**(10), 1296–1308 (2010)

Chapter 30
State-of-the-Art and Future Directions for Predictive Modelling of Offshore Structure Dynamics Using Machine Learning

U. T. Tygesen, K. Worden, T. Rogers, G. Manson, and E. J. Cross

Abstract Ramboll Oil and Gas are leading the field in the development of Structural Health Monitoring Systems (SHMS) for offshore structures. This paper outlines the State-of-the-Art process for predictive maintenance that Ramboll have developed and implemented for offshore structures. This system is one of the first, if not the only one, that creates a maintenance schedule based on knowledge of the structure's current state.

The State-of-the-Art methods of today, as adopted by Ramboll, encompass advanced analysis methods ranging from linear and non-linear system identification, expansion processes, Bayesian FEM updating, wave load calibration, quantification of uncertainties from measured data, damage detection and structural re-assessment analysis to Risk- and Reliability-Based Inspection Planning (RBI) analysis.

The paper will be the first in a series of papers that will outline various promising methods contributing to an even better understanding of the issues at stake in the offshore structures context.

Keywords Linear and nonlinear system identification · FEM updating · Modal expansion · Wave load calibration · Digital twin · Uncertainties · Predictive maintenance · Machine learning · Grey-box

30.1 Introduction

Close to 600 offshore oil and gas platforms in the North Sea have either already exceeded their design lifetime or are currently approaching the end of their life within the next few years. The industry is facing huge investments in infrastructure in the near future for maintaining today's oil and gas production. In Denmark alone, investments for a major upgrade/reinforcement of the Danish oilfield infrastructure have been estimated at a cost of close to $5 billion for maintaining Danish oil and gas production in the future. The alternative to upgrading of the fields and reducing production is decommissioning of the platforms, and the cost for shut down and removing the platforms is estimated at $2.5 billion. The availability of reliable and enhanced methods to provide the information for decision making is of paramount importance for economic profitability, environmental protection and human safety for future operation.

Recent reviews performed by the International Oil & Gas Producers Association [1] of the industry methods and practices for structural design and re-assessment reveal potential inconsistencies in the methods adopted and even knowledge gaps. All structures shall be operated within required safety levels as specified by codes and standards, so it is of the utmost importance that the real uncertainties associated with the design methods are correctly quantified, so that the risks of operating the platform are controlled throughout the lifetime of the structure, either for the original design, for the reinforced/upgraded design, or for other extended lifetime activities performed for ageing platforms.

Throughout the past 15 years, Ramboll have developed novel methods for assessment of the real/measured uncertainties associated with the design and analysis of platforms and today Ramboll is considered to be among the leading experts within application of Structural Health Monitoring Systems (SHMS) for condition-based lifetime extension and for quantification of the real model prediction uncertainties. The uncertainties form the basis for assessing the time dependent Probability of Failure (PoF) for the platforms, i.e. the safety level of the platforms, and hence are the key factor for both verification of extended lifetime of the platforms, but also for reduction of maintenance costs during the years of production. Reductions in maintenance costs are achieved by linking the quantification of measured uncertainties with Risk- and Reliability-based

U. T. Tygesen (✉)
Ramboll Oil and Gas, Esbjerg, Denmark
e-mail: UTT@ramboll.com

K. Worden · T. Rogers · G. Manson · E. J. Cross
Dynamics Research Group, Department of Mechanical Engineering, University of Sheffield, Sheffield, UK

S. Pakzad (ed.), *Dynamics of Civil Structures, Volume 2*, Conference Proceedings of the Society for Experimental Mechanics Series, https://doi.org/10.1007/978-3-319-74421-6_30

Inspection Planning (RBI) methods. Projects on large offshore structures have been performed for which measured site specific uncertainties have formed the basis for RBI with the result of considerable cost savings for future platform maintenance cost.

30.2 Ramboll Predictive Maintenance

An overview of the Ramboll Predictive Maintenance Process in shown in Fig. 30.1, the process can be split into two stages. The first is the calibration of a Finite Element (FE) model with regard to the measured data from the structure of interest. The second stage is where decisions, as to the maintenance and operation of the structure, are made Fig. 30.2.

The calibration phase of the process can be seen as the combination of two Finite Element Model Update (FEMU) steps. The beginning of the calibration method is to build a high fidelity FE model, which is achieved using Ramboll's ROSAP package [2]. After a period of data acquisition on the structure including measurement of accelerations, strains,

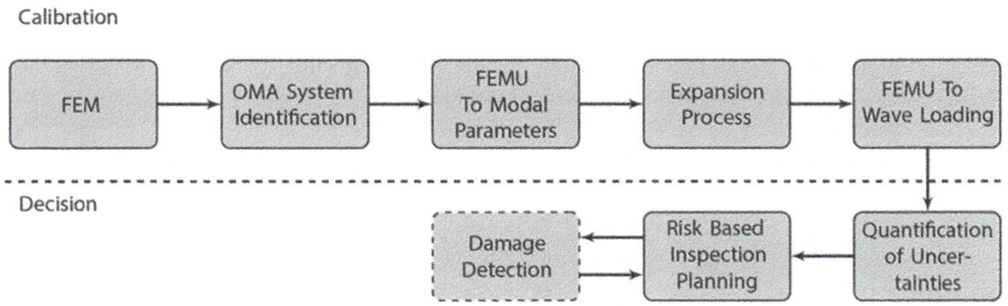

Fig. 30.1 Flow diagram giving a top level view of the Ramboll Predictive Maintenance Process, the process of data to decision can be clearly seen in the methodology

Fig. 30.2 Typical sensor layout for an SHM system

wave characteristics (via wave radar), and GPS measurements; the modal parameters of the structure can be determined using the Stochastic Subspace Identification (SSI) method for Operational Modal Analysis (OMA).

The first FEMU step is the update of the FE model to the measured modal parameters based on a sensitivity analysis and prior information in the form of a Bayesian parameter update. Once satisfied that the FE model sufficiently represents the dynamic behaviour of the structure, a modal expansion process can be used to determine the strains/stresses across the structure. A compensation has to be made for the quasi-static response of the structure with relation to the wave loading. The predictions obtained from the expansion process and the measured sea states are then used in the second round of FEMU to calibrate the parameters of the wave loading.

Throughout the calibration process, it is possible to quantify the uncertainty introduced in the model, the measurements, and the updating steps with respect to fatigue damage accumulation experienced by the structure. The predictions of the fatigue damage accumulation and their uncertainties, summarised as a bias value and a Coefficient of Variation (CoV) are used in a risk based inspection planning (RBI) system. The addition of online damage detection to this is a key step towards a robust predictive maintenance system informed by data. This forms the decision half of the process which is not discussed in full in this work, rather, insight to the calibration methods are presented.

The key first step to providing a predictive maintenance strategy for Ramboll's clients is the acquisition of high-fidelity, information-rich data. Typically a Structural Health Monitoring System (SHMS) used by Ramboll will be based on a range of sensors, such as accelerometers, strain gauges, GPS's, anemometers and wave radars of different types.

Initial work has been carried out to ensure that the base data logged from monitoring systems are of high quality. A number of activities for quality assurance of the data collected are performed, ranging from statistical interrogation to further improvement of data accuracy of platform displacement data. Additionally, algorithms for combining signals from GPS and accelerometers for improved displacement measurements [3], laboratory experiments [4], and full scale experiments on offshore structures [5] are used to validate the principles used.

30.3 System Identification

Normally, a prediction model used for design purposes is thought to be a good representation of the behaviour of the real structure. Design according to the prediction model normally results in the required safety level of the structure. For some cases, however, one needs to know the actual state of a structure in terms of safety. This could be the case for a number of conditions ranging from design verification of structures that are very sensitive to the real dynamic characteristic of the structure or verification of structures that have been exposed to changes during the lifetime of the structure. The changes could be due to a number of reasons such as changes in subsidence, scour, topside weights, corrosion, fatigue cracks or other material degradation.

A number of different methods for system identification are adopted. The choice of the optimal method is dependent on the specific type of structure considered. The preferred method today is SSI as presented in Ref. [6]. For other methods used reference can be made to [7–10]. Most of the methods referred to commonly can fit under the group of Operational Modal Analysis (OMA) methods.

The only measurements required for system identification based on SSI are the displacements $X(t)$ as measured by the SHM system in a few selected locations on the Topside of the Jacket structure. Measurement of the loading $f(t)$ is not required for extraction of the modal parameters. The modal parameters can be extracted continuously for constant monitoring of structural changes in real-time. For clarity, the modal parameters determined from linear system identification will be referred to as 'experimental modal parameters' to distinguish them from the 'analytical modal parameters' represented by the FE model.

State Space based equations have also been used for other purposes than system identification of modal parameters. In references [11, 12] the application of Kalman filters to be used for wave load identification is presented. The advantage of Kalman filter based methods is that these methods allow for analysis of both linear and some nonlinear system behaviour. In addition, they offer the possibility to quantify both measurement uncertainties and prediction model uncertainties from measured and predicted data.

30.4 FEM Updating of Modal Parameters

The typical FE model input parameters that influence the modal parameters are: mass, centre of gravity (CoG), element stiffness, Local Joint Stiffness (LJF), soil stiffness, damping properties, etc.

Several methods for FEM updating are adopted. Here, for the first FEMU step, the Bayesian Parameter Estimation algorithm is presented. In Ref. [13] a more detailed description of the method is presented for updating of FE model against results attained from experimental modal analysis (EMA) in terms of updating to the Frequency Response Function (FRF). Typically, measured FRFs are not available for offshore structures; therefore, the presented method concerns an update with respect to parameters identified by OMA.

The input parameters required for FEM updating are:

- From SHM system: Modal parameters from system identification.
- From FEM model: Predicted modal parameters from FEM.
- From FEM model: Sensitivity matrix for the selected parameters to be updated (mass, COG, stiffness's, etc.).

The updating is performed using the Bayesian Parameter Estimator algorithm:

$$\{\Delta P\} = \left([C_p] + [S]^T [C_R] [S]\right)^{-1} [S]^T [C_R] \{\Delta R\} . \tag{30.1}$$

Where:

$\{\Delta P\}$: Resulting parameter corrections (updated FEM input parameters).
$\{\Delta R\}$: Response difference (difference between predicted and measured modal parameters).
$[C_p]$: Confidence matrix for parameters to be updated (uncertainty of FEM input parameters).
$[C_R]$:Confidence matrix for response (uncertainty of modal parameters).
$[S]$: Sensitivity matrix, produced through perturbation.

The result from the FEM updating is that the analytical modal parameters now are close to the experimental modal parameters. This is the crucial step in generating a True *Digital Twin* of the real structure. In case of nonlinear system behaviour, the Bayesian Estimator method allows for multi-FEM updating. The issue will not be discussed here.

An example outlining the method in principle for Markov Chain Monte Carlo (MCMC) based FEM updating for offshore structures is presented in Ref. [10]. The advantages of the MCMC based methods include more advanced estimation of both measurement uncertainties and model uncertainties.

During the process of FEMU, valuable information for the design engineer is gained from the generation of the sensitivity matrix. Important information about the characteristics of a structure can be observed. Typically, it can be observed that the natural frequencies are very sensitive to Young's modulus, whereas the mode shapes are not. However, as the prior uncertainty of the Young's modulus is very low, it is typically not changed during an FEM updating process. Updating of masses typically only effects the natural frequencies, but not the mode shapes. Conversely, CoG for masses can affect both natural frequencies and mode shapes (especially in torsional modes).

An interesting trend for FEMU of offshore structures is that the soil/foundation parameters commonly affect the natural frequencies, but not the first two bending mode shapes (two horizontal directions). It does, however, have a significant effect on the torsional mode shapes. This fact typically allows for direct calibration of the soil/foundation properties.

30.5 Expansion Process

The purpose of the expansion process is to determine stresses and strains in all elements and joints of the structure above and below the Mean Sea water Level (MSL) from displacement measurements of the structure in only a limited number of sensor locations above MSL. Typically, for ease of installation, the sensors are mounted on the Topside structure.

For the simplest case, the System Equivalent Reduction Expansion Process (SEREP) method can be used for expanding the data [14]. As the name indicates, the SEREP method originally was developed for reduction of the number of degrees of freedom when comparing analytical and measured data. However, the method here is adopted for expansion purposes.

The basic equation for expansion of the measured displacements, $X(t)$ can be performed by multiplying the updated analytical mode shapes, Φ with the modal coordinates $q(t)$:

$$X(t) = \Phi \cdot q(t) \tag{30.2}$$

The modal coordinates, $q(t)$ are computed independently at each time step t, through a least-squares solution. Strain (and analogue stresses) are now calculated from the determined modal coordinates, $q(t)$:

$$\varepsilon(t) = \Phi_\varepsilon \cdot q(t) \tag{30.3}$$

where the mode shapes, Φ_ε are now expressed in terms of normalised strain/stress values calculated from the updated FE model, instead of normalised displacements.

As an alternative to modelling dynamic system behaviour by modal analysis (eigenvalue vectors), in some cases dynamic analysis can be performed by use of Load-Dependent Ritz vectors (LDR). For a description of the LDR method, reference is made to e.g. [15]. The advantage of the LDR method is that both the static (quasi-static) behaviour and the dynamic modes shapes can be analysed. One short-coming of the LDR can be the approximate solutions attained for the dynamic mode shapes.

In Ref. [16], an example of an adopted method for expanding both in the dynamic mode shapes and in the quasi-static Ritz vectors (deflection mode shapes) is described for an experimental laboratory test of an offshore platform – Valdemar VBA. The initial methodology relies on splitting the measured response signal in a low pass filtered (LP) and a high pass filtered (HP) response signal. An alternative approach is through the use of wavelet filters. Ramboll uses Wavelet filters for different purposes ranging from removal of unwanted contributions from natural frequency responses in laboratory wave loading experiment tests to special identification of nonstationary processes.

Figures 30.3 and 30.4 show an example of expansion results from an offshore jacket structure where the expansion successfully recreates measured behaviour.

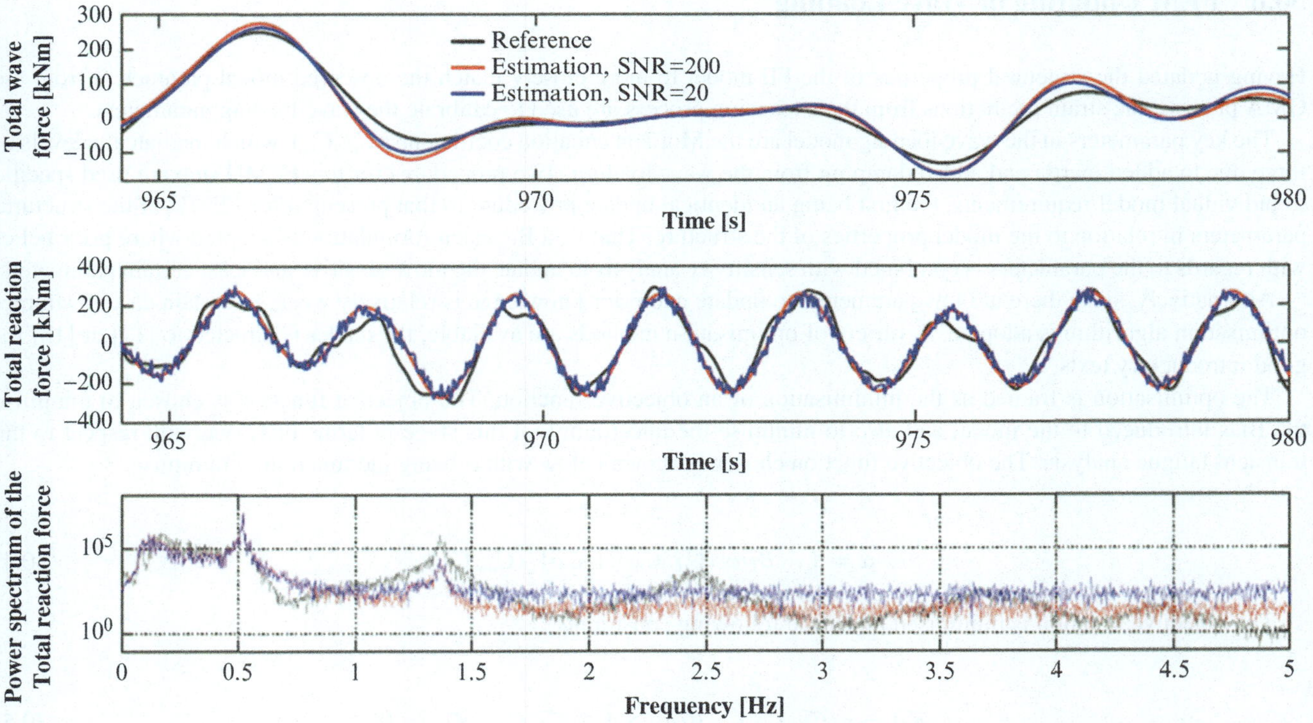

Fig. 30.3 Example of wave load identification using Kalman filter. Top figure show wave force estimated by Kalman filter versus reference total wave loading force acting on an offshore structure in the time domain. Middle figure show the same, but for the structure reaction forces, instead of total load forces. Bottom curve show the results in the frequency domain. Curves are shown for high and low measurement noise levels

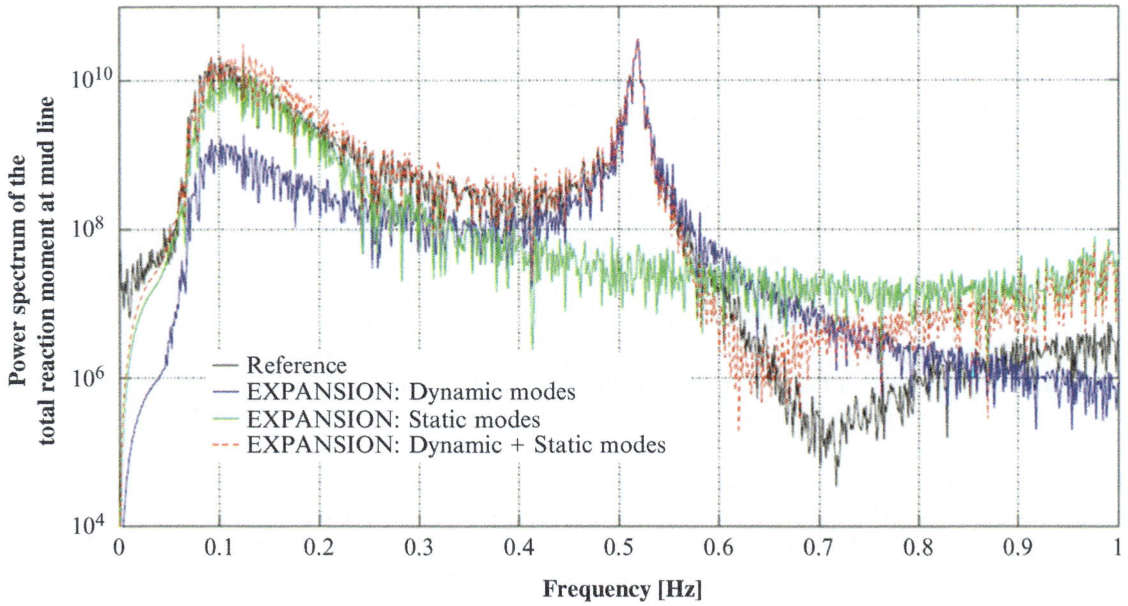

Fig. 30.4 Expansion showing the results from using only dynamic modes in the expansion (blue), using only the quasi-static deformation modes (green) and using both the dynamic modes and the quasi-static modes in the expansion (red). The measured data is shown as the black curve. It is seen that it is important to expand both in the quasi-static and the dynamic modes

30.6 FEM Updating of Wave Loading

Having updated the structural properties of the FE model to more closely match the observed modal parameters from the OMA process, the strain predictions from the expansion process are used to calibrate the wave loading parameters.

The key parameters in the wave loading model are the Morison equation coefficients (C_d, C_m), which mediate the levels of force due to added inertia and added damping from the wave loading. Two procedures for this FEMU step are used specific to individual model requirements, the first being an identical update procedure to that presented for FEMU of the structural parameters in relation to the model properties of the structure. That is, a Bayesian formulation is adopted where prior belief with regards to the parameters is combined with sensitivity analysis to update the most sensitive and least certain parameters.

Alternatively, since there are few parameters to update and prior knowledge is relatively weak, in certain cases a simpler optimisation algorithm is adopted. A variety of optimisation methods are available, the reader is directed to [17] or [18] for good introductory texts.

The optimisation is framed as the minimisation of an objective function. The objective function is chosen to minimise the Bias introduced to the model and also to minimise the uncertainty (at this stage in terms of CoV), with respect to the transient fatigue analysis. The objective function chosen is shown below with α being the function to minimise.

$$\alpha = 1 - Bias\,(C_d, C_m) + CoV\,(C_d, C_m) \tag{30.4}$$

The objective function is minimised under the following constraints.

$$Kd = r\,(C_d, C_m)\,, \quad Bias \geq 1, \quad C_d > 0, \quad C_m > 0, \tag{30.5}$$

where Kd is the FE model analysis (e.g. transient fatigue analysis) to be solved as a function of the wave load parameters during the optimisation.

The Bias and Coefficient of Variation (CoV) are calculated for the ratios of measured and predicted stress ranges as a function of Rain Flow Count (RFC) bins for the full measurement period. The purpose of the wave load calibration is to minimize the Bias, i.e. arrive at a Bias value as close as possible to 1.0. Bias values below 1.0 mean a conservative result,

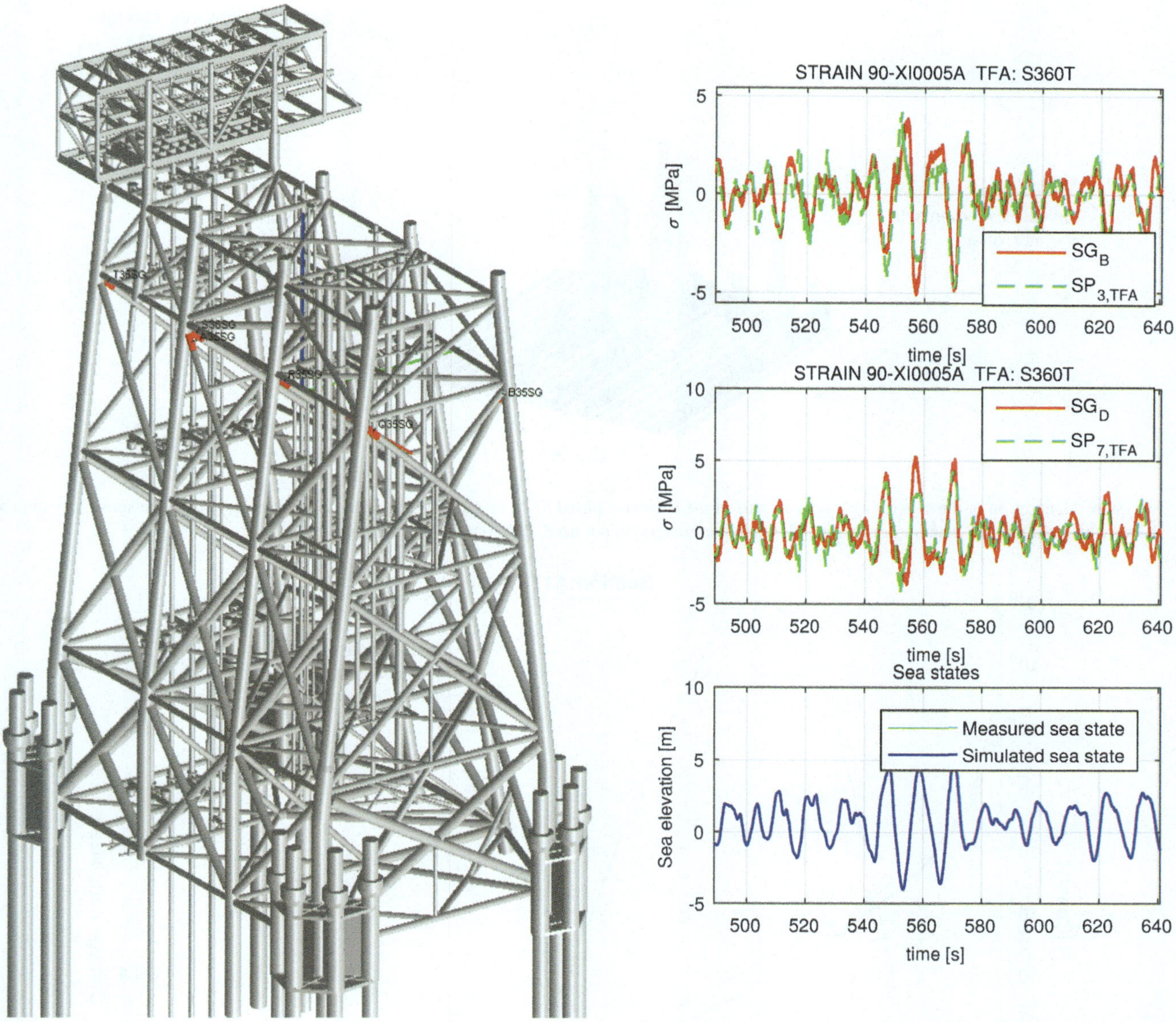

Fig. 30.5 Example of expansion of strains/stresses in braces in the splash zone of a jacket structure from data from only a few sensors at the Topside (incl. wave radars) and compared to strains/stresses measured directly with strain gauges (blind strain gauges). Even the complicated in-plane and out-of-plane local member bending moment phenomena are correctly expanded

while Bias values above 1.0 will not be allowed as a result, i.e. this being a non-conservative calibration. At the same time, the CoV is minimised. The CoV value in this concept is related to the uncertainty of the prediction model.

The optimisation is also constrained to realistic physical values for the parameters and to improve computation speed such that, C_d lies in the range e.g. (0.8, 1.5) and C_m in (1.4, 2.0). The solution to this optimisation problem allows a more accurate representation of the wave loading experienced by the structure which reduces uncertainty in the amount of fatigue the structure has experienced.

A summary of some of the results obtained in the wave loading FEMU step can be seen in Figs. 30.5, 30.6, 30.7, and 30.8. In Fig. 30.6 the sensitivity to the parameters (C_d, C_m) for each member informs the update step and the improvements from the updated values are also shown.

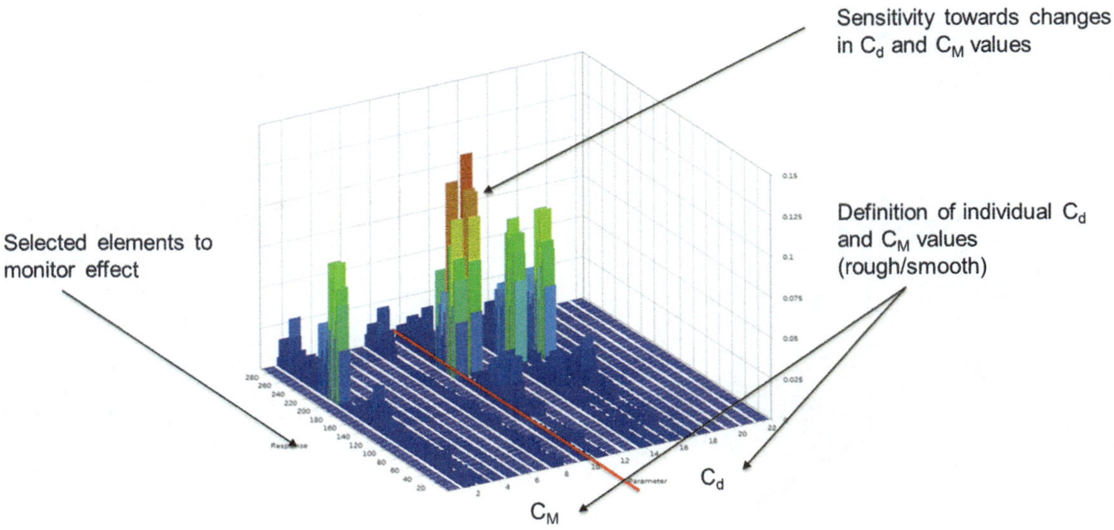

Fig. 30.6 Generation of the sensitivity matrix as part of the Bayesian based FEM updating of the wave loading parameters. The sensitivity matrix shows which parameters that are dominant and which parameters do not need to be part of the calibration

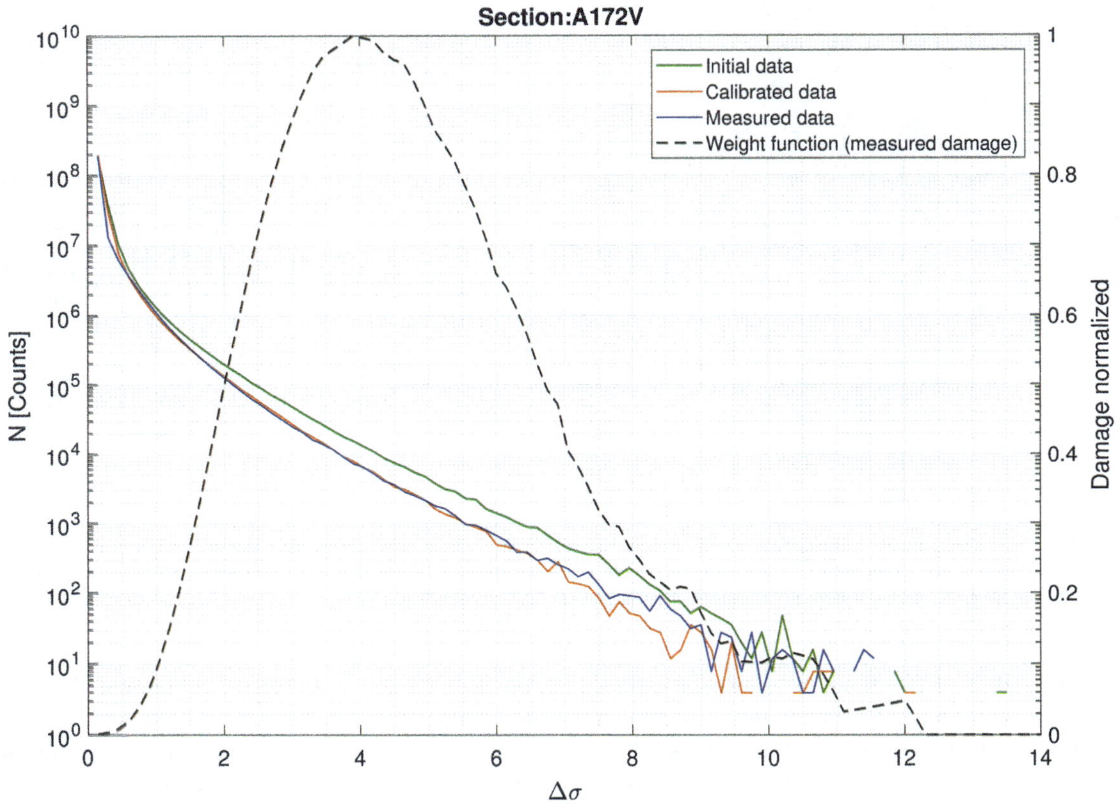

Fig. 30.7 Results from FEM updating of the wave load parameters in terms of Rain Flow Count stress range history curves, comparing the initial model performance with the updated model performance against the measured curves

30.7 Discussion

This paper has outlined a number of the steps in Ramboll's process for predictive maintenance. The main focus of this paper has been on the creation of a validated model of the structure, including an estimate of the wave loading, updated with information from the measured sea state.

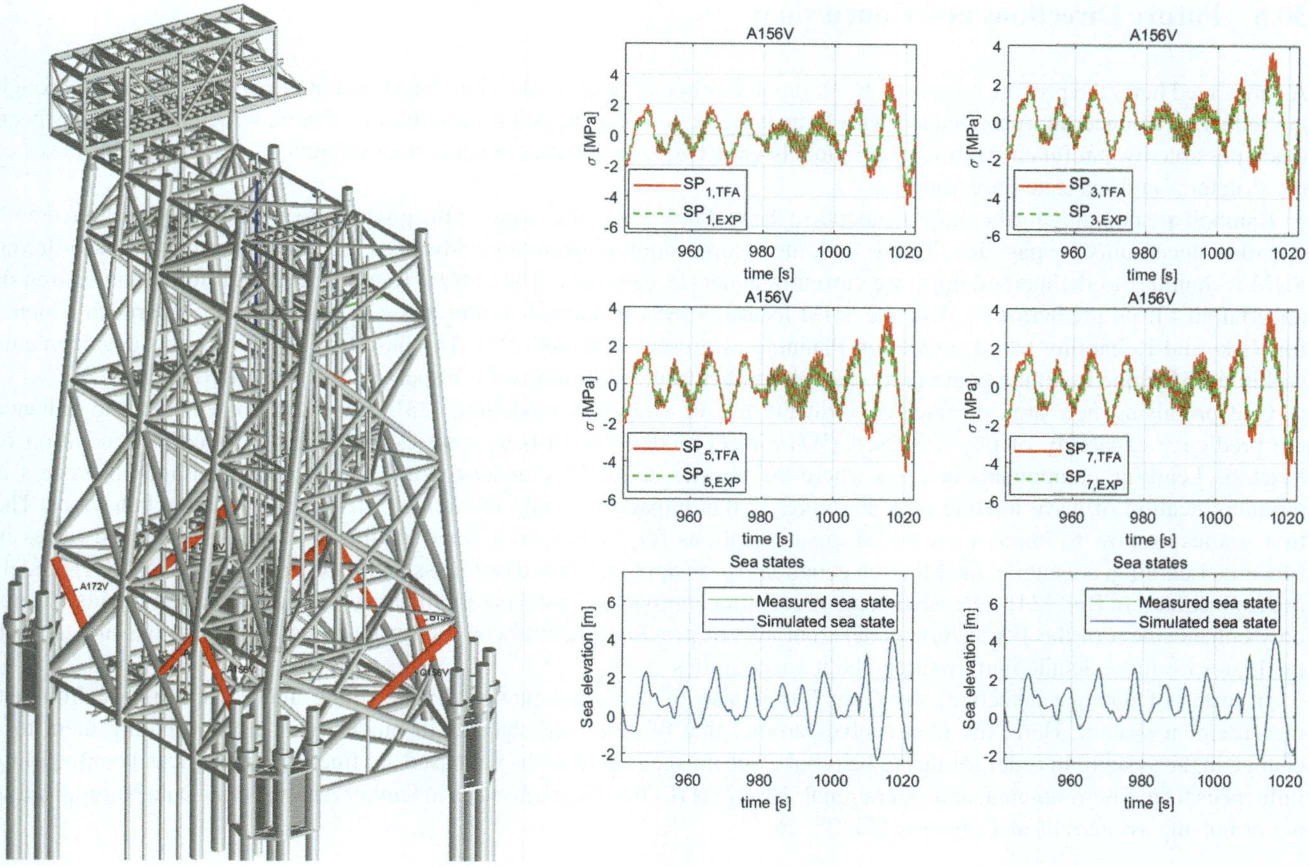

Fig. 30.8 Curves showing the performance of the updated FE model in term of stresses in the time domain (time histories). The green stress curves are the stresses expanded in the bottom braces of the Jacket based on displacement data measured at the Topside in a few locations. The red stress curves are the stresses in the bottom braces calculated by the updated FE model with measured wave load input from wave radars

Not presented above, are the next stages in the process, which include quantification of uncertainties and the Risk- and Reliability-based Inspection Planning informed by fatigue calculations based on the validated model. The FEMU of the prediction models results in improvement of the prediction models, but the question remains: by how much?

The purpose of the uncertainty assessment is to quantify how well the improved prediction model performs. The experience from all performed projects up to now is that the uncertainties are reduced by implementation of the presented procedures. Any reduction in uncertainties results in cost reductions for maintenance.

The scope for the uncertainty assessment is from SHM based data to identify and quantify all contributing factors; here follow examples of uncertainties that is quantified:

- Uncertainty from the updated FEM analysis to predict the structural response in terms of element stress ranges when exposed to a measured sea state.
- Uncertainty from the calibration of the FEM analysis when only based on a limited range of measured sea states.
- Uncertainty of the year-to-year sea state variations for short term periods. The term is relevant for measurement periods less than the full design life of the platform. Typically, the value is calculated for a 3 year period, but any period can be selected. The period is related to the time to first inspection (RBI), not related to the length of the measurement period.
- Uncertainty from measurement noise. The uncertainty covers all kinds of sensors used.
- Uncertainty from the expansion process using an updated FEM to calculate stress ranges for all joints and members in the jacket structure from measurement of responses in only a limited number of locations.

The specific method for quantification of each of the identified contributions from the SHM measurements cannot be presented in the present paper, but will follow in separate papers.

With access to measured structural response and a high-fidelity *True Digital Twin*, online damage detection also becomes more feasible. Ramboll are investigating a number of avenues, including fully data-based approaches [19, 20], as well as those that make inferences from an FE model and the SSI process [21].

30.8 Future Directions and Conclusions

As presented here, the process employed by Ramboll has been found to add value for clients in the offshore industry through the reduction in uncertainty associated with fatigue predictions for the full design life of offshore structures. This has been made possible by careful calibration of FE models via a two-step updating process with reference to measured responses of the structures and sea state observations.

Ramboll are pursuing further improvements to the method, where the stages of the process are more closely integrated with regard to uncertainty propagation. To this end, the incorporation of even more advanced Machine Learning and data-driven SHM techniques to the methodology are currently under development. The present Ramboll approach for quantification of uncertainties from prediction models and SHM measurements is defined so that it is consistent with the approach adopted for Risk- and Reliability-based Inspection Planning as presented in Ref. [22]. The introduction of the latest developments within the Machine Learning community may also affect future approaches for inspection planning (RBI).

One promising new area of research is in the use of *Grey-Box* modelling [23]. This development aims to enhance the predictive capability of physics-based (*White-Box*) models (multi-body physics, FEM, etc.) through the addition of Machine Learning components in cases where the physics is not fully understood. A key step in the Ramboll process is the identification of wave loading on a structure, as this impacts not only the fatigue life but the dynamic behaviour. The first studies on how to improve on the Morison equations for prediction of wave loading by using the latest advances in Machine Learning, enhancing the Morison equations by an approach based on Gaussian Process NARX (GP-NARX) models is demonstrated in Ref. [24]. The advantage of adapting approaches based on nonlinear system identification methods is not only enhancement of the *White-Box* models, but also of providing natural confidence intervals for predictions and as such facilitates for more detailed information about uncertainties.

In terms of damage detection, the *Grey Box* is also of particular interest for detecting abnormal response from the measurement system. Here, the idea evolves around that of detecting changes in model error and attributing them to a change in the system. In order for this to be robust, one must ensure that the data used are free from outliers and confounding influences from Environmental and Operational Variations (EOV). Here advances in Outlier Analysis and Data Normalisation are promising avenues of investigation [24, 25, 26].

Acknowledgement The authors wish to thank the oil and gas operators in the Danish North Sea: Maersk Oil, Hess Denmark and DONG Energy for their participation in several projects forming the basis for the developed methods as of today. TR specially wishes to thank Ramboll for providing financial support for this work.

References

1. OGP Standard Committee: Reliability of offshore structures – current design and potential inconsistencies, OGP report no. 486. International Association of Oil and Gas Producers (OGP, IOGP) (Mar 2014)
2. Ramboll Oil and Gas: ROSAP, program ROSA, structural analysis, user's guide. Ramboll Offshore Structural Analysis Program Package (ROSAP), Rev. 5.1 (Feb 2017)
3. Hansen, J.B., Brincker, R., Knudsen, M.B., Tygesen, U.: Combining GPS and integrated sensor signals. In: International Operational Modal Analysis Conference, Istanbul, Turkey (2011)
4. Skafte, A., Tygesen, U., Brincker, R.: Expansion of mode shapes and responses on the offshore platform Valdemar. In: International Modal Analysis Conference (IMAC), Orlando, FL, USA (2014)
5. Dascotte, E., Strobbe, J., Tygesen, U.T.: Continuous stress monitoring of large structures. In: International Operational Modal Analysis Conference (IOMAC), Guimaraes, Portugal (2013)
6. Brincker, R., Andersen, P.: Understanding stochastic subspace identification. In: Proceedings of the 24th International Modal Analysis Conference (IMAC), St. Louis, MO, USA (2006)
7. Peeters, B., Van der Auweraer, H., Guillaume, P., Leuridan, J.: The PolyMAX frequency-domain method: a new standard for modal parameter estimation Shock Vib. **11**, 395–409 (2004.) IOS Press
8. Brincker, R., Zhang, L., Andersen, P.: Modal identification from ambient response using frequency domain decomposition. In: Proceedings of the 18th International Modal Analysis Conference (IMAC), San Antonio, TX, USA, pp. 625–630 (2000)
9. Zhang, L., Brincker, R., Andersen, P.: An overview of operational modal analysis: major development and issues. In: Proceedings of the 1st International Operational Modal Analysis Conference (IOMAC), Copenhagen, Denmark (2005)
10. Green, P.L., Tygesen, U.T., Stevanovic, N.: Bayesian modelling of offshore platforms. In: The Society for Experimental Mechanics (SEM), International Modal Analysis Conference (IMAC), Model Validation and Uncertainty Quantification, Orlando, FL, USA (2016)
11. Perisic, N., Kirkegaard, P.H., Tygesen, U.T.: Load identification of offshore platform for fatigue life estimation. In: International Modal Analysis Conference (IMAC), Orlando, FL, USA (2014)
12. Perisic, N., Tygesen, U.T.: Cost-effective load monitoring methods for fatigue life estimation of offshore platform. In: Proceedings from the ASME 2014 33rd International Conference on Ocean, Offshore and Artic Engineering (OMAE), San Francisco, CA, USA (2014)

13. Lauwagie, T., Guggenberger, J., Strobbe, J., Dascotte, E.: Model updating using operational data. In: International Conference on Noise and Vibration Engineering (ISMA), Leuven, Belgium (2010)
14. O'Callahan, J., Avitabile, P., Riemer, R.: System Equivalent Reduction Expansion Process (SEREP). In: Proceeding of the 7th International Modal Analysis Conference (IMAC), pp. 29–37 (1989)
15. Sohn, H., Law, K.H.: Extraction of Ritz vectors from vibration test data. Mech. Syst. Signal Process. **15**, 231–226 (2001)
16. Skafte, A., Kristoffersen, J., Vestermark, J., Tygesen, U.T., Brincker, R.: Experimental study of strain prediction on wave induced structures using modal decomposition and quasi static Ritz vectors. J. Eng. Struct. **136**, 261–276 (2017.) Elsevier
17. Fletcher, R.: Practical Methods of Optimization, 2nd edn. John Wiley & Sons (2013). https://doi.org/10.1002/9781118723203
18. Simon, D.: Evolutionary Optimization Algorithms. John Wiley & Sons, Inc., Hoboken, New Jersey (2013)
19. Ulriksen, M.D., Tcherniak, D., Hansen, L.M., Johansen, R.J., Damkilde, L., Frøyd, L.: In-situ damage localization for a wind turbine blade through outlier analysis of SDDLV-induced stress resultants. Struct. Health Monit. **16**, 745–761 (2017)
20. Ulriksen, M.D., Damkilde, L.: Structural damage localization by outlier analysis of signal-processed mode shapes: Analytical and experimental validation. Mech. Sys. Signal Proc., 68–69 (February), pp. 1–14. https://doi.org/10.1016/j.ymssp.2015.07.021
21. Dohler, M., Hille, F.: Subspace-based damage detection on steel frame structure under changing excitation. In: International Modal Analysis Conference (IMAC), Orlando, FL, USA (2014)
22. DNVGL-RP-C210: Probabilistic Methods for Planning of Inspection Planning for Fatigue Cracks in Offshore Structures. DNV-GL Recommended Practice, Edition (Nov 2015)
23. Rogers, T., Holmes, G.R., Cross, E.J., Worden, K.: On a Grey Box modelling framework for nonlinear system identification. In: Special Topics in Structural Dynamics, vol. 6, pp. 167–178. Springer Link (Mar 2017)
24. Worden, K., Rogers, T., Cross, E.J.: Identification of nonlinear wave forces using Gaussian process NARX models. In: Nonlinear Dynamics, vol. 1, pp. 203–221. Springer Link (May 2017)
25. Dervilis, N., Cross, E.J., Barthorpe, R.J., Worden, K.: Robust methods of inclusive outlier analysis for structural health monitoring. J. Sound Vib. **333**, 5181–5195 (2014)
26. Cross, E.J., Worden, K., Chen, Q.: Cointegration: a novel approach for the removal of environmental trends in structural health monitoring data. Proc. R. Soc. A. **467**, 2712–2732 (2011). https://doi.org/10.1098/rspa.2011.0023

Chapter 31
Structural Identification of a Five-Story Reinforced Concrete Office Building in Nepal

Mehdi M. Akhlaghi, Supratik Bose, Babak Moaveni, and Andreas Stavridis

Abstract This study is focused on system identification, finite element (FE) modeling, and FE model updating of the five-story masonry-infilled reinforced concrete building of the National Society of Earthquake Technology (NSET), located in Kathmandu, Nepal. The structure, shown in Fig. 31.1, was subjected to a series of ground motions during the 2015 Gorkha Earthquake and its aftershocks.

Keywords System identification · Structural health monitoring · Model updating · Nonlinear finite element modeling · Gorkha earthquake

In September 2016, some of the authors visited the building and recorded its ambient vibration using an array of 16 accelerometers (Fig. 31.1). The building is also equipped with a continuous monitoring system that includes sensors at the basement and the fourth floor. Response of the building was recorded with this system for one of the moderate aftershocks in 2015. A system identification study is performed to extract the modal parameters of the structure including the natural frequencies, mode shapes and damping ratios using the peak-picking and covariance-driven stochastic subspace identification methods [2] from the measured ambient data. Results of these two system identification studies, i.e. natural frequencies and mode shape values, are presented in Table 31.1.

In order to simulate the behavior of the building during the earthquake, a 3D finite element (FE) model of the building is developed using struts for modeling the effect of the masonry infills (Fig. 31.2). These struts are calibrated using a novel approach, which is based on a recently proposed simplified analytical tool for such structures [3, 4]. Model is validated with the identified modal parameters obtained from the recorded ambient vibration data. This validated numerical model is then used to simulate response of the structure subjected to a moderate aftershock and the results are compared with their identified counterparts.

Figure 31.3 compares the first four identified mode shapes of the structure with those from the initial FE model. Natural frequencies and MAC values of this initial FE model are also compared with the identified values in Table 31.2.

M. M. Akhlaghi (✉) · B. Moaveni
Department of Civil and Environmental Engineering, Tufts University, Medford, MA, USA
e-mail: mehdi.akhlaghi@tufts.edu

S. Bose · A. Stavridis
Department of Civil, Structural and Environmental Engineering, University at Buffalo, Buffalo, NY, USA

S. Pakzad (ed.), *Dynamics of Civil Structures, Volume 2*, Conference Proceedings of the Society for Experimental Mechanics Series, https://doi.org/10.1007/978-3-319-74421-6_31

Fig. 31.1 Street view of the NSET building [1]

Table 31.1 Natural frequencies and MAC values for the first 4 identified modes

	Mode 1	Mode 2	Mode 3	Mode 4
Peak picking frequencies (Hz)	4.25	4.53	13.70	15.18
NExT ERA frequencies (Hz)	4.26	4.56	13.73	15.16
MAC values	0.997	0.977	0.994	0.917

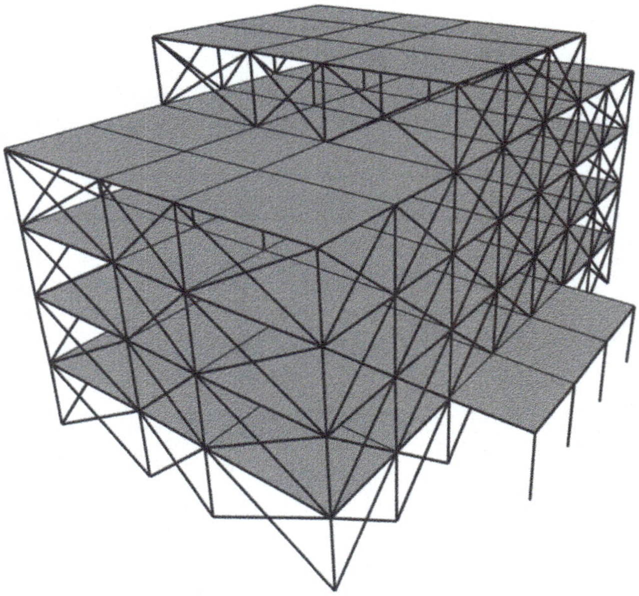

Fig. 31.2 Finite element model of the building

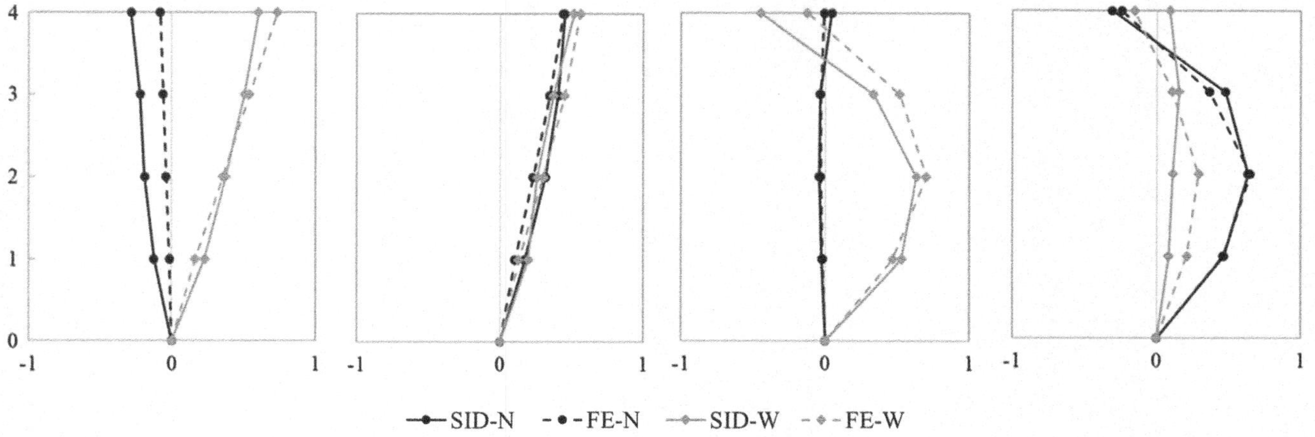

Fig. 31.3 Modes shapes for the identified and the initial FE model

Table 31.2 Natural frequencies and MAC values for the identified and the initial FE model

	Mode 1	Mode 2	Mode 3	Mode 4
NExT ERA frequencies (Hz)	4.26	4.56	13.73	15.16
Initial FE model (Hz)	4.29	4.52	13.70	15.15
MAC values	0.882	0.970	0.855	0.875

Acknowledgements Partial support of this study by the National Science Foundation Grant 1254338 is gratefully acknowledged. The opinions, findings, and conclusions expressed in this paper are those of the authors and do not necessarily represent the views of the sponsors and organizations involved in this project.

References

1. http://peer.nset.org.np. Downloaded on 27 Oct 2017
2. Caicedo, J.M., Dyke, S.J., Johnson, E.A.: Natural excitation technique and eigensystem realization algorithm for phase I of the IASC-ASCE benchmark problem: simulated data. J. Eng. Mech. **130**, 49–60 (2004)
3. Martin Tempestti, J., Stavridis, A.: Simplified method to assess the lateral resistance of infilled reinforced concrete frames. In: Proceedings of the 16th World Conference in Earthquake Engineering, Santiago, Chile (Jan 2017)
4. Bose, S., Nozari, A., Stavridis, A., Moaveni, B.: Nonlinear modeling of the seismic performance of a building at Sankhu Nepal during the 2015 Nepal Earthquake. In: Proceedings of the 16th World Conference in Earthquake Engineering, Santiago, Chile (Jan 2017)

Chapter 32
Structural Identification for Dynamic Strain Estimation in Wind Turbine Towers

Mansure Nabiyan, Hamed Ebrahimian, Babak Moaveni, and Faramarz Khoshnoudian

Abstract Fatigue is a common issue in steel structures such as wind turbine towers, which is caused by cyclic wind and wave excitations. Therefore, estimation of the remaining fatigue life of the structural and foundation system is of concern. For this purpose, continuous monitoring of the structure is necessary to obtain strain data at fatigue critical points. Since installing and maintaining strain sensors in critical underwater location is difficult, strain data is often available only from a few sensors at accessible locations. Using these sparse sensors, the strain time histories at fatigue critical points can be estimated using estimation techniques. These techniques can identify the structural system using limited measured response data and a system model. In this paper, we implement a model updating approach followed by modal expansion to estimate the strain time history at critical points in a numerical case study representing an offshore wind turbine tower. The acceleration response of the structure is simulated using a finite element model and polluted with Gaussian white noise to represent measurements. The measurements are then used for model updating and strain estimation. The accuracy of the methods and their robustness to the measurement noise and model uncertainty are investigated. The estimated strain response time histories can later be used as input to an appropriate fatigue damage model to estimate the current state of fatigue damage in the system.

Keywords System identification · Modal parameters · Model updating · Strain estimation · Modal expansion

32.1 Introduction

Offshore wind turbines are subjected to continuous cyclic wind and wave excitations and therefore, fatigue monitoring is essential to guide maintenance and prevent unwarned failures. Fatigue monitoring requires the knowledge of strain response at critical points of the structure. There are several approaches in the literature which use data from a sparse set of sensors to estimate strain at monitoring points. Some methods utilize the modal decomposition and expansion technique to predict full-field strain response of the structure [1–3]. Iliopoulos et al. [4] applied this method to a monopile offshore wind turbine using only acceleration measurements. Skafte et al. [5] divided the measured data into two parts, including the data with low frequency due to the wave effect on the structure and the data with high frequency due to the dynamic properties. After expanding both low and high frequency signals, they obtained full strain time history. Other approaches make use of Kalman filtering method for state estimation [6, 7]. This algorithm was first implemented for fatigue analysis of structural systems by Papadimitriou et al. [8]. Maes et al. [9] showed the application of Kalman filter for the strain estimation of an offshore monopile wind turbine. They also demonstrated that the Kalman filter algorithm can account for different types of input excitation acting on the structure in the operational condition. A dual implementation of the Kalman filter is proposed in [10] to estimate jointly the force excitation and the state of a linear state-space model using sparse acceleration measurements. In most of these state estimation applications, the input excitation was assumed as a zero-mean Gaussian white noise. To overcome the uncertainty of the input excitation, joint input-state estimation algorithms have been developed to simultaneously estimate both the input and model parameters in the updating process [11–13]. Maes et al. [14] compared

M. Nabiyan · F. Khoshnoudian
School of Civil Engineering, Amirkabir University of Technology, Tehran, Iran

H. Ebrahimian
Department of Mechanical & Civil Engineering, California Institute of Technology, Pasadena, CA, USA

B. Moaveni (✉)
Department of Civil and Environmental Engineering, Tufts University, Medford, MA, USA
e-mail: babak.moaveni@tufts.edu

S. Pakzad (ed.), *Dynamics of Civil Structures, Volume 2*, Conference Proceedings of the Society for Experimental Mechanics Series, https://doi.org/10.1007/978-3-319-74421-6_32

the Kalman filter algorithm, joint input-state estimation algorithm, and a modal expansion algorithm, for the estimation of dynamic strains in the tower of an offshore monopile wind turbine. They concluded that for this application, all three methods yield similar results.

This paper presents a numerical application of strain estimation using a limited set of simulated noise-polluted acceleration data. The acceleration data is first used for identification of modal parameters and model updating. Strains are then estimated through modal decomposition and expansion method, using the calibrated finite element model. The aim is to estimate the strain time history at the base of a wind turbine tower which is a critical fatigue point and evaluate the accuracy of this estimate.

32.2 Strain Estimation Process

The process of strain estimation from measured ambient acceleration response can be summarized as follows:

- Estimating the modal parameters through operational modal analysis.
- Updating a finite element (FE) model based on the identified natural frequencies and mode shapes.
- Predicting the nodal displacement response at all degrees-of-freedom (DOFs) using the modal expansion method.
- Calculating section curvature and estimating strain time history at the point of interest.

32.2.1 Modal Parameter Identification and Model Updating

Modal parameters are identified as a first step using an operational (output-only) modal analysis method from the measured output data. Note that for wind turbines, measuring the wind force is difficult if not impossible. In this paper, the natural excitation technique (NExT) [15] and the eigensystem realization algorithm (ERA) [16] are used to extract the modal parameters. Then, a least-square optimization is employed to update certain parameters of an initial FE model of the structure. The considered updating parameters in this work are flexural stiffness of different structural components. The updating process minimizes the discrepancy between the identified modal parameters (both natural frequency and mode shapes) from data and those from the FE model. The calibrated FE model then provides complete mode shapes to be utilized in the modal expansion algorithm described below.

32.2.2 Modal Expansion Method for Predicting Response at Unmeasured DOFs

Consider the dynamic equation of motion describing a linear time-invariant structural system:

$$\mathbf{M}\ddot{\boldsymbol{u}}(t) + \mathbf{C}\dot{\boldsymbol{u}}(t) + \mathbf{K}\boldsymbol{u}(t) = \boldsymbol{f}(t) \tag{32.1}$$

where **M**,**C**, and **K** are the mass, damping and stiffness matrices, respectively. $\boldsymbol{u}(t)$ is the nodal displacement vector and $\boldsymbol{f}(t)$ represents the force vector. Through modal decomposition [17], the displacement response time history can be written as a linear combination of mode shape vectors and displacement modal coordinate vector of $\boldsymbol{q}(t)$:

$$\boldsymbol{u}(t) = \sum_{j=1}^{n} \varphi_j q_j(t) = \boldsymbol{\phi}\boldsymbol{q}(t) \tag{32.2}$$

where $\boldsymbol{\phi} = [\varphi_1\ \varphi_2 \ldots \varphi_n]$ and n is the number of mode shapes considered. Usually, high frequency mode shapes have smaller contribution to the structural response than the lower mode shapes and therefore their effects can be neglected. In this paper, a limited number of lower modes (n_{modes}) are used. The mode shape matrix can be written as

$$\boldsymbol{\phi} = \begin{pmatrix} \boldsymbol{\phi}^{meas.}_{m\times n_{modes}} \\ \boldsymbol{\phi}^{pred.}_{p\times n_{modes}} \end{pmatrix}_{n\times n_{modes}} \tag{32.3}$$

in which $\boldsymbol{\phi}^{meas.}_{m\times n_{modes}}$ and $\boldsymbol{\phi}^{pred.}_{p\times n_{modes}}$ are partitions of mode shape matrix corresponding to the measured DOFs (m) and the predicted DOFs (p), respectively. For the measured DOFs, Eq. (32.2) can be written as follows.

$$\boldsymbol{u}^{meas.}(t) = \boldsymbol{\phi}^{meas.}\boldsymbol{q}(t) \tag{32.4}$$

Assuming $n_{modes} < m$, the displacement modal coordinate can be obtained from the measured displacement response vector as

$$\boldsymbol{q}(t) = \left(\left(\boldsymbol{\phi}^{meas.}\right)^{\boldsymbol{T}}\boldsymbol{\phi}^{meas.}\right)^{-\boldsymbol{1}}\left(\boldsymbol{\phi}^{meas.}\right)^{\boldsymbol{T}}\boldsymbol{u}^{meas.}(t) = \left(\boldsymbol{\phi}^{meas.}\right)^{\dagger}\boldsymbol{u}^{meas.}(t) \tag{32.5}$$

where $(\boldsymbol{\phi}^{meas.})^{\dagger}$ is pseudo-inverse of $\boldsymbol{\phi}^{meas.}$. Accordingly, the nodal displacement at the predicted DOFs can be obtained as

$$\boldsymbol{u}^{prid.}(t) = \boldsymbol{\phi}^{prid.}\left(\boldsymbol{\phi}^{meas.}\right)^{\dagger}\boldsymbol{u}^{meas.}(t) \tag{32.6}$$

Note that $\boldsymbol{\phi}^{prid.}$ can be obtained from a calibrated finite element model. Similarly, Eq. (32.6) can be written for predicting acceleration response time history at the unmeasured DOFs.

$$\boldsymbol{a}^{prid.}(t) = \boldsymbol{\phi}^{prid.}\left(\boldsymbol{\phi}^{meas.}\right)^{\dagger}\boldsymbol{a}^{meas.}(t) \tag{32.7}$$

32.2.3 Strain Estimation

The transverse displacement field $u(x, t)$ can be computed using the element nodal displacements (including rotations) and beam shape functions [18] as

$$u\,(x, t) = \boldsymbol{N_e}(x)\overline{\boldsymbol{u}}_e(t) \tag{32.8}$$

where $\overline{\boldsymbol{u}}_e(t)$ is the element nodal displacement vector ignoring axial deformations, i.e.,

$$\overline{\boldsymbol{u}}_e(t) = \begin{pmatrix} u_1(t) \\ \theta_1(t) \\ u_2(t) \\ \theta_2(t) \end{pmatrix} \tag{32.9}$$

and $\boldsymbol{N_e}(x)$ is the row vector of beam shape functions:

$$\boldsymbol{N_e}(x) = \left[\left(1-\frac{x}{l_e}\right)^2\left(1+\frac{2x}{l_e}\right), l_e\left(1-\frac{x}{l_e}\right)^2\left(\frac{x}{l_e}\right), \left(\frac{x}{l_e}\right)^2\left(3-\frac{2x}{l_e}\right), -l_e\left(\frac{x}{l_e}\right)^2\left(1-\frac{x}{l_e}\right)\right] \tag{32.10}$$

with l_e is the length of the element as depicted in Fig. 32.1.

Second derivative of Eq. (32.8) with respect to x provides section curvature ($\kappa(x)$) and is obtained as

$$\kappa\,(x, t) = u''\,(x, t) = \boldsymbol{N_e}''(x)\overline{\boldsymbol{u}}_e(t) \tag{32.11}$$

where $\boldsymbol{N_e}''(x)$ is the second derivative of $\boldsymbol{N_e}(x)$. Assuming small deflections, the linear relationship between bending moment and curvature of an Euler-Bernoulli beam is expressed as

$$M\,(x, t) = E(x)\,I(x)\,\kappa\,(x, t) \tag{32.12}$$

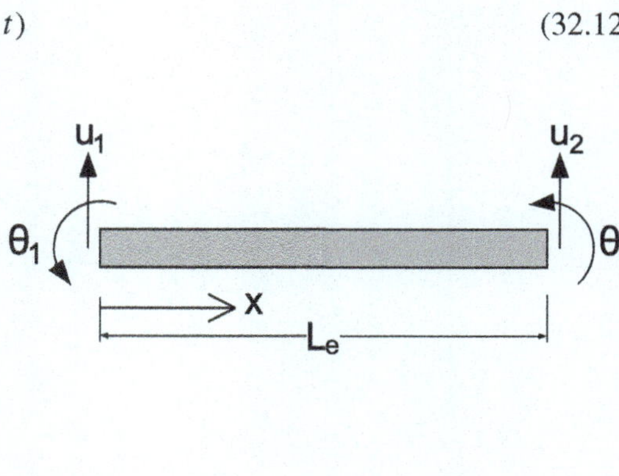

Fig. 32.1 Element end displacements considered in formulation

where E and I are the Young modulus and second moment of inertia of the cross section obtained using the model updating process. The flexural strain due to the bending moment is then computed according to the following equation

$$\varepsilon(x,t) = \frac{M(x,t)\,r(x)}{E(x)\,I(x)} \tag{32.13}$$

where $r(x)$ is the distance between the elastic neutral axis and the outer fiber of the section. Substituting Eq. (32.12) into Eq. (32.13), the maximum section strain time history can be obtained as

$$\varepsilon(x,t) = \kappa(x,t)\;r(x) \tag{32.14}$$

Therefore, the strain time history at each section (x) can be calculated using the curvature. It is worth noting that this method will be part of an output-only system identification method, therefore it does not rely on the knowledge of the input excitation.

32.3 Numerical Application to a Wind Turbine Model

The proposed approach is applied for estimation of strain at the base of a wind turbine tower. The numerical model is considered as a steel cantilevered vertical structure with the 100 m height and a tapered hollow circular cross-section with the wall thickness of 0.04 m and the outer diameter 5.0 m at the base and 2.5 m at the top (similar to the structure that Hernandez et al. considered in Ref. [19]). The considered structure is shown in Fig. 32.2. The FE model includes 21 nodes and 20 beam elements across the height. The elastic Young's modulus of steel is considered to be 210,000 MPa. Simulated acceleration responses at nodes of 4, 7, 10, 13, 16, and 19 are considered as the measured data (location of considered accelerometers are shown Fig. 32.2). The acceleration response time history is simulated when a non-stationary wind load is applied at the top of the tower (node 20) as shown in Fig. 32.3. The simulated data are then polluted with 5% noise to signal ratio Gaussian white noise (i.e., the RMS of the noise is selected as 5% of the RMS of the acceleration response time history).

The acceleration response time history at all nodes are predicted using the reviewed modal expansion approach. The results for six nodes are compared with the exact value and shown in Fig. 32.4. In offshore wind turbines, predicting strain time history at the mudline (the base of the structure) is necessary to estimate the remaining fatigue life. Hence, Eq. (32.14) will be used at $x = 0$. The estimated strain time history at the base is shown in Fig. 32.5. For comparison, the exact strain time

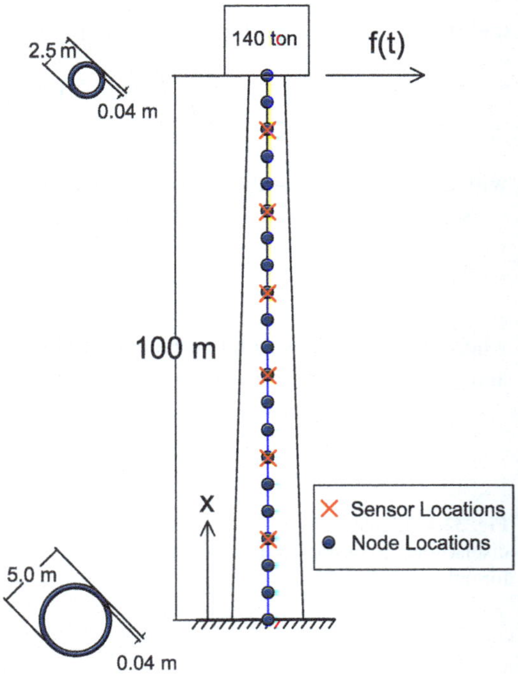

Fig. 32.2 The simulated wind turbine tower

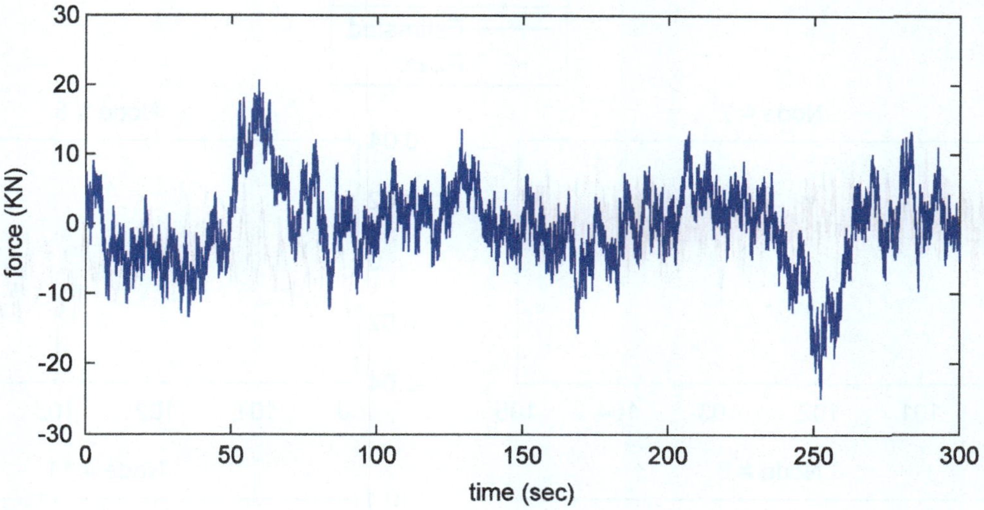

Fig. 32.3 The wind load time history

history is also illustrated in this figure. As can be seen, the estimated strain time history is in good agreement with the actual (true) time history. Furthermore, the error time history which is the difference between the exact and the estimated strain at each instance of t is computed. The maximum absolute value and the RMS (root mean square) of the error are 1.51e-6 and 3.50e-7 for the case of considering the noise effect of 5% in the acceleration response.

32.4 Conclusions

In this paper, a modal decomposition and expansion method is applied for strain estimation, using the simulated data for a wind turbine pole model. This method can identify the strain data at location of interest using limited measured data and a FE model. Mode shape vectors identified using model updating process, are fed into the modal expansion algorithm to estimate the nodal displacement response at all degrees-of-freedom (DOFs). The strain time history at the base is then estimated by calculating curvature at the base section. Comparison the strain time history at the base to the actual quantity reveals a very high estimation accuracy. These results show that the algorithm can be successfully applied for the strain estimation. In the future steps of this work, the effects of model uncertainty, including material and geometry nonlinearity will be investigated.

Acknowledgements Partial support of this study by the National Science Foundation Grant 1254338 is gratefully acknowledged. The opinions, findings, and conclusions expressed in this paper are those of the authors and do not necessarily represent the views of the sponsors and organizations involved in this project.

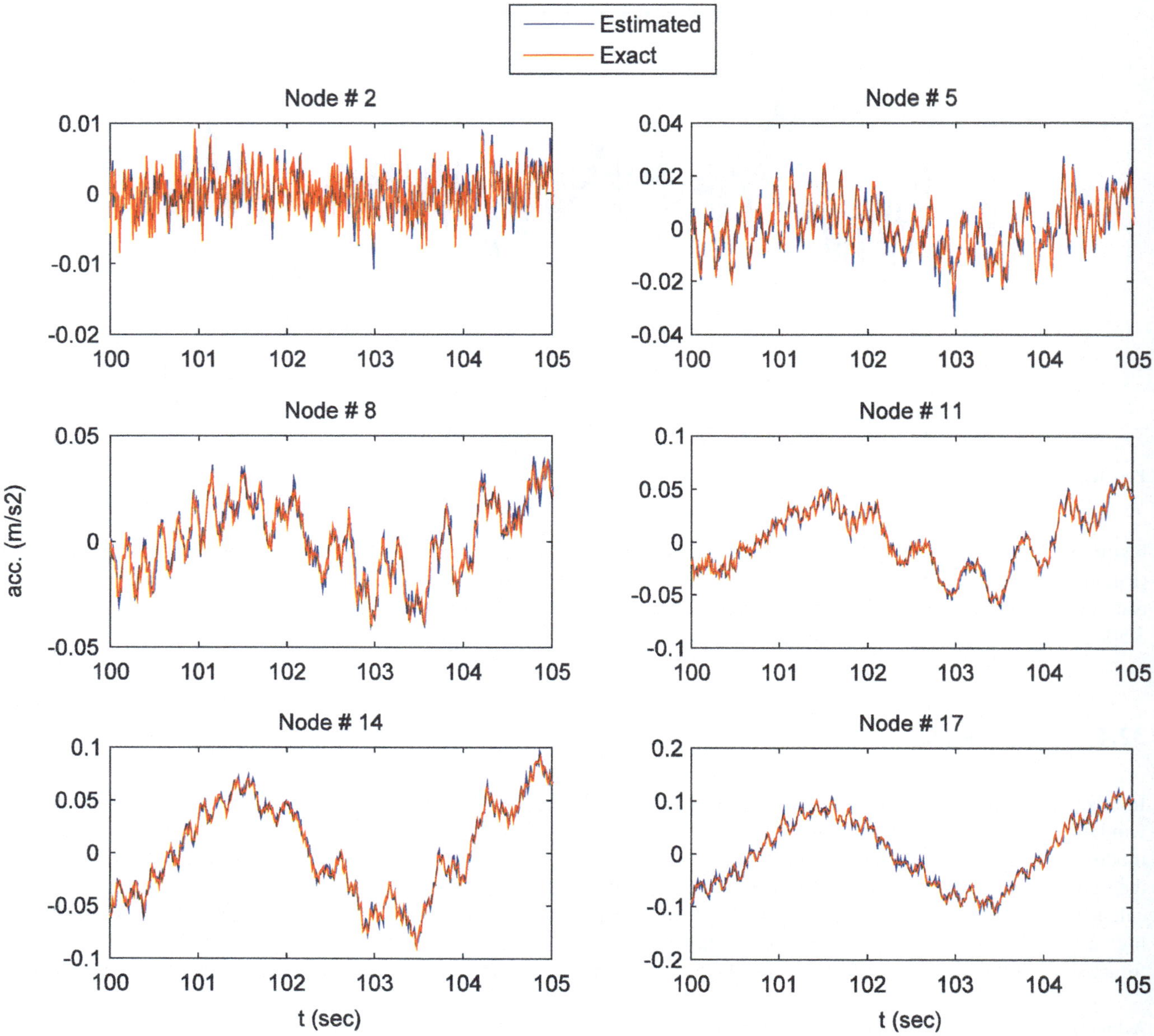

Fig. 32.4 The acceleration time histories at nodes 2, 5, 8, 11, 14 and 17 (blue: estimated, red: exact)

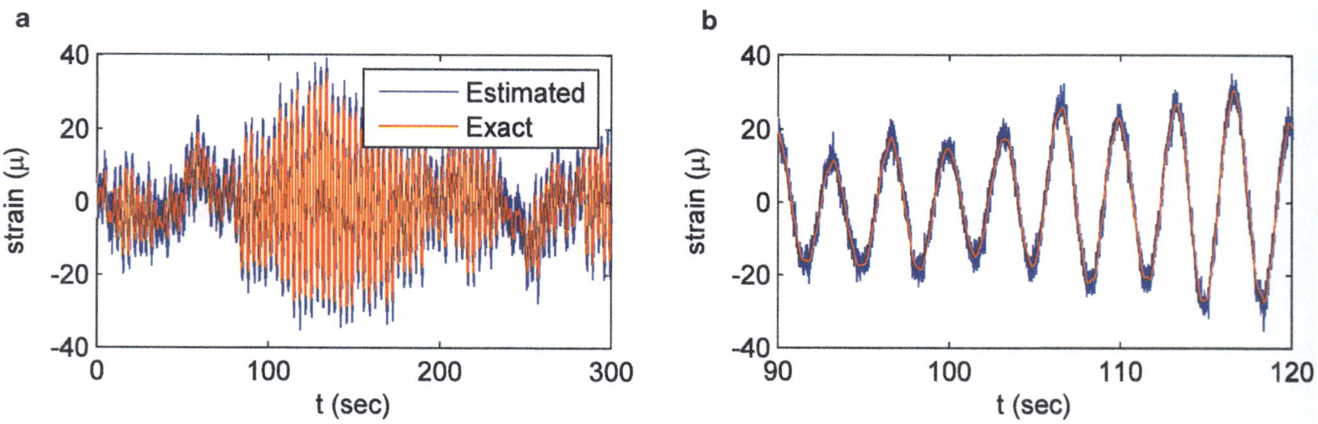

Fig. 32.5 The strain time history (**a**) and zoom-in (**b**) at the base

References

1. Skafte, A., Tygesen, U.T., Brincker, R.: Expansion of mode shapes and responses on the offshore platform Valdemar. In: Dynamics of Civil Structures, vol. 4, pp. 35–41. Springer, Cham (2014)
2. Iliopoulos, A., et al.: Prediction of dynamic strains on a monopile offshore wind turbine using virtual sensors. In: Journal of Physics: Conference Series. IOP Publishing, USA (2015)
3. Iliopoulos, A., et al.: A modal decomposition and expansion approach for prediction of dynamic responses on a monopile offshore wind turbine using a limited number of vibration sensors. Mech. Syst. Signal Process. **68**, 84–104 (2016)
4. Iliopoulos, A.N., et al.: Continuous fatigue assessment of offshore wind turbines using a stress prediction technique. In: SPIE Smart Structures and Materials + Nondestructive Evaluation and Health Monitoring. SPIE, UK (2014)
5. Skafte, A., et al.: Experimental study of strain prediction on wave induced structures using modal decomposition and quasi static Ritz vectors. Eng. Struct. **136**(Supplement C), 261–276 (2017)
6. Smyth, A., Wu, M.: Multi-rate Kalman filtering for the data fusion of displacement and acceleration response measurements in dynamic system monitoring. Mech. Syst. Signal Process. **21**(2), 706–723 (2007)
7. Palanisamy, R.P., et al.: Experimental validation of Kalman filter-based strain estimation in structures subjected to non-zero mean input. Smart Struct. Syst. **15**(2), 489–503 (2015)
8. Papadimitriou, C., et al.: Fatigue predictions in entire body of metallic structures from a limited number of vibration sensors using Kalman filtering. Struct. Control Health Monit. **18**(5), 554–573 (2011)
9. Maes, K., et al.: Continuous strain prediction for fatigue assessment of an offshore wind turbine using Kalman filtering techniques. In: 2015 IEEE Workshop on Environmental, Energy and Structural Monitoring Systems (EESMS). IEEE, Trento (2015)
10. Azam, S.E., Chatzi, E., Papadimitriou, C.: A dual Kalman filter approach for state estimation via output-only acceleration measurements. Mech. Syst. Signal Process. **60**, 866–886 (2015)
11. Estrada, I., Real, E., Mirambell, E.: General behaviour and effect of rigid and non-rigid end post in stainless steel plate girders loaded in shear. Part II: extended numerical study and design proposal. J. Constr. Steel Res. **63**(7), 985–996 (2007)
12. Van der Male, P., Lourens, E.: Operational vibration-based response estimation for offshore wind lattice structures. In: Structural Health Monitoring and Damage Detection, vol. 7, pp. 83–96. Springer, Cham (2015)
13. Niu, Y., Klinkov, M., Fritzen, C.: Online force reconstruction using an unknown-input Kalman filter approach. In: Proceedings of the 8th International Conference on Structural Dynamics. EURODYN, Leuven (2011)
14. Maes, K., et al.: Dynamic strain estimation for fatigue assessment of an offshore monopile wind turbine using filtering and modal expansion algorithms. Mech. Syst. Signal Process. **76**, 592–611 (2016)
15. James III, G.H., Carne, T.G., Lauffer, J.P.: The Natural Excitation Technique (NExT) for Modal Parameter Extraction from Operating Wind Turbines. Sandia National Labs., Albuquerque, NM (1993)
16. Juang, J.-N., Pappa, R.S.: An eigensystem realization algorithm for modal parameter identification and model reduction. J. Guid. **8**(5), 620–627 (1985)
17. Heylen, W., Sas, P.: Modal Analysis Theory and Testing. Katholieke Universteit Leuven, Departement Werktuigkunde, Leuven, Belgium (2006)
18. Gan, B.S.: Finite element formulation of beam elements. In: An Isogeometric Approach to Beam Structures, pp. 61–126. Springer, Cham (2018)
19. Hernandez, E.M., Bernal, D., Caracoglia, L.: On-line monitoring of wind-induced stresses and fatigue damage in instrumented structures. Struct. Control Health Monit. **20**(10), 1291–1302 (2013)

Chapter 33
Modal Parameter Identification from Measurements of Vehicle-Bridge Interaction

Yi Liu, John MacDonald, and Dario Di Maio

Abstract The rise of output-only modal analysis has offered an economical and efficient way to identify modal parameters of civil engineering structures, namely natural frequencies, damping ratios and mode shapes. However, since the forcing is unknown, it is not possible to directly estimate modal masses, and estimates of damping ratios may be inaccurate. With the advancement of wireless sensor networks both vehicle and bridge responses can be simultaneously measured. This offers the possibility of estimating true Frequency Response Functions (FRFs), since the vehicle acceleration gives an estimate of the force input to the bridge. Hence in principle it is possible to estimate modal masses and more accurate damping ratios. However, the spatial and temporal variation of the moving load from a passing vehicle gives challenges to this idea and precludes the direct use of existing single-input-multiple-output (SIMO) system identification methods. Even if the system is treated as a multiple-input-multiple-output (MIMO) one, the inputs are highly correlated so existing methods for these systems are not applicable either. For this reason, a two-stage strategy is proposed to modify an existing method to solve this moving load problem.

Keywords System identification · Input-output · Moving vehicle · Mode superposition · Modal mass

The rise of output-only modal analysis has offered an economical and efficient way to identify modal parameters of civil engineering structures, namely natural frequencies, damping ratios and mode shapes. However, since the forcing is unknown, it is not possible to directly estimate modal masses, and estimates of damping ratios may be inaccurate. With the advancement of wireless sensor networks both vehicle and bridge responses can be simultaneously measured. This offers the possibility of estimating true Frequency Response Functions (FRFs), since the vehicle acceleration gives an estimate of the force input to the bridge. Hence in principle it is possible to estimate modal masses and more accurate damping ratios. However, the spatial and temporal variation of the moving load from a passing vehicle gives challenges to this idea and precludes the direct use of existing single-input-multiple-output (SIMO) system identification methods. Even if the system is treated as a multiple-input-multiple-output (MIMO) one, the inputs are highly correlated so existing methods for these systems are not applicable either. For this reason, a two-stage strategy is proposed to modify an existing method to solve this moving load problem.

Firstly, an output-only system identification method is performed on the measured acceleration responses of the bridge to extract the mode shapes. Secondly, the mode superposition technique is used to transform the measured input and output responses into modal space, decomposing the system into a series of Single-Input-Single-Output (SISO) systems, each represented by an FRF. Using the theoretical accelerance, the parameter estimates for each mode (natural frequency, damping ratio and modal mass) can then be obtained using an optimisation procedure.

The feasibility of this methodology has already been demonstrated by simulating the response of a simply supported beam with a single load moving with constant velocity [1]. The moving load was modelled as a constant weight plus coloured noise representing the effect of road roughness. Records of individual vehicle passages were considered, including both the period that the vehicle was on the bridge and the free decay of the structure after the vehicle has left. In this simple case, five modes of the bridge were considered. For the natural frequencies and modal masses of the beam and assigned damping ratios of each mode, time histories of the bridge response were calculated to give simulated data with known modal parameters to test the proposed system identification technique.

Y. Liu (✉) · J. MacDonald
Department of Civil Engineering, University of Bristol, Bristol, UK
e-mail: yl15193@brsitol.ac.uk

D. Di Maio
Department of Mechanical Engineering, University of Bristol, Bristol, UK

S. Pakzad (ed.), *Dynamics of Civil Structures, Volume 2*, Conference Proceedings of the Society for Experimental Mechanics Series, https://doi.org/10.1007/978-3-319-74421-6_33

In the first stage of the system identification, an output-only system method, based on the decomposition of the Power Spectral Density (PSD) matrix using Singular Value Decomposition (SVD) technique, i.e., Frequency Domain Decomposition (FDD), is used to identify the mode shape of the bridge. Note that, since the response of the bridge is induced by a moving load, the application of the FDD to this non-stationary signal is only approximate. Nevertheless, preliminary analysis shows that the method works well, even with 10% or more added noise. This first stage of mode shape identification is vital to the proposed method. As in forward analysis, the orthogonality of the mode shapes with respect to the mass and stiffness matrix can also be used to tackle the inverse problem to decouple the Multiple Degree of the Freedom (MDOF) system. The measured or simulated bridge responses at a series of discrete locations are decoupled into modal responses, while the physical load, varying in space and time, is converted into modal forces. Hence the bridge is then effectively considered as a series of Single-Degree-of-Freedom (SDOF) systems.

Once the generalised force and generalised acceleration of each bridge mode is available, the FRFs for the system can be easily calculated based on the Discrete Fourier Transform (DFT). This 'measured' accelerance of the bridge and its theoretical counterpart can be used to form an objective function, thus an optimization procedure can be applied to the extract the unknown modal parameters of the bridge, namely, natural frequencies, modal damping ratios and modal masses.

The previous work described above has successfully proved the feasibility of the proposed method [1], but the it hasn't considered the noise effect. In the recent study the robustness of which has been proved by adding different levels of the noise to the input and output response of the system. A statistical analysis has also conducted by running the analysis on 200 simulated time histories of vehicle loads and bridge responses. The results show that:

1. With 0.3% noise added to the input and 5% noise added to the output, the mean values of the three modal parameters for each mode are very close the true values used in the forward analysis. For the natural frequencies, the errors in the mean estimates are less than 0.04% for all modes except for the fourth mode with the highest damping ratio, which had an error of 0.35%. Damping ratios are much more difficult to estimate accurately, but very encouraging results were obtained with the proposed methods, with the mean errors for all five modes being less than 7%. The largest error was for the fifth mode. Accordingly, the largest percentage error for the modal mass occurred in the fifth mode (19%) as well. One possible explanation is that, this fifth mode received less power than the lower modes from the moving load excitation, and since the radius of the Nyquist plot is mainly determined by the ratio between the damping ratio and modal mass, they share the same level of impact to the optimization evaluated around the resonance frequency. Therefore, when the damping is relatively small than other modes, a small variation in the identified damping ratio could result in a large deflection in modal mass, for it has larger scale than the damping ratio, and the standard deviation is about one-tenth of their dimensions. This is also the reason that the modal mass the percentage errors are larger than the damping ratios'. However, the overall results are satisfactory, for the first three modes, they are less than 4.2%, the lowest one is only 0.8% (the third mode). While for the forth mode, the slightly larger damping ratio results in a more scattered data set, the percentage error is about 15.2%, which is consistent with the identified natural frequency and the damping ratio.

   Apart from this, a comparison work has done to compare with the conventional vehicle excitation free vibration analysis. The results indicate that without the noise, using only the free vibration response of the bridge can generate high accuracy natural frequencies and damping ratios for each mode. When 5% noise level is added to the same sequences, the average results for the natural frequencies are still satisfactory, with less than 0.5% percentage error for all five modes. However, the percentage errors for the damping ratios are all over 16%, the largest one, which is the fourth mode, has up to 72.7% percentage error. And the standard deviations are a factor of two larger than the results identified by the proposed method. Hence it is justifiable to say that the proposed method is quite effective and robust.
2. The effect of the noise levels on the input and outputs has different impact. The accuracy of the identified results is largely affected by the noise level added to the input signal. As long as the noise level of the input is kept at a very low level, the noise level for the output signal can be increased up to as high as 50% with good results still being achievable. To be more specific, when 0.1% and 50% noise levels are added to the moving load and the bridge output response, respectively, the identified natural frequencies have less than 0.1% percentage errors (except for the first mode which is 0.5%), while the damping ratios have the highest percentage error (37%) in first mode, but with the rest of the percentage errors remained under 12%. As for the modal mass, it has the same pattern, the first mode is the worst one with 22.7% percentage error, while the higher modes have no more than 10% percentage errors.
3. Since this the method uses a simple vehicle excitation, with now readily available wireless sensor technology, actual experiments could be conducted bridges easily and economically, using multiple vehicle passages to obtain results with low errors due to averaging.

   The previous work also did not consider true vehicle-bridge interaction, the bridge loading being defined by a prescribed moving force. Considering the vehicle dynamics, in practice there will be greater bridge excitation around natural frequencies of the vehicle. This effect is investigated in the current work. The moving load model is replaced with a quarter car model and the system of equations of the vehicle-bridge system is non-dimensionalised. In this more realistic model, when the vehicle crosses the bridge, it fully interacts with it and an appropriate road roughness model is incorporated to give vertical excitation to the vehicle. Simulated data based on this system is used as input to the same system identification procedure as described above. The aim is to test the proposed method for a more realistic scenario. Results for different vehicle-bridge natural frequency ratios will be presented. Also, since the mode shapes of the bridge play a vital role in the method, the influence of the accuracy of the identified mode shapes will be discussed.

Reference

1. Liu, Y., Macdonald, J., Di Maio, D.: Identification of modal parameters based on moving load excitation. Procedia Eng. **199**, 960–965 (2017)

Chapter 34
Bridge Structural Identification Using Moving Vehicle Acceleration Measurements

Soheil Sadeghi Eshkevari and Shamim Pakzad

Abstract Identification of dynamic characteristics of structures is a desired objective for existing infrastructure and has been accounted as a serious challenge for civil engineers. In this research, a structural identification method is proposed, which is capable of identifying dynamics of structures using sensor data inside vehicles passing over a bridge. The methodology utilizes a special type of identification algorithm facilitated by Expectation Maximization (STRIDEX) that is capable of identifying systems using mobile data networks. In this study, it is assumed that the mobile sensor measurements are the accelerations inside rigid vehicles and are primarily a mixtures of accelerations caused by the road roughness and bridge dynamic acceleration. With this regard, a stochastic State-Space model represents the equation of motion for a linear dynamic vehicle-bridge system consisting of an impure input. The observation vector is treated as a linear mixture of two sources that are not known. Therefore, the problem turns to a Blind Source Separation (BSS) procedure that is aiming to draw out the bridge vibrations from the mixture. An algorithm called Second Order Blind Identification (SOBI) has been utilized for source separation and validated using simulation. The entire algorithm, including both SOBI and STRIDEX acting together, could successfully identify natural frequencies and mode shapes of a numerical bridge model.

Keywords Expectation Maximization · Blind Source Separation · System Identification · Output Only Algorithms · Structural Health Monitoring

34.1 Introduction

System identification is a process that extracts mechanical properties of a dynamic system from different types of easy-to-access and cheap data channels such as accelerometer sensors. In recent years, many structural health monitoring (SHM) algorithms have been customized to utilization of wireless sensors, which form a flexible and efficient type of sensors in practice [1–5]. Wireless sensors are able to collect dense data and enjoying an increasingly more powerful on-board microprocessor, they are able to communicate with the control nodes more efficiently [6, 7].

System identification procedure is usually laid on a mathematical framework [8]. The governing equation of motion for dynamic systems is a second order differential equation in which the system's degrees of freedom are included as independent variables as shown in Eq. 34.1:

$$M\ddot{U}[n] + C\dot{U}[n] + KU[n] = F \tag{34.1}$$

Because of the inherent complexity of the conventional form of the equation of motion, in system identification application State Space modal is utilized as in [9]:

$$x_k = Ax_{k-1} + n_k \tag{34.2}$$

$$y_k = Cx_k + v_k \tag{34.3}$$

$$x_1 \sim N\left(\bar{\mu}, \bar{V}\right) \tag{34.4}$$

S. Sadeghi Eshkevari (✉) · S. Pakzad
Department of Civil and Environmental Engineering, Lehigh University, Bethlehem, PA, USA
e-mail: ses516@lehigh.edu

S. Pakzad (ed.), *Dynamics of Civil Structures, Volume 2*, Conference Proceedings of the Society for Experimental Mechanics Series, https://doi.org/10.1007/978-3-319-74421-6_34

$$\eta_k \sim N(0, Q) \tag{34.5}$$

$$\nu_k \sim N(0, R) \tag{34.6}$$

where A is the state matrix, C is the observation matrix, x_k is the state vector which is not observable, and y_k is the observation vector which is the collected data at the accessible channels. η_k and ν_k are systemic and sensing noises respectively, which for simplicity are assumed to be uncorrelated Gaussian white noise with covariance matrices as Q and R [10].

In the case of structural engineering application, it is often not possible to accurately measure the input of the dynamic system, e.g. loads. For example, in case of bridge system identification, an accurate description of traffic load or wind load is extremely challenging. As a result, identification problems of output-only systems have been studied extensively [11, 12].

In general, there are numerous types of system identification methods, which are proper for different applications. In SHM, two general approaches are common: frequency domain and time domain identifications. Time domain identification algorithms such as Eigensystem Realization Analysis (ERA) [13], ERA using exogenous excitations (ERA-NExT), ERA using Observer Kalman Filter Identification (ERA-OKID) [12] and SubSpace Identification (SSI) [14] are a few examples of algorithms for real-time monitoring and asymptotically converge to the more accurate results as the time passes. In this study, STRIDEX [15], a more recent algorithm which is not only capable of time-domain identification, but also is appropriate for mobile sensing networks is utilized.

Structural Identification using Expectation Maximization (STRIDE) [16] is a novel state space-based procedure that uses Expectation Maximization (EM) algorithm [17] to iteratively maximize the estimation of the unobserved state vector and unknown state-space matrices and is developed for output-only identification. The algorithm defines a super parameter, which consists of all state space unknowns as shown in Eq. 34.7 and then assuming a given set of unobserved states, maximizes the log-likelihood function of the state space model (Maximization Step).

$$\Psi = \left(\bar{\mu}, \bar{V}, A, Q, C, R\right) \tag{34.7}$$

$$\begin{aligned}\log L\left(\Psi | x, y\right) = &-\frac{pN}{2}\ln\left(2\pi\right) - \ln\left|\bar{V}\right| - \frac{1}{2}(x_1 - \bar{\mu})^T \bar{V}^{-1}\left(x_1 - \bar{\mu}\right) - \frac{KN}{2}\ln\left(2\pi\right) - \frac{K}{2}\ln|R| \\ &- \frac{1}{2}\sum_{k=1}^{K}\left(y_k - Cx_k\right)^T R^{-1}\left(y_k - Cx_k\right) - \frac{(K-1)\,pN}{2}\ln\left(2\pi\right) - \frac{K-1}{2}\ln|Q| \\ &- \frac{1}{2}\sum_{k=2}^{K}\left(x_k - Ax_{k-1}\right)^T Q^{-1}\left(x_k - Ax_{k-1}\right).\end{aligned} \tag{34.8}$$

In order to estimate unobserved states, in the Expectation Step of the algorithm, STRIDE implements Kalman Filter and Rauch-Tung-Striebel (RTS) smoother on the input data. STRIDE is an asymptotically converging algorithm that finds a local maximum. However, to capture the highest local maximum (global maximum), initialization of the super parameter is crucial. The algorithm has shown to be computationally rapid and inexpensive in [16].

A distinctive feature of STRIDE is its ability to upgrade for mobile sensing networks. STRIDEX is different formulation of STRIDE in which the position of mobile sensors are assumed known at each time step. Consequently, the new state-space equations for the mobile sensing problem are:

$$x_k = Ax_{k-1} + n_k \tag{34.9}$$

$$y_k = \Omega_k C x_k + \upsilon_k \tag{34.10}$$

As indicated in the governing equations, despite the time varying observation matrix $\Omega_k C$, this set of equations can still be solved with EM, knowing that mode shape regression matrix Ω_k is given for each time step. The mode shape regression matrix (MSR) is a transformation matrix that converts moving measurements at an arbitrary set of sensing nodes to the measurements of prefixed Virtual Probing Locations (VPL) that are stationary.

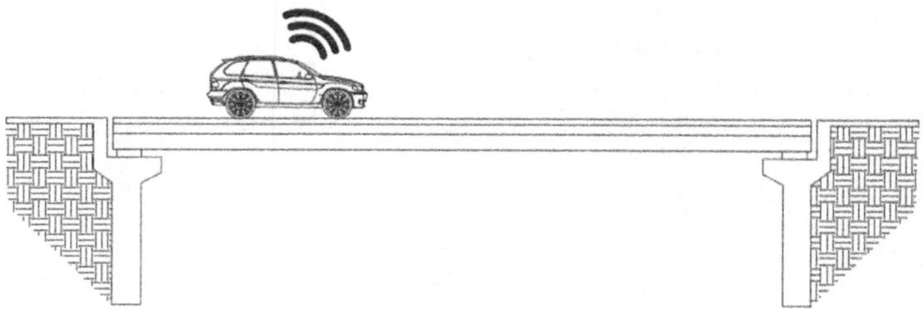

Fig. 34.1 Schematic idea of the mobile sensing

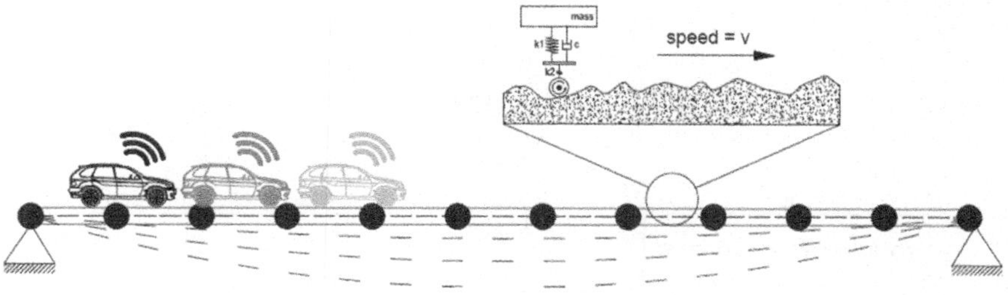

Fig. 34.2 Generic figure of the numerical model

One of the most significant challenges for utilizing STRIDEX for mobile sensing is the intense contamination of the structural vibration caused by physical distractions. Specifically, for bridge health monitoring application, sensing vehicles are passing over a pavement with a certain roughness (Figs. 34.1 and 34.2), whose spatial profile produces oscillating patterns that affect the bridge vibration measurements [3]. In other words, the acceleration time history that is recorded by the moving sensors is not collecting only the bridge dynamic response, but also the vibrations induced by the surface roughness. Therefore, in order to feed the identification algorithm by a valid input, a blind source separation (BSS) of mixed data should be used. In the literature, many algorithms have been proposed for separating mixed signals to their initial sources, each of which has some limitations. Among these, Independent Component Analysis (ICA) [18, 19] and Second Order Blind Identification (SOBI) [20, 21] has been practiced more often and performed better for civil engineering applications.

In the BSS literature, ICA is one of the commonly used and thoroughly evaluated algorithms [18]. In ICA, the principal assumption is that the sources are statistically independent and at least one of them is non-Gaussian. These two assumptions lead the algorithm to find the independent components of a mixed signal. Note that the independent sources may not be the sources that are actually mixed, but the most independent sources that can be pulled out of the mixtures, regardless of their physical interpretation. Moreover, the algorithm guarantees the most statistically possible independent sources, which is much stronger than uncorrelatedness. The algorithm tries to maximize non-Gaussianity by maximizing corresponding quantitative metrics, such as the forth order cumulant or Kurtosis.

Despite ICA's great ability to separate mixed data into its original sources, it is implicitly clear that it does not consider the sequence of samples by parameters such as autocorrelations or spectral densities [21]. In fact, it treats a signal as data points without orders in a way that the sequence is not of a great importance. However, in case of mixed signals, especially when they are contaminated versions of structural dynamic responses, samples order is indeed significant, since their temporal correlation make up the response spectrum from which the identification can be performed. To address this drawback, SOBI algorithm is utilized, which is a more robust alternative that benefits, whenever possible, of the temporal structure of the sources for facilitating their separation [20]. SOBI is more suitable for mixed signals with sources of different spectral contents, which is often the case in structural dynamics [21]. In this algorithm a weaker version of independency, which is covariance - a second order statistical moment - is of interest. As a result, SOBI is not only more desirable because of its ability of capturing temporal sequences for structural dynamics vibrations, but also is a computationally less expensive algorithm than ICA, since it does not consider higher than the second order independencies.

In this study, a numerical model of a bridge is made in OpenSees and the mobile sensor records are simulated. Primarily, these measures are not manipulated by the road profile roughness, so a sample generator MATLAB code has been developed to produce mixed signals. In the second phase, the polluted mixtures are fed through the SOBI algorithm to find both roughness-caused and the dynamic response vibrations of the bridge. Finally, a group of separated dynamic responses is input to the STRIDEX to identify natural frequencies and the mode shapes. Detailed explanation of each part will be given in the following sections. A flow chart of the proposed structural system identification algorithm is demonstrated below (Fig. 34.3).

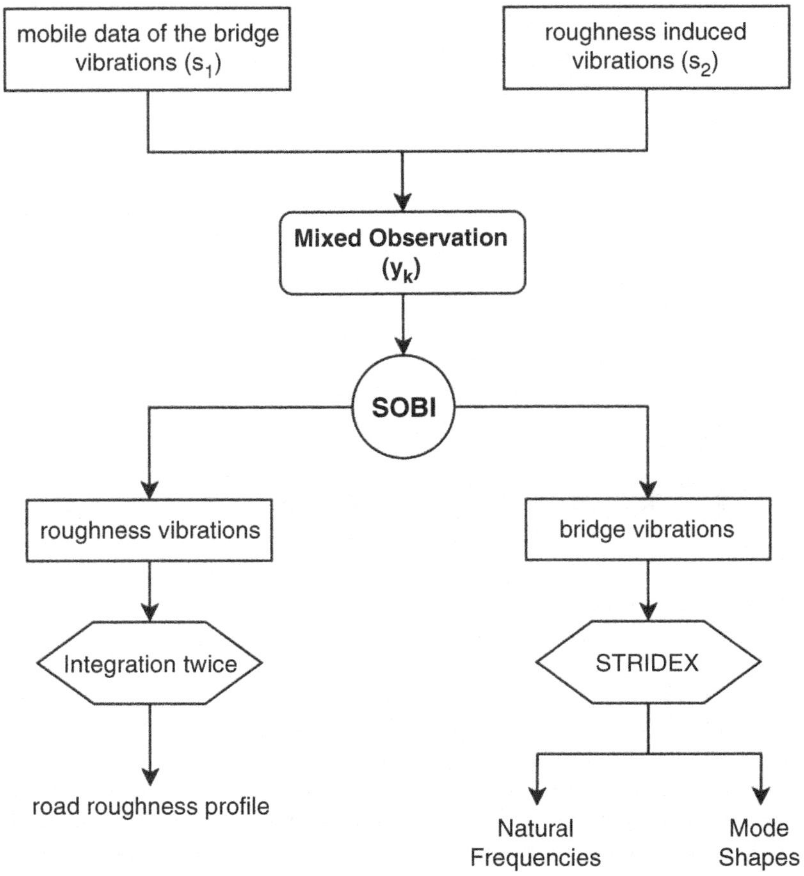

Fig. 34.3 Flow-chart of the proposed identification algorithm

34.2 Problem Statement

A 2D model of a 120 m long simply supported, single span bridge is built in OpenSees. The span is divided be 4800 pieces to make a 4801 degrees of freedom system and lateral accelerations are recorded at all DOFs for 114 sec. A set of nine wind load time histories is generated using spectral representation method [22] and applied on nine DOFs evenly spaced along the span. The wind intensity is mild and is supposed to act like an ambient random loading on the structural system.

After collecting data of the dense matrix function of accelerations at each DOFs per time, mobile sensor vectors has to be produced. Regarding that, based on the sensing vehicle speed and at each time step, one sample is chosen from a column that is attributed to the location of vehicle at that specific time [23]. The mobile sensing vector generation has been illustrated in the figure below (Fig. 34.4).

After composing mobile sensing data, it is needed to linearly add roughness-induced vibrations to these data to generate mixed signal samples. The road roughness profile is commonly assumed to be a zero mean stationary Gaussian random process [24]. In the present study, the following power spectral density (PSD) function for the roughness profile is utilized [25, 26]:

$$\varphi(n) = \varphi(n_0)\left(\frac{n}{n_0}\right)^{-2}; \quad (n_1 < n < n_2) \tag{34.11}$$

where n = spatial frequency (cycle/m); n_0 = discontinuity frequency of 1/2π (cycle/m); $\varphi(n_0)$ = road roughness coefficient (RRC) (m^3/cycle) dependent to the road condition (ISO 1995 gives five categories of very good, good, fair, poor, and very poor). Here, a good road condition with $\varphi(n_0) = 120e-6$ is selected and the road roughness profile is generated with spectral representation technique using Eq. 34.12:

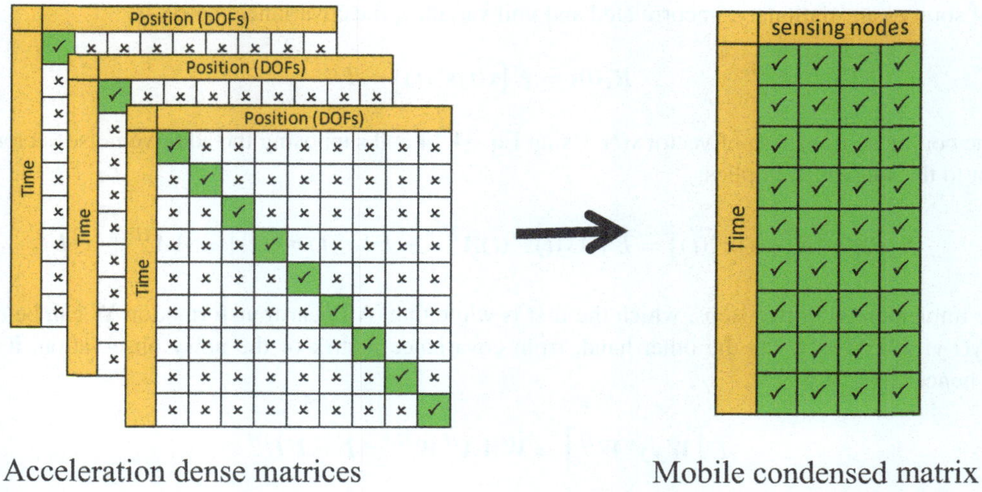

Fig. 34.4 Mobile sensing data generation using acceleration dense matrix

$$r(x) = \sum_{k=1}^{N} \sqrt{2\varphi(n_k)\,\Delta n} \cos(2\pi n_k x + \theta_k) \tag{34.12}$$

where N = spatial frequency depending on the length of the road, θ_k = random phase angle uniformly distributed from 0 to 2π; φ = PSD function for the road surface; n_k = wave number (cycle/m). Note that the generated random function is not a function of time, and as a result, cannot be directly added to the mobile sensor records. Moreover, since the mobile sensors collect accelerations, the roughness displacement profile has to be differentiated twice to give accelerations caused by the roughness. The spatial profile can be transformed to a time series signal with respect to the moving sensor's speed. In this numerical analysis, the sensor's velocity is tuned in a way that for each time step, sensor proceeds one DOF further and collect data on that point. That implies the mobile sensing vector can be acquired by diagonalization of the submatrix of the dense matrix.

At this point, since the generation of the mixed signal samples is completed, the goal is to identify the bridge structural system using these mixed signals that are intensively disturbed by the roughness-induced vibrations. In this study, the approach to solve this problem is to use SOBI method as a successful blind source separation algorithm for distinctive spectral sources. In the next section, SOBI is explained and implemented on the mixed data to draw out bridge dynamic vibrations from the mixed signals using which STRIDEX can identify the modal properties.

34.3 Second Order Blind Identification (SOBI)

In general, a mixed signal is not necessarily a weighted sum of several sources and may be a convolution or nonlinear mixture of the sources. However, in the present application, it is assumed that the mobile sensors seat on rigid vehicles that do not convolve linear mixture of roughness-caused and bridge dynamics vibrations. Therefore, the noisy mixed signals can be expressed as:

$$x_i(t) = \sum_{j=1}^{n} a_{ij} s_j(t) + \sigma_i(t), \qquad i = 1, 2, \ldots, n \tag{34.13}$$

and in the matrix notation:

$$\boldsymbol{x}(t) = \boldsymbol{A}\boldsymbol{s}(t) + \boldsymbol{\sigma}(t) = \boldsymbol{y}(t) + \boldsymbol{\sigma}(t) \tag{34.14}$$

where $\boldsymbol{A}$ is the mixing matrix and $\boldsymbol{\sigma}(t)$ is an uncorrelated noise vector. $\boldsymbol{s}(t)$ is also a vector of sources at time t and apparently, $\boldsymbol{x}(t)$ is the mixed observation. It is assumed that the mixture matrix is an unknown constant matrix that is to be estimated based on the observations. Note that it is a common assumption that the sources are unit variance random processes, so $E\left[s_i^2(t)\right] = 1$for $i = 1, 2, \ldots, n$.

As a result, if sources are stationary, uncorrelated and unit variance, the covariance matrix is:

$$\boldsymbol{R}_s(0) = E\left[\boldsymbol{s}(t)\boldsymbol{s}^*(t)\right] = \boldsymbol{I} \tag{34.15}$$

where $s^*(t)$ is the conjugate transpose of vector $s(t)$. Using Eq. 34.14 and supposing the additive noise vector is uncorrelated and independent to the sources, it implies:

$$\boldsymbol{R}_x(0) = E\left[\boldsymbol{x}(t)\boldsymbol{x}^*(t)\right] = E\left[\boldsymbol{A}\boldsymbol{s}(t)\boldsymbol{s}^*(t)\boldsymbol{A}^H\right] + E\left[\boldsymbol{\sigma}(t)\boldsymbol{\sigma}^*(t)\right] = \boldsymbol{A}\boldsymbol{A}^H + \sigma^2\boldsymbol{I} \tag{34.16}$$

SOBI can be implemented in two steps, which the first is whitening. A linear transformation **W** can be easily calculated such that: $E[\boldsymbol{W}\boldsymbol{y}(t)\boldsymbol{y}^*(t)\boldsymbol{W}^H] = \boldsymbol{I}$. On the other hand, from covariance matrix of the noisy observation, it is obvious that: $E[\boldsymbol{y}\boldsymbol{y}^*] = \boldsymbol{A}\boldsymbol{A}^H$, hence:

$$E\left[\boldsymbol{W}\boldsymbol{y}\boldsymbol{y}^*\boldsymbol{W}^H\right] = \boldsymbol{W}\boldsymbol{A}\boldsymbol{A}^H\boldsymbol{W}^H = \boldsymbol{I} = \boldsymbol{U}\boldsymbol{U}^H \tag{34.17}$$

So, using the whitening matrix $\boldsymbol{W}$, a unitary matrix $\boldsymbol{U}$ can be found such that: $\boldsymbol{U} = \boldsymbol{W}\boldsymbol{A}$. Note that the actual observation is not $\boldsymbol{y}$, but $\boldsymbol{x}$, which has noise effects, thus if say $z(t) = \boldsymbol{W}\boldsymbol{x}(t)$:

$$E\left[zz^*\right] = \boldsymbol{E}\left[\boldsymbol{W}\boldsymbol{x}\boldsymbol{x}^*\boldsymbol{W}^H\right] = \boldsymbol{W}\boldsymbol{R}_{\boldsymbol{x}}(0)\boldsymbol{W}^H \tag{34.18}$$

The second step is to determine the unitary matrix U, considering time lagged observation covariance matrix, as shown below:

$$\boldsymbol{R}_z^w(\tau) = E\left[z(t+\tau)z^*(t)\right] = \boldsymbol{W}E\left[\boldsymbol{x}(t+\tau)\boldsymbol{x}^*(t)\right]\boldsymbol{W}^H = \boldsymbol{W}\boldsymbol{A}E\left[\boldsymbol{s}(t+\tau)\boldsymbol{s}^*(t)\right]\boldsymbol{A}^H\boldsymbol{W}^H, \quad \forall\tau \neq 0 \tag{34.19}$$

which gives $\boldsymbol{R}_z^w(\tau) = \boldsymbol{U}\boldsymbol{R}_s(\tau)\boldsymbol{U}^H$. The key idea to find $\boldsymbol{U}$ is that since it is a unitary matrix, and it was previously assumed that the sources are uncorrelated, implying that $\boldsymbol{R}_s(\tau)$ is diagonal, therefore it is realized that any whitened covariance matrix $\boldsymbol{R}_z^w(\tau)$ is diagonalizable by the unitary transform $\boldsymbol{U}$. To determine such a matrix, it is just required to decompose the time-lagged covariance matrix of the whitened observation $\boldsymbol{R}_z^w(\tau)$ and accordingly, $\boldsymbol{A}$ can be found from $\boldsymbol{U} = \boldsymbol{W}\boldsymbol{A}$ and finally, sources as $\boldsymbol{x}(t) = \boldsymbol{A}\boldsymbol{s}(t)$. The blind source separation using SOBI is completed here. The entire SOBI calculation is coded through a built-in MATLAB toolbox for SOBI [27] and the package has been used in the present study.

After the separation procedure on the mixed signals via SOBI, the separated source which is recognized as the bridge vibration is fed to the STRIDEX algorithm and the modal properties are extracted.

34.4 Numerical Modeling of a 4800-DOF Beam

To evaluate the proposed procedure for structural identification, the aforementioned simply supported beam is assessed as a simplified bridge model. In a mobile sensing procedure, the dynamic response of the bridge for six moving sensors according to a specific movement path as shown in Fig. 34.5 are simulated via the dense matrix. Note that for each sensor, the regional distance is scanned twice to collect more data. Sensor placement is a sensitive parameter that affects the identification quality. In this study, the same placement as used to validate STRIDEX in [15] has been considered. However, thorough discussions on sensor placement can also be found in [32, 32] for different types of structures. The experimental results of various sensor placements are also shown in [32] and effects are discussed.

The sensors' data are dynamic responses of the structure, which are not realistic measurements, since they are not affected by the roughness-induced vibrations. Therefore, a time signal of vertical accelerations caused by road roughness is generated via spectral representation technique and added up with the bridge acceleration in a synchronized fashion. Finally a Gaussian noise is added to consider measurement errors of the sensors. The PSDs for both pure and mixed signals are presented in Fig. 34.6. As illustrated, the frequency content of the bridge vibration is clearly indicator of the natural frequencies of the model via the location of the sharp spikes on the PSD. However, this clarity is not observable anymore on the mixed signal PSD. Correspondingly, structural identification using mixed signals does not seem a plausible desire, since the input is highly disturbed.

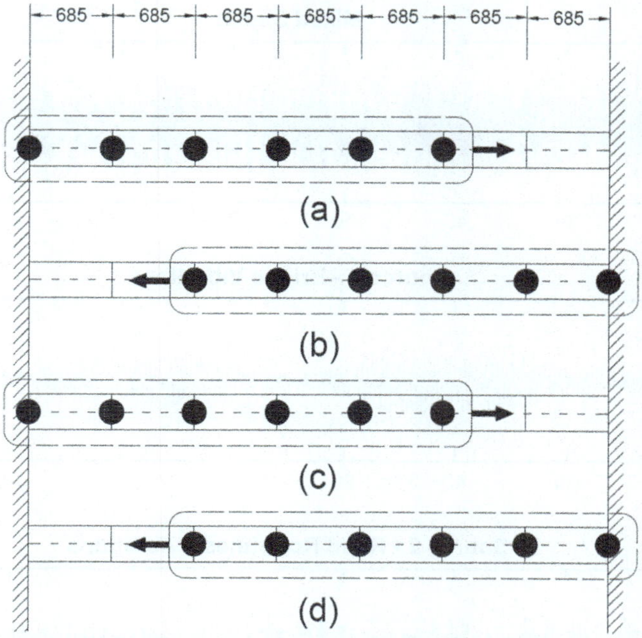

Fig. 34.5 Configuration of moving sensors. a-d showing scanned paths

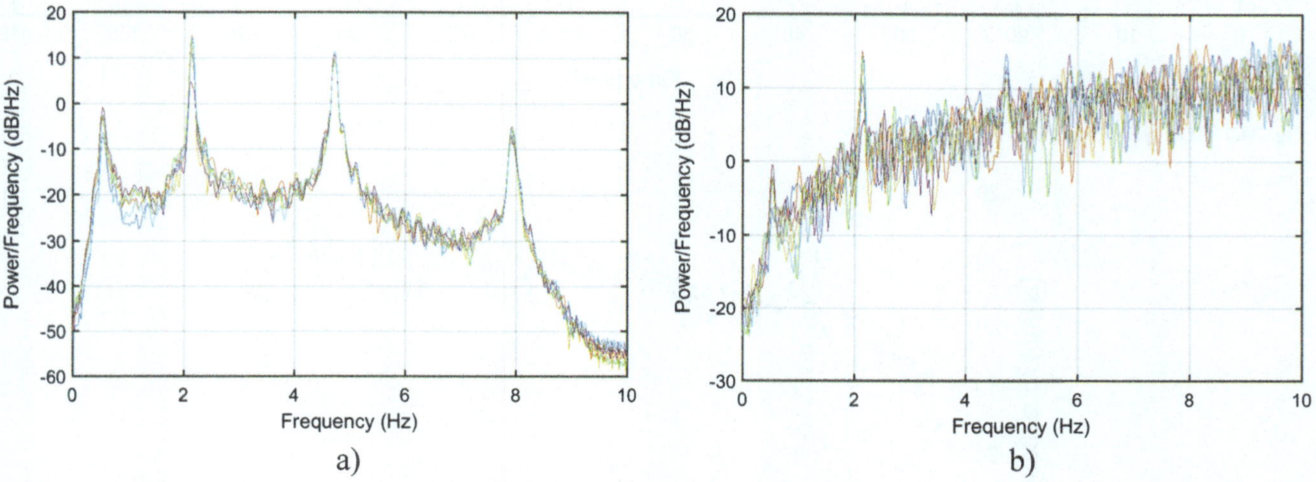

Fig. 34.6 Welch Power Density Spectra (**a**) Bridge vibration only signal (**b**) Mixed signal

After generating samples, the mixed signals are fed to the SOBI algorithm in order to draw out the source signals, which is expected to be bridge dynamic and roughness-caused vibrations. An example of a mixed signal and the corresponding separated sources from SOBI versus time is shown in Fig. 34.7. The spectra for the mixed and source signals are displayed in Fig. 34.8 as well. The first source spectrum, which is correlated to be the bridge vibration, is showing spikes in the same location as the bridge natural frequencies, although these spikes are not visible in the other source and also the mixed spectra. It should be noted that although the first source is not identical to the reference signal in time, it is a linearly transformed version which is practically of the same value, thanks to the SOBI formulations.

On the other hand, if compare the second detected source against the generated roughness-caused acceleration, a significant compliance can be realized, as indicated in Fig. 34.9. This can be promising for the road roughness identification using mobile sensors' records. In fact, since the SOBI is successful to extract roughness-induced part of a mixed signal, the operator can simply find the road roughness pattern by differentiating the mined signal twice and find the roughness profile.

The last phase is to use the implementation of the extracted sources that are correlated to the bridge dynamic responses, to the STRIDEX identification algorithm. For each mobile sensor, one source is mined out as the bridge vibration and fed to the STRIDEX. In this identification algorithm, the accelerations of the mobile sensors at each time steps and locations are

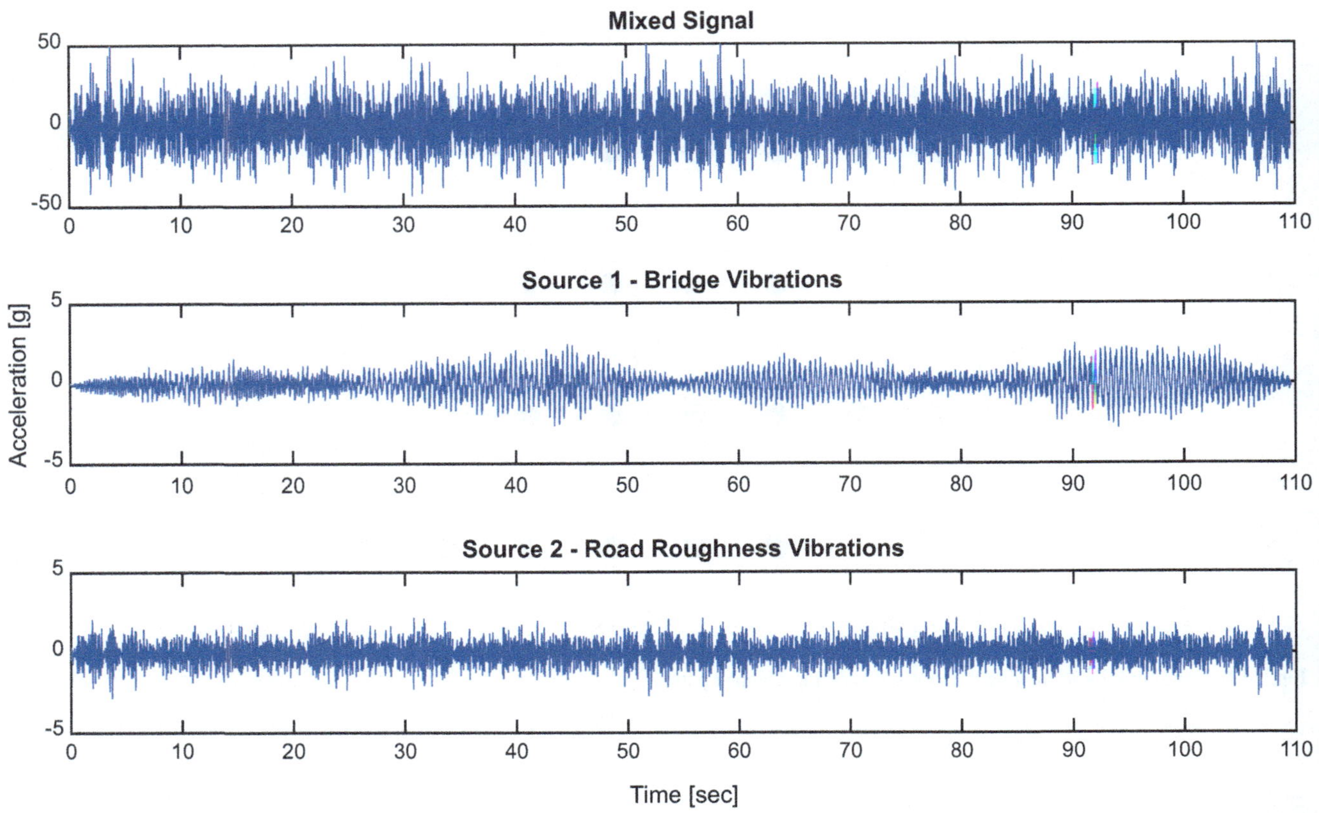

Fig. 34.7 SOBI performance on a mixed signal

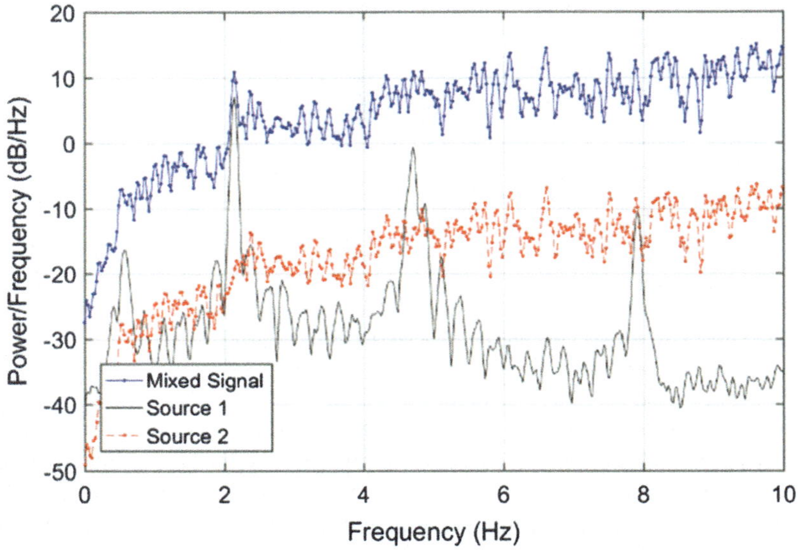

Fig. 34.8 Spectra for mixed and separated signals

mapped to the associated accelerations on preset virtual probing locations (VPL) that are stationary. Therefore, the method is able to enhance the identified modal properties by changing the VPLs, but feeding only one set of data. In this study, three VPLs are considered and their locations are shown in Table 34.1.

The identification algorithm is executed and the results are shown in Fig. 34.10 and Table 34.2. Note that the maximum likelihood-based identification methods can be concluded as a set of sensitivity metrics which are more informative than scalar results shown in this study [32]. However, for simplicity, in the presented evaluation, the representative scalar metrics

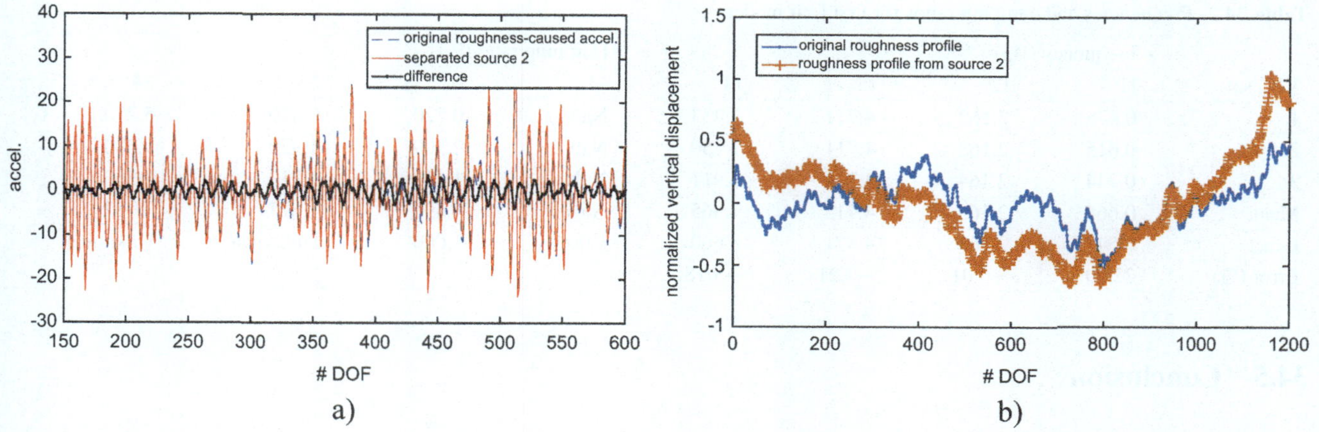

Fig. 34.9 Comparison between generated and separation-derived roughness (**a**) Acceleration difference (**b**) Profile difference

Table 34.1 Virtual probing locations

	DOF label					
VPL set	1	2	3	4	5	6
1	685	1370	2055	2740	3425	4110
2	342	1027	1712	2397	3082	3767
3	1027	1712	2397	3082	3767	4452

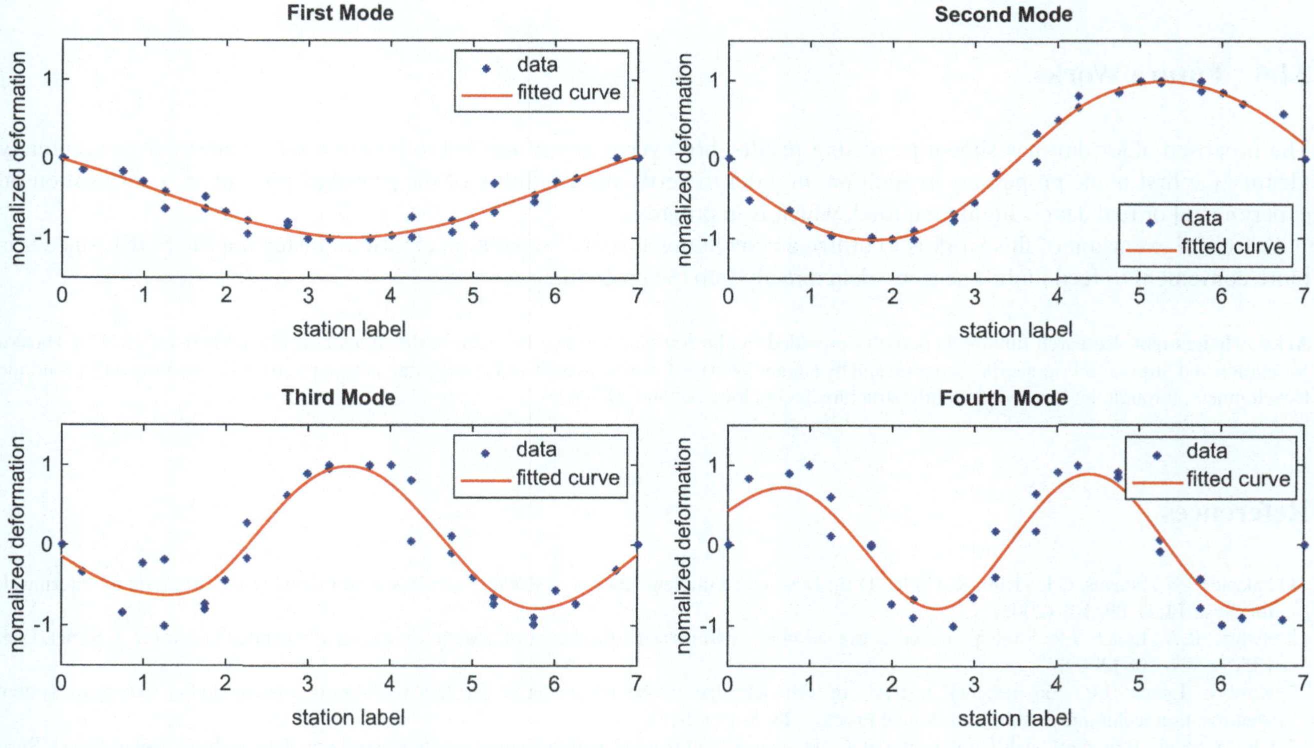

Fig. 34.10 Mode shapes derived by STRIDEX using separated signals

are assessed. As it is shown, the algorithm is able to identify fundamental modes clearly and estimate natural frequencies and mode shapes. In Fig. 34.10, for each mode shape, asterisk points are showing VPL locations for five different VPL sets and the curve is a spline that is best-fitted for these points. The algorithm is also fairly capable to estimate Rayleigh damping ratios, which were tuned in a way to show 1% modal damping on first and second modes. However, the algorithm is not very effective to identify the first mode and its corresponding attributes.

Table 34.2 Frequencies and damping ratios for first four modes

	Frequency (Hz)				Damping ratio (%)			
VPL set	f1	f2	f3	f4	ξ1	ξ2	ξ3	ξ4
1	0.638	2.162	4.711	7.933	NaN	0.720	0.870	5.270
2	0.615	2.168	4.711	7.750	NaN	2.300	1.950	9.580
3	0.744	2.164	4.722	7.911	NaN	1.450	2.210	8.360
Mean	0.666	2.165	4.715	7.865	NaN	1.490	1.677	7.737
Exact	0.541	2.165	4.871	8.660	1.000	1.000	Unknown	Unknown
Error (%)	22.99	−0.01	−3.21	−9.18				

34.5 Conclusion

In this study, a new approach for structural system identification using mobile sensing networks has been proposed. The idea is to separate mixed signals collected by mobile sensors to the producing sources, one of which is the bridge dynamic response. The separated signal may not be identical to the producing source signal, however it is closely related and inherit the same temporal characteristics.

The numerical modeling and validation have exhibited an acceptable performance for the proposed procedure. The Second Order Blind Identification (SOBI) was successful to separate linearly mixed signals. Moreover, the roughness part of the separated sources is also highly correlated to the original roughness profile. This observation offers that the proposed algorithm is capable to have two-folded applications on both bridge and pavement identification simultaneously, without any preliminary information needed.

34.6 Future Works

The proposed algorithm has shown promising results, however it is still needed to be more robust in order to accurately identify the first mode properties. In addition, in order to verify the eligibility of the proposed procedure, a verification on experimental or real data is highly required, which is in progress.

A fruitful extension of this work is to utilize a convolutional source separation algorithm instead of the SOBI, since it is more convenient to feed mobile sensors' data directly into the procedure.

Acknowledgement Research funding is partially provided by the National Science Foundation through Grant No. CMMI-1351537 by Hazard Mitigation and Structural Engineering program and by a grant from the Commonwealth of Pennsylvania, Department of Community and Economic Development, through the Pennsylvania Infrastructure Technology Alliance (PITA).

References

1. Pakzad, S.N., Fenves, G.L., Kim, S., Culler, D.E.: Design and implementation of scalable wireless sensor network for structural monitoring. J. Inf. Syst. **14**(1), 89–101 (2008)
2. Swartz, R.A., Lynch, J.P.: Strategic network utilization in a wireless structural control system for seismically excited structures. J. Struct. Eng. **135**(5), 597–608 (2009)
3. Kim, J., Lynch, J.P.: Experimental analysis of vehiclebridge interaction using a wireless monitoring system and a two-stage system identification technique. Mech. Syst. Signal Process. **28**, 3–19 (2012)
4. Cho, S., et al.: Structural health monitoring of a cable-stayed bridge using wireless smart sensor technology: data analyses. Smart Struct. Syst. **6**(5–6), 461–480 (2010)
5. Dorvash, S., Pakzad, S.N., Cheng, L.: An iterative modal identification algorithm for structural health monitoring using wireless sensor networks. Earthq. Spectra. **29**(2), 339–365 (2013)
6. Sohn, H., Farrar, C.R., Hemez, F.M., Czarnecki, J.J.: A Review of Structural Health Review of Structural Health Monitoring Literature 1996–2001. Los Alamos National Laboratory, Los Alamos, NM (2002)
7. Farrar, C.R.: Structural health monitoring: technological advances to practical implementations. Proc. IEEE. **104**(8), 1508–1512 (2016)
8. Juang, J.-N.: Applied System Identification, 1st edn. Prentice Hall, New York (1994)
9. Peeters, B., De Roeck, G.: One-year monitoring of the Z 24-bridge: environmental effects versus damage events. Earthq. Eng. Struct. Dyn. **30**(2), 149–171 (2001)

10. Matarazzo, T.J., Pakzad, S.N.: Direct state-space models for time-varying sensor networks. Proc. 10th Int. Structural Health Monitoring 2015. **7**, 59–65 (2015)
11. Chang, M., Asce, S.M., Pakzad, S.N., Asce, A.M.: Observer Kalman filter Identification for output-only systems using interactive structural modal identification toolsuite. J. Bridg. Eng. **19**, 1–11 (2014)
12. Chang, M., Pakzad, S.N.: Observer Kalman filter identification for output-only systems using interactive structural modal identification toolsuite. J. Bridg. Eng. **19**(5), 4014002 (2013)
13. Juang, J.-N., Pappa, R.S.: An eigensystem realization algorithm for modal parameter identification and model reduction. J. Guid. **8**(5), 620–627 (1985)
14. Van Overschee, P., De Moor, B.L.: Subspace Identification for Linear Systems: Theory—Implementation—Applications. Springer Science & Business Media, Berlin (2012)
15. Matarazzo, T.J., Pakzad, S.N.: Scalable structural modal identification using dynamic sensor network data with STRIDEX. Comput. Civ. Infrastruct. Eng. **33**(1), 4–20 (2018)
16. Matarazzo, T.J., Pakzad, S.N.: STRIDE for structural identification using expectation maximization: iterative output-only method for modal identification. J. Eng. Mech. **142**(4), (2016)
17. Shumway, R.H., Stoffer, D.S.: Time Series Analysis and Its Applications, vol. 97. Springer, Cham (2011)
18. Oja, E., Hyva, A.: Independent component analysis: algorithms and applications. Neural Netw. **13**, 411–430 (2000)
19. Kerschen, G.Ã., Poncelet, F., Golinval, J.: Physical interpretation of independent component analysis in structural dynamics. Mech. Syst. Signal Process. **21**, 1561–1575 (2007)
20. Belouchrani, A., Abed-Meraim, K., Cardoso, J.-F., Moulines, E.: A blind source separation technique using second-order statistics. IEEE Trans. Signal Process. **45**(2), 434–444 (1997)
21. Poncelet, F., Kerschen, G., Golinval, J., Verhelst, D.: Output-only modal analysis using blind source separation techniques. Mech. Syst. Signal Process. **21**, 2335–2358 (2007)
22. Shinozuka, M., Deodatis, G.: Simulation of stochastic processes by spectral representation. Appl. Mech. Rev. **44**(4), 191–204 (1991)
23. Matarazzo, T.J., Pakzad, S.N.: Truncated physical model for dynamic sensor networks with applications in high-resolution mobile sensing and BIGDATA. J. Eng. Mech. **142**(5), 4016019 (2016)
24. González, A., O'brien, E.J., Li, Y.-Y., Cashell, K.: The use of vehicle acceleration measurements to estimate road roughness. Veh. Syst. Dyn. **46**(6), 483–499 (2008)
25. Kong, X., Cai, C.S., Kong, B.: Numerically extracting bridge modal properties from dynamic responses of moving vehicles. J. Eng. Mech. **142**(2011), 4016025 (2016)
26. Huang, D., Wang, T.-L.: Impact analysis of cable-stayed bridges. Comput. Struct. **43**(5), 897–908 (1992)
27. McNeill, S.: Blind Modal Identification (BMID) toolbox. MATLAB (2011)
28. McNeill, S.I., Zimmerman, D.C.: A framework for blind modal identification using joint; approximate diagonalization: Mechanical Systems and Signal Processing. **22**(7), 1526–1548 (2008)
29. Chang, M., Pakzad, S.N.: Optimal sensor placement for modal Identi fi cation of bridge systems considering number of sensing nodes. J. Bridg. Eng. **19**(6), 1–10 (2014)
30. Chang, M., Pakzad, S.N.: Optimal sensor configuration for flexible structures with multi-dimensional mode shapes. Smart Mater. Struct. **24**(5), 55012 (2015)
31. Valeti, B., Matarazzo, T.J., Pakzad, S.N.: Experimental study on wireless mobile sensor configurations for output-only modal identification of a beam testbed. In: Sensors and Instrumentation, vol. 5, pp. 71–77. Springer, Cham (2017)
32. Matarazzo, T.J., Pakzad, S.N.: Sensitivity metrics for maximum likelihood system identification. ASCE-ASME J. Risk Uncertain. Eng. Syst. Part A Civ. Eng. **2**, B4015002 (2015)

Chapter 35
NDE of Additively Manufactured Parts via Directly Bonded and Mechanically Attached Electromechanical Impedance Sensors

C. Tenney, M. Albakri, C. B. Williams, and P. Tarazaga

Abstract Additive Manufacturing (AM) allows increased complexity which poses challenges to quality-control (QC) and non-destructive evaluation (NDE) of manufactured parts. The lack of simple, reliable, and inexpensive methods for NDE of AM parts is a significant obstacle to wider adoption of AM parts.

Electromechanical impedance measurements have been investigated as a means to detect manufacturing defects in AM parts. Impedance-based NDE utilizes piezoelectric wafers as collocated sensors and actuators. Taking advantage of the coupled electromechanical characteristics of piezoelectric materials, the mechanical characteristics of the part under test can be inferred from the electrical impedance of the piezoelectric wafer. Previous efforts have used piezoelectric wafers bonded directly to the part under test, which imposes several challenges regarding the applicability and robustness of the technique. This paper investigates the use of an instrumented clamp as a solution for measuring the electromechanical impedance of the part under test. The effectiveness of this approach in detecting manufacturing defects is compared to directly bonded wafers.

Keywords Electromechanical Impedance · Non-Destructive Evaluation · Additive Manufacturing · Piezoelectrics · Manufacturing Defects

35.1 Introduction

Additive Manufacturing (AM) processes' layer-wise approach to fabrication provides a means to greatly increase part complexity and thus functionality. However, this added geometric complexity also poses challenges to quality-control (QC) and non-destructive evaluation (NDE) of manufactured parts. Current methods, such as x-ray computed tomography (CT), permit detailed imaging of complex parts [1], but can be slow and expensive. The lack of simple, reliable, and inexpensive methods for NDE of AM parts is a significant obstacle to their wider adoption.

Electromechanical impedance measurements have been investigated as a means to detect manufacturing defects in AM parts [2]. Impedance-based NDE utilizes piezoelectric wafers as collocated sensors and actuators. Taking advantage of the coupled electromechanical characteristics of piezoelectric materials, the mechanical characteristics of the part under test can be inferred from the electrical impedance of the piezoelectric wafer. Previous efforts have used piezoelectric wafers bonded directly to the part under test, which imposes several challenges regarding the applicability and robustness of the technique [2, 3].

This paper investigates the use of an instrumented clamp that can be temporarily affixed onto a part for measurement of its electromechanical impedance. Moving towards a mechanically attached solution would eliminate the need to wait for a bonding agent to cure, and would also address issues related to removing the sensor following testing. However, a mechanical

C. Tenney (✉)
Vibrations and Adaptive Structures Testing Laboratory (VAST), Design, Research & Education for Additive Manufacturing Systems (DREAMS) Lab, Virginia Polytechnic Institute and State University (Virginia Tech), Blacksburg, VA, USA
e-mail: charten@vt.edu

M. Albakri · P. Tarazaga
Vibrations and Adaptive Structures Testing Laboratory (VAST), Virginia Polytechnic Institute and State University (Virginia Tech), Blacksburg, VA, USA

C. B. Williams
Design, Research & Education for Additive Manufacturing Systems (DREAMS) Lab, Virginia Polytechnic Institute and State University (Virginia Tech), Blacksburg, VA, USA

S. Pakzad (ed.), *Dynamics of Civil Structures, Volume 2*, Conference Proceedings of the Society for Experimental Mechanics Series, https://doi.org/10.1007/978-3-319-74421-6_35

attachment mechanism makes the sensor larger and heavier, which could affect the sensitivity of the measurement. To address this potential issue, the effectiveness of this approach in detecting manufacturing defects is compared to directly bonded wafers.

In order to evaluate the use of the instrumented clamp, this paper proceeds as follows. First, a theoretical overview of impedance-based NDE is presented. Next, the design and fabrication of test specimens, instrumentation of the clamp and specimens, and the testing parameters are laid out. Then the results of testing are presented and analyzed. Finally, some concluding remarks and suggestions for future work are made.

35.2 Impedance-Based Non-destructive Evaluation

Impedance-based NDE examines the dynamic vibrational response of a test object and compares that response to some established baseline. The rationale for this comparison is that the changes in mass, stiffness, and damping that result from manufacturing defects will manifest as changes in the object's dynamic response, affecting the frequency and magnitude of the object's resonances.

Interrogation of the dynamic response is accomplished using piezoelectric transducers as collocated sensors and actuators. Since the mechanical and electric properties of a piezoelectric transducer are coupled, its electrical impedance can be related to the mechanical properties of the transducer and the object to which it is bonded. In this work, the piezoelectric transducer takes two forms: a macro fiber composite (MFC) patch that consists of lead zirconate titanate (PZT) strands adhered within two layers of polyimide film [4], as well as a monolithic PZT wafer.

Instrumenting the test object is done by bonding the piezoelectric transducer to a free surface, as shown in Fig. 35.1, then electrically exciting the transducer with an applied voltage. For the monolithic wafers, voltage applied through the thickness causes response in the 31 mode: electrical stimulus is applied in the thickness (3-direction), and strain is developed along the length (1-direction). In the MFC patch being used the response is in the 33-mode, but due to the orientation of the fibers and electrodes the excitation applied to the bonded surface is similar. For the monolithic wafer, we can solve analytically for the electrical impedance of the piezoelectric in terms of material properties and geometry. A similar derivation could be done for the MFC patch.

First, the constitutive equations for linear piezoelectricity that capture the 31 mode are shown in Eq. 35.1 [5]

$$\varepsilon_{11} = s_{11}^{E}\sigma_{11} + d_{13}E_3 \tag{35.1}$$

$$D_3 = \left(d^T\right)_{31}\sigma_{11} + \epsilon_{33}^{\sigma}E_3$$

where ε_{11} is the normal strain in the 1-direction, s_{11}^{E} is the complex elastic compliance constant measured at constant electric field, σ_{11} is the normal stress in the 1-direction, d_{13} and $(d^T)_{31}$ are piezoelectric constants, E_3 is the electric field strength in the 3-direction, D_3 is the charge displacement in the 3-direction, and ϵ_{33}^{σ} is the complex permittivity in the 3-direction measured at constant stress.

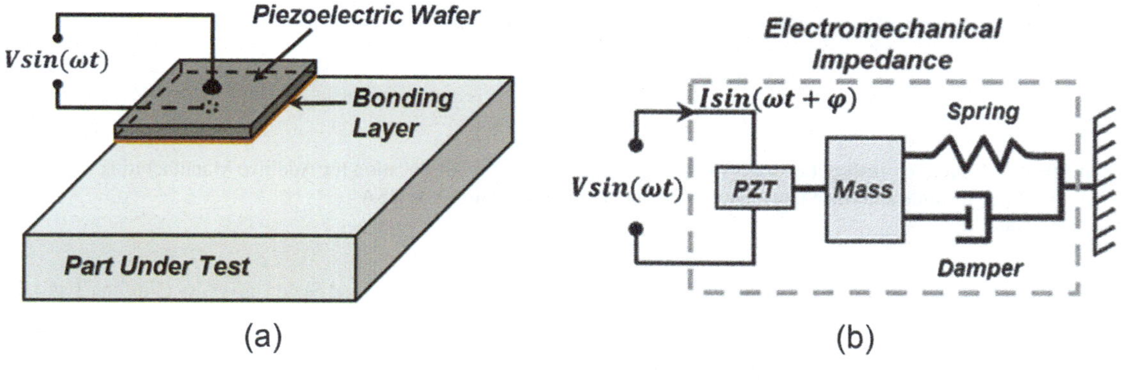

Fig. 35.1 (**a**) a piezoelectric transducer bonded to a test object, (**b**) an abstracted system diagram for the arrangement in (**a**) when excited at a particular frequency [3]

Table 35.1 Process parameters

Machine model	Stratasys Fortus 450mc	*Nozzle temperature*	355 °C
Model material	Stratasys Nylon 12	*Build chamber temperature*	120 °C
Support material	Stratasys SR-110	*Layer height*	0.01″ (0.254 mm)

For a test object, the dynamic response to excitation at any particular frequency can be approximated by a single degree of freedom system as shown in Fig. 35.1. The parameters of the system can be written m_r, k_r, and ζ_r, denoting mass, stiffness, and damping, respectively.

Then, assuming the piezoelectric transducer is perfectly bonded to the test object, Eq. 35.2 shows the electrical impedance of the transducer written in terms of the properties of the piezoelectric and of the test object [6, 7].

$$Z(\omega) = \left[i\omega \frac{bl}{h} \left(\frac{d_{11}^2}{s_{11}^E} \left(\frac{\tan(kl)}{kl} \left(\frac{Z_{pzt}}{Z_{pzt} + Z_{st}} \right) - 1 \right) + \epsilon_{33}^{\sigma} \right) \right]^{-1} \tag{35.2}$$

where $Z_{pzt} = -i\,(bh/l)\left(s_{11}^E \omega \tan(kl)/kl\right)^{-1}$ is the piezoelectric transducer impedance under short-circuit conditions, $Z_{st} = 2\zeta_r(k_r m_r)^{1/2} + i(m_r\omega^2 - k_r)/\omega$ is the mechanical impedance of the test object, $k = \omega\left(\rho s_{11}^E\right)^{1/2}$ is the wavenumber, ρ, b, h, and $2l$ are the piezoelectric density, width, thickness, and length, respectively.

In this study, a baseline impedance signature will be established, then compared to the impedance signature of a damaged test object. Differences between the two signatures will be interpreted as damage. This procedure has been carried out in many other studies, including the ones conducted by the authors of this paper [2, 3]. In order to quantitatively compare the signatures, Root-Mean-Square Deviation (RMSD) and the Correlation Coefficient (r) will be used as damage metrics. These metrics can be calculated using Eqs. 35.3 and 35.4.

$$RMSD = \sqrt{\sum \frac{(Z_D - Z_{BL})^2}{Z_{BL}^2}} \tag{35.3}$$

$$r = 1 - \left| \frac{n\sum Z_D Z_{BL} - \sum Z_D \sum Z_{BL}}{\sqrt{\left[n\sum Z_D^2 - \left(\sum Z_D\right)^2\right]\left[n\sum Z_{BL}^2 - \left(\sum Z_{BL}\right)^2\right]}} \right| \tag{35.4}$$

where Z_D is the impedance signature of the part being tested, Z_{BL} is the baseline impedance signature, and n is the number of data points in each impedance signature. As defined above, these metrics converge to zero when two signatures are identical, and increase as the signatures become less similar.

35.3 Test Specimen and Clamping Device Specifications

The test specimens used in this study, shown in Fig. 35.2, are rectangular beams measuring 72.5 × 7.25 × 5 mm, not including the 1 mm height of the L-shaped rails used to align the piezoelectric transducers. Each specimen is fabricated using a material extrusion additive manufacturing process (also referred to as "Fused Filament Fabrication", FFF). This process builds up a three dimensional structure by extruding filament through a heated nozzle. The machine model and process parameters can be found in Table 35.1. The part material, nylon, is intended to be representative of a common plastic type found in a range of polymer-based additive manufacturing (AM) processes. Additionally, the parts were fabricated with solid infill.

Following fabrication, the parts were post-processed by soaking them in a basic solution that dissolves the support material, according the material supplier's specifications. All support material had been removed after 2 h. Though it might have been preferable to fabricate the parts with no support material at all, the toolpath software for the Stratasys Fortus (Insight 10.1) automatically adds a raft of support material to every part. Finally, the parts were rinsed in fresh water, dried by hand, and then left to thoroughly air-dry for 24 h.

The clamping mechanism, shown in Fig. 35.2, used in this study is a small, metal, c-shaped clamp with a threaded rod that can be advanced to adjust the clamping force. The clamp fits within a volume of $105 \times 65 \times 60$ mm when the threaded rod is retracted to minimize the height of the clamp. When fully extended, the clamp will accept an object with a height of approximately 73 mm.

In preparation for testing, an MFC patch was bonded to the clamp, and a monolithic PZT wafer was bonded to one of the beam specimens. On the clamp, a Smart Materials M-2814-P1 MFC patch was bonded using cyanoacrylate[1] to the base, as shown in Fig. 35.3. This MFC patch provides exposed soldering points to connect to the interior electrodes. On the side of the clamp, a strain gage was bonded similarly to monitor the strain due to bending. This strain will be used to infer and monitor the clamping force.

On the beam specimen, a 12.7×6.35 mm PZT-5H wafer was bonded using cyanoacrylate to the area defined by the L-shaped rails described previously. This can be seen in Fig. 35.4. The wafer's top and bottom surfaces are nickel-coated to serve as electrodes. This arrangement leaves one surface inaccessible after bonding the wafer to the beam. To allow access, a short section of copper tape with a conductive adhesive was applied to the wafer before bonding.

35.4 Experimental Setup

Two configurations are considered in this paper. The first will be referred to as the "bonded configuration": a monolithic PZT wafer is bonded directly to a beam specimen, and then the wafer-and-beam system is suspended by monofilament. The second will be referred to as the "mechanically attached configuration": an MFC patch is bonded to the clamping mechanism, an un-instrumented-beam is clamped within it, and then the whole assembly is suspended by monofilament. By suspending each configuration, a free boundary condition is simulated.

Once the beam specimens and the clamp had been instrumented, impedance signatures were taken, as shown in Fig. 35.4a. For both cases, the impedance signature was measured using a Keysight E4990A impedance analyzer: a device that precisely

(a) (b)

Fig. 35.2 (**a**) a rendering of a beam specimen, (**b**) the clamping mechanism

(a) (b)

Fig. 35.3 (**a**) the MFC bonded to the bottom of the clamp, (**b**) the strain gage bonded to the side of the clamp

[1]Known by its trade name: Super Glue.

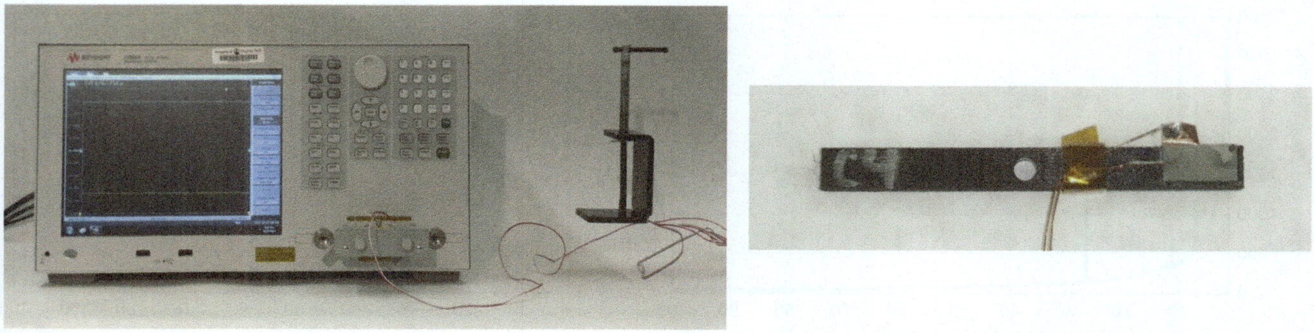

(a) (b)

Fig. 35.4 (**a**) the suspended clamped beam specimen with the MFC patch connected to the impedance analyzer, (**b**) the free-beam with bonded PZT wafer (rectangle at right end) and damage applied (circle at center span)

delivers a voltage signal and measures current in order to determine electrical impedance. The signature was sampled at 1 Hz increments from 1000 Hz to 100 kHz. The excitation at each frequency was a 1 V peak-to-peak sinusoidal signal. Before taking each measurement, the excitation was applied to the test object to allow it time to settle. Without time to settle, the impedance signature has many peaks and troughs near any resonance of the system; a parsimonious explanation for this is that the shift from one frequency to the next was creating broadband excitation that needed to be damped out before the response at precisely one frequency could be accurately measured. For the mechanically attached configuration, it was observed that approximately 100–1000 ms was required to avoid spurious results, depending on the excitation frequency. The bonded configuration alone was observed to require only about 10 ms, presumably due to the higher damping of the plastic as compared to the metal of the clamp.

For the mechanically attached configuration, a Vishay 2110A strain gage conditioner was used to monitor the clamping force. The clamping force is relevant to the impedance signature because strains in a material affect the wave propagation speed within it, and therefore shift its resonant frequencies. The output gain of the conditioner was adjusted so that the output voltage matched the output of a force transducer being compressed by the clamp. A ball bearing was added to the end of the clamp's threaded rod in order to minimize contact area and thereby avoid torqueing the test object. For this study, the clamping force was initially adjusted to 150 N (34 lb), though over the course of testing this value fell to approximately 140 N (32 lb), possibly due to gradual deformation of the beam over hours.

35.5 Procedure

In this study, each beam specimen is being compared to itself: an impedance signature is recorded with the beam specimen in its defect-free state, then again after the application of damage. Because the specimen is being compared to itself—and because the impedance signature has been found to be repeatable over time [8]—a single measurement with the beam in its defect-free state is used as the baseline for comparison.

The damage applied to each beam specimen was made large due to account for the unknown damage-sensitivity of the instrumented clamp. In both the clamped and free configurations, the damage introduced was a drilled hole of diameter 3.9 mm, located equidistant from both ends of the beam as shown in Fig. 35.4b. After the damage had been applied, a second impedance measurement was taken.

35.6 Results and Discussion

In this section, the collected impedance signatures are presented and analyzed. First, the baseline measurements of the mechanically attached and directly bonded configurations are presented and regions of interest are identified. Then, the baseline and damaged signatures are compared.

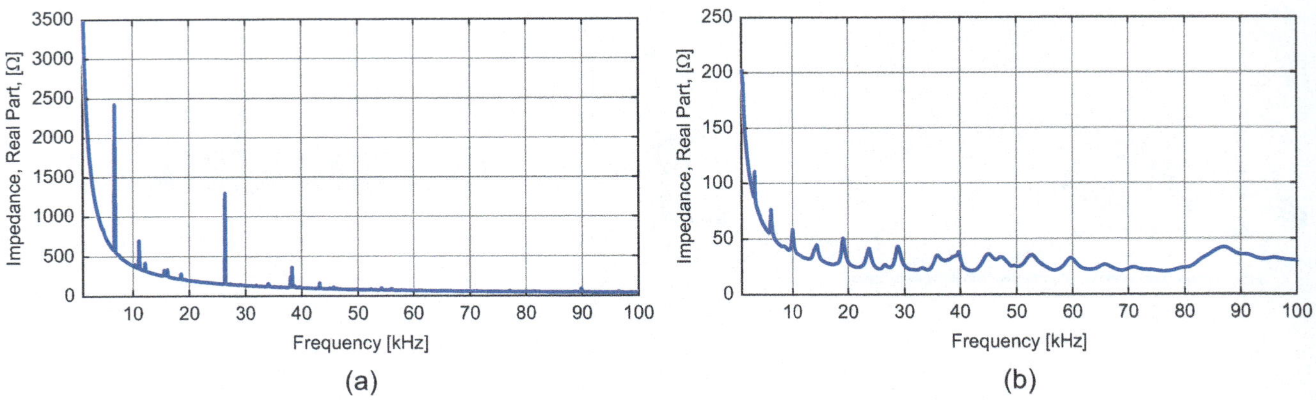

Fig. 35.5 The real part of the impedance signature for (**a**) mechanically attached and (**b**) directly bonded configurations

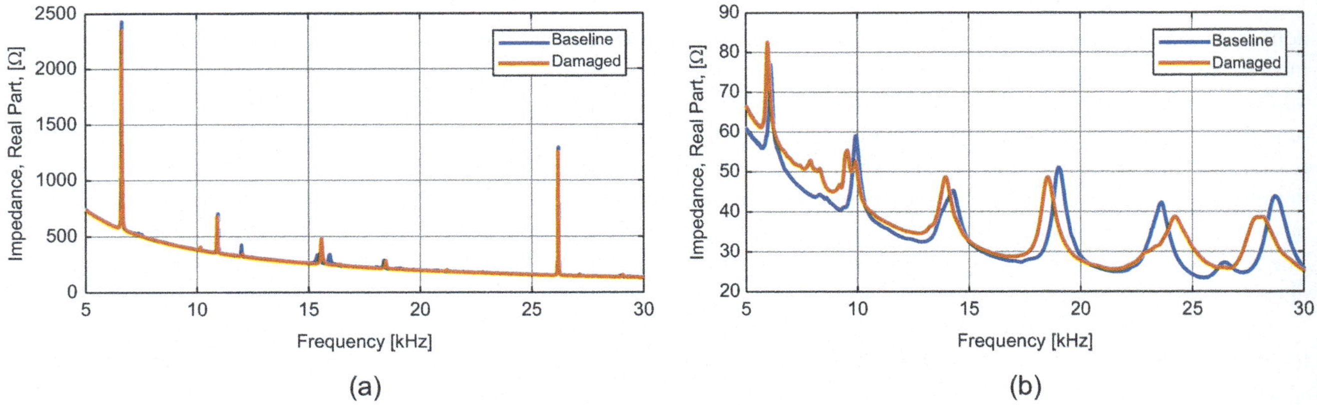

Fig. 35.6 Comparing baseline to damage: (**a**) the real part of the mechanically attached impedance signatures, and the same for (**b**) the directly bonded signatures

35.6.1 Baseline Measurements

A baseline is established by obtaining an impedance signature from damage-free specimens. In Fig. 35.5, the real part of the impedance signature for both the attached and bonded configurations are presented over the full frequency range measured (1–100 kHz). The qualitative response of each is similar: both have an overall decaying trend punctuated with peaks in similar frequency ranges. The individual peaks don't all align, since these are completely different structures. However, we should expect to see something of the beam's response in the response of the clamp-beam system.

Some major differences between the two plots can be attributed to the differences in the configurations. The difference in overall magnitude is due to the different piezoelectric transducers used in each test. The clamp was instrumented with an MFC patch while the beam was instrumented with a PZT wafer. Due to geometry and construction, the MFC has a greater magnitude response regardless of the structure to which it is bonded. The sharpness of the peaks is related to the damping of the structure; the sharp peaks of the mechanically attached configuration are consistent with a mostly-metal structure, while the rounded peaks of the bonded configuration are consistent with a more strongly damped, mostly-plastic system. Since both signatures are rich in peaks around the 5–30 kHz region, regions of interest will be drawn from this range.

35.6.2 Comparison to Damaged Specimens

Next, impedance signatures are obtained from the specimens after damage has been applied. In Fig. 35.6, the impedance signatures of the damaged attached-configuration and damaged bonded-configuration are compared to their respective baselines over the 5–30 kHz range described in the previous subsection.

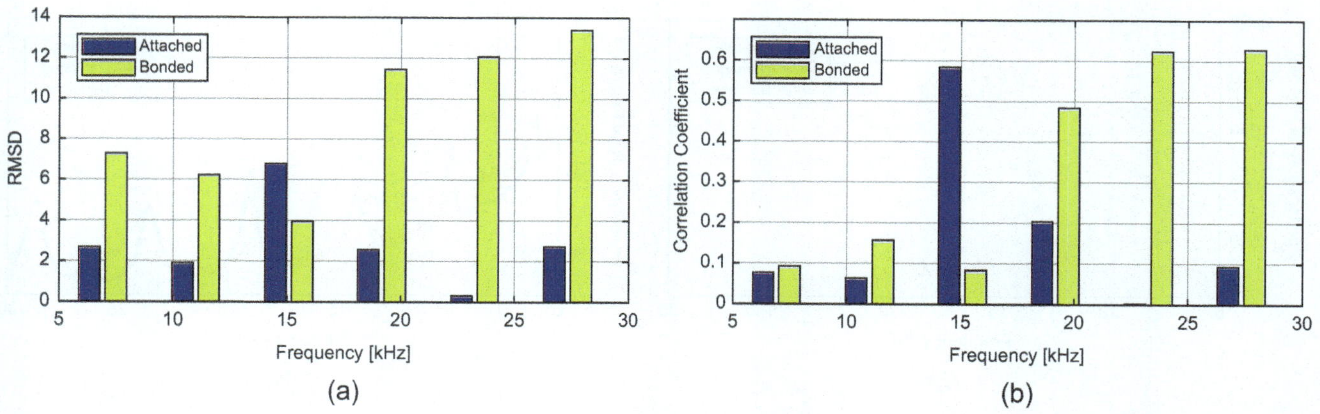

Fig. 35.7 Damage metrics calculated over the frequency range of interest: (**a**) RMSD, (**b**) correlation coefficient (r)

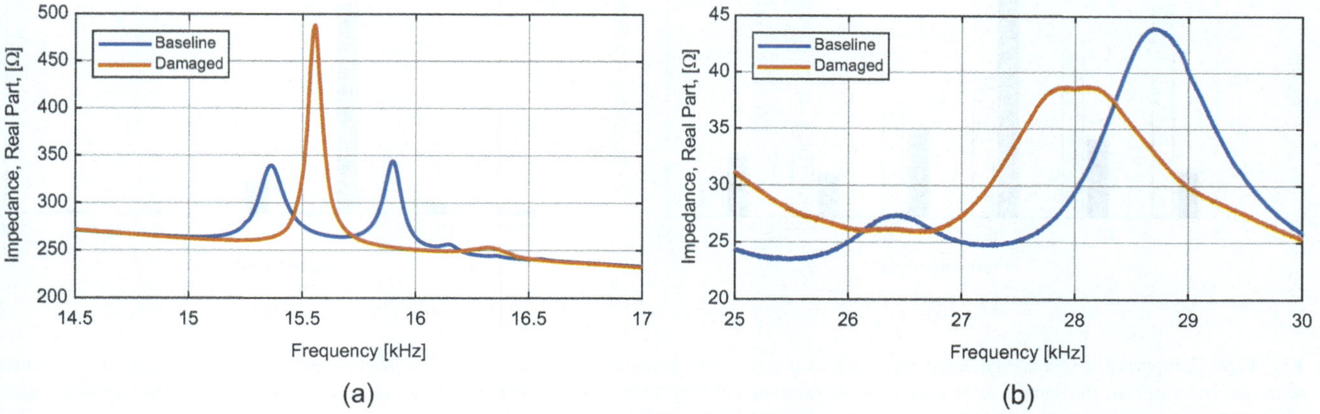

Fig. 35.8 Examining the regions with highest correlation coefficient damage metrics: (**a**) a range within the mechanically attached signature, and (**b**) a range within the directly bonded signature

Looking at the effects of damage application in each configuration, the bonded configuration shows a clear frequency shift across all peaks in the region. In the mechanically attached case, magnification is needed to clearly see the frequency shifting, provided later in Fig. 35.8. This is consistent with the fact that the damage introduced is much larger in comparison with the size of the bonded configuration—the beam alone—than in comparison to the attached configuration—beam and clamp together. Moreover, the transducer is both closer to the damage and more directly attached to the damaged object in the bonded configuration. But critically, in both cases, the damage can be clearly seen with appropriate magnification. In Fig. 35.7, the damage metrics described in Sect. 36.2 are calculated in 4 kHz bins for the frequency range of interest.

Another interesting result is the distribution of the damage throughout the signature. In the attached case, the damage metrics spike in one bin, while the damage is more evenly distributed throughout the range in the bonded case. This can be explained by considering the attached signature to be dominated by the dynamic response of the clamp. The damage to the beam specimen only affects the clamp if the particular mode of vibration is applying force through the beam. A torsional mode of the clamp or a bending mode in the bottom plate might have very low sensitivity to changes in the beam, for example. This is behavior that deserves further investigation. Figure 35.8 shows the range with the highest correlation coefficient damage metric in both configurations, to provide a point of comparison

Finally, in order to demonstrate the measurement repeatability of the impedance signature, an additional measurement was taken for the damaged specimens. The beam specimens used for the second measurement are the same specimens used for the first measurement in each configuration. In Fig. 35.9, the baseline is compared to the two damaged signatures for each configuration, and the damage metrics are calculated. By inspection of the impedance signature, it is clear which two signatures come from the post-damage case. Additionally, the damage metrics are much smaller than the values shown in Fig. 35.7, indicating a small amount of change. These small changes may be due to temperature variations, fatigue at the electrical contacts, or some other source, but they are small enough that they are easily distinguishable from the damage introduced above.

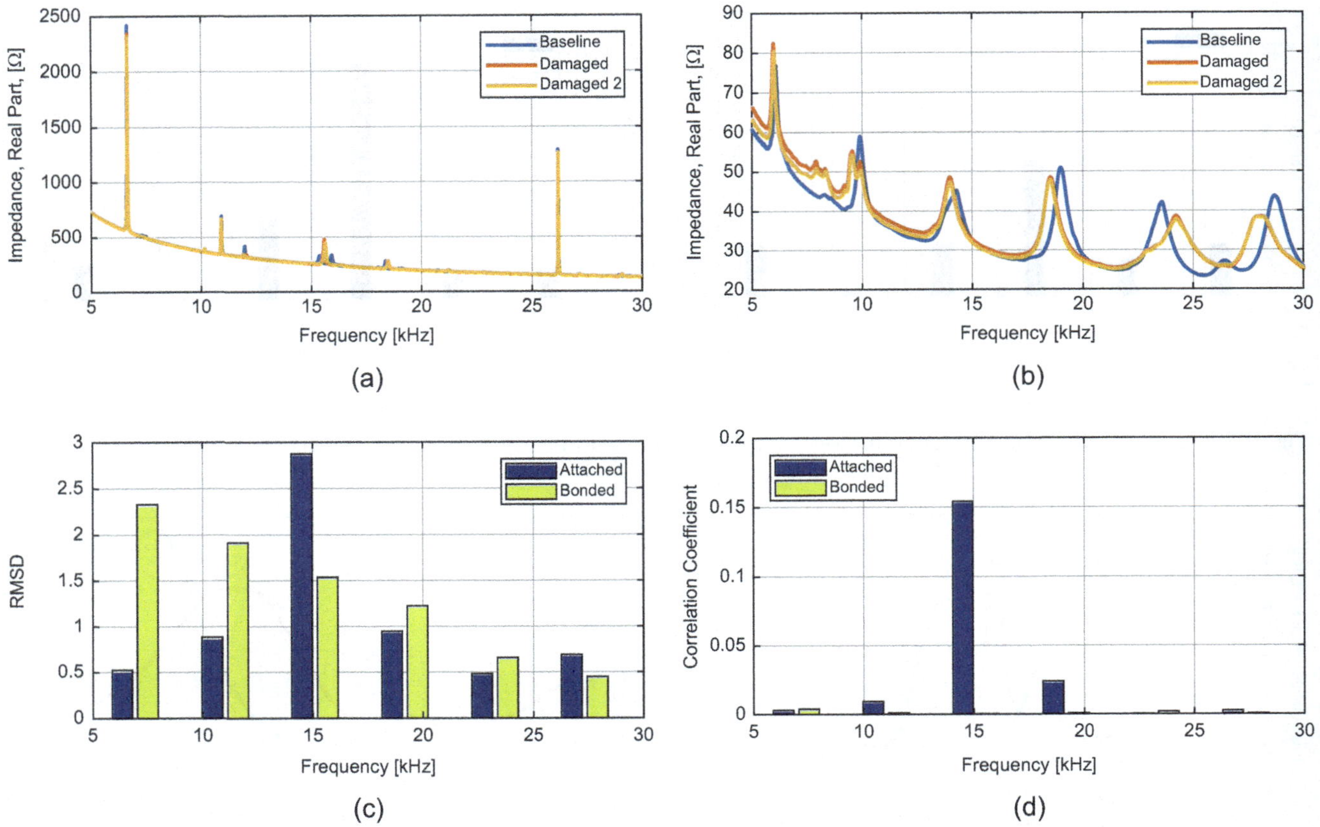

Fig. 35.9 Comparing a second damaged signature to the first. (**a**) impedance for the attached configuration, two damaged signatures plotted with one baseline, (**b**) the same for the bonded configuration, (**c**) RMSD values comparing 'Damaged' and 'Damaged 2' for each configuration, (**d**) correlation coefficient values comparing 'Damaged' and 'Damaged 2' for each configuration

35.7 Conclusions

In this paper, the plausibility of mechanically attached piezoelectric transducers for damage detection in AM parts is examined and compared to directly bonded transducers. The test specimens were rectangular beams, and damage was applied to the beams at mid-span for each of the two test configurations: directly bonded and mechanically attached. For both configurations, the electrical impedance of the attached/bonded transducer was measured over a 1–100 kHz range. Then, damage was applied and the measurement was repeated. Examining the measurements, it was found that both configurations had a dense area of resonant peaks in the 5–35 kHz range. For this range, root-mean-square and correlation damage metrics were calculated comparing the damaged measurement to the un-damaged measurement baseline.

By inspection of the impedance signatures and based on the values of the damage metrics, it was determined that both the mechanically attached and directly bonded configurations are capable of detecting damage to beam specimens at the scale employed in this work. However, it was noted that damage was registered at several resonant peaks in the bonded configuration, while fewer peaks clearly reflected the effects of damage for the attached configuration. This indicates that some resonances of the clamp are more sensitive to damage in the beam specimen than others.

Going forward there is much room for characterization of the effect of clamping orientation and force, as well as the effect of using specimens of different sizes, materials, and geometries. Overall though, mechanically attaching a specimen to an instrumented clamp is found to have potential as an alternative to directly bonding piezoelectric transducers for electromechanical impedance measurements.

35.8 Acknowledgement

This material is based upon work supported by the National Science Foundation under Grant Number CMMI-1635356. Any opinions, findings, and conclusions or recommendations expressed in this material are those of the author(s) and do not necessarily reflect the views of the National Science Foundation.

Additionally, the authors thank Joseph Kubalak for fabricating the beam specimens used in this paper, and for his help in accurately describing the process.

References

1. Du Plessis, A., Le Roux, S.G., Els, J., Booysen, G., Blaine, D.C.: Application of microCT to the non-destructive testing of an additive manufactured titanium component. Case Stud. Nondestruct. Test. Eval. **4**, 1–7 (2015)
2. Albakri, M.I., Sturm, L.D., Williams, C.B., Tarazaga, P.A.: Impedance-based non-destructive evaluation of additively manufactured parts. Rapid Prototyp. J. **23**(3), 589–601 (2017)
3. Tenney, C., Albakri, M., Kubalak, J., Sturm, L., Williams, C., Tarazaga, P.: Internal porosity detection in additively manufactured parts via electromechanical impedance measurements. In: Smart Materials, Adaptive Structures and Intelligent Systems. American Society of Mechanical Engineers, New York (2017)
4. Smart Material: Macro fiber composite (MFC) datasheet. p. 8, (2017)
5. Meitzler, A.H., Tiersten, H.F., Warner, A.W.: An American National Standard IEEE Standard on Piezoelectricity. IEEE, New York (1987)
6. Park, G., Sohn, H., Farrar, C.R., Inman, D.J.: Overview of piezoelectric impedance-based health monitoring and path forward. Shock Vib. Dig. **35**(6), 451–463 (2003)
7. Liang, C., Sun, F.P., Rogers, C.A.: Coupled electro-mechanical analysis of adaptive material systems-determination of the actuator power consumption and system energy transfer. J. Intell. Mater. Syst. Struct. **8**(4), 335–343 (1997)
8. Bhalla, S., Surendra, A., Naidu, K., Wee, C.: Practical issues in the implementation of electro-mechanical impedance technique for NDE. SPIE. **4935**, 484–494 (2002)

Chapter 36
Classification of Human Walking Patterns through Singular Value Decomposition Projection

Ellis Kessler and Pablo A. Tarazaga

Abstract Sensing of structural vibration provides a rich source of information that can be used in structural health monitoring and impact/fault localization among other applications. In this paper, acceleration measurements from vibration sensors (accelerometers), installed in an operational smart building (Virginia Tech's Goodwin Hall), are used to classify footsteps of different kinds from building occupants. Goodwin Hall is a 160,000 square foot five story building instrumented with over 200 accelerometers mounted to the building's structure. Singular value decomposition (SVD) projection is used to classify measured data into categories seeded with training data. Contrary to the black box machine learning approach, the SVD framework allows classification parameters to be easily modified and their effects visualized to be understood. Better understanding of the classification problem and its dominant parameters will allow the development of more accurate and robust algorithms for classification of a wide variety of signals.

Keywords Gait · Singular value decomposition · Classification · Smart infrastructure · User identification

36.1 Introduction and Motivation

Classification of building occupants using non-invasive sensors has applications including retail analytics [1], improved security [2], energy savings [3], and emergency response [4]. In this paper a study of classification of walking patterns using underfloor mounted accelerometers will be presented. Utilizing building instrumentation for tasks such as classification has the opportunity to become more common as buildings are instrumented in the future.

Previous work on this subject from Bales et al. [5] used machine learning techniques to achieve a maximum of 88% accuracy in gender classification from this same data set. While the results are impressive, machine learning essentially acts as a black box. There is little intuition for why the techniques achieve the results they do, or how the inputs could be adjusted to achieve even better results. In this work, authors will take a step back from machine learning and take a simpler approach to classification. Through a simpler approach the phenomenon of classification will be studied more in depth in order to find what features of a step pattern lead to classification. Finally, work here in will leverage this knowledge in an attempt to increase the accuracy of classification.

36.2 Goodwin Hall and Walking Data Set

The data set central to this work was taken in the fourth floor of Goodwin Hall. Goodwin Hall is a 160,000 ft^2 building at Virginia Tech with over 200 accelerometers placed on the buildings structure. During a time of low building use, subjects walked down a hallway while the acceleration from 14 accelerometers under the hallway recorded data at 51,200 Hz. 15 different subjects walked down the hall while their gender (male or female) and shoe sole type (hard sole or soft sole) were noted. A total of 120 trials were run. Figure 36.1 shows an example of one accelerometer's signal from a single trial.

Each step is visible in Fig. 36.1 as a spike and subsequent decay back to noise. Every trial resulted in somewhere between 20 and 30 steps, although not all sensors were able to detect all steps. As seen in Fig. 36.1, noise plays a significant role in the acceleration signal.

E. Kessler (✉) · P. A. Tarazaga
Department of Mechanical Engineering, Virginia Tech Smart Infrastructure Lab (VTSIL), Blacksburg, VA, USA
e-mail: ellisk1@vt.edu

S. Pakzad (ed.), *Dynamics of Civil Structures, Volume 2*, Conference Proceedings of the Society for Experimental Mechanics Series, https://doi.org/10.1007/978-3-319-74421-6_36

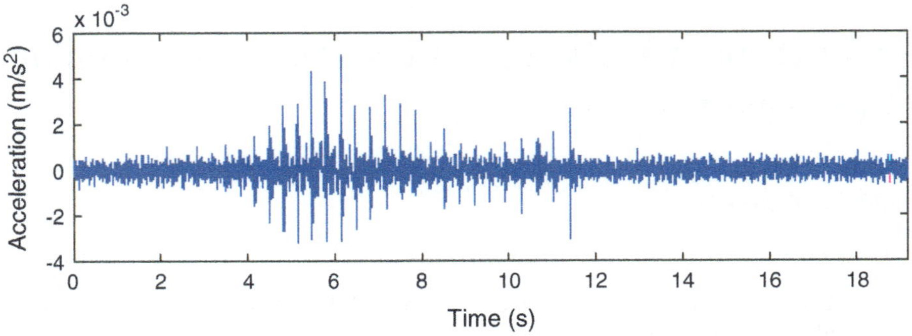

Fig. 36.1 Example of one accelerometer's recordings of a single trial. This trial was Female A with hard soled shoes

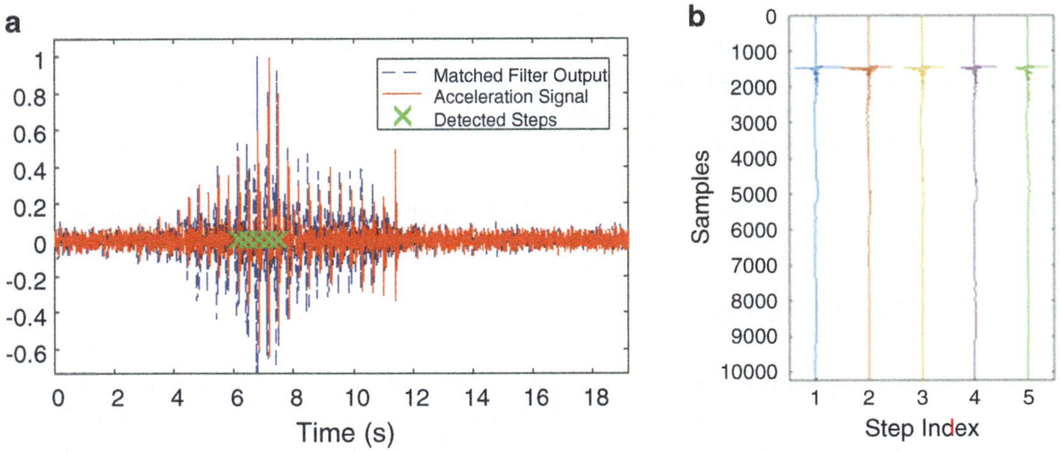

Fig. 36.2 (**a**) Matched filter showing where steps were detected with green xs. (**b**) A representation of the matrix A built up where each step is a column and steps are peak aligned

36.3 Singular Value Decomposition Projection Comparison

The SVD classification framework used in this work is similar to that used by Kasarda et al. [2], and is expanded in this work specifically for footsteps from full walking trials performed in the Goodwin building.

Our classification approach will start from the assumption that there is a defining similarity within a class (such as gender) which will allow steps from subjects in different classes to be differentiated. In order to single out these features we will use singular value decomposition (SVD) to find the dominant singular vectors of a snapshot matrix A. In order to build the matrix A, steps were identified from the acceleration signal $\alpha(t)$ using a matched filter. The output of the matched filter, $y(t)$ see in Eq. (36.1), depends on how well the signal matches a template $z(t)$, $0 \leq t \leq T$, and in the continuous domain is calculated as:

$$y(t) = \int_{-\infty}^{\infty} \alpha(\tau)\, z\,(t_s - t + \tau)\, d\tau \tag{36.1}$$

For this work two different templates were used, one for hard soled shoes and one for soft soled shoes. The process of using a matched filter has known probability of detection and probability of false alarms which depend on the decision threshold γ_d, signal energy E_s and noise variance σ_N^2 [6]. Figure 36.2a shows an example of how steps were detected. After detecting where a step is in the acceleration signal, the step was defined as $s_i \in \mathbb{R}^{10240 \times 1}$ where the peak of the step is at index 1450. A visual representation of five columns of the snapshot matrix A is shown in Fig. 36.2b where the acceleration signal of each step is plotted vertically as it is placed in the matrix A.

For each class we want to identify, a matrix $A \in \mathbb{R}^{l \times n_s}$ is created where each column is a single step, l is the length of a step (10,240 samples), and n_s is the number of steps detected for that class. A can be decomposed through the SVD,

$$A = \left[s_1 \; \dots \; s_{n_s}\right] = U\Sigma V^T = \sum_{j=1}^{n} \sigma_j u_j v_j^T \tag{36.2}$$

where $n =$ rank (A) is the number of linearly independent columns of A. In our case since $n_s \ll l$ and each step in A is distinct, we expect that the rank of the matrix will be equal to the number of steps detected $n = n_s$.

After finding the SVD of the snapshot matrices, trial steps can be compared to the singular vectors from each snapshot. Trial matrices consist of a single step from an unknown source: $x \in \mathbb{R}^{l \times 1}$. The relative error of this trial step with each class is calculated by projecting x onto the dominant singular vectors of each snapshot matrix. For the projection, the first r singular vectors are used. These dominant singular vectors correspond to an order lower than n which captures the important features of each class. For this work, $r = 20$ was chosen for all comparisons. The projection onto these singular vectors is denoted P_x and is calculated as:

$$P_x := \sum_{j=1}^{r} u_j u_j^T x. \tag{36.3}$$

Once the projections are calculated, the relative error is found by comparing the norm of the error in the projection with the norm of the original trial step:

$$e = \frac{\|x - P_x\|}{\|x\|}. \tag{36.4}$$

This error will be 0 if the projection is exactly the original, which is the case if x is completely contained in the space of the first r singular vectors of U. The error reaches a maximum of 1 if the trial is not in the space of the singular vectors at all.

36.4 Classification Results

As a survey of how the classification approach performs, five trials were chosen to begin with. Female A with hard soles, Female D with soft soles, Male B with hard soles, Male A with soft soles, and Female A with soft soles. This set includes each combination of gender and sole type, along with one subject wearing each type of shoes.

First, 20 steps from each of these trials were compared to the snapshots from all five trials. Figure 36.3 shows the errors from all of these comparisons. The actual subject is shown on the bottom, while the error with each comparison is shown as a colored rectangle. Blue means an error of zero, while red means an error of one. The bottom part of the figure shows the minimum of each comparison, giving the best guess for which trial the test step came from. Correct classifications are in black, while incorrect are shown in red.

As seen in Fig. 36.3, 91 out of the 100 classifications yielded the correct classification. These results, however, were aided by the fact that there were only five possibilities, and each snapshot matrix only had one person's signature in them. To further explore the accuracy of classification, snapshot matrices were made for other comparison classes as well: shoe sole type, gender, etc. The accuracy for each of these classes is shown in Table 36.1.

As seen in Table 36.1 the SVD classification method is most accurate with shoe sole type, achieving an accuracy of 100 percent. The next two best classifications also included the sole type, while the least accurate classification was the gender with an accuracy of 74%. Higher accuracy in classifying sole type is also seen in Fig. 36.3 where for hard sole trials, there is a more clear definition in color between the correct subject, and other similar sole type comparisons. For soft sole trials on the other hand, all other soft sole trials are very similar shades showing that the errors are almost the same.

In an effort to investigate why sole type was easier to classify than gender, first the singular value decay for snapshot matrices was plotted in Fig. 36.4. There is not a noticeable difference between sole type or gender. All singular values decay quickly at first and then quickly level off around a minimum normalized value of 10^{-2}. The singular value decay will be limited by the noise in the acceleration signal.

The projection based classification framework offers another way to investigate why classification is working or struggling. The error, defined as $x - P_x$, of the projection of a single step onto multiple snapshot matrices gives insight into what features are getting matched by the singular vectors and which are not. Figure 36.5 shows various steps compared to different snapshot matrices.

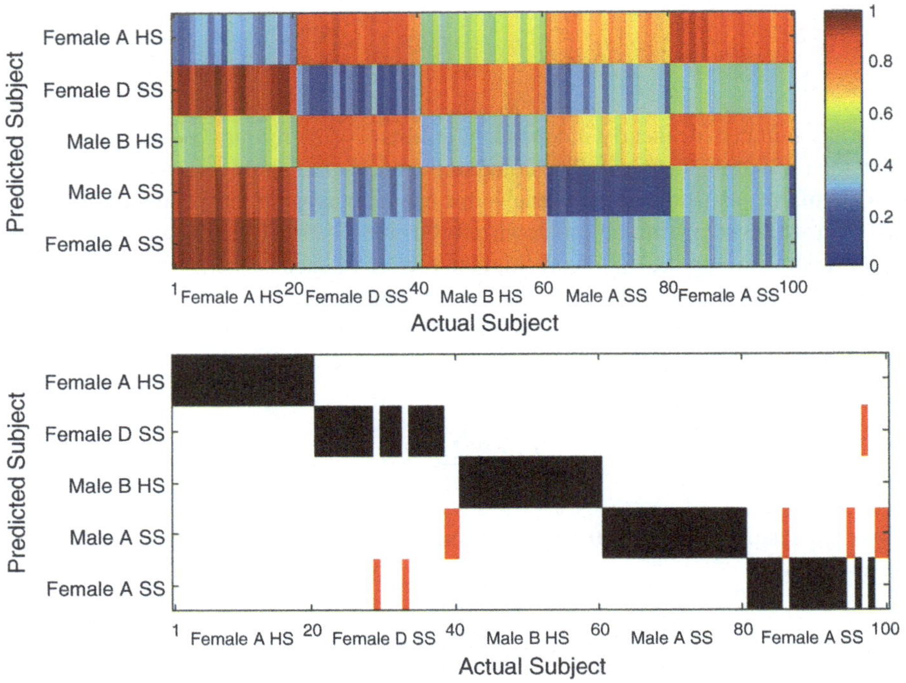

Fig. 36.3 *Top:* Errors of each comparison. The actual subject is noted along the x axis, while the comparison subject is on the y axis. *Bottom:* The minimum error used as a classification method. Correct classifications are in black while incorrect are in red

Table 36.1 Accuracy of classification for various classes

Accuracy (%)	Class name	Classes				
100	Sole	HS	SS	–	–	–
97	Sole & gender	F-Hs	F-SS	M-Hs	M-SS	–
91	Sole & subject	1-Hs	2-SS	3-Hs	4-SS	1-SS
88	Subject	1	2	3	4	–
74	Gender	F	M	–	–	–

The results from Fig. 36.3 are shown in "Sole & Subject" while each other class represents a combination of the five trials chosen
1 = Subj 1 2 = Subj 2 3 = Subj 3 4 = Subj 4 F = Female M = Male HS = Hard Sole SS = Soft Sole

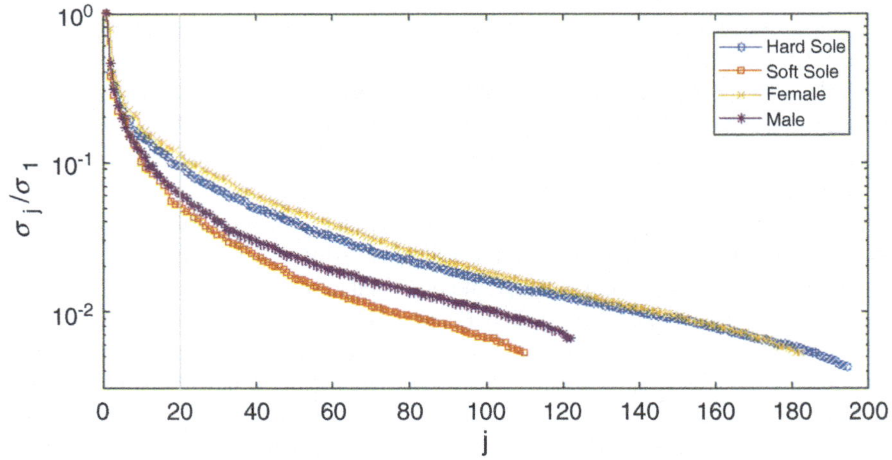

Fig. 36.4 The singular value decay for each snapshot matrix

From the Fig. 36.5a, b it is straight forward to see why the sole type is easily classified. A trial with a soft sole projected onto the hard-soled snapshot matrix (5a in red) gives large errors right around the time of the step with high frequency

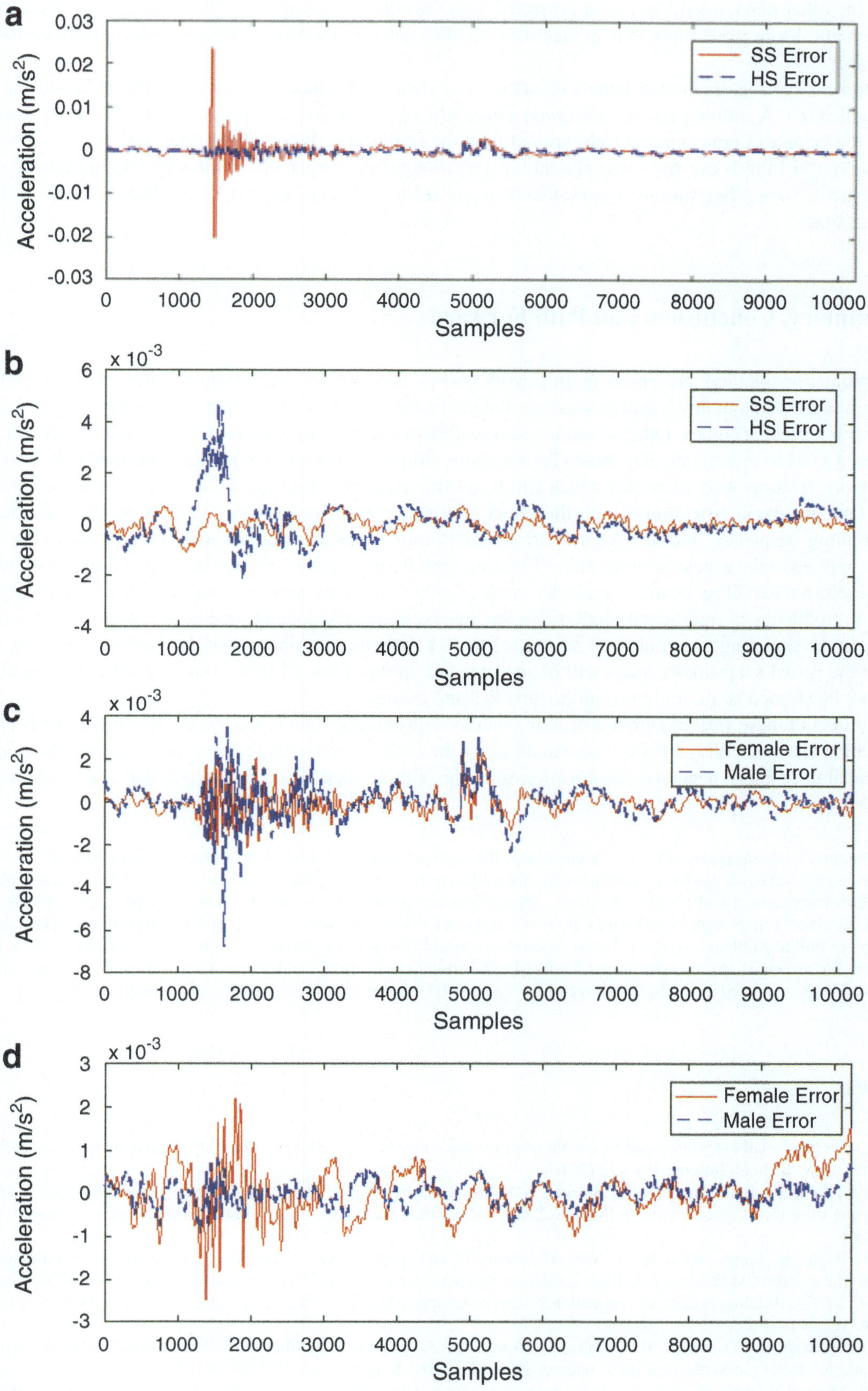

Fig. 36.5 Errors ($x - P_x$) for four cases: (**a**) Hard soled trial and (**b**) soft soled trial both compared with hard soled and soft soled snapshot matrices. (**c**) Female trial and (**d**) male trial both compared with female and male snapshot matrices

content. On the other hand, a hard sole trial projected onto the soft-soled snapshot (5b in blue) gives large errors with low frequency content. From these plots, it is evident that the difference in frequency content makes classification of sole type straightforward.

The differences between male and female errors are less defined. Neither error gives a clear indication of what feature leads to classification. A similar pattern is observed where when a female step was projected onto the male snapshot matrices (5c in blue) the error was concentrated right around the time of the step. When a male step was projected onto the female snapshot matrix (5d in red) the error was spread out for a longer time after the initial step and appears to contain lower frequency content. These observations suggest that there are features that allow gender classification that may be exploitable to increase accuracy.

36.5 Summary, Conclusion, and Path Forward

The SVD projection method presented in this work shows promise for classifying human walking patterns through acceleration signals. Without any signal processing, the method is already successful at classifying the type of shoe worn by the subject with perfect accuracy. Other classifications were less accurate, ranging between 74 and 97 percent, however this method offered tools to investigate *why* these classifications struggled compared to a black box approach. By examining the projection errors, features were identified which will be used to attempt to improve accuracy for various classification types.

Moving forward, parametric studies into the effect of various signal processing methods on the classification such as filtering, sampling frequency, and how signals are aligned in the snapshot matrices will be carried out. For this work, no filtering was applied to the acceleration samples. Filtering high frequency content may be able to limit the effect of shoe sole type, and therefore make classification of gender more accurate. The data used was sampled at a frequency of 51,200 Hz. This sampling frequency is significantly high since the highest frequency content seen in any steps was only around 1 kHz, suggesting that the sampling rate could be lowered below 10 kHz without losing any of the dynamics. Finally, the alignment of signals in the actual snapshot matrices will be investigated. In this work all steps were aligned at their peak value, while they could not be aligned at all or aligned at the time the step occurs.

Once the effect of these various signal processing methods on classification is known, the lessons learned could be applied back to machine learning. With a better understanding of the underlying signal features, new machine learning application should be capable of more accurate classification than this simple method, and also provide more accurate results than previous work.

Acknowledgements Dr. Tarazaga would like to acknowledge the financial support of the John R. Jones Faculty Fellowship.

The authors wish to acknowledge the support as well as the collaborative efforts provided by our sponsors, VTI Instruments, PCB Piezotronics, Inc.; Dytran Instruments, Inc.; and Oregano Systems. The authors are particularly appreciative for the support provided by the College of Engineering at Virginia Tech through Dean Richard Benson and Associate Dean Ed Nelson as well as VT Capital Project Manager, Todd Shelton, and VT University Building Official, William Hinson. The authors would also like to acknowledge Gilbane, Inc. and in particular, David Childress and Eric Hotek. We are especially thankful to the Student Engineering Council (SEC) at Virginia Tech and their financial commitment to this project. The work was conducted under the patronage of the Virginia Tech Smart Infrastructure Laboratory and its members.

References

1. Farshidi, A.: The new retail experience and its unaddressed privacy concerns: how RFID and mobile location analytics are collecting customer information. J. Law. Technol. Internet. **7**(1), 15 (2016)
2. Kasarda, M.E., Tarazaga, P.A., Embree, M., Gugercin, S., Woolard, A., Joyce, B.S., Hamilton, J.: Detection and identification of firearms upon discharge using floor-based accelerometers. IMAC XXXIV A Conference and Exposition on Structural Dynamics, Orlando, Florida, January 25–28 (2016)
3. Agarwal, Y., Balaji, B., Gupta, R., Lyles, J., Wei, M., Weng, T.: Occupancy-driven energy management for smart building automation. In: Proceedings of the 2nd ACM Workshop on Embedded Sensing Systems for Energy-Efficiency in Building, pp. 1–6. ACM, New York (2010)
4. Pan, S. et al.: BOES: Building occupancy estimation system using sparse ambient vibration monitoring. in Proc. SPIE, San Diego, CA, USA, Art. no. 90611O (2014)
5. Bales, D.B., Tarazaga, P.A., Kasarda, M.E., Batra, D., Woolard, A.G., Poston, J.D., Malladi, V.V.N.S.: Gender classification of walkers via underfloor accelerometer measurements. IEEE Internet Things J. (2016). https://doi.org/10.1109/JIOT.2016.2582723
6. Poston, J.D., Michael Buehrer, R., Tarazaga, P.A.: Indoor footstep localization from structural dynamics instrumentation. Mech. Syst. Signal Process. **88**, 224–239 (2017)

Chapter 37
Dynamic Characterization of a Prestressed Concrete Bridge by Strain and Acceleration Measurements

Kirk A. Grimmelsman

Abstract Structural health monitoring systems for bridges typically employ a mixture of different types of sensors to measure and track long-term structural performance under various environmental and service live loads. A significant challenge in designing such monitoring systems is to optimize the quantities and locations of the different sensor types to achieve the monitoring objectives while minimizing instrumentation costs. The final instrumentation design is usually a compromise between an ideal instrumentation plan and one that incorporates economic and practical considerations and constraints. It follows that there is merit in exploring the use of different types of sensors to accomplish more than one characterization objective. This paper presents a study to explore the limits of using both strain gage and accelerometer measurements from a prestressed concrete highway bridge for the dynamic characterization of the structure. The bridge specimen used for this study is primarily instrumented with strain gages to measure the stresses induced by heavy trucks. A smaller array of six accelerometers is also included specifically to measure characterize the vibration response of the bridge due to different truck loads, configurations and crossing speeds. The accelerometers are installed along the transverse and longitudinal centerlines of the bridge span. Strain gages are installed on each beam in the bridge cross section at 28%, 32% and 50% of the longitudinal span length locations. The monitoring system operates in a triggered acquisition mode in which the strain gages and accelerometers are both sampled at 200 Hz during truck crossing events. Strain and vibration measurements from truck crossing events are individually processed and evaluated to identify the dynamic characteristics, and the combination of the two types of measurements for dynamic characterization are evaluated and discussed.

Keywords Strain · Vibration · Prestressed Concrete · Bridges · Structural Monitoring

37.1 Introduction

Structural monitoring programs are most frequently implemented for highway bridges to meet one or more of the following objectives: (1) evaluating in-situ performance characteristics such as load distribution, composite structural behavior, and dynamic characteristics in support of refined structural analyses for load ratings and repair or retrofit designs, (2) diagnosing and evaluating specific serviceability or performance problems such as fatigue, excessive displacements or vibrations, (3) documenting the in-situ state of stress and performance following potential damage events such as superload or other heavy load crossings, following actual damage events, or before and after the implementation of major repairs or retrofits, and (4) research related to evaluating the performance of new materials, structural details or construction methods, and methods for monitoring and diagnosing long-term structural health and performance. In each of the cases, the particular sensing technologies employed and the types, quantities and locations of the sensors installed on the structure will depend on the specific data needs for the monitoring program and also the scale and structural complexity of the bridge being monitored.

Short-to-medium span length steel or concrete multi-beam highway bridges are monitored using strain gages for most of the objectives listed above. Accelerometers are less commonly employed in structural monitoring practice for these types of structures, and usually only for specific objectives related to vibration serviceability, in-situ characterization of the dynamic characteristics, or for structural health monitoring research. A combination of strain gages and accelerometers may be used in some non-research monitoring applications, but the numbers of accelerometers typically employed is not sufficient for an in-depth dynamic characterization of the structure. Given the state of practice with respect to bridge monitoring applications,

K. A. Grimmelsman (✉)
Intelligent Infrastructure Systems, Philadelphia, PA, USA
e-mail: kgrimmelsman@iisengineering.com

S. Pakzad (ed.), *Dynamics of Civil Structures, Volume 2*, Conference Proceedings of the Society for Experimental Mechanics Series, https://doi.org/10.1007/978-3-319-74421-6_37

there is merit in exploring the capabilities and limitations of using strain gages, either exclusively or in combination with a limited number of accelerometers, to obtain a dynamic characterization of the structure being monitored in addition to the primary objective of the instrumentation.

Dynamic characterization of bridges by strain measurements may been accomplished through strain modal testing and analysis theory [1–3]. One of the most significant challenges to this approach is that the measured strains will be very small for higher modes [1, 3]. The very small strains associated with higher modes can be difficult to resolve using conventional foil resistance type strain gages. Strain sensors with much higher sensitivities such as semiconductor strain gages or with sub-microstrain resolutions such as fiber optic strain gages [4, 5] are often recommended for strain modal testing and analysis. However; high resolution strain measurements are often not necessary for most bridge monitoring applications since the focus is often on evaluating large stresses. There are few examples in the literature where strain modal testing has been explored for operating, full-scale bridges using any of the available strain gage technologies. There have been some recent efforts related to strain modal testing in the laboratory with full-scale bridge components [5] and small-scale models of bridge structures [6, 7]. Operating bridges are subject to many complexities and uncertainties in their loadings and responses that may significantly impact the capabilities and utility of strain modal testing in the field, and these are difficult to quantify and assess through laboratory testing applications.

Given the lack of strain modal analysis applications for full-scale, operating highway bridges, a study was initiated to evaluate the capabilities and limitations associated with dynamic analysis of strain measurements recorded from a representative multi-beam highway bridge instrumented with a representative type of strain gage used in bridge monitoring applications and subject to actual operating highway truck traffic loads. The study bridge was also instrumented with a small number of accelerometers to assess dynamic characteristics of the truck loading events and bridge responses, such as correlating peak-to-rms vibration levels with truck weights and characteristics. The accelerometers were not specifically included to characterize and monitor the dynamic properties of the bridge over the long term, although a limited potential for accomplishing this is possible with these sensors.

This paper presents and discusses some preliminary dynamic analysis results obtained from strain measurements recorded from the prestressed concrete multi-beam highway bridge under actual operational, loading and field-measurement conditions. The focus of this paper is a comparison of the natural frequencies identified from high-speed strain gage measurements and accelerometer measurements recorded for truck live loads crossing the structure. A five-year-long monitoring program is currently underway for this bridge structure, and this study is performed to determine if there is sufficient merit for conducting more in-depth analysis of the strain gages to evaluate the long-term dynamic characteristics of the structure in conjunction with measuring and evaluating the environmental and live load related stresses and other bridge performance characteristics.

37.2 Overview of the Bridge and Monitoring System

The bridge evaluated for this study carries two westbound lanes of Interstate-84 traffic over a rural access road in eastern Oregon. Approximately 2000 trucks cross the bridge each day. The bridge crossing consists of three simple spans of prestressed concrete beams and has a 30° right forward skew. The middle span of the bridge crossing was instrumented and is currently being monitored for the five-year-long study. The middle span of the bridge has a total length of 24028.4 mm measured between the centerlines of the bearings and consists of 6 AASHTO Type III prestressed concrete beams spaced on 2254 mm centers. The bridge has a 205-mm-thick, cast-in-place reinforced concrete deck that is continuous over the interior bents. The deck continuity detail is not designed to provide live load continuity across the adjacent spans, it is considered a link-slab that is used instead of deck joints at these locations. The width of the bridge deck is 12,484 mm measured between the toes of the barriers, and the deck is striped for two 3658-mm-wide traffic lanes, a 3048-mm-wide right shoulder and an 1830 mm-wide left shoulder.

The middle span of the bridge crossing was instrumented with a strain gages on the prestressed beams and deck underside, vertical accelerometers, and linear displacement sensors and tiltmeters to measure and track the bearing movements and beam end rotations. The strain gages and accelerometers were installed along transverse instrumentation lines as shown in Fig. 37.1. Strain gages were installed on the bottom flanges and webs of each beam in the section; however, only the bottom flange strain gage measurements from Sections B, C and D (18 total gages) were evaluated with the 5 accelerometers on the span for this study. The strain gages used on the bridge are Geokon Model 4151 vibrating wire strain gages that have a range

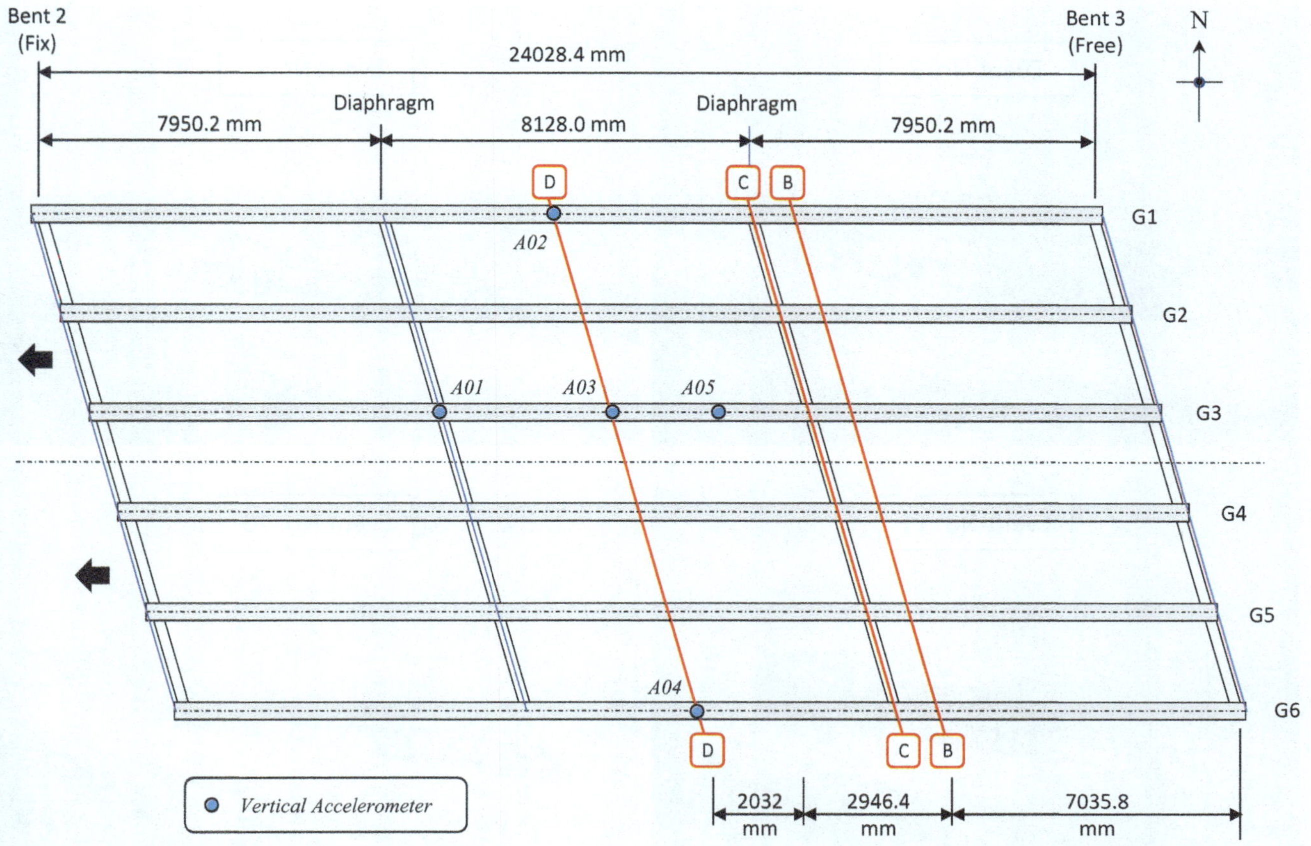

Fig. 37.1 Sensor locations on the test bridge

of 3000 microstrain and a resolution of 0.4 microstrain according to the manufacturer's specifications. The accelerometers are Model 4312 M3 from TE Connectivity with a range of ±2g and a frequency range of DC to 200 Hz. The vibrating wire strain gages were used for the project since they can measure and report the dynamic live load strains (200 Hz), the quasi-static thermal strains (1 Hz) and the temperature of the sensor (1 Hz) simultaneously. The live load strain and accelerometer measurements are automatically recorded at 200 Hz for 8 s whenever a truck crosses the bridge. The strain and accelerometer measurements are all synchronized through simultaneous sampling. An overview camera also takes a snapshot of the traffic on the bridge during each truck crossing event. The test site also includes a pavement Weigh-in-Motion (WIM) system located in both traffic lanes of the roadway just upstream from the bridge location. The WIM system records the weight and geometric characteristics (e.g. number of axles, axle spacings, speed, etc) of every truck crossing the bridge. An all-in-one weather station records local weather conditions at the site including wind speed and direction, ambient temperature, relative humidity, solar radiation, and precipitation.

37.3 Data Analysis and Results

The primary objective of this study was to evaluate the natural frequencies obtained for the accelerometers and the bottom flange strain gages due to truck loads. The specific data analyzed for this study consisted of 42 truck crossing events recorded over an approximately 2.5-h-timeframe. The trucks that produced the recorded measurements had various axle configurations and weights, and all were crossing in the right (slow) traffic lane. Some representative images of the trucks that produced the measurement data analyzed for this paper are shown in Fig. 37.2.

Fig. 37.2 Representative truck images from the 42 truck crossing events

Power spectral density (PSD) plots were generated for the accelerometers and the bottom flange strain gages from the total measurement record for each truck event using Welch's method with an FFT block size of 1024 points, a 50% overlap, and a hamming window. The resulting frequency resolution of the accelerometer and strain measurements was 0.195 Hz. The PSD plots for the strain gages are dominated by low frequency content related to the quasi-static truck weight component. The strain gage measurements were subsequently high pass filtered using a cutoff frequency of 3 Hz, which was below the fundamental mode of the bridge span identified from the accelerometer measurements, and the resulting filtered data was used for frequency analysis. Representative PSD plots for the accelerometers and the strain gages for different truck crossing events are shown in Figs. 37.3 and 37.4, respectively.

Natural frequencies were identified from the accelerometer and strain PSDs by peak-picking. The PSDs for each individual strain gage and accelerometer were first added together to obtain an averaged frequency spectrum for each sensor from the 42 distinct truck events. The resulting averaged PSD spectra for the accelerometers is shown in Fig. 37.5. Similar plots were obtained for each strain gage channel. Peaks in the PSD spectra for acceleration and strain channels were identified in the frequency band between 3 Hz and 25 Hz using a minimum peak distance criterion of at least 3 frequency lines. The strain and accelerometer channels were each evaluated separately in case a sensor was located at a nodal point for one of the modes. The peak frequencies that were consistently identified across multiple accelerometer and strain gage channels were selected as the natural frequencies of the bridge. The natural frequencies identified from the accelerometer and strain gage channels are summarized in Table 37.1.

The natural frequencies identified from the acceleration and strain measurements show good agreement with each other except the last two modes. The natural frequencies were consistently identified from peaks in the averaged PSDs for most of the accelerometer channels, but not as consistently from the averaged PSDs for the strain gage channels. In some instances, there is only a single frequency line (0.195 Hz) difference between the location of the peaks for different strain gage channels. Some strain gage channels did not show a peak at frequencies identified from other strain gage channels and the accelerometer channels. This may be a result of the frequency resolution of the measurements which was controlled by the 8 s records.

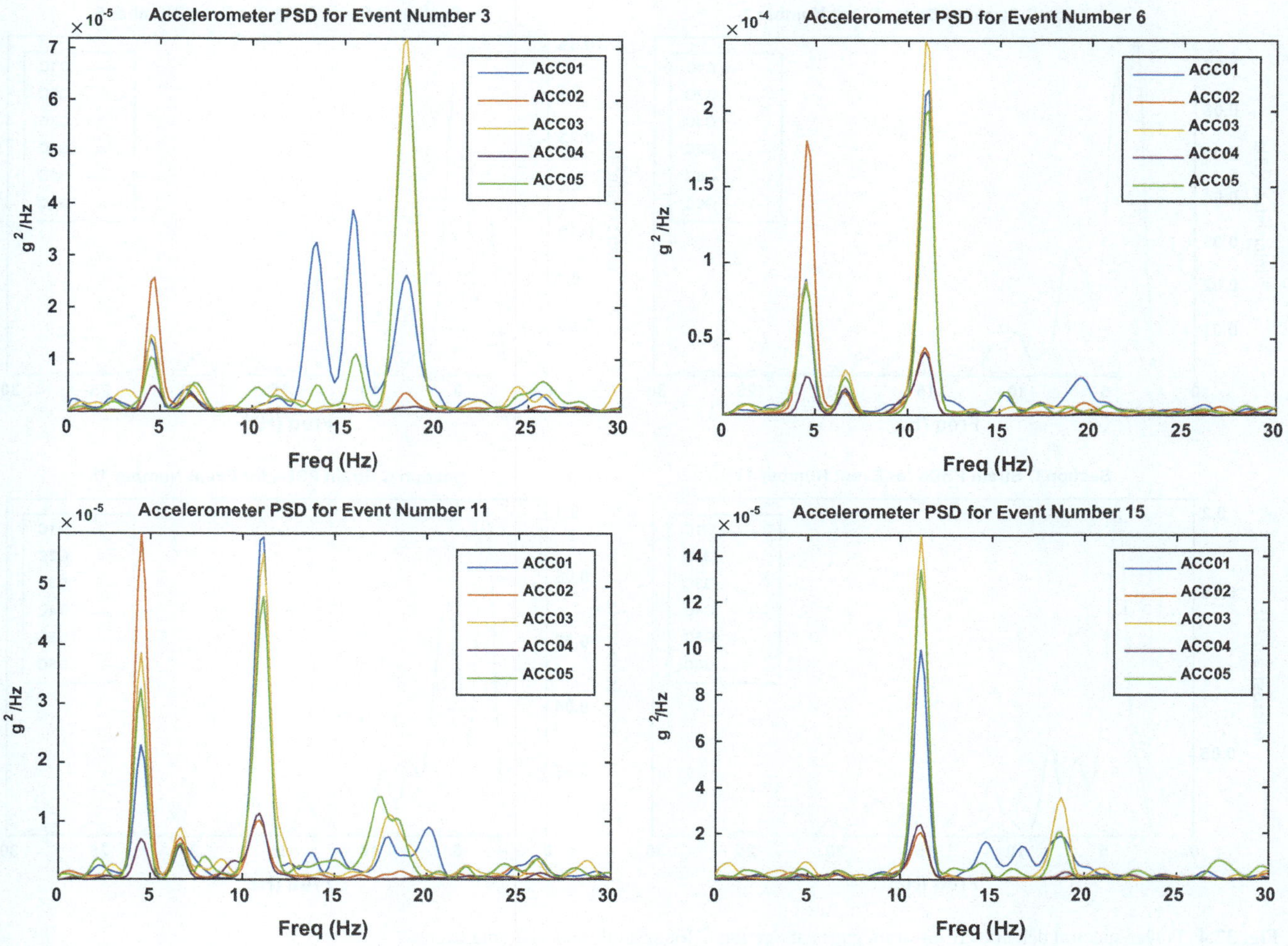

Fig. 37.3 Power spectral densities from accelerometers for several truck crossing events

At higher frequencies, the natural frequencies identified from the acceleration and strain measurements begin to diverge. It is expected that it may be difficult to identify at higher frequencies since the modal displacements (and the corresponding bending strains) associated with these modes will be very small.

Since no attempt was made to extract mode shape vectors from this measurement data, it was not clear if the frequencies identified were all associated with actual vibration modes of the structure, or if some were related to noise or other errors. A preliminary FE model of the bridge was developed, and modal analysis was performed to provide a baseline for comparison with the natural frequencies identified from the accelerometers and the strain gages. A total of 10 modes were identified from the preliminary FEM in the frequency band from 4 to 25 Hz. The FEM natural frequencies and mode shapes for Mode 1 through Mode 6 are shown in Fig. 37.6 and for Mode 7 through Mode 10 in Fig. 37.7. Some of the FEM modes are very close to the natural frequencies identified from the accelerometers and strain gages, while other FEM modes have no experimental identified counterpart. Mode shapes for the accelerometer measurements need to be extracted, or at least the spatial phase information from the sensors at each identified peak for comparison with the FEM mode shapes. The mode shape analysis will be completed in the future for this data.

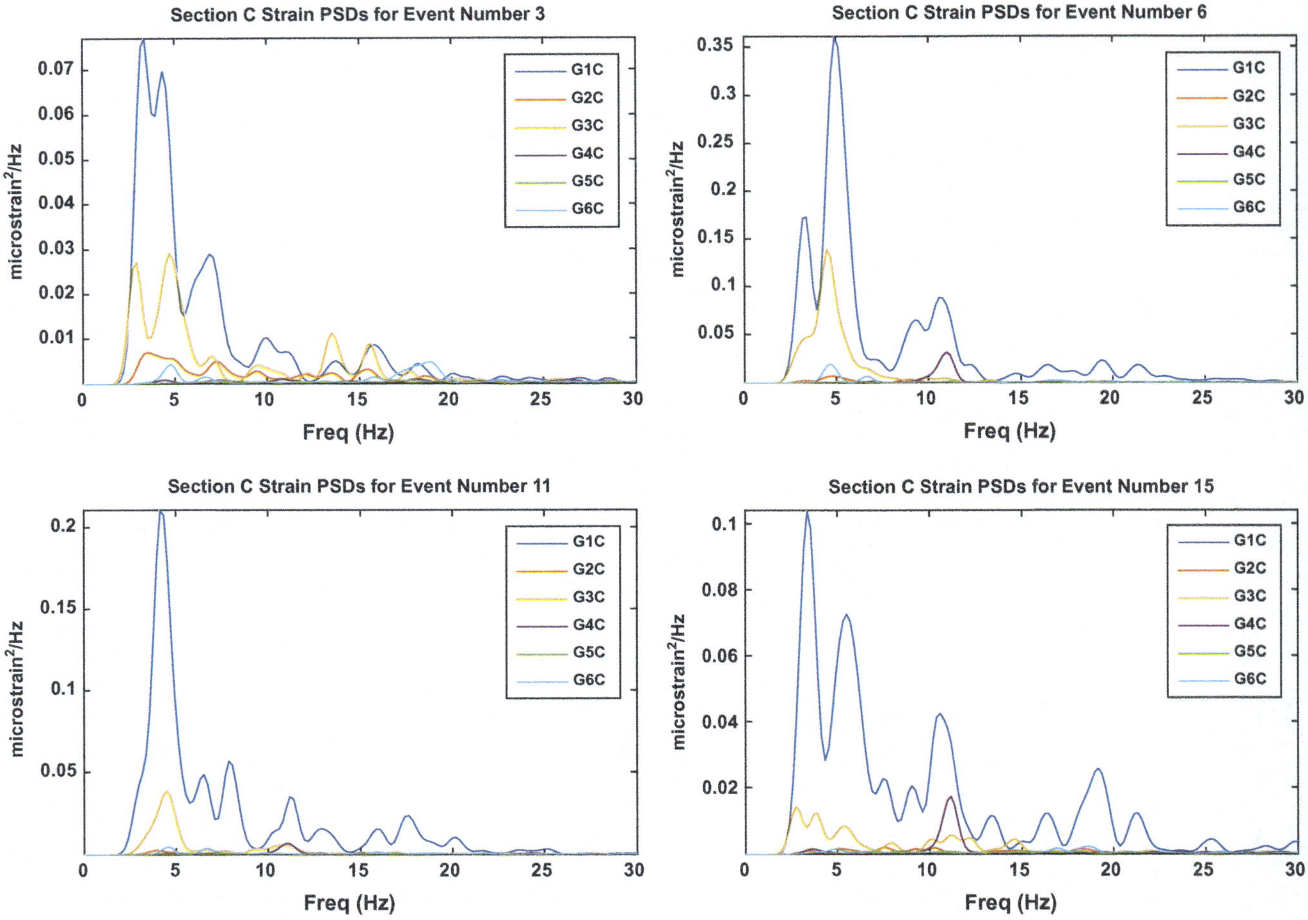

Fig. 37.4 Power spectral densities from strain gages at Section C for several truck crossing events

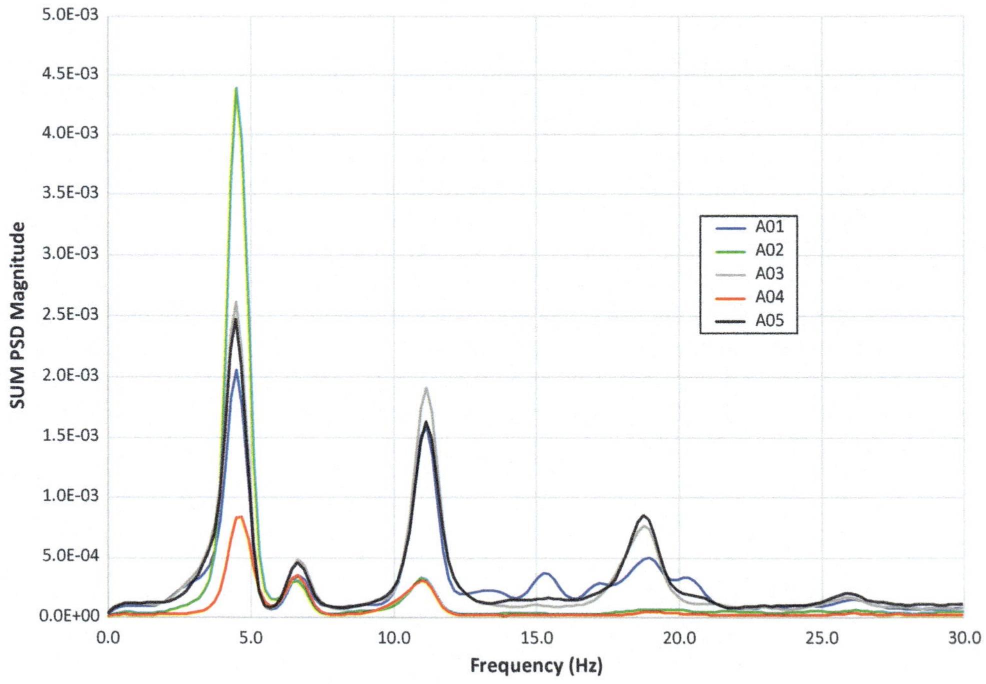

Fig. 37.5 Averaged power spectral densities for accelerometers (42 truck crossing events)

Table 37.1 Natural frequencies (Hz) identified from accelerometer and strain gage channels

Mode	Accelerometer channels	Strain gage channels
1	4.492	4.492
2	6.641	6.641
3	11.133	11.133
4	18.750	18.750
5	22.852	22.656
6	24.219	25.000

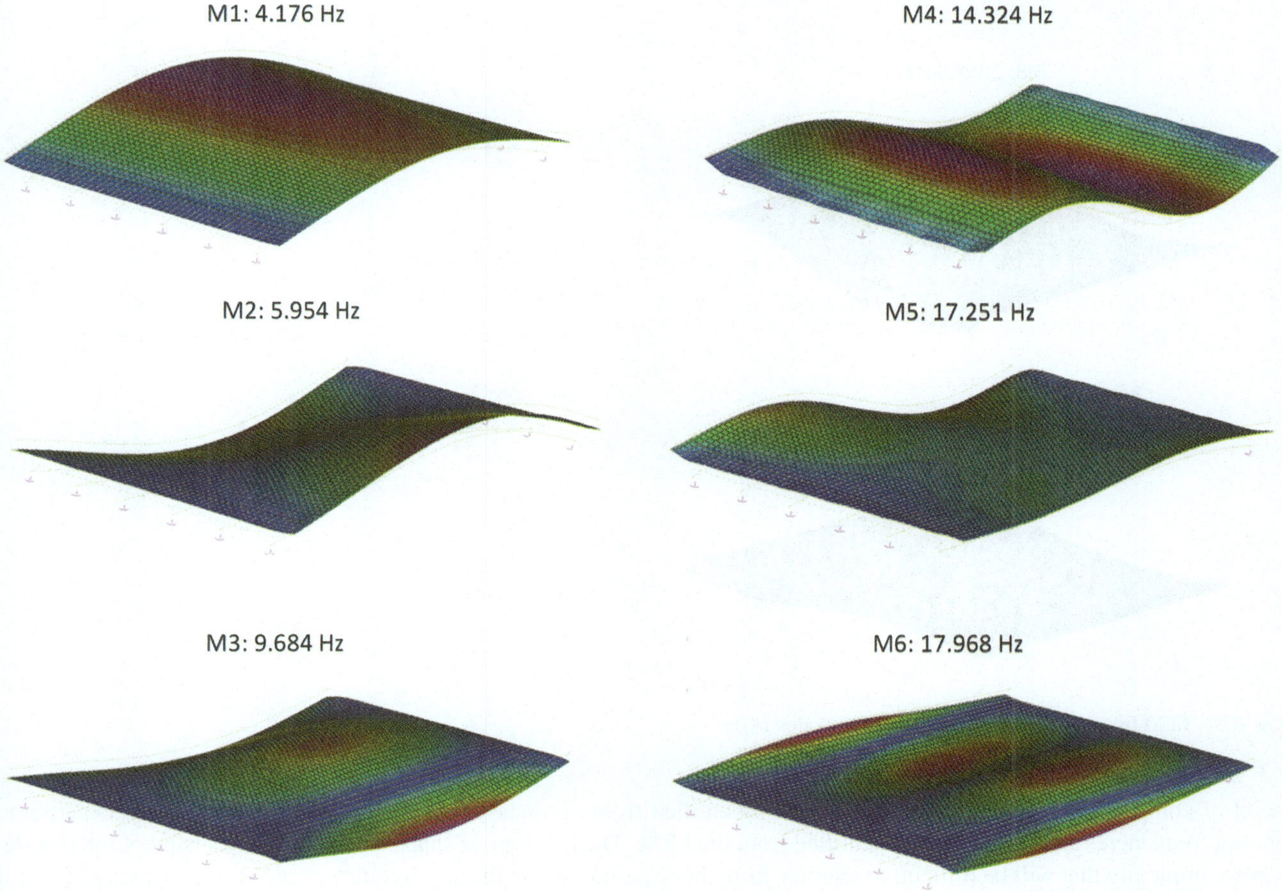

Fig. 37.6 FEM frequencies and mode shapes for modes 1–6

37.4 Conclusions and Future Work

Acceleration and strain measurements from a full-scale, operating prestressed concrete multi-beam highway bridge were analyzed to evaluate and compare natural frequencies. The measurements were recorded from the bridge for a total of 42 different truck crossing events. The strain gages used in the study were vibrating wire type sensors with a strain resolution of 0.4 microstrain. Natural frequencies were identified by peak picking from the averaged power spectral densities computed for each of 5 accelerometers and 18 bottom flange strain gages for the 42 truck events.

A total of six natural frequencies were identified from the accelerometers and the strain gages in the frequency range of 4–25 Hz. There was very good agreement between the two sensor types for the first 4 natural frequencies, and the results began to diverge for the final 2 natural frequencies. The natural frequencies were easier to identify from peak picking with the accelerometers than the strain gages, and were generally more consistently identified across the accelerometer channels than across the strain gage channels. Generally, the strain gages proved capable for identifying natural frequencies that were found from the accelerometer measurements.

The mode shapes associated with each natural frequency were not identified in this study, but will be evaluated in the future. Natural frequencies and mode shapes for the first 10 modes were identified from a preliminary FEM to serve as a

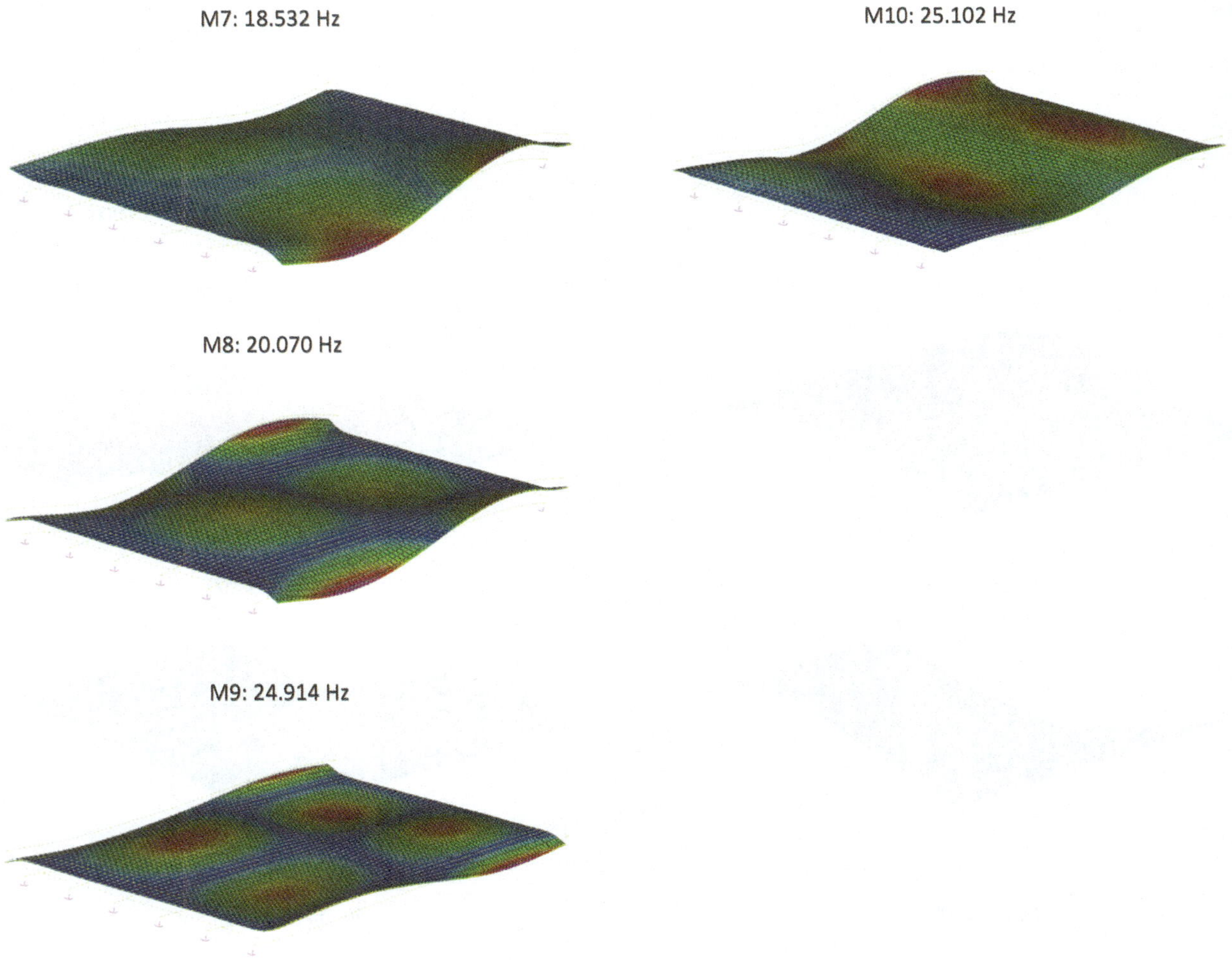

Fig. 37.7 FEM frequencies and mode shapes for modes 7–10

point of comparison with the natural frequencies identified from the measurements. In general, some of the experimental natural frequencies are close to those identified from the FEM. The FEM mode shapes in the frequency band of interest show some complexity that will be difficult to identify from the locations where the accelerometers and strain gages are physically installed on the structure. An important limitation of this study is that the strain gage locations were not designed with the intent of identifying dynamic properties of the structure. A better identification might be possible if the strain gages were more spatially distributed across the superstructure; however, the strain gage locations were representative of what is used for many routine bridge monitoring applications to evaluate stresses.

The frequency analysis performed in this study did indicate that a minimum number of accelerometers are very helpful for gaining confidence in the frequency results obtained from the strain gages. The strain gage measurements were dominated by quasi-static truck load effects, and the dynamic response from these gages was very small. The unfiltered average PSD spectrums for the accelerometers showed very clear peaks at the natural frequencies identified for the bridge. The natural frequency peaks for the filtered average PSD spectra for the strain gage channels did not contain the same frequency peaks for all channels.

The bridge monitoring system will continue to collect measurement data for a total of 5 years. Currently, the monitoring system only collects dynamic acceleration and strain measurements during truck crossing events. Additional and longer measurement records should be periodically collected from the system to identify natural frequencies and compare with the truck measurements. Since this preliminary analysis indicated that many of the bridge's natural frequencies can be identified from the strain measurements, the strain gage and accelerometer data should be continuously analyzed for natural frequencies for all truck crossing events. These results may reveal potential relationships with the various measured characteristics of the trucks. Long-term changes in the identified natural frequencies could potentially be correlated with changes in the measured stresses, composite behavior and load distribution characteristics which are also being tracked for each truck event.

References

1. Bernasconi, O., Ewins, D.J.: Application of strain modal testing to real structures. In: Proceedings of the 7th International Modal Analysis Conference, pp. 1453–1464. Society for Experimental Mechanics, Bethel, CT (1989)
2. Debao, L., Hongcheng, Z., Wang, B.: The principle and techniques of experimental strain modal analysis. In: Proceedings of the 7th International Modal Analysis Conference, pp. 1285–1289. Society for Experimental Mechanics, Bethel, CT (1989)
3. Vari, L.M., Heyns, P.S.: Using strain modal testing. In: Proceedings of the 12th International Conference on Modal Analysis, pp. 1264–1270. Society for Experimental Mechanics, Bethel, CT (1994)
4. Marques dos Santos, F.L., Peeters, B., Van der Vorst, R., Desmet, W., Sandoval Goes, L.C.: The use of mixed/strain acceleration measurements for modal analysis. In: Proceedings of the 9th International Conference on Structural Dynamics, pp. 2247–2254. EURODYN, Rome (2014)
5. Anastasopoulos, D., De Smedt, M., De Roeck, G., Vandewalle, L., Reynders, E.: Modal strain identification using sub-microstrain FBG data from a pre-stressed concrete beam during progressive damage testing. In: Proceedings of the 10th International Conference on Structural Dynamics, pp. 1846–1851. EURDODYN, Rome (2017)
6. Wang, T., Celik, O., Catbas, F.N., Zhang, L.M.: A frequency and spatial domain decomposition method for operational strain modal analysis and its application. Eng. Struct. **114**, 104–112 (2016)
7. Skafte, A., Aenlle, M.L., Brincker, R.: A general procedure for estimating dynamic displacements using strain measurements and operational modal analysis. Smart Mater. Struct. **25**(2), 025020 (2016)

Chapter 38
Evaluation of a New Energy-Based Human Tracking Method in a Smart Building Using Floor Vibration Measurements

Sa'ed Alajlouni and Pablo A. Tarazaga

Abstract Tracking occupants in an indoor environment has applications in intruder detection, emergency response and evacuation (e.g., locating an occupant in a burning building), and energy saving (through activity-based control of building lighting and HVAC system). In this document, we show that tracking occupants in an indoor environment can be done using the floor vibration caused by occupant footstep impacts. In order to track an occupant, each footstep impact location must first be estimated. For that purpose, we evaluate the performance of a newly developed energy-based localization (multilateration) method for the case of localizing occupant footsteps in a real-life operational smart building. The new method is based on the fact that the energy of the impact-generated wave will be attenuated as the wave travels away from the impact location. Localization is achieved using a network of vibration sensors (accelerometers) placed underneath the walking floor, which provides a non-intrusive and tamper-proof localization system. The new method has small computational time and requires a relatively small sensor data sampling rate. It is anticipated that the new method will have a smaller footstep localization error compared to conventional time of flight/arrival methods. Occupant tracking experiments show that the new method has a promisingly small localization error.

Keywords Occupant tracking · Underfloor accelerometers · Sensor networks · Floor vibrations · Multilateration

38.1 Introduction

An energy source (e.g., an earthquake, vocal cords vibrating, a transmitting antenna, etc.) generates waves (e.g., seismic waves, acoustic waves, or electromagnetic waves) that propagate in a certain medium (the waveguide). If a sensor network is deployed in the waveguide, measuring some feature of the incident traveling wave, then it is possible to estimate the source location using these sensor measurements.

Traditionally, attempts to solve the source localization problem were mainly done using the time difference of wave arrival (TDOA) between each sensor and a chosen reference sensor. Conventional TDOA localization methods estimate the source location assuming that the impact-generated wave travels, undistorted in shape, with a single propagation speed.

Unfortunately, when considering the problem of impact localization (e.g., footstep impact) on a floor instrumented with an accelerometer sensor network, conventional TDOA methods suffer immensely producing location estimates with high error variance and occasionally large error values (see the experimental impact localization results in [1–4]). The reduced accuracy is due to the fact that the floor is a dispersive waveguide [5–7]. Dispersion causes the floor wave to distort with the traveled distance as a result of different wave frequency components having different propagation speeds.

Methods exist in the literature that adapt conventional TDOA to improve localization in dispersive waveguides [8–13]. These methods are computationally expensive in general, and all of them—except for the method in [13]—have been tested on isotropic plates rather than a real building floor.

Due to the dispersion-caused wave distortion, in addition to the high computational cost of current TDOA-based methods, an alternative approach to impact localization in dispersive waveguides was introduced in [14]. The new approach utilizes the fact that an impact generates waves propagating along the floor, where the wave energy is attenuated with the traveled distance. The method in [14] was evaluated using hammer impacts on a concrete floor, where the floor was instrumented with

S. Alajlouni (✉)
Department of Electrical and Computer Engineering, Virginia Tech, Blacksburg, VA, USA
e-mail: saed13@vt.edu

P. A. Tarazaga
Department of Mechanical Engineering, Virginia Tech Smart Infrastructure Lab (VTSIL), Blacksburg, VA, USA
e-mail: ptarazag@vt.edu

S. Pakzad (ed.), *Dynamics of Civil Structures, Volume 2*, Conference Proceedings of the Society for Experimental Mechanics Series, https://doi.org/10.1007/978-3-319-74421-6_38

an underfloor accelerometer sensor network. The method was shown to be computationally cheap and relatively accurate (compared to hammer localization experiments using conventional TDOA, where error was larger with a larger variance [1–4]).

It is interesting to evaluate the accuracy of the method in [14] using real occupant footsteps—rather than hammer impacts—which will be the focus of this presentation. Therefore, in this document, the method in [14] is evaluated using occupant walking experiments. In particular, two occupants walked into a corridor on a floor instrumented with accelerometers. Both test subjects walked a total of 162 × 2 footsteps, where the localization algorithm in [14] was used to estimate the location of each footstep.

The remainder of this document is organized as follows: Sect. 38.2 briefly presents the localization algorithm procedure following the presentation in [14]. The reader interested in more details is referred to [14]. Then, Sect. 38.3 reports the results of the occupant footstep localization experiments. Finally, Sect. 38.4 briefly states the conclusions.

38.2 Methodology

Consider a floor instrumented with an accelerometer sensor network. This localization system must be able to detect, then localize, a footstep impact. The detection algorithm employed is a peak-picking algorithm that determines if an impact is present based on an averaged sum of all sensor readings. After a footstep is detected, the average signal power, denoted Q_i is calculated for each sensor as follows [14]

$$Q_i(t_{0i}, t_{pi}) = \frac{1}{t_{pi} - t_{0i}} \int_{t_{0i}}^{t_{pi}} z_i^2(t)dt \,, \tag{38.1}$$

where z_i denotes the ith sensor reading ($i = 1, 2, \ldots, N$), t_{pi} denotes the time instance when the peak wave envelope passes the ith sensor (determined by the peak-picking algorithm), and t_{0i} denotes the beginning (arrival) time of the footstep wave determined by a threshold crossing criterion, going back in time from t_{pi}.

Q_i is assumed to derease exponentially with sensor-source distance, d_i, according to the relation: $Q_i = Q_s e^{\beta d_i}$, where β is the floor-dependent attenuation parameter that must be known a priori (usually estimated during an offline calibration phase), and where Q_s is an unknown quantity. To eliminate Q_s, a reference sensor is chosen, and energy ratios are calculated with respect to the reference sensor as follows

$$Q_{iR} \stackrel{\text{def}}{=} \frac{Q_i}{Q_R} = e^{\beta(d_i - d_R)} \,, \tag{38.2}$$

where, $d_i = \|\boldsymbol{r}_i - \boldsymbol{r}_s\|$ ($\boldsymbol{r}_i$ is the ith sensor location vector, and $\boldsymbol{r}_s$ is the unknown, to be estimated, footstep location). Equation (38.2) generates $N - 1$ nonlinear equations. These equations are then solved simultaneously for $\boldsymbol{r}_s$ using nonlinear programming, where the obtained solution is denoted $\hat{\boldsymbol{r}}_s$. Finally, localization error–denoted ϵ–is calculated such that $\epsilon = \hat{\boldsymbol{r}}_s - \boldsymbol{r}_s$.

38.3 Experimental Results

Two occupants walked, individually during separate times, on an instrumented floor inside a smart building (*Goodwin Hall*; located inside the campus of Virginia Tech), and the proposed localization method was used to estimate their footstep locations. Each test subject stepped 162 footsteps while traversing the path shown in Fig. 38.1. Data from 11 accelerometers were used for localization while being sampled at a rate of 1024 samples per second.

The root mean squared error (RMSE $\stackrel{def}{=} \sqrt{\sum_{k=1}^{162} \|\epsilon(k)\|_2^2}$) was used to evaluate the accuracy of the localization algorithm. Roughly 2/3 of all footsteps have location error distances $\|\epsilon(k)\|$ ($k = 1, 2, \ldots, 162$) that are less than or equal to the calculated RMSE.

RMSE values were found to be 1.31 meters and 1.38 meters for the first and second test subjects, respectively. These RMSE values were obtained given that $\beta = -0.45$ (corresponding to a wave power attenuation of -1.95 dB/m). This value of β was determined, using a parameter sweep, simply as the value yielding the least overall RMSE. Therefore, the reported RMSE values represent a best case localization scenario. It is worth mentioning that, in practice, β is estimated during a calibration phase as explained in [14], which produces an estimate of β that almost always will not yield the least possible RMSE.

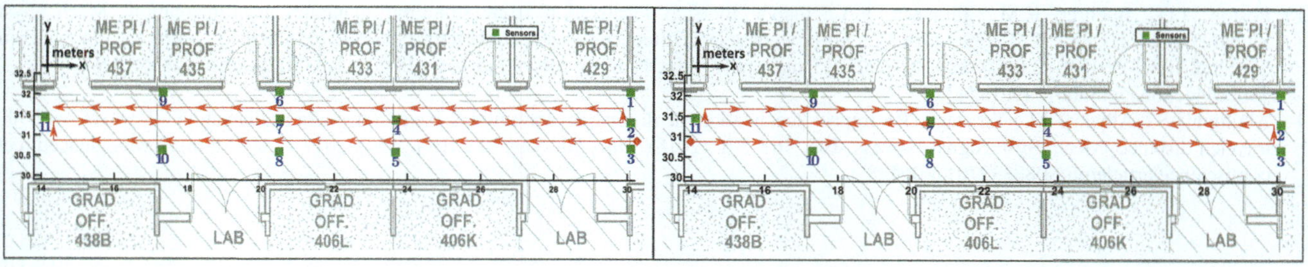

(a) (b)

Fig. 38.1 Walking path on the 4th level of Goodwin Hall. (**a**) Part 1 of the walking path. (**b**) Part 2 of the walking path

38.4 Conclusion

Reported RMSE values motivate future improvement of the method. In particular, future work will investigate ways of improving accuracy and removing the need to estimate β before running the algorithm.

Acknowledgements The authors are thankful for the support and collaborative efforts provided by our sponsors VTI Instruments, PCB Piezotronics, Inc.; Dytran Instruments, Inc.; and Oregano Systems. The authors are particularly thankful for the support provided by the College of Engineering at Virginia Tech through Ed Nelson and Dean Richard Benson as well as Capital Project Manager Todd Shelton. The authors would also like to acknowledge the collaboration with Gilbane, Inc.; in particular members David Childress and Eric Hotek. We are especially thankful to the Hashemite University of Jordan and the Student Engineering Council at Virginia Tech for their financial support. The authors would also like to recognize the support provided by the John R. Jones III faculty fellowship. The work was conducted under the patronage of the Virginia Tech Smart Infrastructure Laboratory and its members.

References

1. Poston, J.D., Buehrer, R.M., Woolard, A.G., Tarazaga, P.A.: Indoor positioning from vibration localization in smart buildings. In: 2016 IEEE/ION Position, Location and Navigation Symposium (PLANS), pp. 366–372. IEEE (2016)
2. Woolard, A.G., Phoenix, A.A., Tarazaga, P.A.: Assessment of large error time-differences for localization in a plate simulation. In: Dynamics of Coupled Structures, vol. 4, pp. 369–376. Springer (2016)
3. Poston, J.D., Schloemann, J., Buehrer, R.M., Malladi, V.S., Woolard, A.G., Tarazaga, P.A.: Towards indoor localization of pedestrians via smart building vibration sensing. In: 2015 International Conference on Location and GNSS (ICL-GNSS), pp. 1–6. IEEE (2015)
4. Schloemann, J., Malladi, V.V.N.S., Woolard, A.G., Hamilton, J.M., Buehrer, R.M., Tarazaga, P.A.: Vibration event localization in an instrumented building. In: De Clerck, J. (ed.) Experimental Techniques, Rotating Machinery, and Acoustics, **8**, pp. 265–271. Springer International Publishing, Cham (2015)
5. Bahroun, R., Michel, O., Frassati, F., Carmona, M., Lacoume, J.: New algorithm for footstep localization using seismic sensors in an indoor environment. J. Sound Vib. **333**, 1046–1066 (2014)
6. Ziola, S.M., Gorman, M.R.: Source location in thin plates using crosscorrelation. J. Acoust. Soc. Am. **90**, 2551–2556 (1991)
7. Graff, K.F.: Wave Motion in Elastic Solids. Courier Corporation, Oxford (1975)
8. De Marchi, L., Marzani, A., Speciale, N., Viola, E.: A passive monitoring technique based on dispersion compensation to locate impacts in plate-like structures. Smart Mater. Struct. **20**, 035021 (2011). http://iopscience.iop.org.ezproxy.lib.vt.edu/article/10.1088/0964-1726/20/3/035021/pdf
9. Perelli, A., De Marchi, L., Marzani, A., Speciale, N.: Acoustic emission localization in plates with dispersion and reverberations using sparse PZT sensors in passive mode. Smart Mater. Struct. **21**(2), 025010 (2012). http://iopscience.iop.org.ezproxy.lib.vt.edu/article/10.1088/0964-1726/21/2/025010/pdf
10. Gaul, L., Hurlebaus, S.: Identification of the impact location on a plate using wavelets. Mech. Syst. Signal Process. **12**(6), 783–795 (1998). <Go to ISI>: WOS:000077054200004
11. Ciampa, F., Meo, M.: Acoustic emission source localization and velocity determination of the fundamental mode a0 using wavelet analysis and a newton-based optimization technique. Smart Mater. Struct. **19**(4), 045027 (2010). http://iopscience.iop.org.ezproxy.lib.vt.edu/article/10.1088/0964-1726/19/4/045027/pdf
12. Hyunjo, J., Young-Su, J.: Fracture source location in thin plates using the wavelet transform of dispersive waves. IEEE Trans. Ultrason. Ferroelectr. Freq. Control **47**(3), 612–619 (2000). http://ieeexplore.ieee.org/ielx5/58/18204/00842048.pdf?tp=&arnumber=842048&isnumber=18204

13. Poston, J.D., Buehrer, R.M., Tarazaga, P.A.: Indoor footstep localization from structural dynamics instrumentation. Mech. Syst. Signal Process. **88**, 224–239 (2017). http://www.sciencedirect.com/science/article/pii/S0888327016305015. http://ac.els-cdn.com/S0888327016305015/1-s2.0-S0888327016305015-main.pdf?_tid=c8aab7a8-c0ba-11e6-b364-00000aacb35f&acdnat=1481582094_380ce34ce8b635ef8a0b7549c1a74390
14. Alajlouni, S., Albakri, M., Tarazaga, P.: Impact localization in dispersive waveguides based on energy-attenuation of waves with the traveled distance. Mech. Syst. Signal Process. **105**, 361–376 (2018). https://doi.org/10.1016/j.ymssp.2017.12.007. http://www.sciencedirect.com/science/article/pii/S0888327017306428

Chapter 39
Innovative Sensing by Using Deep Learning Framework

Nur Sila Gulgec, Martin Takáč, and Shamim Pakzad

Abstract Structures experience large vibrations and stress variations during their life cycles. This causes reduction in their load-carrying capacity which is the main design criteria for many structures. Therefore, it is important to accurately establish the performance of structures after construction that often needs full-field strain or stress measurements. Many traditional inspection methods collect strain measurements by using wired strain gauges. These strain gauges carry a high installation cost and have high power demand. In contrast, this paper introduces a new methodology to replace this high cost with utilizing inexpensive data coming from wireless sensor networks. The study proposes to collect acceleration responses coming from a structure and give them as an input to deep learning framework to estimate the stress or strain responses. The obtained stress or strain time series then can be used in many applications to better understand the conditions of the structures. In this paper, designed deep learning architecture consists of multi-layer neural networks and Long Short-Term Memory (LSTM). The network achieves to learn the relationship between input and output by exploiting the temporal dependencies of them. In the evaluation of the method, a three-story steel building is simulated by using various dynamic wind and earthquake loading scenarios. The acceleration time histories under these loading cases are utilized to predict the stress time series. The learned architecture is tested on acceleration time series that the structure has never experienced.

Keywords Structural health monitoring · Long short-term memory · Recurrent neural networks · Deep neural network

39.1 Introduction

As sensor networks today provide the opportunity to collect an enormous amount of data from any structure, Structural Health Monitoring (SHM) applications start posing a BIGDATA problem [1, 2]. Deep Neural Networks (deep learning or DNN) is an ideal state-of-the-art set of techniques for exploiting the opportunities hidden in BIGDATA [3]. Deep learning algorithms are designed such that they can learn from data. Therefore, deep learning is ideally suited to use large representative training datasets to learn complex features. During this learning process, they build a model which is then used to make data-driven predictions or decisions.

Many traditional SHM and condition assessment methods need full-field strain or stress measurements to be used in remaining fatigue life estimation, assessment of loading conditions, corrosion detection, composite material testing and structural design check. Large scale deployment of wired strain gauges, however, poses a fundamental limitation: they are expensive and laboriously impractical as more spatial information is desired [4]. An important, but relatively new method to measure the strain field is that of indirect monitoring [5–7]. Indirect sensing approaches first and foremost eliminate the installation costs associated with wiring and also provide a robust way to access critical locations in structures. One of the popular indirect methods for measuring kinematic quantities (displacement, strain) is Digital Image Correlation (DIC) which resolves relative movement using a reference image [8, 9]. The drawback of DIC system is the high cost of the equipment and data storage for a long and continuous dynamic monitoring protocol.

N. S. Gulgec (✉) · S. Pakzad
Department of Civil and Environmental Engineering, Lehigh University, Bethlehem, PA, USA
e-mail: nsg214@lehigh.edu

M. Takáč
Department of Industrial and Systems Engineering, Lehigh University, Bethlehem, PA, USA

S. Pakzad (ed.), *Dynamics of Civil Structures, Volume 2*, Conference Proceedings of the Society for Experimental Mechanics Series, https://doi.org/10.1007/978-3-319-74421-6_39

Acceleration measurements are another form of data that SHM applications rely on. Acceleration data can be collected relatively inexpensively by the means of fixed tethered sensors, wireless sensor networks (WSN) or mobile sensing. WSNs have been utilized in a variety of applications that range from low duty-cycle, low-power environmental monitoring applications to high-fidelity applications with accurate measurements, high sampling rate, and lossless communication for monitoring of mechanical and structural systems [10]. The advantages of WSNs include the low cost of installation and equipment, as well as their robustness and quality of data.

Addressing these limitations beg for an innovative sensing strategy where data can be integrated from inexpensive data sources. This paper presents a deep learning based approach using inexpensive data sources to predict stress or strain information of structural systems. To achieve that, the study proposes a deep learning framework comprised of multi-layer networks and Long Short-Term Memory (LSTM). Exploiting the multi-layer networks, architecture maps the complex relation between input and output. Furthermore, it captures the temporal dependencies of sensor data by using LSTM which is the state-of-the-art technique for time series prediction [11], language translation [12], and speech recognition [13].

The rest of the paper is organized as follows. First, a background information on deep learning is provided in Sect. 39.2; then, the proposed methodology with data preparation and training steps are described in Sect. 39.3. In Sect. 39.4, main findings of this study are discussed. Conclusions and future work are presented in Sect. 39.5.

39.2 Background on Deep Learning

Multi-Layer Neural Networks. Multi-layer neural networks are a subfield of machine learning where they build a deep graph mapped from input data to target [14]. The graphs are organized such that they have multiple linear layers activated by nonlinear transformations (e.g. sigmoid, tanh and others) [15]. The multi-layer networks (fully connected layers or FC layers) are composed of neurons and weight parameters (w) where the value of each neuron (s_i) can be computed by a weighted sum of the values of its input nodes (s'_j) activated by the nonlinear function a:

$$s_i = a\Big(\sum_j w_{ij} s'_j\Big). \tag{39.1}$$

Recurrent Neural Networks. Recurrent neural networks (RNN) are a family of deep neural networks for dealing with the sequential data [16]. Unlike multi-layer neural networks, RNNs are able to map target data from the entire history of previous inputs. RNN models capture the dynamics of the sequences with the directed loops in them [17]. A typical RNN architecture is shown in Fig. 39.1, which demonstrates a RNN being unrolled into a full network that has a chain-like structure. At time t, RNN receives the input $x^{(t)}$ and the hidden values from previous state $h^{(t-1)}$. In other words, current decisions are affected by the previous states. Given an input sequence $\{x^{(1)}, x^{(2)}, \ldots, x^{(T)}\}$, RNN updates the hidden node values $\{h^{(1)}, h^{(2)}, \ldots, h^{(T)}\}$ by the following equation:

$$h^{(t)} = a(W^{hx}x^{(t)} + W^{hh}h^{(t-1)} + b_h), \tag{39.2}$$

where W^{hx} denotes the weight matrix between the input and hidden layer, W^{hh} and b_h denote the recurrent weight matrix and the bias vector in hidden layer, respectively. The output of the weighted sum is typically passed through a function a such as sigmoid, tanh, ReLU or others. Optionally, output sequence at time t can be found with the formulation $y^{(t)} = (W^{yh}h^{(t)} + b_y)$.

Long Short-Term Memory (LSTM). In RNNs, the weight and bias parameters are shared along all the hidden layers [18]. Therefore, they may suffer from long-term time dependency problems where the gradients vanish or explode [19]. In order to address these problems, RNNs are improved over these years. Hochreiter and Schmidhuber [20] proposed long short-term memory (LSTM) which showed groundbreaking performance in time series prediction, language translation and video recognition.

LSTM networks are composed of memory cells which contain a chain of recurrent nodes. These memory cells help the network to control the information by adopting input gate, forget gate and output gate [21]. Diagram illustrated in Fig. 39.2 focuses on single memory cell and shows data flow through it. Each memory cell includes input node, internal state, input, forget and output gates. *Input node*, (g), takes the current input and hidden layer at previous time step, then computes the weighted sum followed by tanh function (ϕ). *Input gate*, (i), controls which input to be passed to the memory cell. For

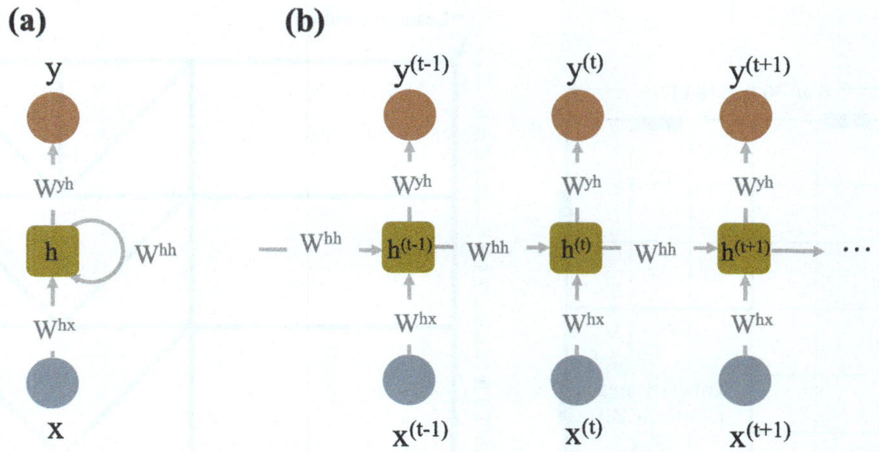

Fig. 39.1 A recurrent neural network (**a**) rolled, (**b**) unrolled

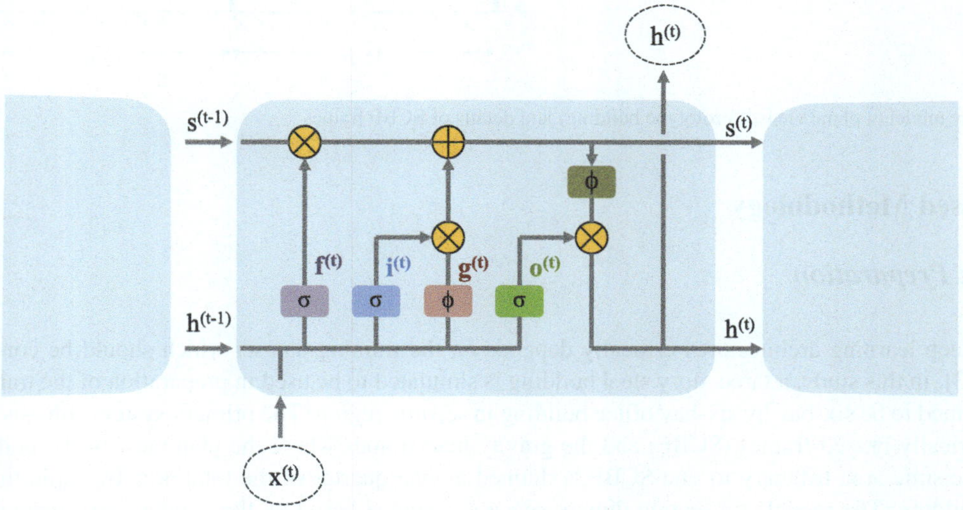

Fig. 39.2 A LSTM memory cell

instance, if the value after sigmoid function (σ) is 0, gate cuts off the input otherwise, it allows data to pass through. *Forget gate*, (f) proposed by Gers et al. (1999) [22] flushes the content if it is necessary. *Internal state*, (s), has a self-connected node that updates itself by forgetting or adding new information. *Output gate*, (o), controls what information to pass the next time step. Equations used in LSTM computations are given as follows:

$$g^{(t)} = \phi(W^{gx}x^{(t)} + W^{gh}h^{(t-1)} + b_g), \tag{39.3a}$$

$$i^{(t)} = \sigma(W^{ix}x^{(t)} + W^{ih}h^{(t-1)} + b_i), \tag{39.3b}$$

$$f^{(t)} = \sigma(W^{fx}x^{(t)} + W^{fh}f^{(t-1)} + b_f), \tag{39.3c}$$

$$o^{(t)} = \sigma(W^{ox}x^{(t)} + W^{oh}h^{(t-1)} + b_o), \tag{39.3d}$$

$$s^{(t)} = g^{(t)} \odot i^{(t)} + s^{(t-1)} \odot f^{(t)}, \tag{39.3e}$$

$$h^{(t)} = \phi(s^{(t)}) \odot o^{(t)}. \tag{39.3f}$$

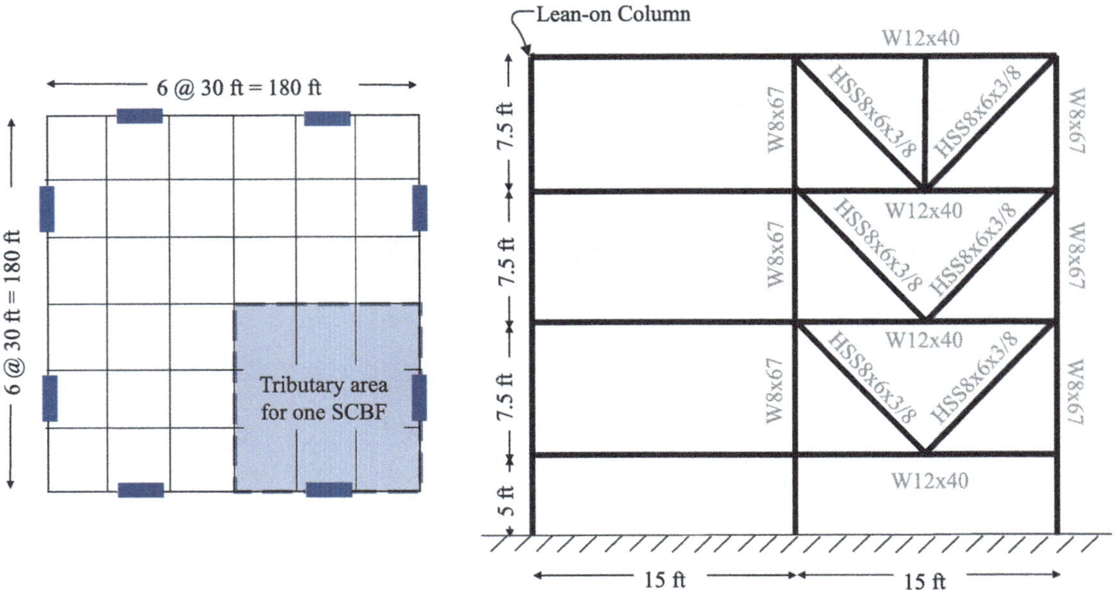

Fig. 39.3 Structure model (a plane view of prototype building) and details of SCBF frames

39.3 Proposed Methodology

39.3.1 Data Preparation

Design of the deep learning architectures critically depends on the training dataset which should be constructed by well-known states [23]. In this study, a three-story steel building is simulated to be used in preparation of the training dataset. The structure is assumed to be six-bay by six-bay office building in seismic region. The primary system consists of eight identical special concentrically braced frames (SCBF) and the gravity load frames where the plan view of the building is shown in Fig. 39.3. The seismic area tributary to one SCBF is defined as one quarter of the total area by exploiting the symmetric layout of the building. The model and section dimensions are designed based on the similar experimental structure model described in Dong et al. [24]. However, damped braced frame (DBF) and moment resisting frame (DBF) are replaced by SCBF for simplicity. The section properties of the designed model is shown in Fig. 39.3.

The building is loaded with different dynamic loading combinations under the effect of wind and earthquake. Twelve different ground motion records are selected from the PEER NGA online database [25] which have the smallest sum of the squared error (SSE) matching with uniform hazard spectrum (UHS). OpenSha is used to generate hazard spectrum [26].

Wind velocity fluctuations are performed by Monte Carlo Simulation [27]. One-dimensional, uni-variate sample functions are created to match the probabilistic characteristics of the wind load. Wind velocity is defined with the Kaimal's Spectrum [28] which can be formulated by the following equation:

$$S_{xx}(w) = \frac{200zu_*^2}{4\pi U(z)\left[1 + \dfrac{50|w|z}{2\pi U(z)}\right]^{5/3}}, \tag{39.4}$$

where $U(z)$ is mean speed at height z, k is Von karman's constant, w is frequency in rad/s, u_* is shear velocity of the flow defined by $u_* = kU(z)/ln(z/z0)$. In this study, roughness length z_0 is adopted as 0.001266 m and mean wind speed height is taken as 8 m/s to simulate wind vibrations. By using the simulated velocities, wind pressure is calculated and dynamically applied to the structure.

The load combinations through 4a to 4d are utilized with 12 ground motions and 6 wind load scenarios. Dynamic responses are collected to have a total of 1000 points with a sampling time of $\Delta_t = 0.05$ s. Total 40 different loading combinations are created. Acceleration ($\ddot{u}(t)$) and stress ($\sigma(t)$) time histories are collected from nine locations as shown in

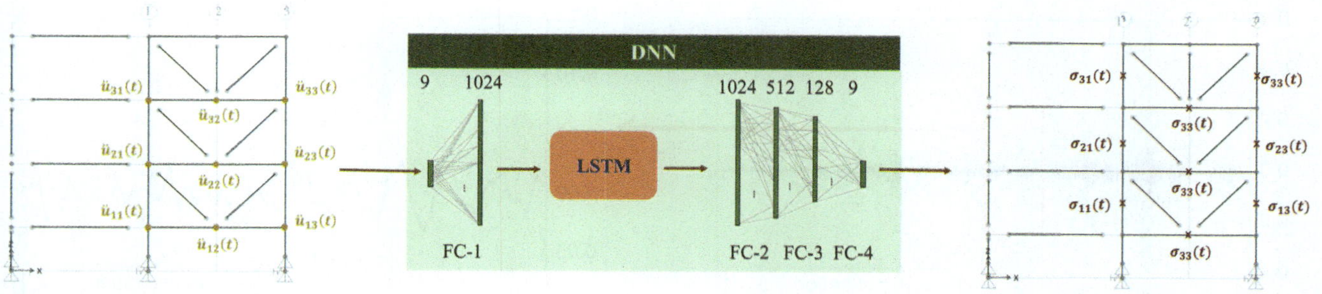

Fig. 39.4 Proposed network topology

Fig. 39.4. The collected responses are distributed to training, validation and testing datasets. Responses are normalized with the overall maximum of sequences.

$$(1.2D + 0.2S_{DS})D + \rho Q_E + L, \tag{39.5a}$$

$$(0.9D + 0.2S_{DS})D + \rho Q_E, \tag{39.5b}$$

$$1.2D + 1.6L + 0.8W, \tag{39.5c}$$

$$1.2D + L + 1.6W, \tag{39.5d}$$

where D and L are dead and live loads calculated based on ASCE7010 design code [29], W is wind load, Q_E is the effect of horizontal seismic forces, S_{DS} is design, 5% damped, spectral response acceleration parameter at short periods which is taken as 1.0 g and ρ is a redundancy factor adopted as 1.

39.3.2 Proposed Network Topology and Training

Let a series of observations $\{\ddot{u}_{11}(t), \ddot{u}_{12}(t), \ddot{u}_{12}(t), \ldots, \ddot{u}_{33}(t)\}$ denotes the acquired acceleration data from the building model. Here the acceleration of the joint of ij sampled at time step t is represented as $\ddot{u}_{ij}(t)$ where i is the number of story and j is the grid number in x direction (as shown in Fig. 39.4). The responses obtained from each joint at the time steps $t = 0, 0.05, \ldots, 50$ for different loading combinations are prepared as tensors (e.g. in the shape of [24, 1000, 9] for training dataset).

The proposed network topology takes the sequences of acceleration data and learns how to predict stress time histories. At time t, model takes input sequences and passes them through the multi-layer neural network size of [1024]. The output of the FC layer is used to feed the LSTM memory cells with size of 1024, which are then followed by three multi-layer network with the sizes of $[1024 - 512 - 128]$. These layers are activated by using tanh() function. The final output of the last DNN layer is used to predict the stress time histories $\{\sigma_{11}(t), \sigma_{12}(t), \sigma_{12}(t), \ldots, \sigma_{33}(t)\}$. The scheme of the proposed architecture can be found in Fig. 39.4.

The loss function is defined as mean squared error of predictions and true values of acceleration sequences. The networks is trained by ADAM optimizer which is an adaptive learning rate algorithm [30] with batch sizes of $N = 24$. In order to favor the short-term dependencies of data, truncated backpropogation through time (BPTT) approach is adopted [31].

39.4 Results

This section presents the performance analysis of the proposed deep learning based methodology. The designed architecture is trained for 10,000 epochs until it overfits the training dataset. By overfitting the training dataset, the capability of proposed architecture is observed and predicted results are compared with the true stress time series. The mean squared error for training process is found to be 0.001. As an example, a performance of one earthquake sample sequence that is collected from Joint$_{31}$ is visualized in Fig. 39.5. In figure, normalized acceleration time series are plotted with both normalized target

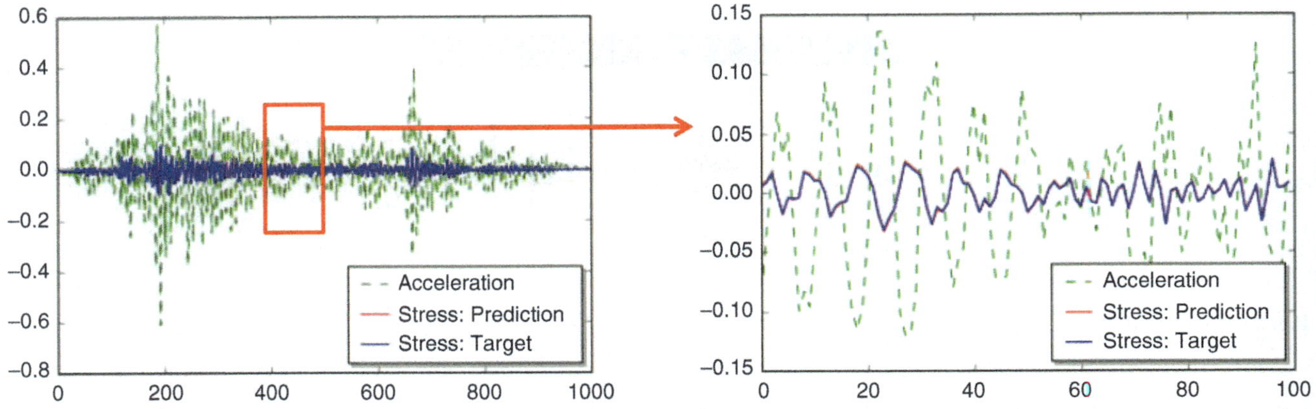

Fig. 39.5 A performance of an example training sample

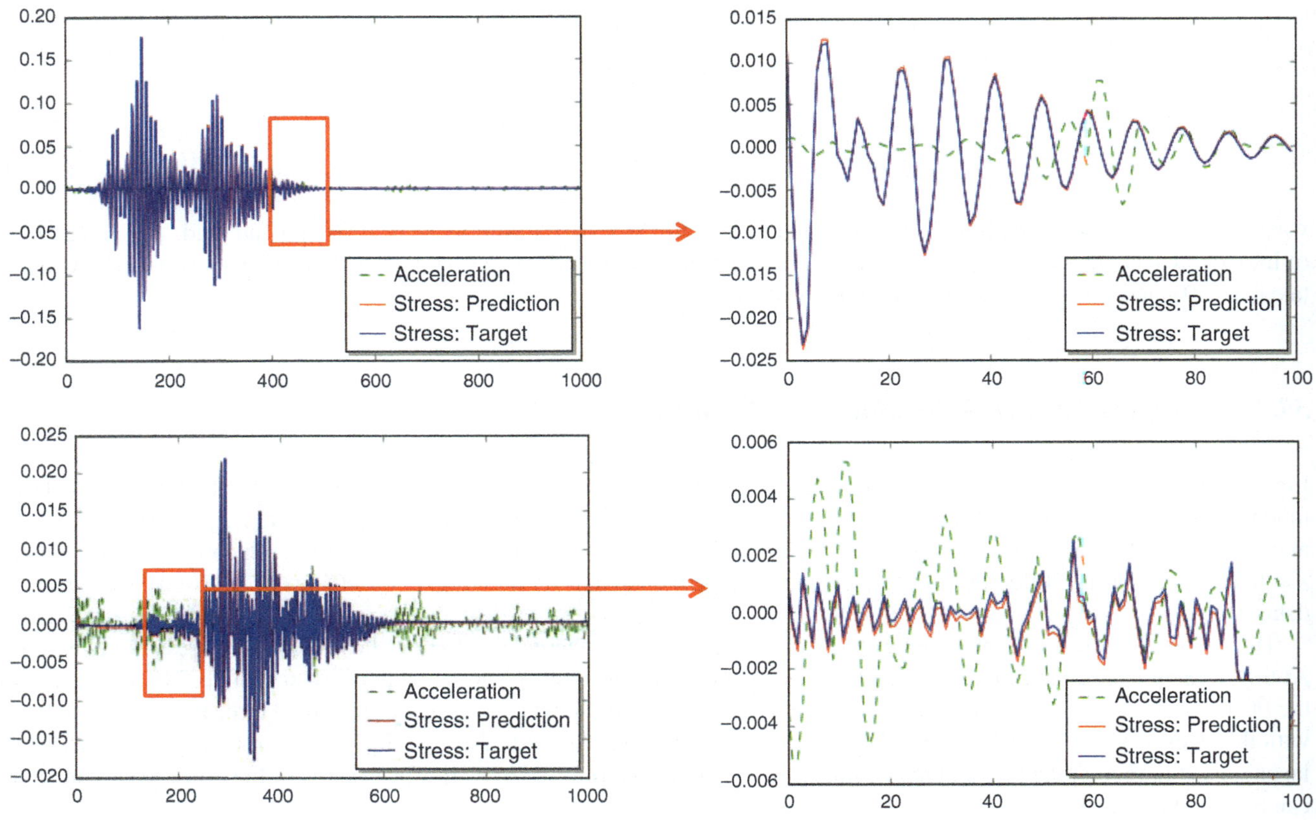

Fig. 39.6 A performance of example testing samples

and predicted stress time series. It is observable that predicted sequence perfectly captures the target sequence. Furthermore, figure shows that although acceleration and stress sequences do not have a relationship that easily noticeable, the designed architecture accurately estimates it by exploiting the LSTM cells.

After training of the network, the model is tested with the earthquake loading scenarios that structure has never been exposed. The mean square error is found to be 0.12. Similar plots are also generated for example earthquake loading cases. The earthquake loads for $Joint_{31}$ and $Joint_{33}$ are presented in Fig. 39.6. These examples shows that even the model is overfitted to training process and unseen loading earthquake cases are used, the model predicts the target sequences almost perfectly. This performance shows that introduced approach is promising for described innovative sensing strategy. The small deviance in true and estimated sequences for $Joint_{33}$ can be reduced by adopting multiple LSTMs in the model or applying a fine tuning.

39.5 Conclusion

This paper introduces a deep learning based platform so the data obtained by using wireless sensor networks can be used to obtain stress or strain information which is necessary for many potential applications including: damage diagnosis, remaining fatigue life estimation, accurate assessment of loading conditions, corrosion detection, composite material testing, and structural design check. The proposed network exploits the temporal modeling of LSTM and nonlinear mapping of FC layers to be able discover temporal dependencies and complex relationships between input and output sequences. Based on the findings of the approach, accurate estimation of stress time series is possible with acceleration acquired from inexpensive sensing system. Results show that stream of prediction values are matching quite well for training samples. The performance of the network needs a little improvement for the load scenarios that structure has never been experienced.

To discover more of abilities of deep neural networks, further research steps are important. The future work aims to extend this work by (i) designing more complex network architecture which considers both temporal and spatial dependencies of data, (ii) using experimental data, (iii) trying different type of data source conversions specific to other structures.

Acknowledgements Research funding is partially provided by the National Science Foundation through Grant No. CMMI-1351537 by Hazard Mitigation and Structural Engineering program, and by a grant from the Commonwealth of Pennsylvania, Department of Community and Economic Development, through the Pennsylvania Infrastructure Technology Alliance (PITA). Martin Takáč was supported by National Science Foundation grant CCF-1618717 and CMMI-1663256.

References

1. Matarazzo, T.J., Shahidi, S.G., Chang, M., Pakzad, S.N.: Are today's SHM procedures suitable for tomorrow's bigdata? In: Structural Health Monitoring and Damage Detection, vol. 7, pp. 59–65. Springer, Heidelberg (2015)
2. Gulgec, N.S., Shahidi, S.G., Pakzad, S.N.: A comparative study of compressive sensing approaches for a structural damage diagnosis. In: Geotechnical and Structural Engineering Congress, pp. 1910–1919. American Society of Civil Engineers, Reston (2016)
3. Gulgec, N.S., Shahidi, G.S., Matarazzo, T.J., Pakzad, S.N.: Current challenges with bigdata analytics in structural health monitoring. In: Structural Health Monitoring & Damage Detection, vol. 7, pp. 79–84. Springer, Berlin (2017)
4. Kim, S., Pakzad, S., Culler, D., Demmel, J., Fenves, G., Glaser, S., Turon, M.: Health monitoring of civil infrastructures using wireless sensor networks. In: Proceedings of the 6th International Conference on Information Processing in Sensor Networks, pp. 254–263. ACM (2007)
5. Giraldo, D.F., Dyke, S.J., Caicedo, J.M.: Damage detection accommodating varying environmental conditions. Struct. Health Monit. **5**(2), 155–172 (2006)
6. González, A., Covián, E., Madera, J.: Determination of bridge natural frequencies using a moving vehicle instrumented with accelerometers and GPS. In: Proceedings of the Ninth International Conference on Computational Structures Technology, CST2008, Athens, 2–5 Sept 2008. Civil-Comp Press (2008)
7. Lederman, G., Wang, Z., Bielak, J., Noh, H., Garrett, JH., Chen, S., Kovacevic, J., Cerda, F., Rizzo, P.: Damage quantification and localization algorithms for indirect SHM of bridges. In: Proceeding of International Conference on Bridge Maintenance, Safety Management, Shanghai (2014)
8. Yoneyama, S., Kitagawa, A., Iwata, S., Tani, K., Kikuta, H.: Bridge deflection measurement using digital image correlation. Exp. Tech. **31**(1), 34–40 (2007)
9. Hild, F., Roux, S.: Digital Image Correlation. Wiley-VCH, Weinheim (2012)
10. Pakzad, S.N., Fenves, G.L., Kim, S., Culler, D.E.: Design and implementation of scalable wireless sensor network for structural monitoring. J. Infrastruct. Syst. **14**(1), 89–101 (2008)
11. Gers, F.A., Eck, D., Schmidhuber, J.: Applying LSTM to time series predictable through time-window approaches. In: Neural Nets WIRN Vietri-01, pp. 193–200. Springer, New York (2002)
12. Sutskever, I., Vinyals, O., Le, Q.V.: Sequence to sequence learning with neural networks. In: Advances in Neural Information Processing Systems, pp. 3104–3112 (2014). arXiv:1409.3215
13. Graves, A., Jaitly, N.: Towards end-to-end speech recognition with recurrent neural networks. In: Proceedings of the 31st International Conference on Machine Learning (ICML-14), pp. 1764–1772 (2014)
14. Gulgec, N.S., Takáč, M., Pakzad, S.N.: Structural damage detection using convolutional neural networks. In: Model Validation and Uncertainty Quantification, vol. 3, pp. 331–337. Springer, Cham (2017)
15. LeCun, Y., Bengio, Y., Hinton, G.: Deep learning. Nature **521**(7553), 436–444 (2015)
16. Chung, J., Gulcehre, C., Cho, K., Bengio, Y.: Empirical evaluation of gated recurrent neural networks on sequence modeling. arXiv preprint arXiv:1412.3555 (2014)
17. Elman, J.L.: Finding structure in time. Cogn. Sci. **14**(2), 179–211 (1990)
18. Lipton, Z.C., Berkowitz, J., Elkan, C.: A critical review of recurrent neural networks for sequence learning. arXiv preprint arXiv:1506.00019 (2015)
19. Bengio, Y., Simard, P., Frasconi, P.: Learning long-term dependencies with gradient descent is difficult. IEEE Trans. Neural Netw. **5**(2), 157–166 (1994)
20. Hochreiter, S., Schmidhuber, J.: Long short-term memory. Neural Comput. **9**(8), 1735–1780 (1997)

21. Graves, A., Mohamed, A.-R., Hinton, G.: Speech recognition with deep recurrent neural networks. In: 2013 IEEE International Conference on Acoustics, Speech and Signal Processing (ICASSP), pp. 6645–6649. IEEE (2013)
22. Gers, F.A., Schmidhuber, J., Cummins, F.: Learning to forget: continual prediction with LSTM. Neural Comput. **12**(10), 2451–2471 (2000)
23. Elkordy, M.F., Chang, K.C., Lee, G.C.: Neural networks trained by analytically simulated damage states. J. Comput. Civ. Eng. **7**(2), 130–145 (1993)
24. Dong, B., Sause, R., Ricles, J.M.: Seismic response and performance of a steel MRF building with nonlinear viscous dampers under DBE and MCE. J. Struct. Eng. **142**(6), 04016023 (2016)
25. PEER NGA: Pacific Engineering Research Database (2008). https://ngawest2.berkeley.edu/
26. Field, E.H., Jordan, T.H., Cornell, C.A.: Opensha: a developing community-modeling environment for seismic hazard analysis. Seismol. Res. Lett., **74**(4), 406–419 (2003)
27. Carassale, L., Solari, G.: Monte carlo simulation of wind velocity fields on complex structures. J. Wind Eng. Ind. Aerodyn. **94**(5), 323–339 (2006)
28. Kaimal, J.C., Wyngaard, J.C., Izumi, Y., Cote, O.R.: Spectral characteristics of surface-layer turbulence. Q. J. R. Meteorol. Soc. **98**(417), 563–589 (1972)
29. American Society of Civil Engineers: Null null: Minimum Design Loads for Buildings and Other Structures, ASCE/SEI 7–10 edn. American Society of Civil Engineers (2013)
30. Kingma, D., Ba, J.: Adam: a method for stochastic optimization. arXiv preprint arXiv:1412.6980 (2014)
31. Werbos, P.J.: Backpropagation through time: what it does and how to do it. Proc. IEEE **78**(10), 1550–1560 (1990)

Chapter 40
The Role of Control-Structure Interaction in Deployable Autonomous Control Systems

K. Goorts and S. Narasimhan

Abstract Modern, innovative structures often rely on auxiliary control devices to suppress excessive vibrations. Currently, most control devices are permanent installations designed specifically for the intended structure and further tuned to a particular structural property. The concept of a deployable, autonomous control system (DACS) is proposed herein where the notion of deployability implies the system can be rapidly implemented on a range of structures while the autonomy aspect facilitates real-time positioning of the system and generation of active control forces. A small-scale prototype targeting short-term control applications of lightweight structures is presented. The prototype consists of an electromagnetic mass damper (EMD), unmanned ground vehicle (UGV) equipped with vision sensors, and computational hardware. The UGV transfers inertial control forces generated by the EMD through the tires which eliminates the need for a rigid connection to the structure and enables rapid deployment for short-term applications. A simultaneous localization and mapping (SLAM) solution utilizing the mobility of the UGV and on-board vision sensors facilitates autonomous positioning of the device at desired locations on the structure. This study investigates the role of control-structure interaction (CSI) by examining the effect of base motion on the position controlled EMD and interaction between the structural response and UGV dynamics. Different controller formulations are presented and comparatively assessed to illustrate the inherent interaction effects. Experimental results confirm CSI has a negligible impact on EMD position tracking while demonstrating the importance of considering UGV dynamics in the controller formulation.

Keywords Control structure interaction · Active control · Autonomous robotics · Mobile vibration control · Lightweight structures

40.1 Introduction

Auxiliary control devices play a key role in enabling innovative structural designs that challenge the existing norms of material usage, size, shape, and flexibility. As structures continue to evolve, so too must the control devices that support them. Most existing structural control implementations consist of passive devices, tuned for a particular structure and hence specific to the particular application. Although reliable, the passive nature of these devices restricts the number of controllable properties and limits adaptability. Active and semi-active devices have been used in particular applications to increase the control effect and adapt to different structural properties [1–3]. Nevertheless, these systems are generally fixed in location which limits their effectiveness to control different modes. By design, these systems are generally only useful for design level events. For example, a tuned mass damper designed for a 5-year return period event may not engage significantly during a 1-year return event. Furthermore, nearly all active and passive systems are permanent installations which may be considered impractical for control of temporary structures (i.e., a lightweight pedestrian bridge for marathon events or short-term access) where the need for control depends on the intended use. These limitations motivate the novel concept of a deployable, autonomous control system (DACS), suitable for use on a range of structures. The notion of deployability implies the system can be rapidly implemented on a range of structures with minimal modification while the autonomy aspect facilitates real-time positioning of the system and generation of active control forces to control multiple properties. Contrary to permanent installations designed for specific applications, a single DACS design could be implemented on a range of structures, as needed, and stored or used elsewhere when not required.

A small-scale prototype DACS has been developed to specifically target temporary control applications of lightweight structures (i.e., up to 1 tonne) [4]. The prototype consists of an electromagnetic mass damper (EMD) mounted on an

K. Goorts (✉) · S. Narasimhan
University of Waterloo, Waterloo, ON, Canada
e-mail: kjgoorts@uwaterloo.ca; sriram.narasimhan@uwaterloo.ca

S. Pakzad (ed.), *Dynamics of Civil Structures, Volume 2*, Conference Proceedings of the Society for Experimental Mechanics Series, https://doi.org/10.1007/978-3-319-74421-6_40

unmanned ground vehicle (UGV) equipped with vision sensors and onboard computational hardware. Active control forces generated by the EMD are transferred through the UGV tires which eliminates the need for a rigid connection to the structure and enables rapid deployment. The relatively wide control bandwidth of the EMD facilitates use on a range of structures with the ability to control various properties. The mobility of the UGV combined with on-board vision sensors allows for autonomous positioning of the device at desired locations through a simultaneous localization and mapping (SLAM) algorithm which addresses the fundamental problem in mobile robotics of navigating an unknown environment [5, 6].

The fundamental challenge associated with the prototype DACS involves characterization of the system dynamics for the purpose of controller design. The inertial forces generated by the EMD are filtered through the inherently non-linear UGV dynamics prior to being transferred to the structure. As such, models of both the EMD dynamics and UGV dynamics are required for controller design. Earlier work on the prototype involved experimental system identification of the UGV and EMD culminating in a preliminary controller design which was later validated on a full-scale pedestrian bridge [4, 7]. Control-structure interaction (CSI), defined as the dynamic coupling between an active control system and structure, directly impacts controller design. Studies on CSI have illustrated the implications of unmodeled effects on the performance and robustness of various active control systems [8, 9]. The role of CSI in the prototype DACS is unique due to the presence of additional UGV dynamics and the fact that there is no rigid connection to the structure. This paper provides an overview of the system modeling and investigates the role of CSI. The effect of base motion on the EMD controller is examined and two controller formulations are comparatively assessed to illustrate the impact of unmodeled UGV dynamics on control performance.

40.2 Prototype System

The three primary components of the prototypical DACS are an UGV equipped with vision sensors, an EMD with auxiliary mass, and on-board compact computational hardware with real-time processing capabilities. Figure 40.1a and b show a schematic of the prototype and a possible implementation for lateral vibration suppression of footbridges respectively.

The UGV utilized in this prototype is a Husky A200 built by Clearpath Robotics [10]. This rugged, skid-steering platform features a payload capacity of 75 kg and 300 mm lug tread tires which provide sufficient static friction in the lateral direction to generate control action up to a peak force of 400 N [4]. The vehicle can reach speeds up to 1 m/s and is well suited for outdoor use with an internal storage compartment. A Parker Automation T3D I-force ironless linear motor positioner powered by a Gemini Gv6K servo drive serves as the EMD. The linear motor consists of a carriage attached to a current carrying coil which moves through a magnetic field created by a series of permanent magnets. An auxiliary mass of 33.6 kg is mounted on the carriage which can reach speeds up to 7 m/s and generate a peak force of 410 N. Contrary to typical EMD implementations which rely on electro-mechanical models of the system to operate in force control, the T3D positioner operates under position control using digital signals.

On-board computational hardware is required for monitoring and processing tasks as well as execution of the active control algorithm. A National Instruments (NI) compact RIO (cRIO) is programmed for the prototype and fits within the

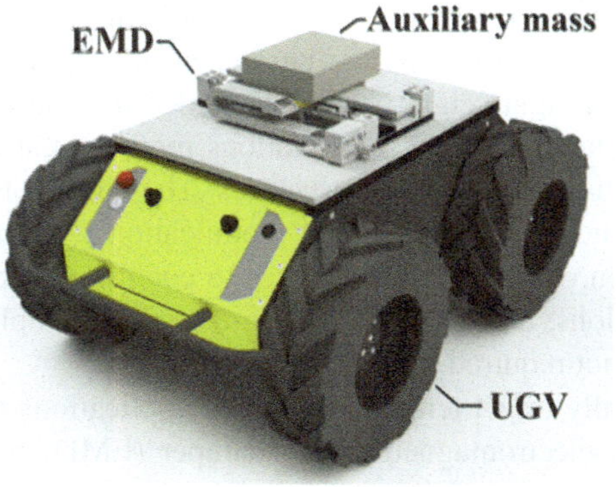

(a)

(b)

Fig. 40.1 Main components and possible implementation of prototypical DACS

UGV storage compartment. The cRIO features a 667 MHz dual-core processor and built-in field programmable gate array (FPGA) for high-speed deterministic data transfer. The cRIO has the capacity to interface directly with the EMD servo drive and UGV controller which facilitates integrated UGV positioning and structural control. A robot operating system (ROS) enabled laptop is used in parallel with the cRIO to drive the UGV, process vision sensor data, and execute SLAM algorithms.

40.3 System Modeling and Controller Design

Generally speaking, active control uses measurements of the structural response (feedback link) and in some cases external excitation (feed-forward link) to determine the command input to the actuator. Most implementations of active control are closed-loop systems with force-feedback; however, the deployable and mobile nature of the prototype removes the ability to measure control forces directly and prevents closed-loop design. Hence, an open-loop controller with built-in compensation for UGV dynamics is required. Provided the UGV dynamics can be sufficiently captured in a linear transfer function H_{UGV}, the following relationship can be derived in the Laplace domain between the EMD inertial force F_{EMD} and active control force transferred to the structure F_a:

$$F_a(s) = H_{UGV}(s)\ F_{EMD}(s) \tag{40.1}$$

In addition, the linear motor positioner is configured to operate in position control using the readily available motor encoder outputs. Although position control offers improved motion profile tracking and higher noise robustness with the use of digital command signals, there is no direct relationship between the inertial force and command input (i.e., position). Modeling the EMD dynamics using a linear transfer function H_{EMD}, yields the following relationship between the EMD inertial force and command position P_{EMD}:

$$F_{EMD}(s) = H_{EMD}(s)\ P_{EMD}(s) \tag{40.2}$$

Substituting Eq. 40.2 into Eq. 40.1 leads to the overall model of the prototype consisting of both the UGV and EMD dynamics:

$$F_a(s) = \{H_{UGV}(s)\}\{H_{EMD}(s)\}\ P_{EMD}(s) \tag{40.3}$$

For open-loop controller design, this model can be inverted and used to compute EMD position commands given desired control forces. It should be noted that a controller design based solely on the above model neglects any interaction effects caused by structural motion below the system. This section describes the system modeling, discusses the role of CSI, and proposes a controller formulation to compensate for the interaction effects.

40.3.1 System Modeling

Characterization of the UGV and EMD dynamics is achieved using experimental identification of linear transfer function models [4]. The validity of the linear models is investigated in a separate study that identifies frequency and force bandwidths for the prototype system [11]. The findings conclude that by restricting the operating frequency to below 4 Hz and capping the force level at 250 N, the root-mean-squared-error (RMSE) between the desired and measured forces will fall below 5%. A brief overview of the identification procedure and resulting models is included below.

40.3.1.1 Modeling EMD Dynamics

Experimental identification of the EMD dynamics captures the effects of closed-loop motor dynamics including position feedback control. The test setup consists of the EMD mounted directly on a shear-type load cell with an accelerometer mounted on the auxiliary mass for measurement redundancy. The EMD is controlled using sine-sweep position time-histories encompassing frequencies from 0.5 to 5.0 Hz. Six different time-histories are used with time-varying amplitudes scaled to achieve constant accelerations, ranging from 1.5 to 9.0 m/s^2, across the span of frequencies. Through curve-fitting to frequency response functions (FRFs) between command position and measured inertial force the following EMD model is derived:

$$H_{EMD}(s) = \frac{3.458s^2 + 1.639s + 1.737}{s^2 + 78.15s + 3459} \tag{40.4}$$

40.3.1.2 Modeling UGV Dynamics

Modeling the UGV dynamics is achieved using two approaches: parametric modeling and experimental modeling. The parametric modeling approach is based on the premise that the UGV can be represented by a single degree-of-freedom (SDOF) system and as such involves estimation of mass, damping, and stiffness parameters. Although stiffness eccentricities and complex damping may limit model accuracy, the motivation behind the parametric approach lies in the ability to use the system parameters directly in controller design. For parameter identification, the prototype is positioned atop four shear-type load cells and subjected to an impulse load. The logarithmic-decrement method is applied to the force measured at the base to estimate the natural frequency and damping ratio. Using a vehicle mass of 50 kg, the damping and stiffness parameters are computed as 337 Ns/m and 80,705 N/m respectively yielding the following parametric UGV model:

$$H_{UGV}{}^{(a)}(s) = \frac{337s + 80705}{50s^2 + 337s + 80705} \tag{40.5}$$

The second approach involves fitting polynomial transfer function models to experimental FRFs. The test setup remains unchanged however in this case, the system is excited using constant amplitude harmonic forces sweeping through the spectral bandwidth of potentially controlled structures (i.e., 0–5 Hz). Six different tests are conducted with increasing force magnitudes up to 150 N after which the UGV begins to slip as the peak force at the base is amplified beyond 400 N. Applying curve-fitting techniques to experimental FRFs between the inertial force of the EMD control force measured at the base yields the following experimental UGV model:

$$H_{UGV}{}^{(b)}(s) = \frac{15.64s + 1485}{0.02s^3 + 1.132s^2 + 33.75s + 1358} \tag{40.6}$$

Note the third-order polynomial in the denominator is a direct result of adding a lead-lag compensator to the estimated second-order system in order to improve the phase performance of the model.

40.3.2 Active Control Strategy

The objective of the active control algorithm is to determine the optimal EMD position command such that the structural response is in some way minimized. An overall controller consisting of a linear-quadratic Gaussian (LQG) algorithm connected in series with the inverse UGV model and inverse EMD model is formulated. In this configuration, referred to as LQG #1, the LQG algorithm is designed based solely on the structural properties and computes the optimal active control force to be applied by the prototype DACS. In order to compute the control input (i.e., EMD position), the controller first compensates for UGV dynamics via the inverse UGV model and subsequently accounts for EMD dynamics through the inverse EMD model. Figure 40.2 provides a schematic block diagram of the LQG #1 controller configuration. Inherent in this formulation is the assumption that the structure and control system can be modeled independently. In other words, it implies the desired control force can be achieved irrespective of the structure motion. For the prototype DACS, it is clear the structural response will engage the UGV dynamics thus directly influencing the generated control force. As such, a controller configuration accounting for this interaction is required.

40.3.3 Role of Control-Structure Interaction

Although the effects of CSI have been documented for a range of active control systems including hydraulic actuators [8] and EMDs [9], the interaction effects in the prototype DACS are unique due to the presence of additional UGV dynamics and use of position control for the EMD. Unlike EMD inertial forces which are relative to the global reference frame, the

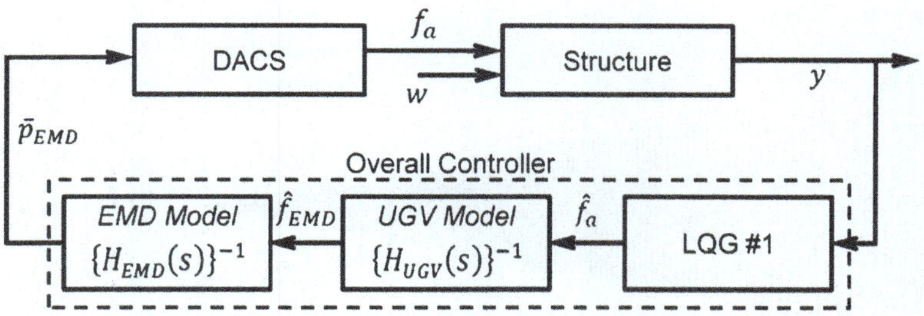

Fig. 40.2 Schematic block diagram of LQG #1 controller configuration

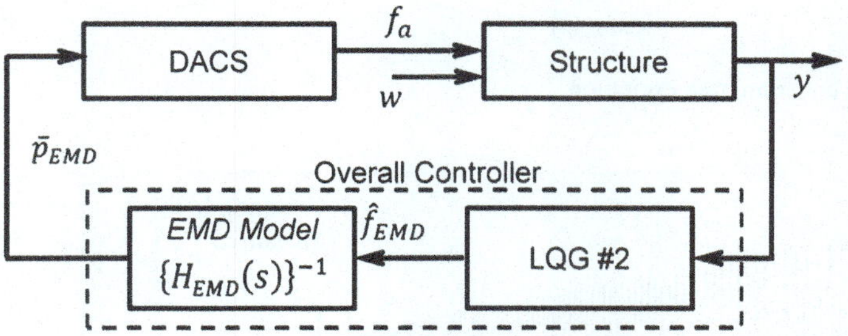

Fig. 40.3 Schematic block diagram of LQG #2 controller configuration

auxiliary mass position is measured relative to the local (i.e., EMD base) reference frame and thus independent of base motion. When subject to base motion, the position feedback algorithm still aims to track the desired position profile relative to the base of the EMD. The position tracking performance under base excitation depends to some extent on the tuning and control authority of the feedback controller but also on the capacity of the linear motor and level of base excitation. Consider the extreme cases: slow base excitation (i.e., low frequency, small amplitude) and fast base excitation (i.e., high frequency, large amplitude). Slow base excitation will generate minimal inertial forces on the auxiliary mass and is unlikely to disrupt position tracking. On the other hand, fast base excitation could generate inertial forces too large and too rapidly for the feedback algorithm to keep up. As a result, the feedback algorithm is expected to maintain a high level of position tracking up to a certain level of base excitation after which the tracking deteriorates and eventually becomes unstable. The assumption is made that position tracking deteriorates outside the control bandwidth of the system and as such the EMD model remains valid for controller design. This assumption is investigated in the following section.

The UGV dynamics are not only excited by the EMD inertial force but also by the structural response (i.e., base motion). As such, the interaction between the structural response and UGV dynamics must be considered in controller design. By idealizing the UGV as a SDOF system, an LQG algorithm is formulated based on the combined model of the structure and UGV. In this configuration, referred to as LQG #2, the output of the LQG algorithm is the desired EMD inertial force to be applied at the top of the UGV. As such, computing the control input requires only the inverse EMD model. The schematic block diagram for LQG #2 is shown in Fig. 40.3.

40.4 Experimental Testing and Results

Experimental testing is used to investigate the position tracking performance of the EMD under base excitation and compare the performance of the two LQG controller designs.

40.4.1 EMD Position Tracking

To assess the position tracking performance of the EMD under base excitation, the EMD is mounted directly on a shear-type load cell atop a shake table. The shake table provides harmonic base excitation at different frequencies and amplitudes while the EMD is programmed to track a 60-s linearly-varying sine-sweep position time-history encompassing frequencies up to 5 Hz. A total of 18 tests are conducted with harmonic base excitation ranging from 1 to 5 Hz with amplitudes from

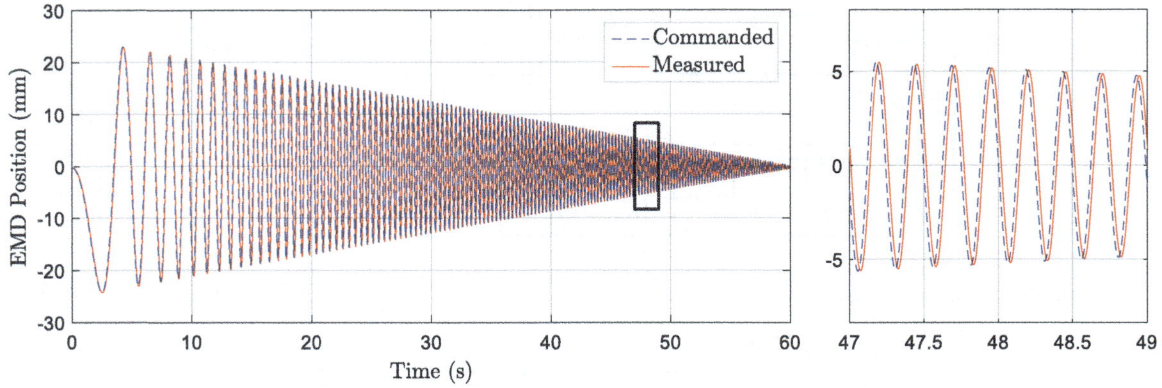

Fig. 40.4 EMD position tracking under no base excitation

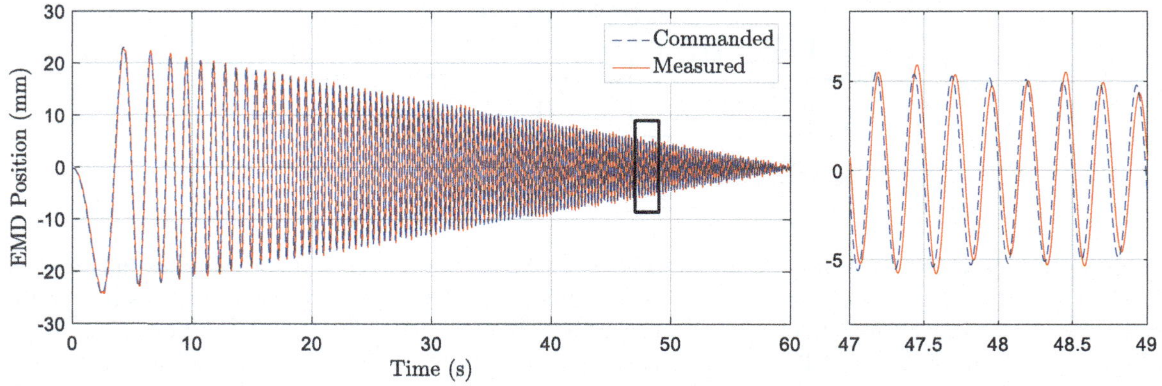

Fig. 40.5 EMD position tracking under harmonic base excitation with peak acceleration of 3.6 m/s^2

2.5 to 25 mm yielding a spectrum of peak base accelerations from 0.2 to 4.9 m/s^2. Figure 40.4 shows the position tracking for the reference case under no base excitation. As shown, the feedback control accurately tracks the desired position with approximately 17 ms of control delay.

Small disruptions in the position tracking first appear when the peak base acceleration exceeds 1.75 m/s^2. These disruptions only occur when the EMD is reaching frequencies beyond 4 Hz. The tracking continues to deteriorate as the peak base acceleration increases and disruptions are evident at lower EMD frequencies. Figure 40.5 shows the EMD position tracking under harmonic base excitation with a peak acceleration of 3.6 m/s^2. Although the system has yet to destabilize, the position tracking at this point begins oscillating as it tries to keep up with the base excitation. The final test with a peak acceleration of 4.9 m/s^2 was unable to finish as the controller destabilized with EMD frequencies under 2 Hz. Overall, the position tracking behaviour is consistent with expectations. The EMD accurately tracks positions under base excitations up to approximately 1.75 m/s^2 where performance first shows signs of disruptions at higher EMD frequencies. Performance continues to decrease until eventually becoming unstable. The measured EMD inertial force at the point when tracking first showed signs of deteriorating is approximately 225 N. This falls well outside the control bandwidth of the prototype system since an EMD inertial force of only 150 N is amplified to approximately 400 N at the base [4]. This result verifies the previous assumption that the EMD model is valid under base excitation within the control bandwidth of the prototype.

40.4.2 LQG Controller Design

Real-time hybrid simulation (RTHS) of a SDOF system controlled using the prototype DACS is used to compare the performance of the two LQG controller design. RTHS is a dynamic testing method that couples numerical analysis with experimental testing [12]. In this study, the SDOF system modeled analytically while the prototype DACS is testing physically on a shake table. The SDOF system is excited by a summation of sinusoids representative of pedestrian loading in

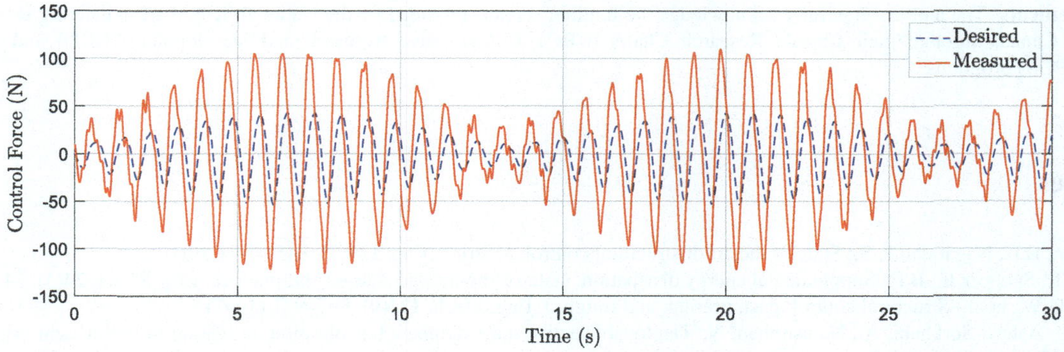

Fig. 40.6 Control performance using LQG #1 control configuration

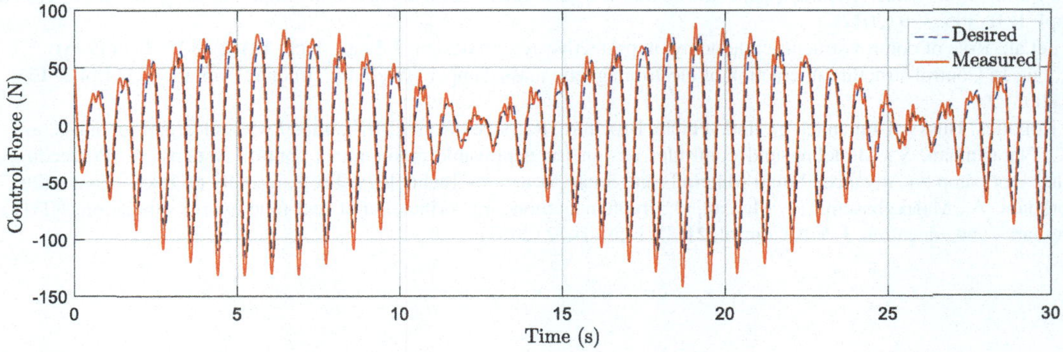

Fig. 40.7 Control performance using LQG #2 control configuration

the first mode of a lightweight bridge. The LQG #1 controller configuration designed independently of the prototype system is implemented first. Figure 40.6 compares the desired and measured control forces. The large discrepancy between desired and measured forces illustrates the interaction effect between the structural response and UGV dynamics. The measured control force is considerably larger than desired primarily due to the additional inertia of the UGV self-weight and dynamics which went unaccounted for in the controller design. In addition to amplitude errors, the interaction also caused a phase shift in the measured force. This phase error is detrimental to the controller effectiveness and in some cases could lead to stability issues.

Figure 40.7 compares the desired and measured control forces using the LQG #2 controller configuration. The control performance is improved considerably, particularly in terms of phase. Although an overshoot of approximately 10% is evident, it is clear that including the UGV dynamics in the controller formulation significantly improves the control performance by accounting for the interaction effects. Errors in the modeling of UGV dynamics or estimation of the UGV states in the LQG algorithm are possible contributors to the force overshoot.

40.5 Conclusions

The concept of a DACS presented herein offers an innovative alternative to permanent installations of auxiliary control devices. The small-scale prototype, consisting of an EMD mounted on a UGV with vision sensors, demonstrates the key advantages including rapid deployment and autonomous, real-time positioning. The primary challenge associated with the prototype DACS involves system modeling of inherently non-linear dynamics and compensating for CSI effects. Experimental modeling is used to characterize the UGV and EMD dynamics in the form of linear transfer functions. These models are used in different LQG controller configurations to illustrate the interaction effects. The effect of base excitation on position control of the EMD was shown to be negligible over the control bandwidth of the prototype system. Furthermore, formulating an LQG controller based on a system that incorporates the UGV portion of the DACS with the structure sufficiently captures the interaction effects that when left unaccounted for are detrimental to control performance.

Acknowledgements The authors gratefully acknowledge the funding support provided by the National Sciences Engineering Research Council (NSERC) of Canada through their Canada Research Chairs (CRC), Collaborative Research and Development (CRD) and Post-Graduate Scholarship (PGS) programs.

References

1. Spencer Jr., B.F., Nagarajaiah, S.: State of the art of structural control. J. Struct. Eng. **129**(7), 845–856 (2003)
2. Soong, T.T., Spencer Jr., B.F.: Supplemental energy dissipation: state-of-the-art and state-of-the-practice. Eng. Struct. **24**(3), 243–259 (2002)
3. Housner, G.W., et al.: Structural control: past, present, and future. J. Eng. Mech. **123**(9), 897–971 (1997)
4. Goorts, K., Ashasi-Sorkhabi, A., Narasimhan, N.: Deployable active mass dampers for vibration mitigation in lightweight bridges. J. Struct. Eng. **143**(12), 04017159 (2017)
5. Thrun, S., Burgard, W., Fox, D.: Probabilistic Robotics. MIT Press, Cambridge, MA (2005)
6. Durrant-Whyte, H., Bailey, T.: Simultaneous localization and mapping: part I. IEEE Robot. Autom. Mag. **13**(2), 99–110 (2006)
7. Goorts, K., Phillips, S., Ashasi-Sorkhabi, A., Narasimhan, N.: Structural control using a deployable autonomous control system. Int. J. Intell. Robot. Appl. **1**(3), 306–326 (2017)
8. Dyke, S.J., et al.: Role of control-structure interaction in protective system design. J. Eng. Mech. **121**(2), 322–338 (1995)
9. Zhang, C., Ou, J.: Control structure interaction of electromagnetic mass damper system for structural vibration control. J. Eng. Mech. **134**(5), 428–437 (2008)
10. Clearpath Robotics: Husky: Unmanned ground vehicle. Brochure. Retrieved from (2017). https://www.clearpathrobotics.com
11. Goorts, K., Narasimhan, S.: Modeling and controller design for deployable autonomous control systems. In: Proceedings of: The 13th International Workshop on Advanced Smart Materials and Smart Structures Technology. The University of Tokyo, Tokyo (2017)
12. Ashasi-Sorkhabi, A., Malekghasemi, H., Mercan, O.: Implementation and verification of real-time hybrid simulation (RTHS) using a shake table for research and education. J. Vib. Control. **21**(8), 1459–1472 (2013)

Chapter 41
Support Vector Machine-Based Face Direction Detection Using an Infrared Array Sensor

Zhangjie Chen, Hanwei Liu, and Ya Wang

Abstract Facing direction detection plays a critical role in human computer interaction, such as face recognition and head pose estimation in biometric identification, spatial microphone/loudspeaker devices, virtual reality games and etc. Currently detection methods are mainly focused on extracting specific patterns of various facial features from the user's optical images, which raises concerns on privacy invasion and these detection techniques do not usually work in the dark environment. To address these concerns, this paper proposes a pervasive solution for a coarse facing direction detection using a low pixel infrared thermopile array sensor. Support vector machine (SVM) method is selected as the classifier. Two methods for feature extraction are compared. One is based on pre-defined features and the other is based on pre-trained convolutional neural network (CNN) model. The detection accuracy resulted from using pre-defined features reaches 87% for identifying five different facing directions up to 1.2 m. However, this accuracy is largely descended when the detection range is increased to 1.8 m. Interestingly, the accuracy resulted from using pre-trained CNN features, however, demonstrates a reliable and steady performance up to 1.8 m. The accuracy keeps above 95% at both detection ranges (1.2 and 1.8 m). This proposed solution leads to many advantages, such as low resolution leading to no intention on privacy invasion, and the low-cost intriguing a potentially large market for human computer interaction in smart home appliances control and computer games.

Keywords Infrared array sensor · Facing direction · SVM · CNN · Human computer interaction

41.1 Introduction

Facing direction detection plays a critical role in human computer interaction (HCI), it is also found to be a natural, simple and effective way of pointing to objects, interaction and communication. Thus, the detection techniques have received significant attention in recent research [1]. Facing direction detection covers a wide range of applications such as video surveillance [2], HCI in computer games [3], driving assistance systems [4], assistive technologies [5], and even suspicious behavior detection [6].

The most popular approach for facing direction detection is computer vision-based. Currently detection methods are mainly focused on extracting specific patterns of various facial features such as hair and skin [7], eye gaze [8]. Murphy-Chutorian and Trivedi presented a static head-pose estimation algorithm and a visual 3-D tracking algorithm based on image processing and pattern recognition for a real-world problem: driver awareness monitoring [9]. Siriteerakul et al. proposed an effective method that tracks changes in head direction with texture detection in orientation space using a low-resolution video. Paone, J. et al. performed baseline facing detection and pose estimation to naturalistic driving study data [10]. Solutions for facing direction detection also include the use of wearable devices such as accelerometers [11], gyro-sensors [12], EEG equipment [13], etc. Video based approaches raise concerns on privacy invasion and these devices do not usually work in the dark environment. Wearable devices have obvious drawbacks as attaching sensors to the subject causes inconvenience.

Z. Chen · Y. Wang (✉)
Department of Mechanical Engineering, Stony Brook University, Stony Brook, NY, USA
e-mail: ya.s.wang@stonybrook.edu

H. Liu
Department of Electrical Engineering, Stony Brook University, Stony Brook, NY, USA

S. Pakzad (ed.), *Dynamics of Civil Structures, Volume 2*, Conference Proceedings of the Society for Experimental Mechanics Series, https://doi.org/10.1007/978-3-319-74421-6_41

To address these arising issues, we evaluate the potential to apply a low-resolution infrared array sensor for coarse facing direction detection. Infrared sensors have the advantages of low-cost and non-privacy invasion. They also have already been widely applied for occupancy estimation [14], moving direction detection [15] etc. In our work, a sensor system building off a single 8 × 8 infrared array sensor is developed to detect facing direction. The final facing detection is achieved by analyzing facing data received from occupancy face and torso.

The rest of the paper is organized as follows. The development of sensor node is described in Sect. 41.2. Two detection algorithms with associated experimental setup are introduced in Sect. 41.3. One is based on Support vector machine (SVM) features classification, and the other one is based on the combined convolutional neural network (CNN) and SVM feature extraction and classification. Results and analysis is described in Sect. 41.4. Finally, Sect. 41.5 draws the conclusions.

41.2 System Design

In our solution, we use a Gird-EYE IR array sensor produced by the Panasonic Corporation, as shown in Fig. 41.1a. It has 8 × 8 pixels with 2.5°C temperature accuracy and I^2C interface. A sample image of Grid-EYE sensor output is shown on Fig. 41.1b. Grid-EYE's 60° field of view (FOV) is comparably wide considering other options with the same price range. Grid-EYE enables three working modes: normal mode, sleep mode, and standby mode, with current consumption 4.5 mA, 0.2 mA, and 0.8 mA, respectively. A Sensor node based on Grid-EYE is developed for the face direction detection. Figure 41.1c illustrate the systematic of the sensor node. The sensor node is powered by a tiny Li-ion battery and can send data to the up-level computer through Bluetooth. The sampling frequency of the sensor node is 10 Hz. The real picture of the sensor node is shown in Fig. 41.1d.

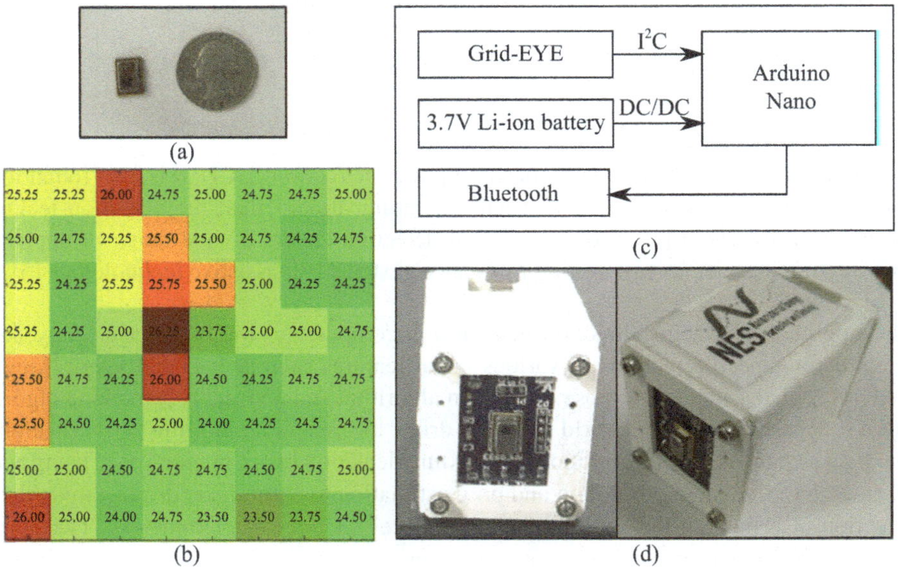

Fig. 41.1 (**a**) Systematic of sensor node (**b**) Grid-EYE sensor (**c**) Example of Grid-EYE 8 × 8 thermal image (**d**) snapshot of Grid EYE sensor node

41.3 Detection Algorithm

41.3.1 System Working Principal

Since a low-resolution IR array sensor is applied in our approach, the detection accuracy is largely depended on the distance between the user and the sensor. A universal algorithm cannot maintain high accuracy of the system. Since our system can be applied to HCI such as computer games and automatic control and the distance between sensor and human during these applications usually remains the same for each user, our system works based on a training process. In other words, every time, the user set the sensor node to a new location, a simple training process need to be performed. Figure 41.2 illustrates the overview of system working principal. The classification of five face directions is based on SVM method. The LIBSVM library [16] is applied for classifier calculation. Two methods are involved in this paper for feature extraction. One is based on pre-defined feature which is also the common method for SVM. Another feature extraction method is based on CNN, a deep learning algorithm. Before the feature extraction process, the 8 × 8 raw image data are converted to 32 × 32 binary image data based on preprocessing. And after classification using SVM, a size three median filter is applied to the final result.

41.3.2 Preprocessing of Data Image

This part introduces the preprocessing algorithms used for the direction detection. A flow chart shown in Fig. 41.3 as well as a visualization of the data shown in Fig. 41.4, both indicate how a raw data of an 8 × 8 temperature matrix from the Grid-EYE sensor is preprocessed step by step.

The first step is to subtract the background temperature from the raw data. Figure 41.4a and b show the photo and raw data received by the Grid-EYE sensor. The purpose of this algorithm is to track the facing direction of the main subject in the FOV. To emphasize the subject and exclude interferences such like other heat resources in the background, background subtraction is taken in the very first step as shown in Fig. 41.4c. Then the 8 × 8 temperature matrix is interpolated to 32 × 32, shown in Fig. 41.4c. Additionally, binarization is implemented before the feature extraction process. Binarization aims to highlight the main body of the subject in the image. The maximum temperature difference of every column is sorted and among which the minimum temperature difference is taken as the threshold for binarization, the final effect after binarization is shown in Fig. 41.4e. Binarization can efficiently reduce redundancy information in the temperature image. The features will then be extracted based on the image with binary data.

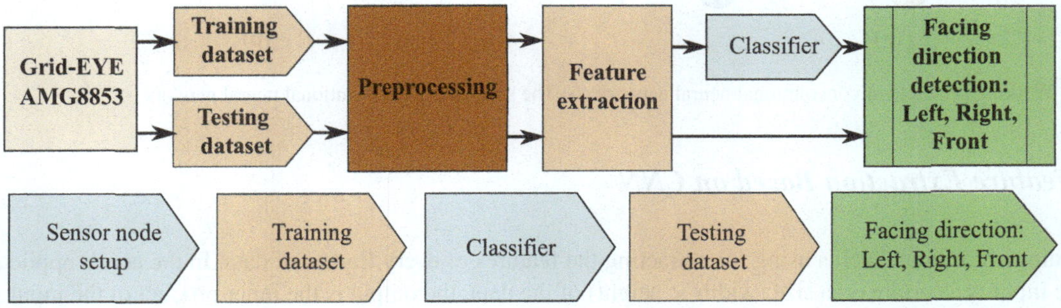

Fig. 41.2 System working principal

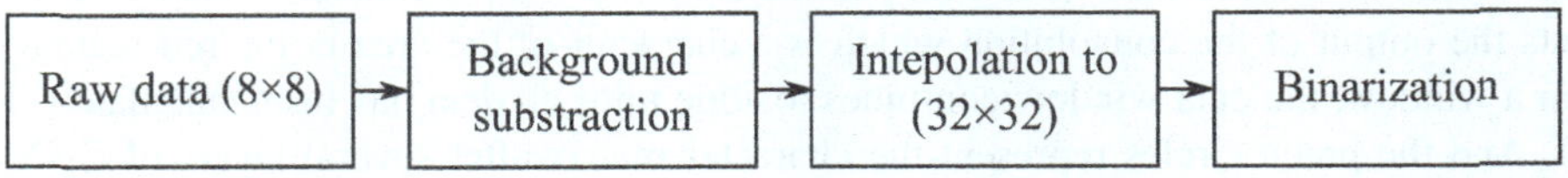

Fig. 41.3 Flow chart of data preprocessing

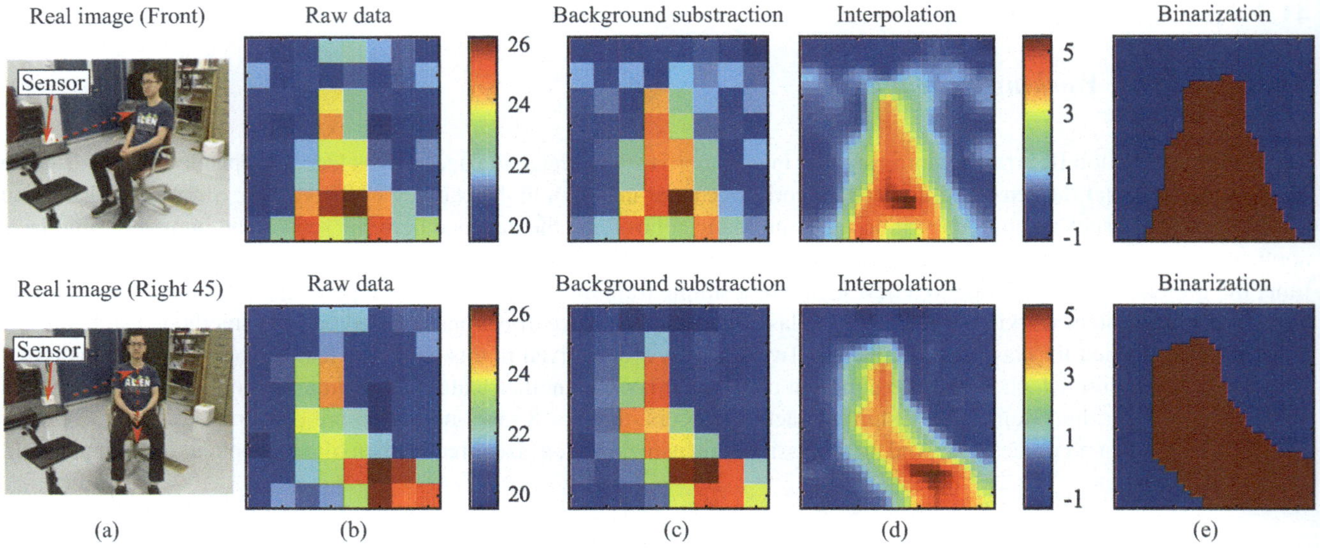

Fig. 41.4 (**a**) Original photo (**b**) Raw data from the IR sensor (**c**) Data after background subtraction; (**d**) Linearly interpolate image to 32 × 32; (**e**) Binarization after taking a threshold

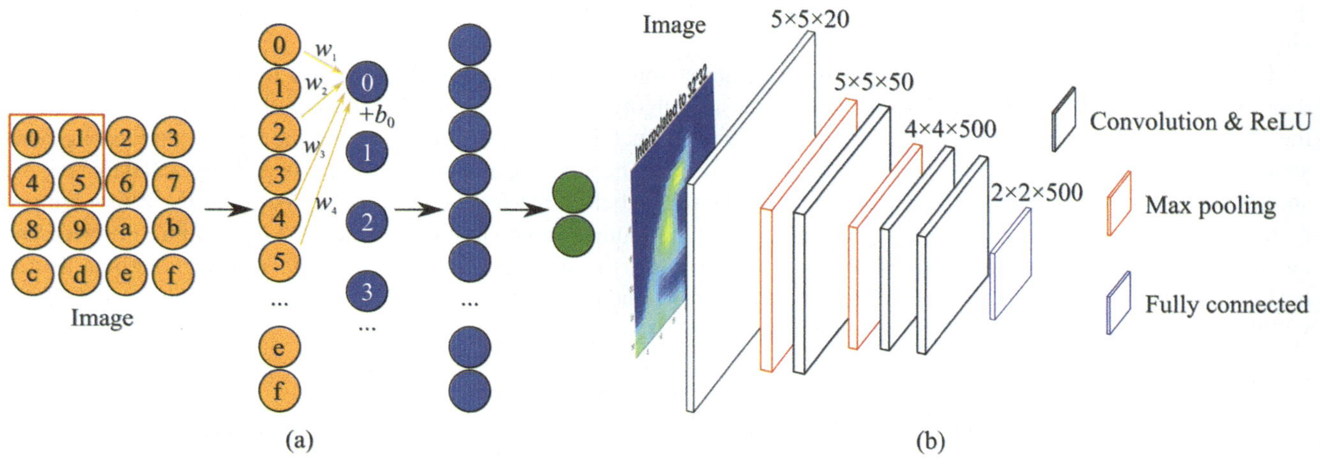

Fig. 41.5 (**a**) Working principle of convolutional neural network (**b**) The structure of convolutional neural network

41.3.3 Feature Extraction Based on CNN

A deep learning algorithm CNN is using for extracting the features of every frames of data. In the actual application process of CNN, the input is a certain area size (width × height) of the data, the output is the inner product of the input and multiple layers of the filter (with a fixed weight of the neurons). Figure 41.5a illustrates the basic working principle of CNN. The red rectangle stands for a filter and the group of yellow circles with label stand for an image. A convolutional calculation is performed on the local input data based on a filter with parameters w_1, w_2, w_3, w_4, determined by the pretrained model. The blue circle represents the output of the convolution which is a character of the area in the red rectangle. After calculating the local data within a window, the data window continues to slide until all data has been calculated. This is considered as a single CNN layer. And the green circles represent the character matrix after several layers of CNN is performed to the original data.

In this experiment, a seven-layer convolutional network is defined shown in Fig. 41.5b, input format is 32 × 32 × 1 image, the first layer is a 5 × 5 × 20 convolutional layer with no padding. The second layer is a max-pool layer, the third layer is a 5 × 5 × 50 convolutional layer, the fourth layer is another max-pool layer, the fifth layer is 4 × 4 × 500 convolutional layer, the sixth layer is a ReLU layer, and the last layer is a 2 × 2 × 500 fully-connected layer which produce a feature of the whole image. This net is pre-trained for recognizing images of characters [17], a characteristic of deep learning that it constructs

Algorithm: Feature extraction

```
R(i, j) ←Data after binarization
   k  ← Size of R(i, j)
   S1 ←area size of the topleft part
   S2 ←area size of the toprightpart
   S3 ←area size of the lowerleft part
   S4 ←area size of the lowerrightpart
M_Co[] ←coordinates of middle point in each row
   For i from 1 to k
      For j from 1 to k
         if R(i, j)> 0
         save(p)←j
         p←p+1
         End if
      End for
         M_Co(1, q)←(save (1)+save(p-1))/2
         M_Co(2, q)←i
   End for
[maxV,maxI ]= max(M_Co(1,:));
[minV,minI]= min(M_Co(1,:));
    Slope←(maxV-minV)/(M_Co(2,maxI)-M_Co(2,minI));
    Feature value1← S1-S2
    Feature value 2← S3-S4
    Feature value 3← Slope
```

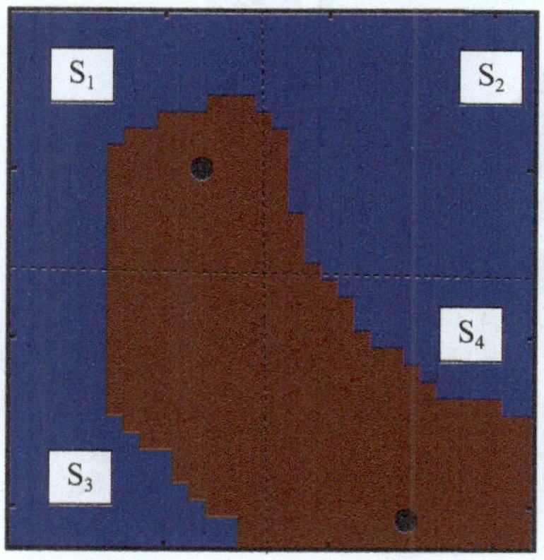

Fig. 41.6 Pre-defined feature extraction algorithm

representations of the data. These representations tend to have a universal value, or at least a value to be applied to an array of problems that transcend the particular task that a model was trained for. By this means, the network can also work for our data, because both cases are to find the contours in the image, and CNN can work more efficiently on the feature extraction of the contour of human.

The seven layers CNN produce a $2 \times 2 \times 500$ feature matrix for every thermal image. We rearrange this matrix to 1×2000 and join all the feature matrix into one file. To minimize the computational complexity and to increase the accuracy, we reduce the dimension of feature matrix by removing the all-zero columns. The motivation behind this is that there are a number of columns that don't have value after binarization, and are taken as redundancy of the feature matrix. This work uses Oxford Visual Geometry Group's convolutional neural network toolbox [18].

41.3.4 Pre-defined Feature Extraction

The preprocessing is performed to get a binarization image, and then three features are introduced to calculate SVM classifier. The algorithm for feature extractions are demonstrated in Fig. 41.6.

Feature 1: Slope of the high temperature area. For each row of the 32×32 data image, the middle point of the high value part will be found first. And then the most left and most right point will be connected to calculate the slope.

Feature 2: Area difference between the lower part of the thermal image. The total area of a 32×32 data image is segmented into four parts, and the area size between the two lower parts will be calculated.

Feature 3: Area difference between the top part of the thermal image. The total area of a 32×32 data image is segmented into four parts, and the area size between the two top parts will be calculated.

41.3.5 Experiment Design and Setup

The experiment is performed in a common laboratory area. Each testing subject/user sits on a chair which can be easily rotated. The sensor node is set in front of the user at a height of 0.75 m. Five facing directions are tested in our experiment: Left 45°, Left 90°, Front, Right 45°, Right 90°. Figure 41.7 shows the snapshot of the experimental environment and five

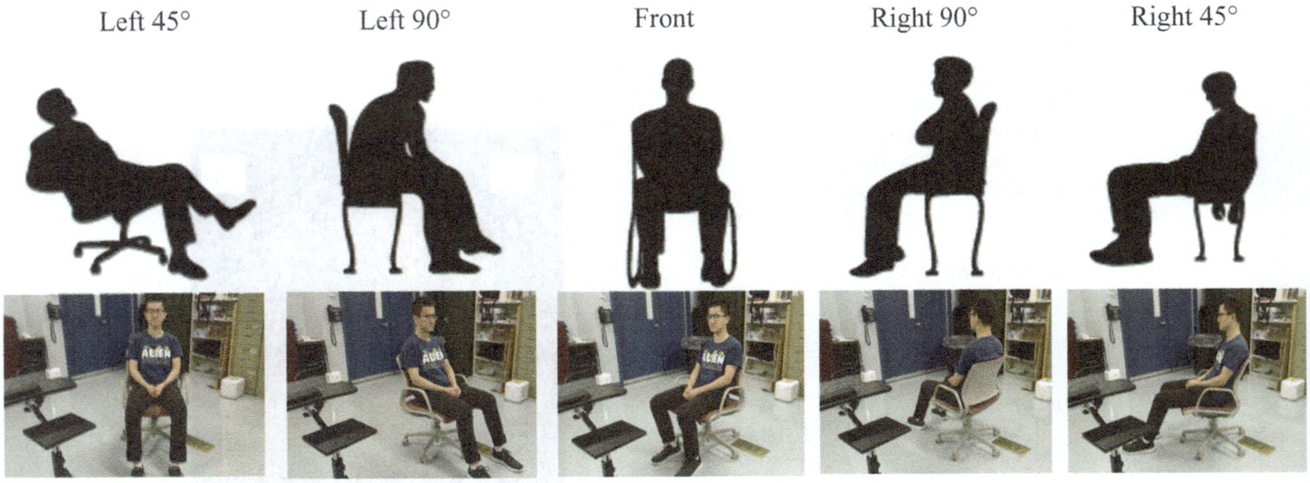

Fig. 41.7 Snapshot of five directions of the experiment setup

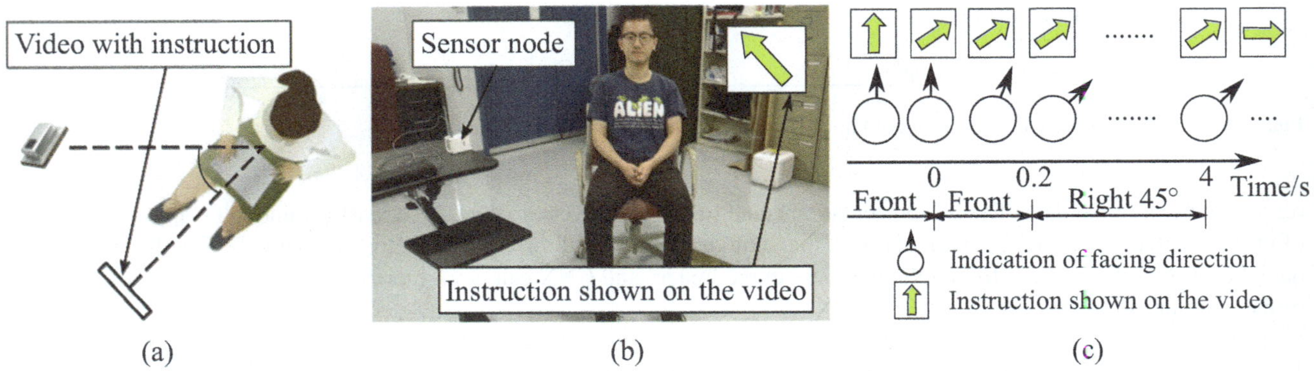

Fig. 41.8 (**a**) Systematic of testing experiment top view (**b**) experiment setup (**c**) example of ground truth for a face rotation process

tested directions. And regarding to the distance from the sensor to the user, two cases are involved in the experiment, 1.2 and 1.8 m. For both distances, most part of the subject body is within the sensor FOV, and the subject facing direction keeps the same as the torso direction. Three users are tested in this experiment.

Training Dataset The training dataset and background features are always captured before further detecting the facing directions. For each user, data images of each facing direction are saved for about 5 s. And only the first 50 frames are included in the training dataset to maintain a balance between different classes. In other words, the training dataset includes 250 frames.

Testing Dataset the testing dataset in this paper is recorded based on the following experimental setup. The subject rotates continuously for eight times, while keeps at a direction for 4 s before rotates to the next direction. When the testing subject is doing the experiment. Another person stands in front of the person holding an IPAD with an instruction video. The video with indication of direction signs is played for the testing subjects. The testing subjects follow the instruction of the video to complete all the testing data recording. And the video recording is used as the ground truth reference.

Detection Accuracy Calculation In reality, the occupancy facing directions always consist of a continuous process. At first, the ground truth definition for our continuous detection results need to be clarified. However, reliable ground truth could be complicated to define, since the rotation pattern for each user are different. Therefore, the accuracy is calculated based on the following approach. Since the rotation process during our experiment is quite quick, and normally completed within about 0.2 s, the rotation process is considered as two frames. For instance, for the period of 'Right 45°' status, normally it includes 40 frames. Then the ground truth of the first two frames are 'Front', and all the rest are 'Right 45°' as shown in Fig. 41.8c.

41.4 Result Analysis

The experiment on 1.2 m is first performed in our work. A feature analysis is conducted to evaluate the performance using pre-defined SVM features. The 3D feature space plotting indicates a high quality in classifier calculation Fig. 41.9a. A highest accuracy of 87.2% is reached by only applying two simple features as shown in Fig. 41.9b.

The same feature set (feature set 1,2) is selected for the experiment at 1.8 m. The detection results at 1.2 m is shown in Fig. 41.10. The detection result based on CNN achieves an accuracy of 98.3%. Since the exact time of facing direction change is not so important, the detection result based on CNN can be considered as 100% since it successfully identifies all the face direction status in correct orders.

Figure 41.11 shows the detection results on a detection range of 1.8 m by comparing the two methods. It's obvious to find that the detection result based on a stand-alone SVM classifier is not so satisfied. The accuracy is only 78.96% when the distance is increased to 1.8 m. And the facing directions of left 45° and left 90° can hardly be differentiated. The features used in the SVM classifier are only based a single thermal image rather than a time sequence. We believe that's why the

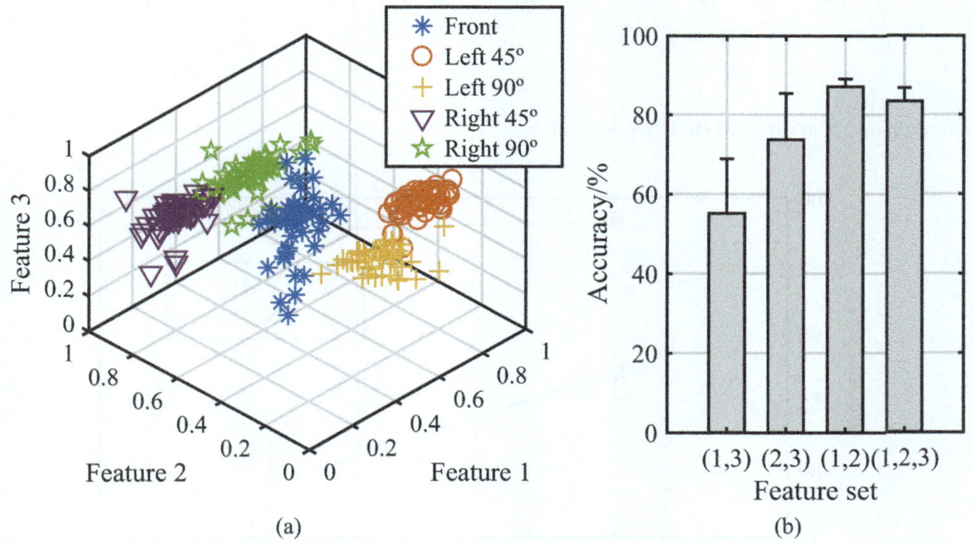

Fig. 41.9 (**a**) Feature plotting (**b**) SVM feature analysis

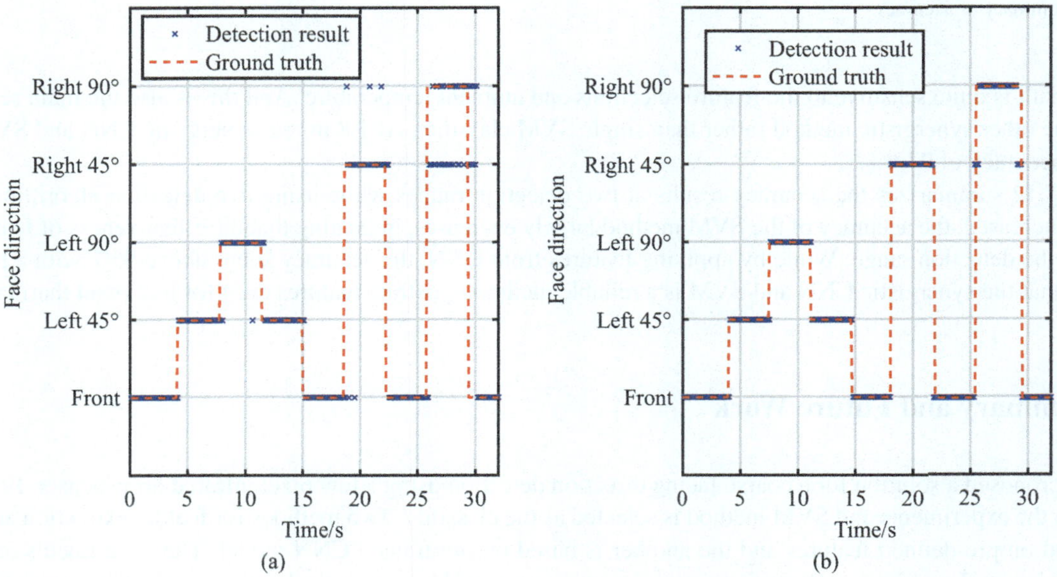

Fig. 41.10 Detection results on 1.2 m (**a**) SVM (**b**) CNN + SVM

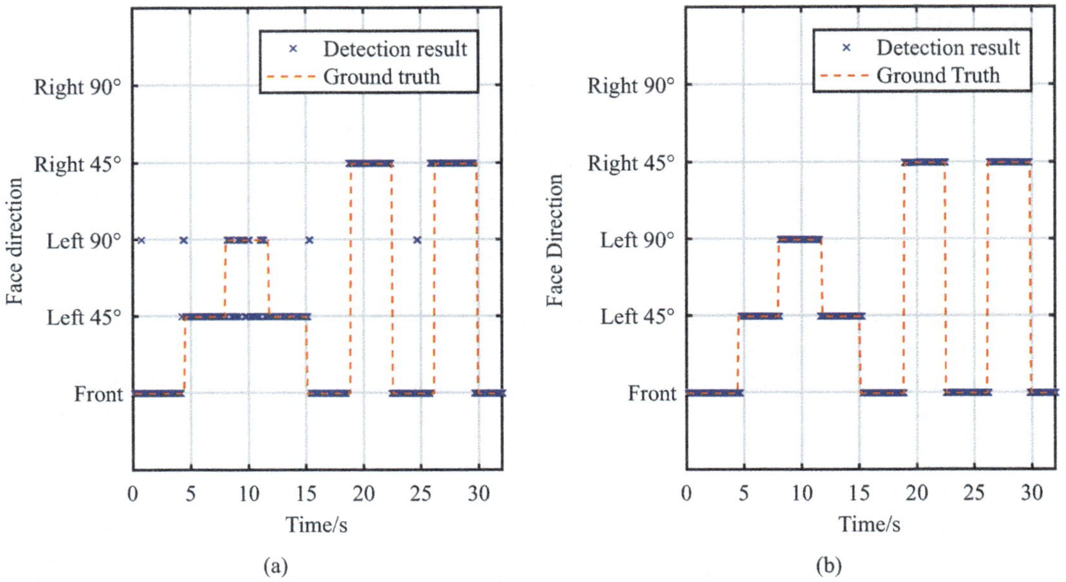

Fig. 41.11 Detection result on 1.8 m (**a**) SVM (**b**) CNN + SVM

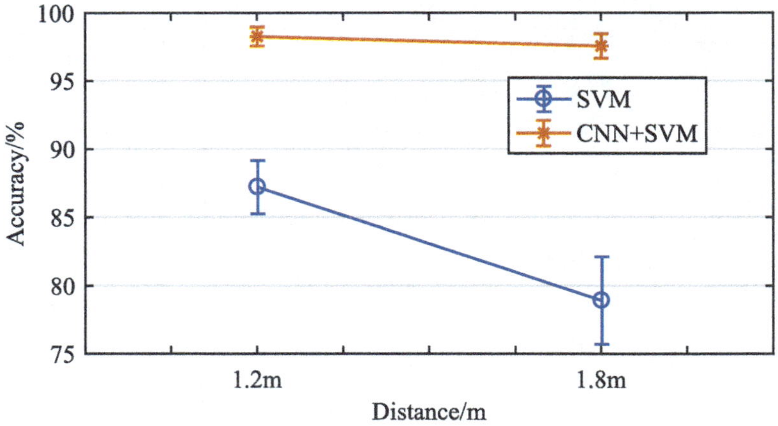

Fig. 41.12 Summary of accuracy

detection result is quite sensitive to the feature selections and ambient temperature. And this is also the main reason that we introduce the other synergistic method rather than single SVM classifier. At 1.8 m, the synergistic CNN and SVM approach reaches an accuracy of 97.6%.

Figure 41.12 summarizes the accuracy results at two detection ranges while using two detection algorithms. When the distance is increased, the accuracy of the SVM method largely decreases, indicating that the effectiveness of features is quite sensitive to the detection range. While by applying features from CNN, the accuracy keeps above 95% with a low variance, suggesting that the synergistic CNN and SVM is a reliable method to extract features even for low-pixel thermal images.

41.5 Summary and Future Work

This paper proposed a solution for a coarse facing direction detection using a low pixel infrared array sensor. Five directions are tested in the experiments and SVM method is selected as the classifier. Two methods for feature extraction are compared. One is based on pre-defined features and the another is based on pre-trained CNN model. The experiments are performed at two distances, 1.2 and 1.8 m. During experiments, most part of the occupancy body is within the sensor FOV for both detection ranges (1.2 and 1.8 m), and thus, provide sufficient data for the detection algorithm. Human face direction remains the same with the human torso during experiment. The accuracy resulted from pre-defined features reaches 87% at a distance of 1.2 m to identify five different directions. However, this accuracy is largely descended below 80% when the distance is increased to 1.8 m. While detection method with feature from CNN demonstrates a reliable performance regardless of the detection ranges. The accuracy keeps above 95% at both distance. The proposed solution leads to many advantages, such as low resolution leading to no intention on privacy invasion, and the low-cost intriguing a potentially large market for HCI in smart home appliances control and computer games.

In this paper, human subjects sit relatively far from the sensor so that most part of the human body is within the sensor FOV. In future work, scenarios with shorter distance that only the human head included in the sensor FOV will be researched. In addition, a distance sensor can be integrated to detect user existence as well as the distance between the user and the sensor. And the distance data can help to update the algorithm for better performance. More realistic scenarios should also be included in the experiments, such as face facing direction different from torso facing direction.

References

1. Al-Rahayfeh, A., Faezipour, M.: Eye tracking and head movement detection: a state-of-art survey. IEEE J. Transl. Eng. Health Med. **1**, 12 (2013)
2. Xie, D., Dang, L., Tong, R.: Video based head detection and tracking surveillance system. In: 2012 9th International Conference on Fuzzy Systems and Knowledge Discovery (FSKD), IEEE (2012)
3. Yim, J., Qiu, E., Graham, T.: Experience in the design and development of a game based on head-tracking input. In: Proceedings of the 2008 Conference on Future Play: Research, Play, Share, ACM (2008)
4. Liu, K., et al.: Attention recognition of drivers based on head pose estimation. In: Vehicle Power and Propulsion Conference, VPPC'08, IEEE (2008)
5. Sasou, A.: Acoustic head orientation estimation applied to powered wheelchair control. In: Second International Conference on Robot Communication and Coordination, ROBOCOMM'09, IEEE (2009)
6. Miyanokoshi, Y., et al.: Suspicious behavior detection based on case-based reasoning using face direction. In: 2006 SICE-ICASE International Joint Conference, pp. 5429–5432 (2006)
7. Ugurlu, Y.: Head posture detection using skin and hair information. In: 2012 21st International Conference on Pattern Recognition (ICPR), IEEE (2012)
8. Al-Rahayfeh, A., Faezipour, M., IEEE: Enhanced eye gaze direction classification using a combination of face detection, CHT and SVM. In: 2013 IEEE Signal Processing in Medicine and Biology Symposium (2013)
9. Murphy-Chutorian, E., Trivedi, M.M.: Head pose estimation and augmented reality tracking: an integrated system and evaluation for monitoring driver awareness. IEEE Trans. Intell. Transp. Syst. **11**(2), 300–311 (2010)
10. Paone, J., et al.: Baseline face detection, head pose estimation, and coarse direction detection for facial data in the SHRP2 naturalistic driving study. In: 2015 IEEE Intelligent Vehicles Symposium, pp. 174–179. IEEE, New York (2015)
11. Manogna, S., Vaishnavi, S., Geethanjali, B.: Head movement based assist system for physically challenged. In: 2010 4th International Conference on Bioinformatics and Biomedical Engineering (iCBBE), IEEE (2010)
12. Kim, S., et al.: Head mouse system based on gyro-and opto-sensors. In: 2010 3rd International Conference on Biomedical Engineering and Informatics (BMEI), IEEE (2010)
13. O'Regan, S., Marnane, W.: Multimodal detection of head-movement artefacts in EEG. J. Neurosci. Methods. **218**(1), 110–120 (2013)
14. Tyndall, A., Cardell-Oliver, R., Keating, A.: Occupancy estimation using a low-pixel count thermal imager. IEEE Sensors J. **16**(10), 3784–3791 (2016)
15. Yun, J., Song, M.: Detecting direction of movement using pyroelectric infrared sensors. IEEE Sensors J. **14**(5), 1482–1489 (2014)
16. Chang, C.-C., Lin, C.-J.: LIBSVM: a library for support vector machines. ACM Trans. Intell. Syst. Technol. **2**(3), 1–27 (2011)
17. http://www.vlfeat.org/matconvnet/pretrained/
18. Vedaldi, A., Zisserman, A.: VGG Convolutional Neural Networks Practical. Available from: http://www.robots.ox.ac.uk/~vgg/practicals/cnn/#vgg-convolutional-neural-networks-practical (2016)

Chapter 42
Estimation of Remaining Useful Life of a Fatigue Damaged Wind Turbine Blade with Particle Filters

Bhavana Valeti and Shamim Pakzad

Abstract Structural maintenance operations in wind energy sector are steering towards condition based maintenance (CBM) which requires prognostic estimates of existing condition of the wind turbine (WT) structural systems that is damage propagation and remaining useful life (RUL). WT blades are highly vulnerable structural components that are subjected to continuous cyclic loads of wind and self weight variation. A method for estimation of RUL of wind turbine blades considering the fatigue mode of failure is proposed in this paper. Stochastic life expectancy methods that use Bayesian updating with measurements of evolving damage for damage propagation estimation have proven to be reliable in RUL estimation. In this study probability density functions for the RUL of WT blades are estimated for diffident initial crack sizes and particle filtering method is used for forecasting the evolution of fatigue damage addressing the non-linearity and uncertainty in crack propagation. The stresses on a numerically modeled life size onshore WT blade subjected to turbulence are used in computing the crack propagation observation data for particle filters.

Keywords Wind turbine blades · Structural health monitoring · Condition based maintenance · Particle filters · Remaining useful life

42.1 Introduction

Wind energy is one of the promising sources of renewable energy available today. The increasing output of wind energy systems, and awareness of public and governments of the need for clean energy from renewable sources added momentum to fast expansion of wind energy sector [1]. This is accompanied by advanced technology, increased WT size and capacities which increased the installation, maintenance and repair costs [2]. The maintenance operations constitute approximately 20–25% of the total levelized cost per kWh of wind power assets [3]. Although preventative and corrective maintenance practices have been popular so far, the wind energy sector is moving towards condition based maintenance (CBM). With CBM, unnecessary labor and pre-mature replacement costs due to scheduled maintenance practices (preventative maintenance) and expensive post-failure repair and replacement costs (corrective maintenance) are minimized by optimizing the maintenance schedule based on the RUL and maintenance, repair and replacement costs [4]. CMB is an optimal maintenance strategy where maintenance activities are determined based on diagnostic and prognostic analyses of the Structural Health Monitoring (SHM) results [5] (See Fig. 42.1) In this paper a prognostic procedure using a dynamic Bayesian estimation method of particle filtering for RUL estimation of WT blades subjected to fatigue damage is presented.

42.2 WT Blade SHM

42.2.1 Damage Hotspots for WT Blades

Among the structural components of the WTs, rotor blades are subjected to the most dynamically varying loads. Blade damage accounts for around 23% of the total failure events accompanied by significant downtime and replacement costs [6, 7]. Being subjected to highly irregular loads caused by a combination turbulent wind flow, varying direction of

---

B. Valeti (✉) · S. Pakzad
Department of Civil and Environmental Engineering, Lehigh University, Bethlehem, PA, USA

S. Pakzad (ed.), *Dynamics of Civil Structures, Volume 2*, Conference Proceedings of the Society for Experimental Mechanics Series, https://doi.org/10.1007/978-3-319-74421-6_42

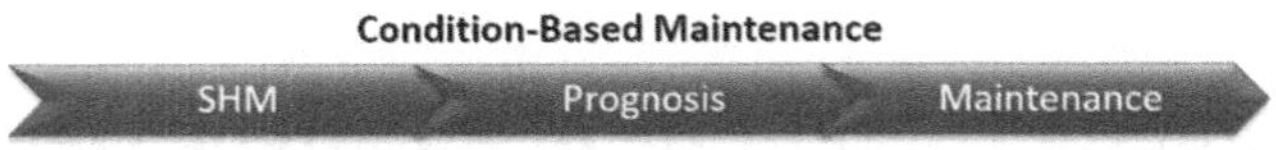

Fig. 42.1 Components of condition based maintenance

gravitational load, centrifugal loads, surface erosion and corrosion from humidity and temperature, the WT blade fatigue is one of the most recurring problems [8, 9].

The WT rotor blades consist of load carrying shear webs or hollow beam laminates for flapwise and edgewise moments along the blade and light weight core to resist torsion and buckling. The laminate frame is made of fiber reinforced polymer composite materials and the core is made of foam or balsa wood and the two are typically bonded by adhesives. The core is covered with composite laminates or gel coats for protection [8, 10]. Although the modern WT blades are designed to withstand the loads for about 20–30 years, there is always a chance for external causes of damage from lightening or an inherent flaw in the complex blade structure initiated during blade manufacture [11]. Multiple studies have identified few hotspots [6] in the WT blade structure with high probabilities for damage. Fatigue crack at the blade root is frequently encountered due to continuous variation of flapwise moments. Cracks at maximum chord location, 30 and 70% of blade length from root are also reported from experiments and simulation studies [12, 13]. Upper spar cap and its flange are also identified as the critical locations depending on the pitch angle of the blade [14]. Delamination, debonding and intralaminar failure are some of the damage modes that need attention in these locations [15].

42.2.2 Damage Detection in WT Blades

Faults in blades if unchecked can grow in size eventually causing a catastrophic damage to the integrity of WT system [6]. Condition monitoring and diagnosis is the pr-requisite for RUL estimation of WT blades. With many advances in structural damage detection and diagnosis [16–18], SHM system should be capable enough to detect and localize the faults at a minimum required stage of evolution such that sufficient time is available to achieve economic CBM. Numerous SHM strategies [9, 19–23] for WT blades have been proposed, studied and tested [24–27].

42.3 Prognosis for RUL Estimation of WT Blades

42.3.1 Prognosis Methods

RUL of a structural component is the time it can last from a particular time during operation [28]. Prognosis is the ability to predict precisely and accurately the RUL of a deteriorating subsystem or a component [29]. The residual life of a structure after fatigue crack initiation depends on speed of crack propagation and the of size of critical defect that can be tolerated by the structure under extreme loading [30]. The fault prognosis step following SHM and diagnosis in the process of condition based monitoring (CBM) helps in making suitable maintenance and operational decisions such as to repair the component, replace the component, continue until the next maintenance schedule or stop operation. The uncertainties from estimated fault or damage characteristics, modeling the fault propagation, material characterization, future loads and measurement error act as impediments in accurate RUL estimation. Due to these uncertainties, the estimation of confidence limits along with RUL is important.

A relatively new approach classifies prognostic methods specifically for RUL estimation into four categories [31] namely (1) Knowledge-based models, (2) Life expectancy models (stochastic and statistical), (3) Artificial Neural Networks and (4) Physical models. The physical models are more accurate but expensive with low range of system applicability compared to the knowledge base methods on the other end of the spectrum. Fatigue, being one of the prime failure modes in WT blades, the ability to asses RUL considering various uncertainties and current damage state is necessary to make confident and safe maintenance decisions. Several studies have been conducted for determining the RUL in WT blades through classical fatigue crack propagation laws [14, 32–34]. Sanchez et al. [35] estimated the RUL for a medium sized WT blade using fatigue stiffness degradation damage model developed for CFRP materials. The damage model was applied on blade root moment sensor information from a WT simulated in a high fidelity simulator FAST v8 (Fatigue, Aerodynamics, Structures and Turbulence) and subjected to different wind speed scenarios. The stochastic life expectancy models consider RUL as a statistically identical independent random variable and define it with a probability density function. This random variable

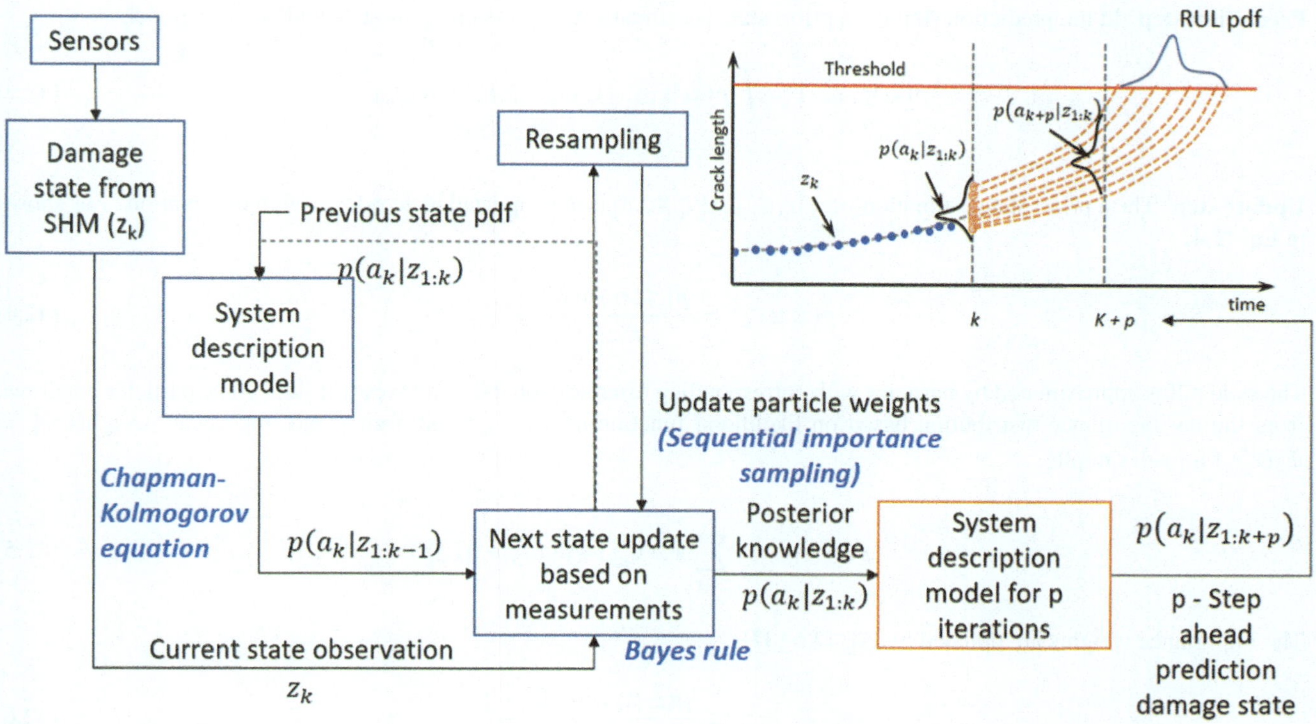

Fig. 42.2 Particle filter algorithm and estimation of RUL

carries the measurement, modeling and loading uncertainties. Dynamic Bayesian network models such as Markov models, Kalman filters and Particle filters are some of the stochastic life expectancy models that are popularly used in engineering prognostics [36]. Nielsen and Sorensen [11] employed hidden Markov models for RUL estimation of WT blades.

42.3.2 *RUL Estimation with Particle Filters*

Particle filtering methods accommodate the nonlinearity in the damage state evolution and non-Gaussian noise in the measurements leading to inaccuracy in prognostic estimates are particularly useful when the posterior distribution is non-standard or multivariate [37–39]. Which makes the method very much suitable for forecasting a nonlinear process such as crack growth [40]. Given a damage propagation is measured until the current time say k, the RUL estimation is performed here in three steps (See Fig. 42.2) [41, 42] *Step 1:* Damage estimation using particle filters upto time $k + 1$, *Step 2:* P-step ahead prediction (x_{k+1} to x_{k+p}) and *Step 3:* Estimating the probability distribution of time to failure, P_{TTF} that is RUL distribution.

42.3.2.1 *Step 1:* State Estimation

Define the damage state progression with a non-linear state equation as a function of previous damage state (Markov model of first order), and observation model as the following state space equations [38, 39, 41] (Eqs. 42.1 and 42.2).

$$x_k = f_x(x_{k-1}, \omega_k) \longleftrightarrow p(x_k|x_{k-1}) \tag{42.1}$$

$$z_k = h_x(x_k, \nu_k) \longleftrightarrow p(z_k|x_k) \tag{42.2}$$

where damage state x_k at time k is described by a first degree Markov process. f_k and h_k are nonlinear state transition function and the measurement model respectively. ω_k and ν_k are non-Gaussian noise vectors of known distributions corresponding to uncertainty in the nonlinear functions of damage state and measurement model respectively. The basic prediction and correction sequence of filtering is employed in particle filters.

Prediction step In the prediction step the a priori state is estimated using the previous state and process model.

$$p(x_k|z_{1:k-1}) = \int p(x_k|x_{1:k-1})p(x_{k-1}|z_{1:k-1})dx_{k-1} \tag{42.3}$$

Update step The a priori state estimation $p(x_k|z_{1:k-1})$ (Eq. 42.3) is now updated to $p(x_k|z_{1:k})$ with observation z_k as shown in Eq. 42.4.

$$p(x_k|z_{1:k}) = \frac{p(z_k|x_k)p(x_k|z_{1:k-1})}{p(z_k|z_{1:k-1})} \tag{42.4}$$

The state pdf is approximated by particles with corresponding discrete probability masses (Eq. 42.5). The particles are drawn from the an importance distribution based on likelihood function $q(x_{0:k}|z_{1:k})$ and normalized importance weights $\tilde{w}_k^i = \tilde{w}_k(x_{0:k}^i)$ for i-th sample.

$$p(x_k|z_{1:k}) = \sum_{i=1}^{N} \tilde{w}_k(x_{0:k}^i) * \delta(x_{0:k} - x_{0:k}^i) \tag{42.5}$$

The importance weights are updated by Eq. 42.6 [37]

$$w_k = w_{k-1}\frac{p(z_k|x_k)p(x_k|x_{k-1})}{q(x_k|x_{0:k-1}, z_{1:k})} \tag{42.6}$$

42.3.2.2 *Step 2:* p-Step Ahead Prediction

Using the current state estimate and updated fault growth model parameters, long term state predictions are generated by a recursive integration process based on importance sampling (Eq. 42.7) [43].

$$p(x_{k+p}|z_{1:k}) = \int p(x_k|z_{0:k}) \prod_{j=k+1}^{k+p} p(x_j|x_{k-1})dx_{k:k+p-1} = \sum_{i=1}^{N} \tilde{w}_k^{(i)} \int \ldots \int p(x_k^i) \prod_{j=k+1}^{k+p} p(x_j|x_{k-1})dx_{k:k+p-1} \tag{42.7}$$

42.3.2.3 *Step 3:* RUL Probability Distribution

For pre-defined upper ($\{H_{up}$) and lower ($\{H_{lb}$) bounds of damage thresholds for failure the probability distribution of time to failure or RUL is given by Eq. 42.8 [41]

$$P_{TTF}(ttf) = \sum_{i=1}^{N} Pr\{H_{lb} \leq x_{ttf}^{(i)} \leq H_{up}\}.\tilde{w}_{ttf}^{(i)} \tag{42.8}$$

42.4 RUL Estimation for WT Blades with Fatigue Damage

The following sections explain implementation of particle filters for RUL estimation of working size 5MW WT blade defined by NREL is assumed to be fatigue damaged at the 30% length of the blade from root. RUL distributions for different initial detected crack lengths are estimated.

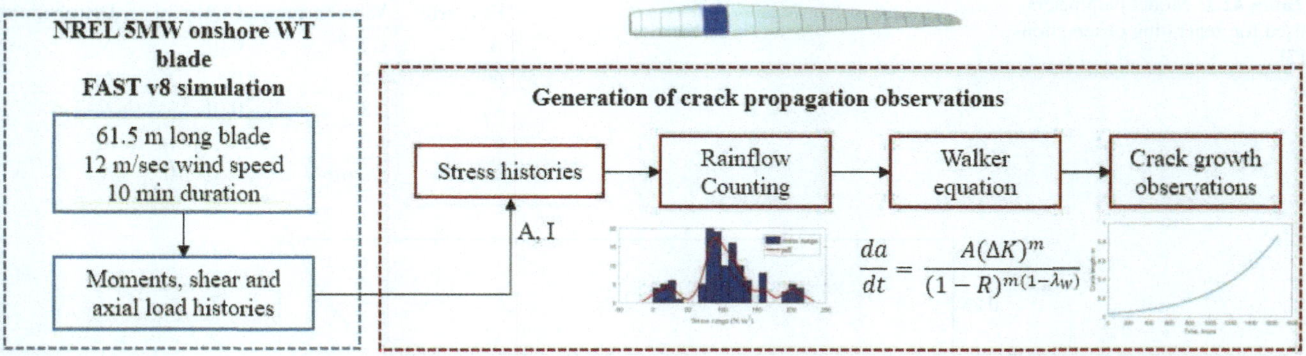

Fig. 42.3 Procedure to generate fatigue crack growth observation data in WT blade

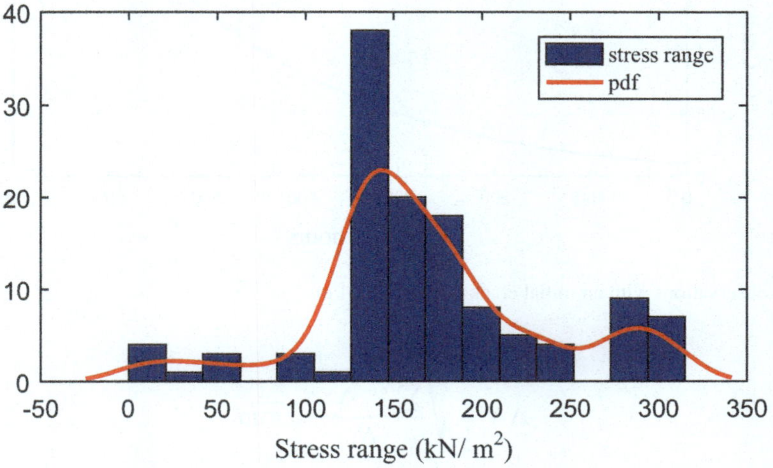

Fig. 42.4 Stress range distribution from rain-flow counting algorithm

42.4.1 Generating Fatigue Crack Propagation Observations

Due to the unavailability of condition monitoring data for fatigue crack propagation observation from real WT blades, data from a simulated WT is used to generate the same. The dynamic response of a 5 MW onshore WT defined by NREL [44, 45] is simulated in FAST v8 [46] a high fidelity aero-elastic computer aided engineering tool that simulates coupled dynamic responses of horizontal axis WTs. The WT with a blade length of 61.5 m is subjected to turbulence with an average wind speed of 12 m/sec for a period of 10 min. The loads due to start and stoppage of WT, sudden gusts and variation of wind speeds between the cut in and cut out speed of the operation are ignored for simplicity and time conservation during simulations. The force response history of the simulated WT blades at multiple locations is stored. This is used to generate axial stress histories using the blade airfoil geometrical properties at desired locations [44] (Fig. 42.3).

With this stress history, crack propagation observations were generated [7] using Wlaker's equation for fatigue crack propagation [47]. For simplicity of crack propagation estimation, homogeneity of the blade materials is assumed. The change in the stress intensity factor ΔK with crack growth is calculated from current crack length a and root mean square stress range (see Fig. 42.4) computed with the Rainflow counting algorithm (Eq. 42.10) [7, 48]. The simulation procedure is summarized in Fig. 42.3.

$$\frac{da}{dt} = \frac{A(\Delta K)^m}{(1-R)^{m(1-\lambda_w)}} \tag{42.9}$$

Where, WT

Table 42.1 Model parameters used for generating observations [7]

Parameter	Value (mean)	COV	Distribution
A	1.2^{-9}	0.01	N
m	1.8	0.1	N
λ_w	0.8	0.01	N
α	1.0	–	–
a	$variable$	1.5×10^{-4}	N

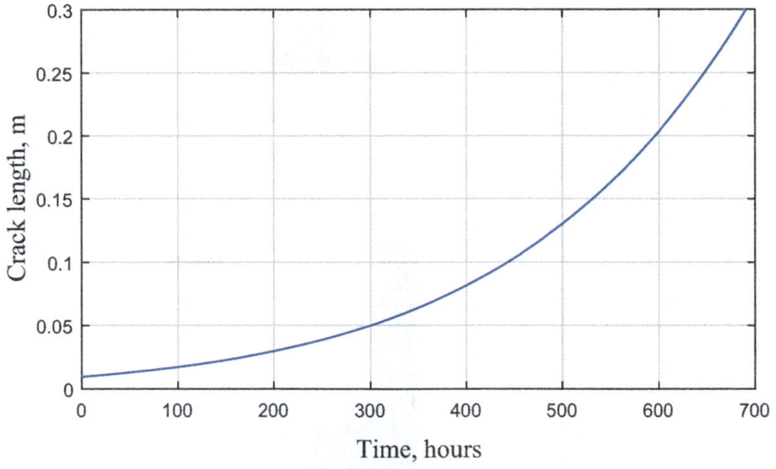

Fig. 42.5 Crack propagation observations with an initial crack length of 0.01 m

$$\Delta K = \sqrt{\frac{\sum_{i=1}^{N} \Delta S_i^2}{N}} \sqrt{\pi a \alpha} \tag{42.10}$$

$$R = \frac{S_{minrms}}{S_{maxrms}} \tag{42.11}$$

A, m, λ_w, α are material properties, R is Stress ratio, S_{minrms} and S_{maxrms} are the RMS values of minimum and maximum stresses in the stress cycles and ΔS is the stress range in each stress cycle.

The fatigue crack growth observation data is computed with the parameters in Table 42.1 at 30% length of the blade from blade root (See Fig. 42.5).

42.4.2 Particle Filter Implementation, Results and Discussion

The fault growth model for the particle filter implementation is defined using the Paris law for fatigue propagation 42.12 and the uncertainty distributions of different parameters are given in Table 42.1.

$$a_k = f(a_{k-1}, \omega_k) = A(\Delta K)^m + a_{k-1} \tag{42.12}$$

The first 50 of the observation samples are used in updating the model parameters that is the first step of the process. This is followed by long-term prediction until the mean of predicted crack length distribution reaches pre-defined threshold (that is assumed to be 0.2 m in this case) as shown in Fig. 42.6. The 5% and 95% percentile boundaries indicate the distribution of crack length a estimation at a particular instant during crack propagation. These bounds widen with larger number of forecast steps indicating higher uncertainty. The filter estimation reaches the assumed damage threshold at after 745 h (approximately 31 days) of initial crack detection. The non- Gaussian distribution of RUL (Fig. 42.7) gives a mean of 700 h. The deviation of prognosis by the filter from measurements cab be reduced by providing more measurements to update the crack propagation model [38]. This is proven by comparing the time from crack detection crossing the damage threshold for 50 measurements (Fig. 42.8) and 100 observations (Fig. 42.9). The time to failure moves closer to the observation. This implies that model

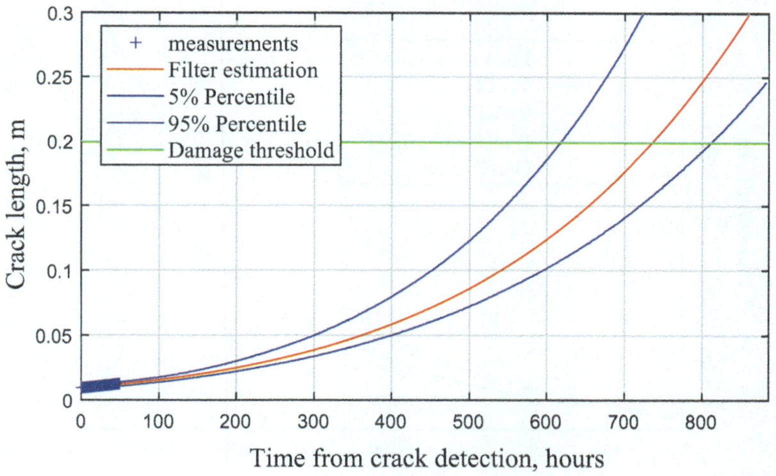

Fig. 42.6 Prognosis by particle filter for 0.01 m initial crack length

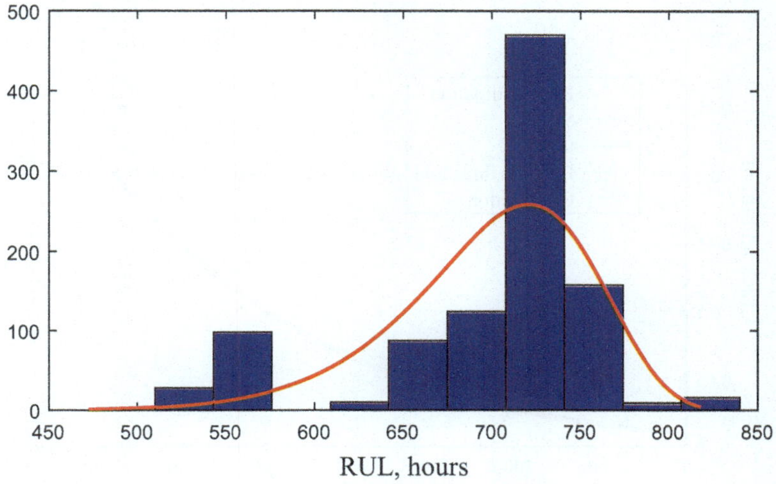

Fig. 42.7 RUL distribution for 0.01 m initial crack length

parameters are improved with higher number of measurements. This can also be inferred from the fact that the damage state distribution has narrowed down for the case with higher number of measurements.

The particle filtering estimation of multiple initial crack lengths is performed at the same location under the same loading conditions. Figure 42.10 shows the fatigue crack prognosis for multiple initial crack lengths we can identify that the mean RUL estimated is lower as the cracks are detected at the later stages of propagation (Fig. 42.11).

42.5 Conclusions

This study is particularly focused on estimation of RUL of the fatigue damaged life size 5 MW NREL WT blades under turbulence loading conditions using the stochastic life-expectancy method of particle filters. Meaningful RUL estimates are achieved for different initial detected crack lengths. Despite the unavailability of sources to validate the estimated RUL in this study, it is necessary to emphasize the effect of simplifications in the measurement simulation method on the estimated RUL. Though the WT blades are made of multiple layers of laminates which have different response to fatigue loading, homogeneity of material was assumed in the measurement simulation. Also, the fatigue cracks initiated at different locations such as between the layers or on the surface behave differently under loading but no such distinction is made in this study. Taking these into account can result in a more reliable model for RUL estimation. It is also established that performance of

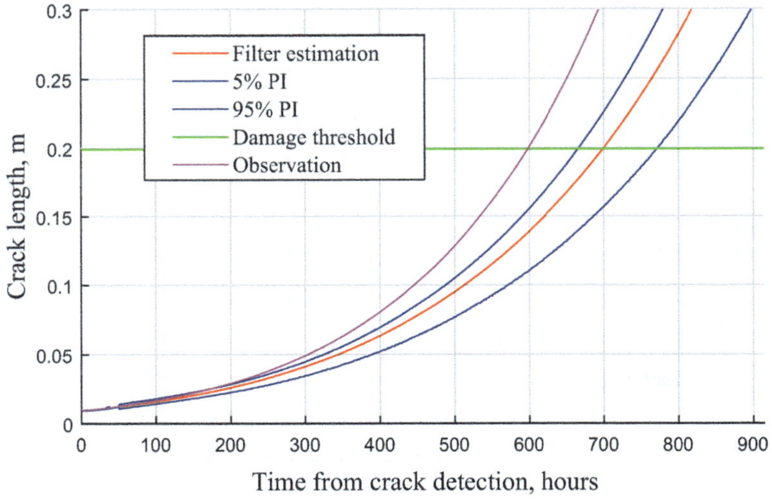

Fig. 42.8 Prognosis by particle filter with 50 measurements for 0.01 m initial crack length

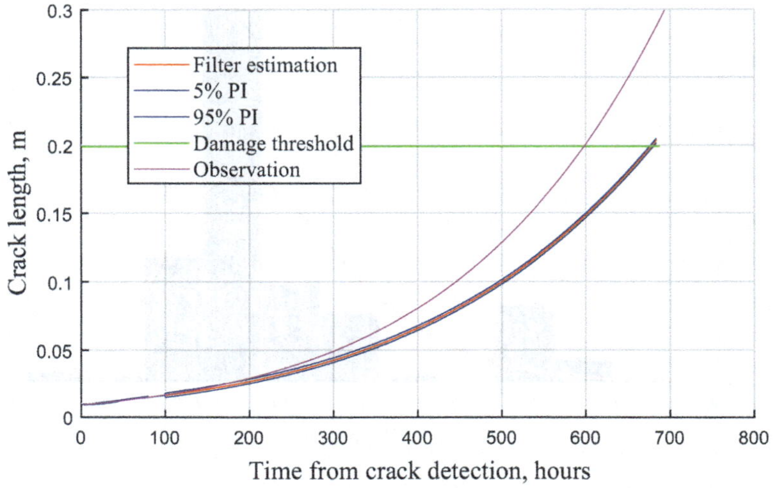

Fig. 42.9 Prognosis by particle filter with 100 measurements 0.01 m initial crack length

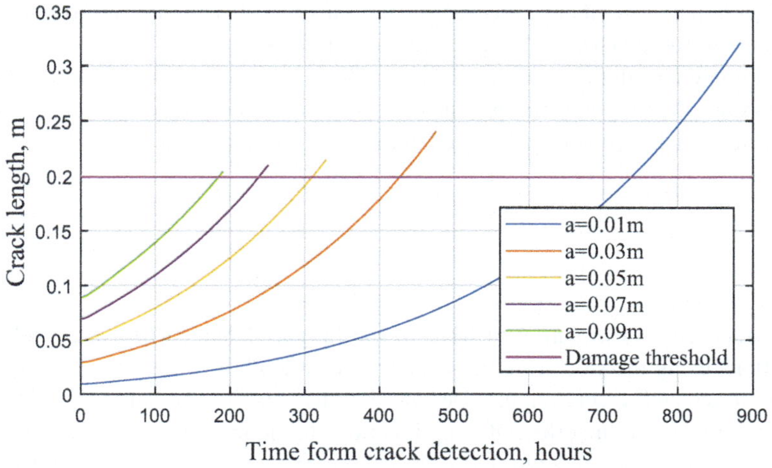

Fig. 42.10 Prognosis by particle filter for different initial crack lengths

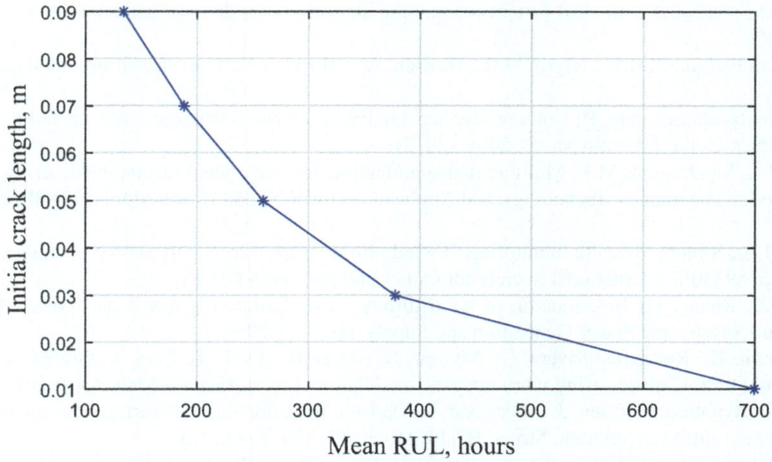

Fig. 42.11 RUL vs initial crack length

particle filters can be improved with higher number of measurements to update the model parameters. Though the study only focused on simple fatigue damage propagation for RUL estimation, influence of other modes of failure due to corrosion, surface erosion and probable sudden and extreme loads also need to be considered. This suggests that addressing uncertainty in the estimation of RUL of a WT blade is paramount not only regarding the material behavior but also the loading and interaction of different failure modes of the blade.

Acknowledgements Research funding is partially provided by the National Science Foundation through Grant No. CMMI-1351537 by Hazard Mitigation and Structural Engineering program, and by a grant from the Commonwealth of Pennsylvania, Department of Community and Economic Development, through the Pennsylvania Infrastructure Technology Alliance (PITA).

References

1. Wymore, M.L., Van Dam, J.E., Ceylan, H., Qiao, D.: A survey of health monitoring systems for wind turbines. Renew. Sust. Energ. Rev. **52**(1069283), 976–990 (2015)
2. Griffin, D.A., Malkin, M.C.: I Introduction, Principal Engineer, Turbine Engineering Group, Senior Engineer, Press Release, Suzlon Energy Unlimited.: Lessons learned from recent blade failures: primary causes and risk-reducing technologies. In: 49th AIAA Aerospace Sciences Meeting Including the New Horizons Forum and Aerospace Exposition, Jan 2011, pp. 1–9
3. El-Thalji, I., Jantunen, E.: On the development of condition based maintenance strategy for offshore wind farm: requirement elicitation process. Energy Proc. **24**, 328–339 (2012)
4. Gupta, A., Lawsirirat, C.: Strategically optimum maintenance of monitoring-enabled multi-component systems using continuous-time jump deterioration models. J. Qual. Maint. Eng. **12**(3), 306–329 (2006)
5. Campos, J.: Development in the application of ICT in condition monitoring and maintenance. Comput. Ind. **60**(1), 1–20 (2009)
6. Ciang, C.C., Lee, J.-R., Bang, H.-J.: Structural health monitoring for a wind turbine system: a review of damage detection methods. Meas. Sci. Technol. **19**(12), 122001 (2008)
7. Florian, M., Sørensen, J.: Wind turbine blade life-time assessment model for preventive planning of operation and maintenance. J. Mar. Sci. Eng. **3**(3), 1027–1040 (2015)
8. Al-Khudairi, O., Hadavinia, H., Little, C., Gillmore, G., Greaves, P., Dyer, K.: Full-scale fatigue testing of a wind turbine blade in flapwise direction and examining the effect of crack propagation on the blade performance. Materials **10**(10), 1152 (2017)
9. Yang, W., Peng, Z., Wei, K., Tian, W.: Structural health monitoring of composite wind turbine blades: challenges, issues and potential solutions. IET Renew. Power Gener. **11**(4), 411–416 (2017)
10. Mcgugan, M., Pereira, G., Sørensen, B.F., Toftegaard, H., Branner, K.: Damage tolerance and structural monitoring for wind turbine blades. Philos. Trans. R. Soc. A **373**(2035) (2015)
11. Nielsen, J.S., Sørensen, J.D.: Bayesian estimation of remaining useful life for wind turbine blades. Energies **10**(5), 664 (2017)
12. Kong, C., Kim, T., Han, D., Sugiyama, Y.: Investigation of fatigue life for a medium scale composite wind turbine blade. Int. J. Fatigue **28**(10), 1382–1388 (2006)
13. Schulz, M.J., Sundaresan, M.J.: Smart sensor system for structural condition monitoring of wind turbines: 30 May 2002–30 Apr 2006, subcontract report NREL/SR-500-40089. Technical report, National Renewable Energy Laboratory (2002)
14. Shokrieh, M.M., Rafiee, R.: Simulation of fatigue failure in a full composite wind turbine blade. Compos. Struct. **74**(3), 332–342 (2006)
15. Lading, L., McGugan, M., Sendrup, P., Rheinländer, J., Rusborg, J.: Fundamentals for remote structural health monitoring of wind turbine blades – a preproject annex B – sensors and non-destructive testing methods for damage detection in wind turbine blades fundamentals for remote structural health monitoring of wind turbine. Technical report, Forskningscenter Risoe, Risoe-R (2002)

16. Yao, R., Pakzad, S.N.: Auto-regressive statistical pattern recognition algorithms for damage detection in civil structures. Mech. Syst. Signal Process. **31**, 355–368 (2012)
17. Shahidi, S.G., Yao, R., Chamberlain, M.B.W., Nigro, M.B., Thorsen, A., Pakzad, S.N.: Data-driven structural damage identification using DIT. In: Imac33 pp. 2–9 (2015)
18. Yao, R., Pakzad, S.N., Venkitasubramaniam, P.: Compressive sensing based structural damage detection and localization using theoretical and metaheuristic statistics. Struct. Control. Health Monit. **24**(4) (2017)
19. Kirikera, G.R., Schulz, M.J., Sundaresan, M.J.: Multiple damage identification on a wind turbine blade using a structural neural system. In: Sensor Systems and Networks: Phenomena, Technology, and Applications for NDE and Health Monitoring. Proceedings of the SPIE, vol. 6530, p. T5300 (2007)
20. Rumsey, M.A., Paquette, J.A.: Structural health monitoring of wind turbine blades. In: Smart Sensor Phenomena, Technology, Networks, and Systems 2008, vol. 6933, p. 69330E. International Society for Optics and Photonics (2008)
21. Zhang, F., Li, Y., Yang, Z., Zhang, L.: Investigation of wind turbine blade monitoring based on optical fiber Brillouin sensor. In: 2009 International Conference on Sustainable Power Generation and Supply, pp. 1–4 (2009)
22. Kim, S., Adams, D.E., Sohn, H., Rodriguez-Rivera, G., Myrent, N., Bond, R., Vitek, J., Carr, S., Grama, A., Meyer, J.J.: Crack detection technique for operating wind turbine blades using vibro-acoustic modulation. Struct. Health Monit. **13**(6), 660–670 (2014)
23. LeBlanc, B., Niezrecki, C., Avitabile, P., Chen, J., Sherwood, J.: Damage detection and full surface characterization of a wind turbine blade using three-dimensional digital image correlation. Struct. Health Monit. **12**, 430–439 (2013)
24. Dutton, A.G., Blanch, M.J., Vionis, P., Lekou, D., van Delft, D.R.V., Joosse, D., Anastassopoulos, P.A., Kouroussis, A., Kossivas, T., Philippidis, T.T., Assimakopoulou, T.P.: Acoustic emission condition monitoring of wind turbine rotor blades: laboratory certification testing to large scale in service deployment. In: European Wind Energy Conference, pp. 1–11 (2003)
25. Kirikera, G.R., Shinde, V., Schulz, M.J., Sundaresan, M.J., Hughes, S., van Dam, J., Nkrumah, F., Grandhi, G., Ghoshal, A.: Monitoring multi-site damage growth during quasi-static testing of a wind turbine blade using a structural neural system. Struct. Health Monit. Int. J. **7**(2), 157–173 (2008)
26. Ou, Y., Dertimanis, V.K., Chatzi, E.N.: Experimental damage detection of a wind turbine blade under varying operational conditions. In: Proceedings of the ISMA, pp. 4183–4198 (2016)
27. Ou, Y., Chatzi, E.N., Dertimanis, V.K., Spiridonakos, M.D.: Vibration-based experimental damage detection of a small-scale wind turbine blade. Struct. Health Monit. **16**(1), 79–96 (2016)
28. Si, X.S., Wang, W., Hu, C.H., Zhou, D.H.: Remaining useful life estimation – a review on the statistical data driven approaches. Eur. J. Oper. Res. **213**(1), 1–14 (2011)
29. Vachtsevanos, G., Lewis, F.L., Roemer, M, Hess, A., Wu, B.: Intelligent Fault Diagnosis and Prognosis for Engineering Systems. Wiley, Hoboken (2006)
30. Benedetti, M., Fontanari, V., Battisti, L.: Structural health monitoring of wind towers: residual fatigue life estimation. Smart Mater. Struct. **22**(4), 045017 (2013)
31. Sikorska, J.Z., Hodkiewicz, M., Ma, L.: Prognostic modelling options for remaining useful life estimation by industry. Mech. Syst. Signal Process. **25**(5), 1803–1836 (2011)
32. Degrieck, J., Van Paepegem, W.: Fatigue damage modelling of fibre-reinforced composite materials: review. Appl. Mech. Rev. **54**(4), 279–300 (2001)
33. Beganovic, N., Njiri, J.G., Rothe, S., Soffker, D.: Application of diagnosis and prognosis to wind turbine system based on fatigue load. In: 2015 IEEE Conference on Prognostics and Health Management (PHM), pp. 1–6. IEEE (2015)
34. Hayat, K., Asif, M., Ali, H.T., Ijaz, H., Mustafa, G.: Fatigue life estimation of large-scale composite wind turbine blades. In: 2015 12th International Bhurban Conference on Applied Sciences and Technology (IBCAST), pp. 60–66 (2015)
35. Sanchez, H., Sankararaman, S., Escobet, T., Puig, V., Frost, S., Goebel, K.: Analysis of two modeling approaches for fatigue estimation and remaining useful life predictions of wind turbine blades. In: Third European PHM Conference (2016)
36. Bartram, G., Mahadevan, S.: Dynamic Bayesian networks for prognosis. In: PHM 2013 – Proceedings of the Annual Conference of the Prognostics and Health Management Society 2013, pp. 167–184 (2013)
37. Arulampalam, M.S., Maskell, S., Gordon, N., Clapp, T.: A tutorial on particle filters for online nonlinear/non-GaussianBayesian tracking. IEEE Trans. Signal Process. **50**(2), 174–188 (2002)
38. Cadini, F., Zio, E., Avram, D.: Monte Carlo-based filtering for fatigue crack growth estimation. Probab. Eng. Mech. **24**(3), 367–373 (2009)
39. Zio, E., Peloni, G.: Particle filtering prognostic estimation of the remaining useful life of nonlinear components. Reliab. Eng. Syst. Saf. **96**(3), 403–409 (2011)
40. Haile, M.A., Riddick, J.C., Assefa, A.H.: Robust particle filters for fatigue crack growth estimation in rotorcraft structures. IEEE Trans. Reliab. **65**(3), 1438–1448 (2016)
41. Orchard, M., Wu, B., Vachtsevanos, G.: A particle filtering frmaework for failure prognosis. In: World Tribology Congress III, vol. 2, pp. 1–2, Washington, DC (2005)
42. An, D., Choi, J.H., Kim, N.H.: Prognostics 101: A tutorial for particle filter-based prognostics algorithm using Matlab. Reliab. Eng. Syst. Saf. 115:161–169 (2013)
43. Orchard, M.E., Vachtsevanos, G.J.: A particle-filtering approach for on-line fault diagnosis and failure prognosis. Trans. Inst. Meas. Control. **31**(3–4), 221–246 (2009)
44. Jonkman, J., Butterfield, S., Musial, W., Scott, G.: Definition of a 5-MW reference wind turbine for offshore system development. Technical report, National Renewable Energy Limited, Golden (2009)
45. Tezduyar, T.E., Sathe, S., Schwaab, M., Conklin, B.S.: Arterial fluid mechanics modeling with the stabilized space – time fluid – structure interaction technique. Int. J. Numer. Methods Fluids **2007**, 601–629 (2008)
46. Jonkma, J., Jonkman, B.: FAST, National Renewable Energy Laboratory. NREL (2016)
47. Stephens, R.I., Fatemi, A., Stephens, R.R., Fuchs, H.O.: Metal Fatigue in Engineering, 2nd edn. Wiley, New York (2001)
48. Downing, S.D., Socie, D.F.: Simple rainflow counting algorithms. Int. J. Fatigue **4**(1), 31–40 (1982)

Chapter 43
A Numerical Investigation of a Gravity-Compensated Nonlinear Energy Sink for the Passive Control of Flooring Systems

J. R. Ramsey and N. E. Wierschem

Abstract Flooring systems are subjected to a variety of human-induced and mechanically-induced loads which can vary in amplitude, frequency, and location. Furthermore, the properties of flooring systems and the acceptable levels of vibration can change during the life of a building as it transitions between multiple different uses. Tuned mass dampers (TMDs) can be effective at controlling floor vibration; however, their effectiveness is limited because TMDs must be tuned and can only effectively control vibrations across a narrow band of frequencies. Recently, a passive mass damper, known as a gravity-compensated nonlinear energy sink (GCNES), was proposed to mitigate vertical vibrations. The unique geometric nonlinearity used to produce this device's stiffness element compensates for the vertical offset resulting from the weight of the device and allows it to dynamically achieve a cubic nonlinearity. This strong nonlinearity allows the GCNES to interact with the flooring system across a broad range of frequencies. In this paper, a numerical model of a flooring system with a GCNES attached is developed. This model is then used to investigate the effectiveness of the GCNES, in comparison to the TMD, at controlling floor vibrations. The results of this study show that, while the TMD is more effective when mitigating excitations at the particular frequency it is tuned to, the GCNES can provide effective vibration control across a wide range of frequencies near the system's resonance point.

Keywords Floor vibration · Mass damper · Nonlinear energy sink · Passive control

43.1 Introduction

Due to the increased use of lightweight materials, long-spans, and open plan floor systems, flooring systems are becoming more susceptible to vibration issues related to a variety of excitation sources [1, 2]. Passive devices, such as the tuned mass damper (TMD), have been implemented to mitigate vertical floor vibration. TMD's, however, are limited due to the fact that they need to be tuned to one specific frequency and multiple TMD's are needed for multiple frequency mitigation [3]. TMDs are also susceptible to detuning caused by issues such as fatigue, damage, and temperature variations which can result in changes in the natural frequencies of the flooring system or TMD [4].

A proposed passive alternative, the nonlinear energy sink (NES), is a different type of mass damper. Unlike the TMD, the NES incorporates an essentially nonlinear stiffness component that, in most cases, can be modelled as a cubic function [5]. With this elastic and repeatable stiffness component, the NES does not have a preferred vibration frequency, giving it a broad operating frequency regime [6]. The main problem when realizing a vertical NES, however, is the vertical gravitational force on the mass of the NES which produces an offset in the at rest position of the device; the device will displace downward due to its self-weight. This offset displacement results in the at rest point being moved from a position where the device has no tangent stiffness to a position where the device has a non-zero tangent stiffness. This can have a large effect on the dynamics of the NES because, instead of possessing a strong non-linearizable nonlinearity, the device will operate like a perturbed linear oscillator at some displacement levels.

In this paper, a NES is proposed using a restoring force mechanism, which was originally studied for application in a vertical vibration isolator [7, 8], that compensates for the effect of gravity while maintaining the required dynamic non-linear stiffness component. A simplified five degree-of-freedom floor model coupled with a gravity-compensated NES (GCNES) is developed and optimized along with a traditional TMD. The acceleration responses of these systems are then analyzed and compared.

J. R. Ramsey · N. E. Wierschem (✉)
The University of Tennessee, Knoxville, TN, USA
e-mail: jramse28@vols.utk.edu; nwiersch@utk.edu

S. Pakzad (ed.), *Dynamics of Civil Structures, Volume 2*, Conference Proceedings of the Society for Experimental Mechanics Series, https://doi.org/10.1007/978-3-319-74421-6_43

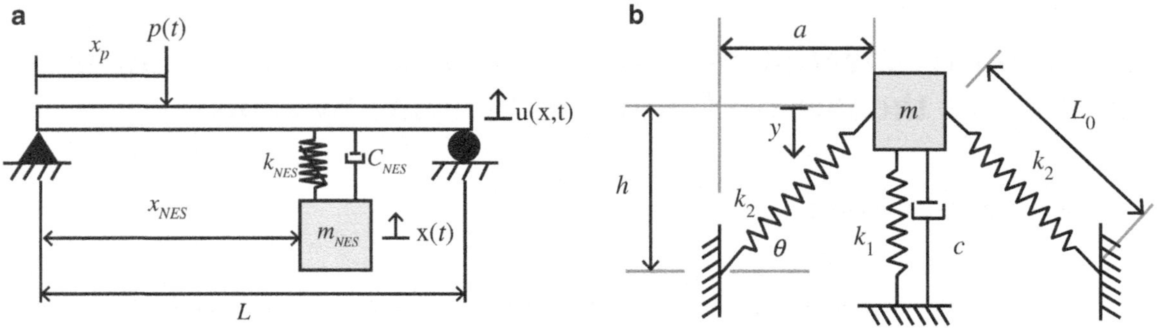

Fig. 43.1 (**a**) Simply supported beam floor model. (**b**) Free body diagram of GCNES

43.2 GCNES and Floor Model

The geometric mechanism utilized by the GNES that provides vertical stiffness and compensates for the vertical effect of gravity is seen in Fig. 43.1b. This device was previously studied as a vertical vibration isolator [7, 8]. The restoring force for this device, in terms of the vertical stiffness was previously derived with this restoring force relative to the configuration where the two oblique springs are completely horizontal [9]. The actual device's restoring force, Eq. (43.1), and its Taylor series cubic approximation, Eq. (43.2), are shown below. The physical properties of this device can be configured such that, along with the acceleration due to gravity, the approximate dynamic restoring force only has a cubic component [9].

$$F_s = k_1\left(\overline{y} + h\right) - 2k_2\left(L_0 - \sqrt{a^2 + \overline{y}^2}\right)\frac{\overline{y}}{\sqrt{a^2 + \overline{y}^2}} \tag{43.1}$$

$$F_{s-approx} = hk_1 + \left(k_1 + 2k_2\frac{a - L_0}{a}\right)\overline{y} - k_2\left(\frac{a - L_0}{a^3} - \frac{1}{a^2}\right)\overline{y}^3 + H.O.T. \tag{43.2}$$

A combination of the assumed modes method and Lagrange's equation was used to develop the considered system's equation of motion (EOM). The flooring system is modelled as a simply supported beam (Length = 144 in, Mass = 25.60 slugs, EI =1.69 × 10^{10} kip-in^2, modal damping = 1%). The assumed beam shape functions were of the form $\sin(n\pi x/L)$ where n = 1,2,3,4, and 5, respectively. The resulting natural frequencies for this system are 32.04, 125.48, 289.45, 525.46, and 804.58 rad/s. The GCNES, located at x_{NES}=108 in, is modelled with the approximated cubic vertical spring and a linear damper coupling a mass (5% of the mass of the beam) to the simplified floor model, as depicted in Fig. 43.1.

43.3 Optimization

A parameter sweep optimization process was used to find the optimal combination of damping and stiffness parameters. The objective function used in this optimization is the minimization of the underlying system's steady state RMS acceleration response at the measurement point (x = 36 in) given a sinusoid loading at (x_p=36 in) with a frequency of 32.04 rad/s (equal to the 1st natural frequency of the structure with a locked NES or TMD). The amplitude of the load used in this optimization was set to a value that scales the peak steady state acceleration amplitude of the underlying structure with a locked NES/TMD to 7.6139 in/s^2 at the measurement point. This amplitude scaling factor was used to scale the beam's response up to a reasonable level of excitation as the NES is an amplitude dependent device. For this optimization, a lower-bound of 100 lb-s/in was used for the TMD and NES damping in order to ensure that the optimized devices did not possess damping values too low to be realistically achieved in a physical device. Figure 43.2 depicts the contour plots produced by the parameter sweep optimization for the NES case and the TMD case. The resulting optimized values produced from the parameter sweep are k_{TMD} = 28650 lb/in and c_{TMD} = 100lb - sec /in for the TMD and k_{NES} = 9.8 ^ 10 lb/in^3 and c_{NES} = 100 lb - sec /infor the NES, respectively.

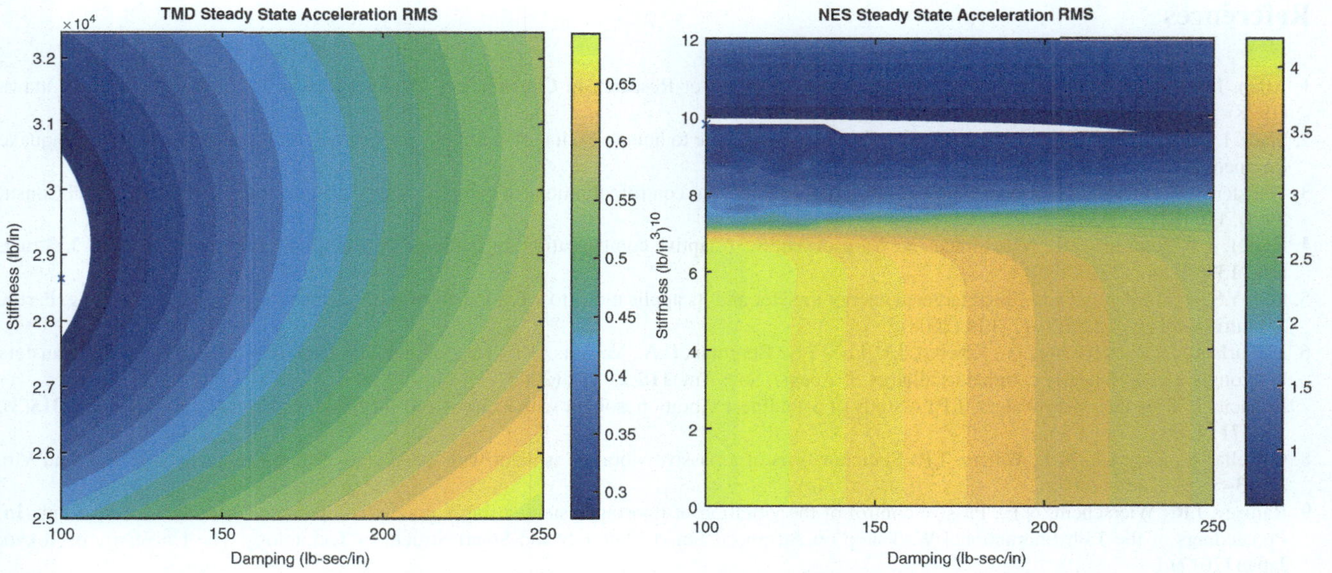

Fig. 43.2 Steady state acceleration RMS contour optimization plots for TMD (Left) and NES (Right)

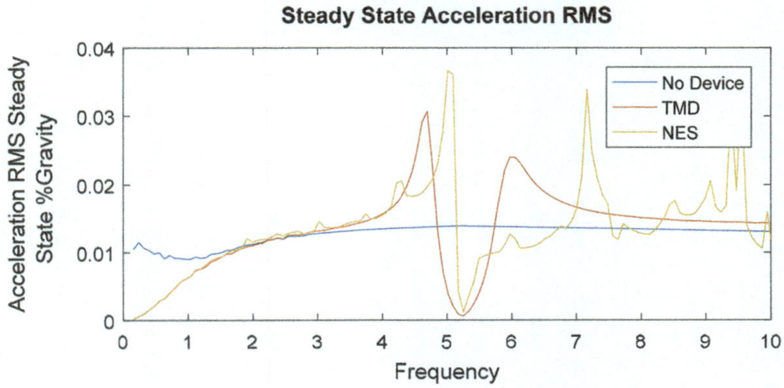

Fig. 43.3 Frequency response curve

43.4 Results

Using the optimized stiffness and damping parameters, the system's frequency response is analyzed with a sine sweep which has a modulating amplitude designed to keep the steady state acceleration response amplitude of the system with the locked TMD/NES at the same level the system was optimized at. Figure 43.3 shows that the TMD significantly reduces the response around the frequency it was optimized at (the 1st mode 5.1 Hz), but produces peaks in the frequency response curve on either side of it. On the other hand, one can see that the GCNES reduces the response at the optimization frequency and a significant range of frequencies above this point; maintaining the first peak, but eliminating the second peak shown in the TMD frequency response curve. While the GCNES creates some additional peaks in the frequency response curve at higher frequencies, this is acceptable as these points are significantly off resonance and require a much higher load amplitude to significantly excite.

43.5 Conclusion

In this paper, a gravity-compensated nonlinear energy sink (GCNES), a mass damper with a geometric stiffness non-linearity that compensates for the effect of gravity, is studied and its ability to mitigate floor vibrations is analyzed. A simplified floor model is derived with both a traditional TMD as well as the proposed GCNES device. The model's optimal damping and stiffness parameters are found through a contour plot optimization process. Overall, the GCNES is shown to be able to better control vibrations at higher frequencies close to the underlying system's resonance frequency.

References

1. Allen, D.E., Pernica, G.: Control of Floor Vibration. Institute for Research in Construction, National Research Council of Canada, Ottawa (1998)
2. Saidi, I., Haritos, N., Gad, E.F., Wilson, J.L.: Floor vibrations due to human excitation-damping perspective. In: Proceedings of the Earthquake Engineering in Australia, pp. 257–264 (2006)
3. Webster, A.C., Vaicaitis, R.: Application of tuned mass dampers to control vibrations of composite floor systems. Eng. J. Am. Inst. Steel Constr. **29**(3), 116–124 (1992)
4. Roffel, A.J., Lourenco, R., Narasimhan, S., Yarusevych, S.: Adaptive compensation for detuning in pendulum tuned mass dampers. J. Struct. Eng. **137**(2), 242–251 (2010)
5. Lee, Y.S., et al.: Passive non-linear targeted energy transfer and its applications to vibration absorption: a review. Proc. Inst. Mech. Eng. Part K J. Multi-Body Dyn. **222**(2), 77–134 (2008)
6. McFarland, D.M., Kerschen, G., Kowtko, J.J., Lee, Y.S., Bergman, L.A., Vakakis, A.F.: Experimental investigation of targeted energy transfers in strongly and nonlinearly coupled oscillators. J. Acoust. Soc. Am. **118**(2), 791 (2005)
7. Kovacic, I., Brennan, M.J., Waters, T.P.: A study of a nonlinear vibration isolator with a quasi-zero stiffness characteristic. J. Sound Vib. **315**(3), 700–711 (2008)
8. Carrella, A., Brennan, M.J., Waters, T.P.: Static analysis of a passive vibration isolator with quasi-zero-stiffness characteristic. J. Sound Vib. **301**(3–5), 678–689 (2007)
9. Ramsey, J.R., Wierschem, N.E.: Passive control of the vibration of flooring systems using a gravity compensated non-linear energy sink. In: Proceedings of the 13th International Workshop on Advanced Smart Materials and Smart Structures Technology. The University of Tokyo, Japan (2017)

Chapter 44
Characterizing Structural Changes to Estimate Walking Gait Balance

Jonathon Fagert, Mostafa Mirshekari, Shijia Pan, Pei Zhang, and Hae Young Noh

Abstract We present a method for improving left-right walking gait balance using structural floor vibration sensing by characterizing changes in structural properties in the sensing area. Understanding and measuring human gait balance can be used to assess overall health status, mobility, and rehabilitation progress. The key research challenge is that structural properties in the sensing area may differ from one footstep location to the next, resulting in inaccurate footstep force and balance estimations. To address this challenge, our method performs sensor selection using the insight that some sensors in the sensor network are in a similar structural region as the footstep location and, therefore, are not as effected by the observed variations in structural properties as the other sensors. We evaluate the performance of our method by conducting uncontrolled real-world walking experiments in a residential structure. This evaluation shows that our method achieves a 1.6X reduction in force estimation error and a 2.4X reduction in balance estimation as compared to the baseline approach.

Keywords Structural vibrations · Footstep ground reaction forces · Sensor selection · Walking gait balance · Structural identification

44.1 Introduction

Walking gait balance is a critical component of health assessment in a number of fields such as physical therapy, elder care, and neurology [1–3]. Existing methods for monitoring gait balance involve visual observation, motion-tracking cameras, pressure sensors, and wearable devices (e.g. accelerometers), but have operational limitations such as specialized staff (observation), line of sight (cameras), dense sensor deployment requirement (pressure sensors), and requiring users to carry a device at all times (wearables) [3–5]. Recently, structural vibration sensing has been introduced to overcome many of the limitations of prior works [6], but may be limited in scenarios where local variations in structural properties (i.e. stiffness) result in increased footstep ground reaction force and walking balance estimation errors.

In this paper, we present a system to estimate walking left-right gait balance (referred to as balance) using structural floor vibrations that is robust to these local variations in structural properties. The primary research challenge addressed through this work is that variations in structural properties at one footstep location can significantly change the amplitude of the measured vibration response as compared to footsteps at adjacent locations, resulting in large footstep force and balance estimation errors.

Our approach utilizes the insight that there exist some sensors in the network that reside in a similar structural region as the footstep location and experience fewer structural effects than the other sensors in the network do. Therefore, we take this structural effect into account during the training phase to select the sensor that has the least training error for each location thereby reducing the errors in footstep ground reaction force estimation and balance estimation. Our method contains three primary components: (1) a footstep detection module, (2) a force-amplitude-distance learning module, and (3) a structure-informed footstep force estimation module. We validate our approach using real-world walking experiments conducted in a residential apartment building.

J. Fagert (✉)
Department of Civil and Environmental Engineering, Carnegie Mellon University, Pittsburgh, PA, USA
e-mail: jfagert@andrew.cmu.edu

M. Mirshekari · H. Y. Noh
Department of Civil and Environmental Engineering, Carnegie Mellon University, Pittsburgh, PA, USA

S. Pan · P. Zhang
Department of Electrical and Computer Engineering, Carnegie Mellon University, Moffett Field, CA, USA

S. Pakzad (ed.), *Dynamics of Civil Structures, Volume 2*, Conference Proceedings of the Society for Experimental Mechanics Series, https://doi.org/10.1007/978-3-319-74421-6_44

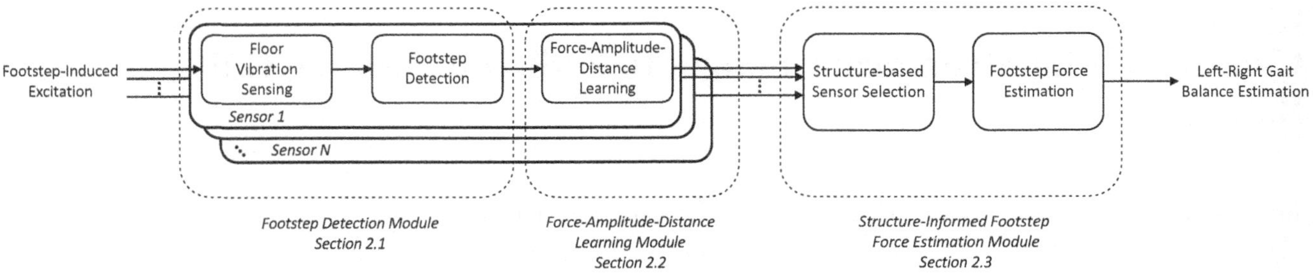

Fig. 44.1 Gait balance estimation approach overview

44.2 Gait Balance Estimation Approach

Our balance estimation approach consists of three main modules: (1) a footstep detection module that isolates individual footstep responses, (2) a force-amplitude-distance learning module that maps footstep forces to signal amplitudes and distances, and (3) a structure-informed footstep force estimation module that selects the sensor with lowest training error to estimate footstep forces and balance. An overview of our system can be found in Fig. 44.1.

44.2.1 Footstep Detection Module

In the first module of our approach, our system senses the floor vibration response due to a footstep-induced excitation using a SM-24 geophone vibration sensor [7] and isolates individual footstep signals. We utilize an anomaly detection algorithm to distinguish the footstep-induced excitations from the ambient noise [8]. Footsteps are identified as vibration responses that exceed an empirically-defined ambient noise threshold. Once identified, individual footstep responses are isolated from the total signal and used for subsequent modules.

44.2.2 Force-Amplitude-Distance Learning Module

In this module, our system estimates the relationship between the input footstep forces, the measured signal amplitude, and the distance between the footstep and the sensor. This relationship is estimated using an adaptation of the method introduced in [6]. Using known footstep forces, peak footstep-induced signal amplitude values, and footstep-sensor distances, our system learns the relationship between footstep forces, amplitude, and distance. This process is completed independently for each sensor located in the region of the footstep.

44.2.3 Structure-Informed Footstep Force Estimation Module

With the final module of our system, we utilize the structural insights described earlier to estimate footstep forces and left-right walking gait balance. With variations in structural properties, the force-amplitude-distance function from the second module may not adequately represent the structural response for effected sensors. Figure 44.2 shows an example of this structure effect. With Sensor 1, the fit curve at a distance of approx. 3.75 m is considerably greater than the ground truth footstep data (actual footstep forces/amplitude). However, when the same step location is considered using Sensor 4, the structural effect is much smaller. Using this observation, this module selects one sensor for each step location that has the smallest training error and uses that sensor to estimate footstep forces for that location. Finally, with estimated footstep forces, we estimate balance by calculating the Symmetry Index (S.I.) which compares consecutive left-foot and right-foot footstep forces [9].

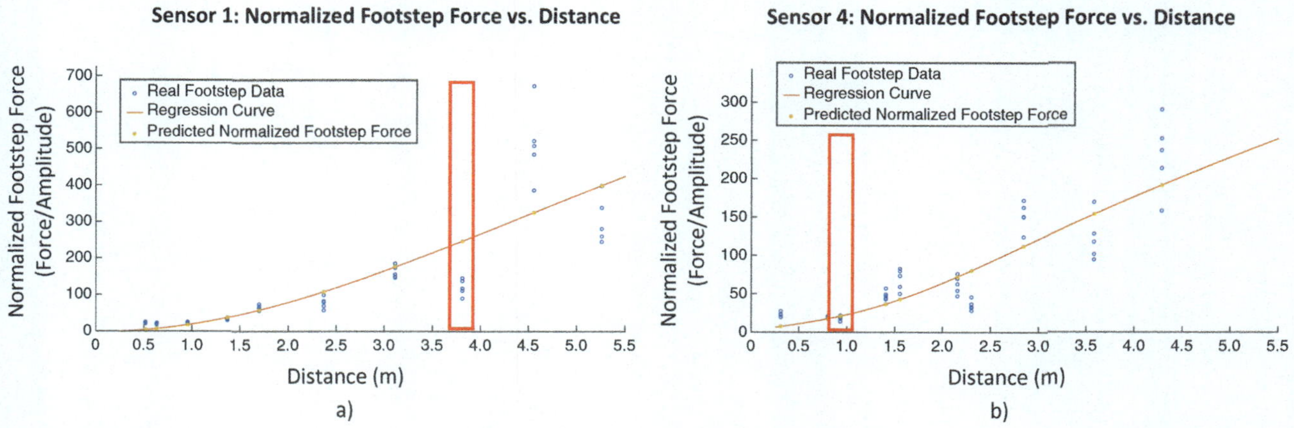

Fig. 44.2 (**a**) at the highlighted location, structural differences result in low accuracy for Sensor 1. (**b**) For the same footstep location, Sensor 4 exhibits much higher accuracy

44.3 Evaluation

To evaluate the validity of our balance estimation approach, we conducted real-world walking experiments in a wood-framed residential apartment structure. For the evaluation, ground truth footstep forces were collected using FlexiForce A401 pressure sensors, [10] and footstep locations were chosen based on the natural stride length of the participant. A total of 50 footsteps were recorded across 10 footstep locations using 4 evenly spaced geophone sensors. Of these 50 footsteps, 30 were randomly chosen for training and 20 were used for evaluation. Our method achieves a 1.6X reduction in footstep force estimation root mean squared error (RMSE) from a baseline approach that averages the force estimation from all four sensors (baseline: 66.5 N (14.95 lbs.), our approach: 40.2 N (9.04 lbs.)). Furthermore, for balance estimation our method achieves a 2.4X reduction in S.I. estimation RMSE from the baseline approach (baseline: 23.4%, our approach: 9.92%).

44.4 Conclusions

In this paper, we introduce a method for estimating left-right gait balance using structural vibrations that incorporates variations in structural properties. To reduce large errors associated with footstep locations where structural properties may have changed, our method selects the sensor that is least effected by these changes and uses it for estimating footstep forces and balance at that location. With this approach, our system achieves a 1.6X reduction in footstep force estimation RMSE and a 2.4X reduction in balance S.I. RMSE from the baseline approach.

References

1. Vaught, S.L.: Gait, balance, and fall prevention. Ochsner J. **3**(2), 94–97 (2001)
2. Furman, J.M., Cass, S.P., Whitney, S.L.: Vestibular Disorders: A Case-study Approach to Diagnosis and Treatment. Oxford University Press, New York (2010)
3. Nnodim, J.O., Yung, R.L.: Balance and its clinical assessment in older adults a review. J. Geriatr. Med. Gerontol. **1**(1), 94–97 (2001)
4. Lindemann, U., Moe-Nilssen, R., Nicolai, S.E., Becker, C., Chiari, L.: Assessment of balance in unsupported standing with elderly inpatients by force plate and accelerometers. Aging Clin. Exp. Res. **24**(1), 37–41 (2012)
5. Mancini, M., Horak, F.B.: The relevance of clinical balance assessment tools to differentiate balance deficits. Eur. J. Phys. Rehabil. Med. **46**(2), 239–248 (2010)
6. Fagert, J., Mirshekari, M., Pan, S., Zhang, P., Noh, H.Y.: Characterizing left-right gait balance using footstep-induced structural vibrations. In: SPIE Smart Structures and Materials+ Nondestructive Evaluation and Health Monitoring, pp. 10168–10169 (2017)
7. I/O Sensor Nederland bv.: SM-24 Geophone Element. P/N 1004117 (2006)
8. Mirshekari, M., Zhang, P., Noh, H.Y. Calibration-free footstep frequency estimation using structural vibration. In: IMAC XXXV A Conference and Exposition on Structural Dynamics. SEM (2017)
9. Vagenas, G., Hoshizaki, B.: A multivariable analysis of lower extremity kinematic asymmetry in running. Int. J. Sport Biomech. **8**(1), 11–29 (1992)
10. Tekscan, Inc.: Flexiforce Standard Model A401 Datasheet (2017)

Chapter 45
Load Rating of a Reinforced Concrete T-Beam Bridge Through Ambient Vibration Testing and Finite Element Model Updating

Abdou K. Ndong, Mehrdad S. Dizaji, Mohamad Alipour, Osman E. Ozbulut, and Devin K. Harris

Abstract As the load demands on highway bridges increases, it is essential that the load rating procedures reliably assess the condition of existing structures. In addition, conventional design office load rating techniques cannot be used for bridges without structural plans, which indicates the need for a more advanced load rating procedure. This paper presents a methodology to compute the live load-carrying capacity of reinforced concrete T-beam bridges, which can be applied for bridges with structural plans or with missing or limited design information. The method involves modal identification of bridge using ambient vibrations and finite element model updating using vibration characteristics for capacity estimation. A simply supported T-beam bridge located in Virginia is selected for field-testing to verify the proposed method. The bridge is composed of five spans of the same length, 12.95 m for each, with a total length of 65.4 m and a width of 8.864 m. A total of nine accelerometers are installed to bridge to collect acceleration data for 15 min at a sampling rate of 500 Hz. The modal properties of the bridge are determined using enhanced frequency domain decomposition technique. The initial finite element model of the bridge is updated such that the modal properties of the bridge match the field measured parameters. The load effects and capacity of the bridge are determined and used to calculate the load rating factor. The rating factors obtained from the proposed method and traditional design office load rating procedures are compared. The results indicate that the proposed method can reveal the reserve capacity of bridges.

Keywords Load rating · Modal analysis · Dynamic testing · Vibrations

45.1 Introduction

Traditionally, vibration testing of full-scale bridges, except notable structures, has seldom conducted by bridge owners due to the cost and potential traffic closures associated with the testing [1]. More recently, ambient vibration testing, which allows the characterization of dynamic properties of bridges under normal operating conditions, has gained attention for the condition assessment of bridges. For bridge structures, passing traffic, wind, wave motions, and walking people are potential causes for ambient vibrations. Ambient vibration testing does not require the measurement of input data. With recent developments in sensor technologies and signal processing techniques, both hardware required to obtain vibration measurements and software designed to provide modal properties of structures without input data have considerably advanced. Therefore, ambient vibration testing has been considered in various studies for structural identification, condition assessment and/or damage detection [2–5]. Coupled with finite element modeling, ambient vibration testing can also provide a more objective load rating procedure for bridges with or without structural plans.

Load rating is an integral part of the process of structural integrity evaluation of existing bridge structures. State departments of transportation and other owner agencies perform load ratings of all their assets to determine the allowable load capacity of the structures under different types of truck loads. The Manual for Bridge Evaluation (MBE) [6] presents an analytical and an empirical approach for load rating and sets forth a standard procedure for each one. Analytical load rating includes establishing the capacity of the members as well as the dead and live load force effects through structural analysis, which can be done either using simplified methods of analysis (such as empirical equations for load distribution factors) or via refined methods of analysis (such as the finite element method). Once the capacity and dead and live force effects are established, the load rating factor for each member can then be established using Eq. (45.1) given below. In this general form

A. K. Ndong · M. S. Dizaji · M. Alipour · O. E. Ozbulut (✉) · D. K. Harris
Department of Civil and Environmental Engineering, University of Virginia, Charlottesville, VA, USA
e-mail: ozbulut@virginia.edu

S. Pakzad (ed.), *Dynamics of Civil Structures, Volume 2*, Conference Proceedings of the Society for Experimental Mechanics Series, https://doi.org/10.1007/978-3-319-74421-6_45

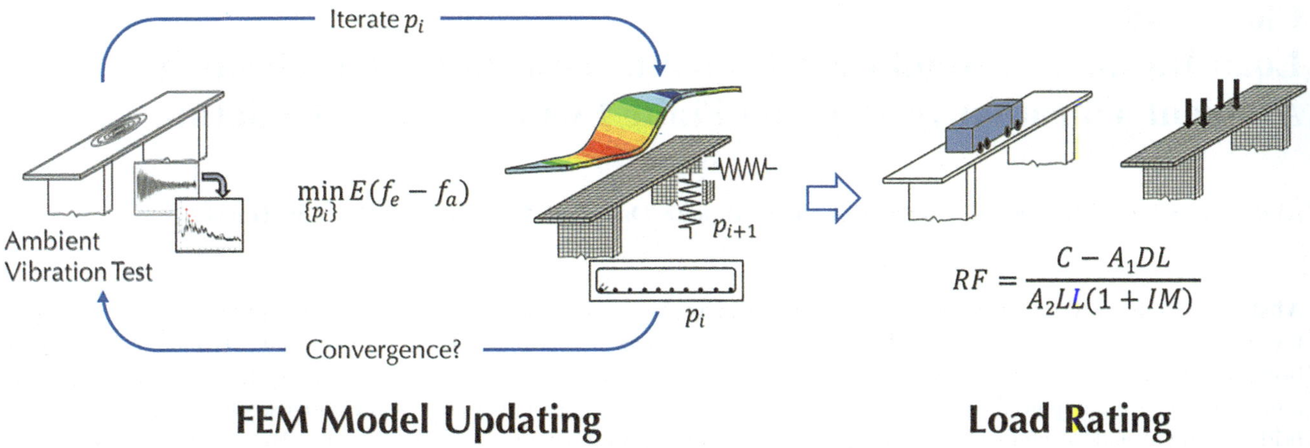

Fig. 45.1 Schematic of load rating using finite element model updating based on an ambient vibration test

of the load rating equation, *C*, *DL* and *LL* represent the member capacity, dead load effect and live load effect, respectively and A_1 and A_2 are factors for dead and live loads.

$$RF = \frac{C - A_1 DL}{A_2 (LL + IM)} \tag{45.1}$$

Finite element analysis can be used to provide a refined method of analysis of load effects especially in cases where simplified analysis is inapplicable, insufficient or results in unsatisfactory ratings. Experimental field testing is another approach that provides the most realistic picture of the structural behavior of the system. This method is especially effective in cases of bridges with complicated load distribution behavior, deteriorated or damaged structures, and bridges with unknown structural plans. MBE identify field testing as an option for cases where the lack of as-built information makes it difficult to establish the make-up of the members or their behavior. Despite the degree of accuracy it provides, the application of the field testing of load rating has been considerably limited by cost, time, test truck requirement, traffic interruption, and safety. Statistics on the methods used for load rating of highway bridges in the US show an estimated 0.6% of the bridges load-rated using field testing [7].

Building upon the principles outlined in the MBE, the analytical and experimental approaches can be combined to create an effective and realistic rating method that aims to leverage the strengths of each one of these approaches while reducing the operational difficulties of load testing. This method relies on the application of finite element model updating based on a simple ambient vibration test. An ambient vibration test is conducted on the bridge and the vibration data is used to update a FE model whose field-calibrated parameters will then be used for load rating as schematically shown in Fig. 45.1. This type of test minimizes traffic disruption by eliminating the need for test trucks or other intentional excitement but provides experimental data that help identify structural unknowns such as damage or unknown material or section properties. This paper investigates the use of this proposed approach on a concrete highway Tee-Beam bridge.

45.2 Bridge Description and Instrumentation

The selected bridge for the field testing is a concrete T-beam bridge, named as Flat Creek and built in 1957. It is a five-span simply-supported bridge located in Rockingham County, Virginia, USA. Each span of the bridge has a total length of 12.954 m and a total width of 8.865 m, and consists of four longitudinal T-beams. Each T-beam has vertical rectangular stem with a width of 0.825 m and a thickness of 0.356 m, and a wide top flange of 0.191 m thick. The wide top flange is the transversely reinforced deck slab and the riding surface for the traffic. The bridge shown is skewed at an angle of 0 degrees to the main road. The ambient vibration testing of one of the spans are conducted in this study.

The dynamic bridge assessment procedure involves the attachment of nine accelerometers underneath of the span of the bridge at 19 measurement points in two set-ups as shown in Fig. 45.2. The connection of the accelerometers to the girders of the deck was performed by means of metallic plates bonded to the surface of the concrete. Note that two common sensors

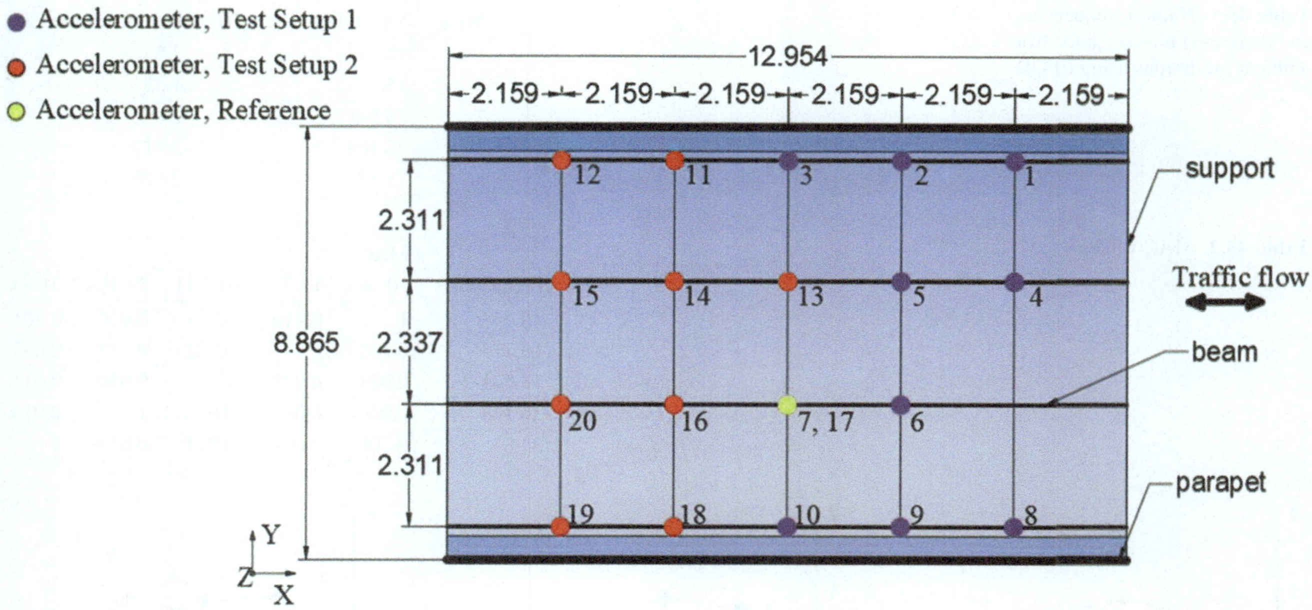

Fig. 45.2 Sensor instrumentation plan for one of the bridge span with two setups (all dimensions are in meters)

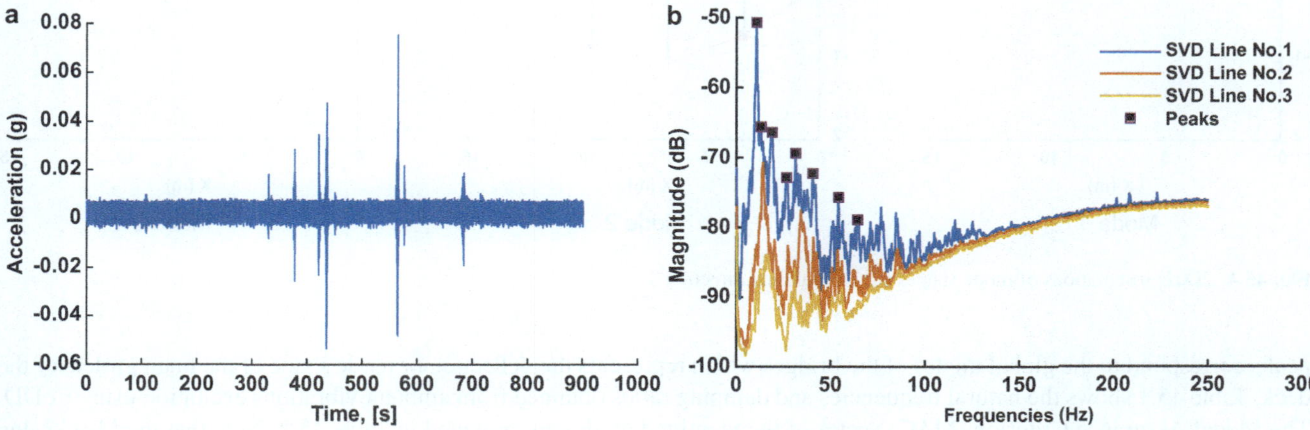

Fig. 45.3 Acceleration response measured by Sensor 7: (**a**) Ambient excitation, (**b**) Singular values for EFDD method

are used as reference in each set-up. The uniaxial accelerometers with a measuring range of ±5 g are used. The response of the bridge to ambient excitations is measured. Ambient vibrations are generated by the passing traffic, wind and walking people and recorded for a total of 15 min. All data is collected with a sampling frequency of 500 Hz.

45.3 Modal Identification

A typical measured acceleration time history recorded by sensor 7 during ambient excitation is shown in Fig. 45.3a. For parameter estimation using the ambient data, the Enhanced Frequency Domain Decomposition (EFDD) method [8], which usually provides improved estimates of modal parameters is used. In this method, the SDOF power density functions are transferred back into the time domain, and the natural frequencies are obtained by calculating number of zero-crossings as a function of time and the damping ratio is estimated from the logarithmic envelope of the corresponding SDOF correlation function using the logarithmic decrement method.

The dynamic properties, such as the natural frequencies, damping ratio and mode shapes are obtained. Figure 45.3b shows the curves of the average normalized singular values of the spectral density matrices of all experimental setups. The marked

Table 45.1 Natural frequencies and damping ratios obtained from ambient excitations using EFDD

Mode	Modal frequency (Hz)	Damping ratio (%)
1	10.743	2.522
2	13.773	1.35
3	18.541	2.856
4	26.168	2.512
5	31.83	2.826

Table 45.2 MAC values

	MAC				
Frequencies	10.743	13.773	18.541	26.168	31.83
10.743	1	0.116	0.027	0.026	0.701
13.773	0.116	1	0.022	0.043	0.093
18.541	0.027	0.022	1	0.016	0.011
26.168	0.026	0.043	0.016	1	0.013
31.83	0.701	0.093	0.011	0.013	1

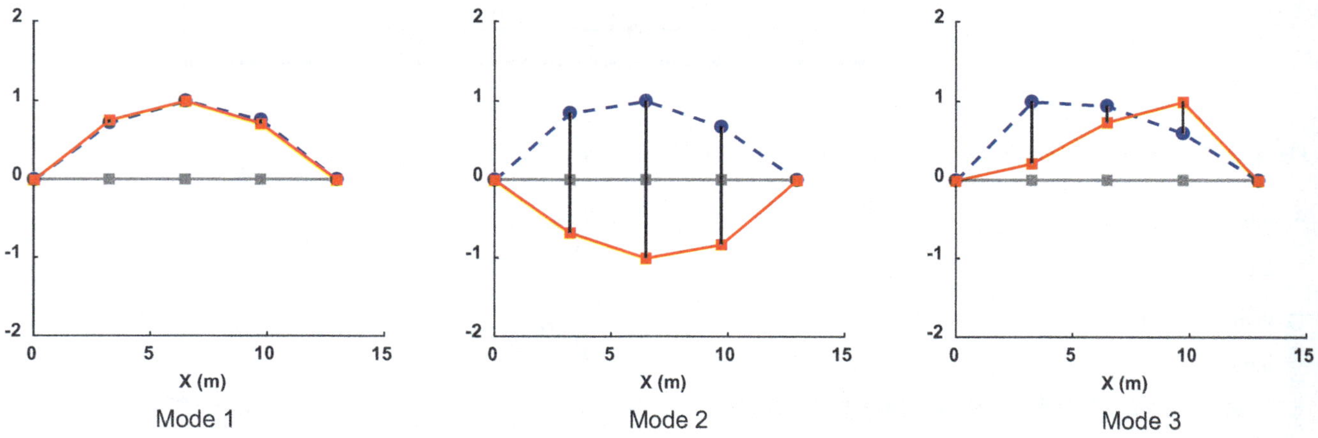

Fig. 45.4 2D representations of mode shapes for longitudinal direction

peaks correspond to the global modes of the bridges which represents the deflection or mode shape of the main girders of the deck. Table 45.1 shows the natural frequencies and damping ratios obtained from ambient vibrations excitation using EFDD. The Modal Assurance Criterion (MAC) vector of the identified modes are provided in Table 45.2. Note that the MAC value calculated for the same modes will have a value of one, while it will be close to zero if the modes are different. Figures 45.4 and 45.5 illustrate 2D and 3D representations of first three modes. The mode shapes in Figs. 45.4 and 45.5 are depicted for the longitudinal direction of the bridge using the sensor data along the sides. The mode shape vector for the left and right sides are represented by the red dashed lines and the blue solid lines, respectively. For the first mode, the natural frequency is 10.73 Hz with a damping ratio of 2.522%. It can be seen from Fig. 45.4 that the mode shape vectors on the two sides are symmetric and coincide with each other for Mode I. For this mode, the deformation of the bridge reaches its maximum at the center and minimum at the ends. This is the typical bending mode of a simply supported beam. The second mode has a natural frequency of 13.73 Hz with a damping ratio of 1.35%. From Fig. 45.4, it can be concluded that this mode is a torsional mode, where two sides have mode shape vectors in the form of an arc and in opposite directions. The third mode has a frequency of 18.541 and a damping ratio of 2.856% and is a bending mode with the two ends that arc upward but having their maximum at the quarter position of the opposite edges.

45.4 Model Updating and Rating Factors

The concept of structural identification can be defined as the process of creating/updating structural model based on experimental observations/data. It aims to bridge the gap between model approximation and the real system behavior through improved simulations. One of the subcomponent of structural identification is Finite Element Model Updating which as

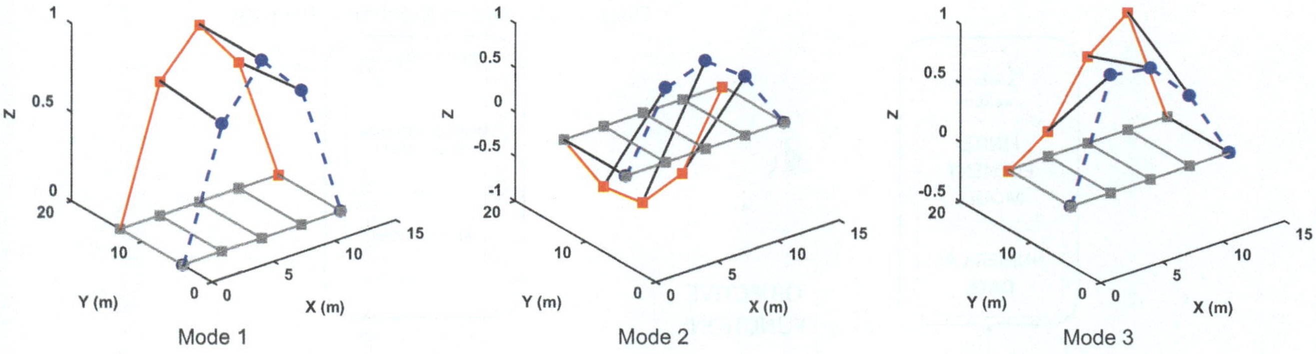

Fig. 45.5 3D representations of mode shapes

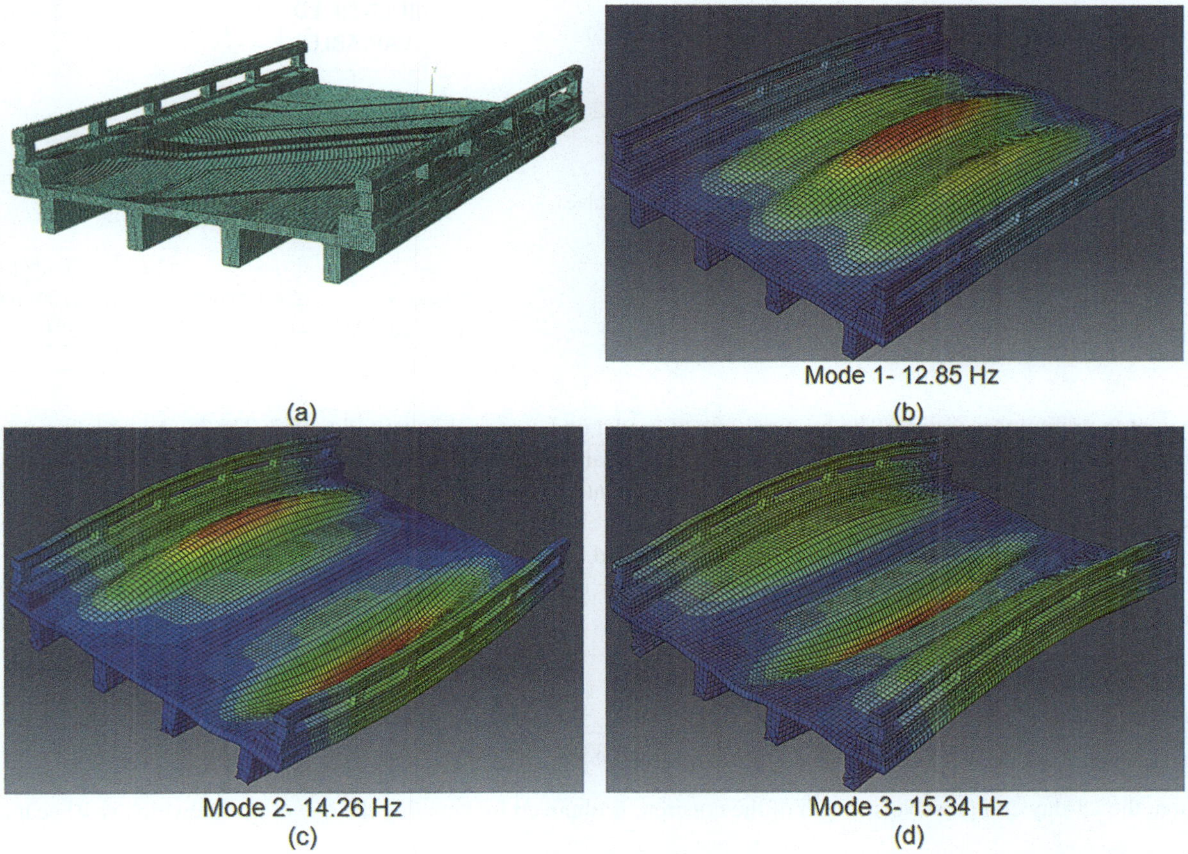

Fig. 45.6 (**a**) A global view of the numerical model of the bridge (**b–d**) natural frequencies and mode shapes obtained from initial FE model.

noted demands the establishing of an initial model that can be updated based on measured data. Updated model reflects the measured data better than the initial model. In this research, initial model is developed in ABAQUS, a robust commercially available finite element software package. ABAQUS allows for the development of an interface with MATLAB, which facilitated the iterative parameter optimization algorithm. The bridge is modeled with solid elements. For modelling of the bridge, particular attention is given to the supports to allow for realistic reproduction of the supports restraint conditions. In the models, a series of linear and rotational springs are used to create the necessary restraint [9, 10]. A global view of the model of the bridge and the first three natural frequencies and mode shapes obtained from the initial model are shown in Fig. 45.6.

The optimization algorithm developed in this investigation incorporated the features of a genetic algorithm and a gradient-based scheme to iterate on the unknown parameters. The Young's Modulus (E_s) is selected as an unknown parameter in the optimization scheme along with restraint stiffness at the support locations, which were determined to exhibit some rigid body

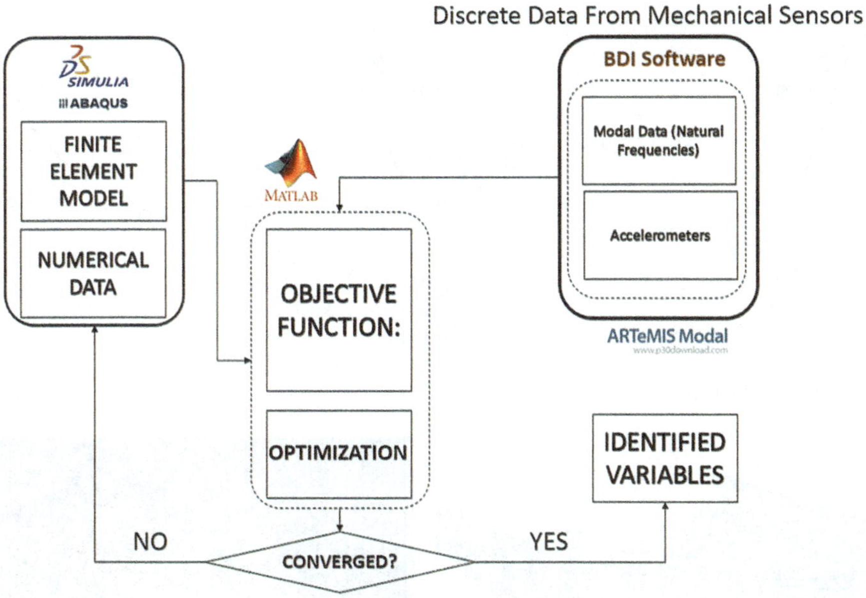

Fig. 45.7 The brief process of finite element model updating of the bridge

Table 45.3 Initial and updated values of unknown parameters

	$E_c(ksi)$	$A_s(in^2)$	$K_x(lb/in)$
Initial range	[2000, 5000]	[0.5, 2]	[100, 100,000]
Updated value	4100	1.0	554

movement in preliminary testing. The brief process of model updating is shown in Fig. 45.7. The model updating procedure minimizes the difference between the natural frequencies obtained in FE model and those obtained from experimental testing.

Leveraging the measured data, the initial FE model is updated using the optimization algorithm to converge on predictions of the beam's Young's Modulus (E_s), the area of the steel reinforcement (A_s) and support stiffness parameters (K_x). Table 45.3 provides the initial ranges selected for each variable and their updated values.

The estimated Young's Modulus result is then used for estimating the ultimate compressive strength of the bridge's concrete:

$$f_c = \left[\frac{E_c}{33\rho^{1.5}}\right]^2 = \left[\frac{4100000}{33 \times 145^{1.5}}\right] = 4.57\ ksi, \tag{45.2}$$

Then, the 28-day compressive strength of the concrete is obtained by knowing the age of bridge which is 40 years:

$$f_c = \frac{4 + 0.85t}{t} f_c = \frac{4 + 0.85 \times 60 \times 365}{60 \times 365} 4.57 = 3.88\ ksi \tag{45.3}$$

Yield strengths of unknown reinforcing steel used in the bridge is estimated by considering the date of bridge construction and the type and shape of reinforcing steel. Table 45.4 provides a list of the type of reinforcing steel and bridge construction's date with corresponding yield strength of reinforcing steel. Based on the age of the bridge, the yield strengths of reinforcing steel is also estimated to be 40 ksi from the table. An elastic modulus of $E_s = 29{,}000$ ksi is used for reinforcing steel material.

Based on the determined values for material properties and reinforcing steel, the bending capacity is computed as follows:

$$f_c = 3.88\ ksi,\ f_y = 40\ ksi, a = \frac{A_s f_y}{\beta f_c b} = \frac{1.0 \times 12 \times 40}{0.85 \times 2.5 \times 91} = 1.56\ in, C = \varphi A_s f_y \left(d - \frac{a}{2}\right) = 1328\ ft.kip/ft \tag{45.4}$$

The model of the bridge subjected to dead loads including structural components is analyzed to determine bending moment DL. For design load rating factor calculation, LRFD design live load of HS-20 according to AASHTO Manual

Table 45.4 Yield strengths of unknown reinforcing steel provided by AASHTO Manual for Bridge Evaluation [6]

Type of reinforcing steel	Yield strength f_y (*ksi*)
Unknown steel constructed prior to 1954	33
Structural grade	36
Billet or intermediate grade, grade 40, or unknown steel constructed during or after 1954	40
Rail or hard grade, grade 50	50
Grade 60	60

for Bridge Evaluation [6] is then applied to the model of bridge to compute bending moment *LL*. Note that a dynamic impact factor of 33% is considered. By considering the obtained value for the capacity, *C*, and load effects, the load rating factor is calculated using Eq. (45.1) as:

$$RF = \frac{1328 - 1.25 \times 312.5}{1.75 \times 215} = 2.47 \tag{45.5}$$

Note that following an analytical load rating procedure that is described in MBE and relies on estimating the actual load capacity by adopting usually conservative assumptions and approximate formulas for load distribution, the load rating factor of the bridge is found to be 1.02. This indicates that the load rating based on vibration testing and FE model updating can reveal the reserve capacity of the bridge, which can avoid posting of structure.

45.5 Conclusions

This paper describes a load rating procedure for reinforced concrete T-beam bridges and an experimental program that illustrates the application of the proposed procedure into an in-service bridge structure. The method first requires the identification of modal parameters of a given structure using an ambient vibration testing. The obtained results are used to calibrate an initial 3D finite element model and then used for load rating. A hybrid optimization procedure which employs a genetic algorithm and a gradient-based optimization scheme is used in the calibration of FE model. It is shown that ambient vibration testing combined with FE model updating provides a relatively cost-effective and non-intrusive method for bridge load rating. The results also reveal that conventional design office rating procedures can be overly conservative.

Acknowledgements This material is based upon the work supported by the Virginia Department of Transportation. The authors would like to thank Dr. Bernard L. Kassner of Virginia Transportation Research Council for his helps in conducting the vibration testing of the bridge.

References

1. Brownjohn, J.M., Moyo, P., Omenzetter, P., Lu, Y.: Assessment of highway bridge upgrading by dynamic testing and finite-element model updating. J. Bridg. Eng. **8**(3), 162–172 (2003)
2. Wang, L., Chan, T.H.: Review of vibration-based damage detection and condition assessment of bridge structures using structural health monitoring. In: QUT Conference Proceedings (2009)
3. Lee, J.J., Yun, C.B.: Damage diagnosis of steel girder bridges using ambient vibration data. Eng. Struct. **28**(6), 912–925 (2006)
4. Gheitasi, A., Ozbulut, O.E., Usmani, S., Alipour, M., Harris, D.K.: Experimental and analytical vibration serviceability assessment of an in-service pedestrian bridge. Case Stud. Nondestruct. Test. Eval. **6**, 79–88 (2016)
5. Gul, M., Catbas, F.N.: Damage assessment with ambient vibration data using a novel time series analysis methodology. J. Struct. Eng. **137**(12), 1518–1526 (2010)
6. AASHTO: The Manual for Bridge Evaluation. American Association of State Highway and Transportation Officials, Washington, DC (2011)
7. Alipour, M., Harris, D.K., Ozbulut, O.E.: Vibration testing for bridge load rating. In: Dynamics of Civil Structures, vol. 2, pp. 175–184. Springer International Publishing, New York (2016)
8. Brincker, R., Ventura, C., Andersen, P.: Damping estimation by frequency domain decomposition. In: 19th International Modal Analysis Conference (2001)
9. Shafiei Dizaji, M., Alipour, M., Harris, D.: Leveraging vision for structural identification – a digital image correlation based approach. In: International Digital Image Correlation Society Conference (iDICs), SEM Fall Conference, Philadelphia, PA, USA (8–11 Nov 2016)
10. Shafiei Dizaji, M., Harris, D., Alipour, M., Ozbulut, O.: En"vision"ing a novel approach for structural health monitoring – a model for full-field structural identification using 3D–digital image correlation. In: The 8th International Conference on Structural Health Monitoring of Intelligent Infrastructure, Bridbane, Australia (5–8 Dec 2017)

Chapter 46
Identifying Modal Characteristics of Reinforced Concrete Bridges Using Smartphones

Abdou K. Ndong, Osman E. Ozbulut, and Devin K. Harris

Abstract This paper explores the use of smartphones as vibration measurement device to identify modal properties of reinforced concrete (RC) bridges. Two in-service RC bridges are instrumented with both conventional accelerometers and smartphone accelerometers. One of the tested bridges is a simply supported RC T-beam bridge structure and the other one is a simply supported RC slab bridge with a skew angle of 15 degrees. The vibration testing includes both traffic-induced ambient excitations and impact hammer excitation. The natural frequencies of the bridges are identified from operational modal analysis using the data obtained from both conventional and smartphone accelerometers at a single point and a peak-picking technique. In addition, the modal properties of two bridges are extracted using data obtained from a dense network of sensors and by employing enhanced frequency domain decomposition method. To assess the correlation between the modal properties identified from smartphone data and data obtained from traditional sensor, statistical analyses are conducted. Results show that there is a good agreement between the modal characteristics extracted from smartphone and reference sensor data as well as those obtained from a dense instrumentation.

Keywords Modal identification · Vibration measurement · Ambient and impact · Accelerometers · Phones

46.1 Introduction

As the fundamental of dynamic analysis and finite element model updating, modal parameter identification plays an important role in structural health monitoring. For civil engineering structures that are usually large and complex, operational modal analysis (OMA) is an important technique used to estimate the modal parameters. OMA was developed in the 1970s [1] and has attracted significant attention in the field of engineering owing to its use in marine platforms, buildings, towers, bridges, and other structures [2]. This technique is based on the measurement of the vibration response under normal operating conditions and extracts modal characteristics such as natural frequencies, mode shapes and damping ratios without requiring the measurement of input forces. The excitation for OMA may come from various sources, such as wind, vehicular traffic, and the operation of machines on the structure [3]. The modal parameter estimation in OMA can be performed either in time domain using the correlation functions among output responses or in frequency domain using the output power spectral density functions.

To overcome the expense and the time required to set the instrumentation of traditional accelerometers, new generation smartphones can actually be used to identify the modal parameters of existing large structures such as bridges and buildings. Smartphones can perform such a task since they are equipped with an internal accelerometer sensor, which can accurately measure vibration response in three directions at relatively high acquisition rates. Though the sampling rate is usually low compared to the data obtained from traditional accelerometers and data acquisition systems, a few recent studies have shown that smartphones can provide good estimates for the modal properties of structures [4, 5].

In this study, operational modal analysis of two bridge structures, one reinforced concrete slab bridge and one reinforced concrete Tee-beam bridge, using smartphone measurements are explored. These bridges are both located in the city of Richmond, in Virginia, USA. The bridges are instrumented with both conventional accelerometers and smartphones during vibration tastings. The testing procedure primarily consists of exciting the structure through passing traffic or an impact hammer, measuring the acceleration response using both smartphones and traditional accelerometers, and analyzing the obtained data for modal identification. The data analysis consists of obtaining the Power Spectral Density (PSD), which describes the power present in a signal as a function of frequency, of the measured signals and extracting natural frequencies

A. K. Ndong · O. E. Ozbulut (✉) · D. K. Harris
Department of Civil and Environmental Engineering, University of Virginia, Charlottesville, VA, USA
e-mail: akhaloo@gmu.edu

S. Pakzad (ed.), *Dynamics of Civil Structures, Volume 2*, Conference Proceedings of the Society for Experimental Mechanics Series, https://doi.org/10.1007/978-3-319-74421-6_46

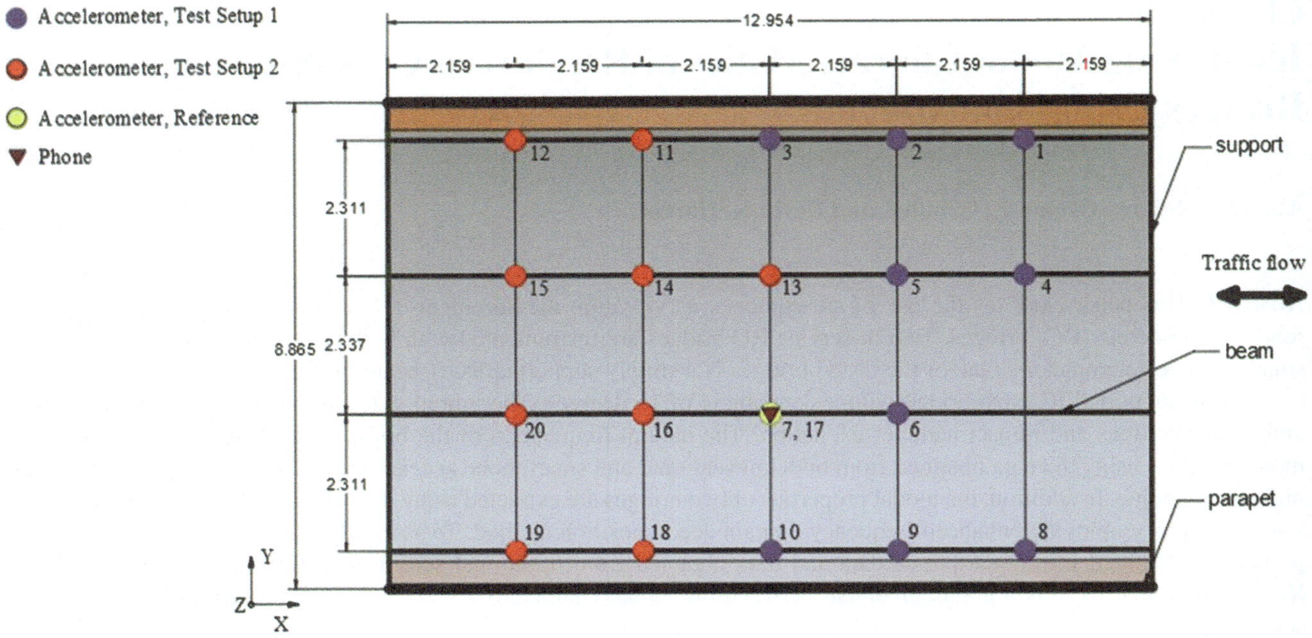

Fig. 46.1 Sensor instrumentation plan for one of the T-beam bridge span with two setups (all dimensions are in meters)

of the tested structures. The modal parameters obtained from both traditional accelerometer and smartphone measurements as well as those obtained from full instrumentation of bridges are compared. In addition, to better interpret the obtained results, statistical modelling techniques are used to further analyze the results.

46.2 Description and Instrumentation of Tested Bridges

46.2.1 T-Beam Bridge

The first bridge for the field testing is a concrete T-beam bridge, named as Flat Creek and built in 1957. It is a five-span simply-supported bridge located in Rockingham County, Virginia, USA. Each span of the bridge has a total length of 12.954 m and a total width of 8.865 m, and consists of four longitudinal T-beams. Each T-beam has vertical rectangular stem with a width of 0.825 m and a thickness of 0.356 m, and a wide top flange of 0.191 m thick. The wide top flange is the transversely reinforced deck slab and the riding surface for the traffic. The bridge shown is skewed at an angle of 0 degrees to the main road. The ambient vibration testing of one of the spans are conducted in this study.

The dynamic bridge assessment procedure involves the attachment of nine accelerometers underneath of the span of the bridge at 19 measurement points in two set-ups as shown in Fig. 46.1. The connection of the accelerometers to the girders of the deck is performed by means of metallic plates bonded to the surface of the concrete. Note that two common sensors are used as reference in each set-up. The uniaxial accelerometers with a measuring range of ±5 g are used. The response of the bridge to ambient excitations is measured. Ambient vibrations are generated by the passing traffic, wind and walking people and recorded for a total of 15 min. For impact excitation, a large sledge impulse hammer with a force capacity of 22.2 kN is used. Three impact locations are chosen and excited by the impact hammer for five repetitions and for a duration of 15 s. Data is collected with a sampling frequency of 500 Hz. The ambient vibration is simultaneously recorded for 15 min.

46.2.2 Slab Bridge

The second bridge selected for the field testing is a concrete slab bridge, named as Smacks Creek and built in 1964. The three-span simply-supported bridge is located in Amelia County, Virginia, USA. Each span of the bridge has a total length of

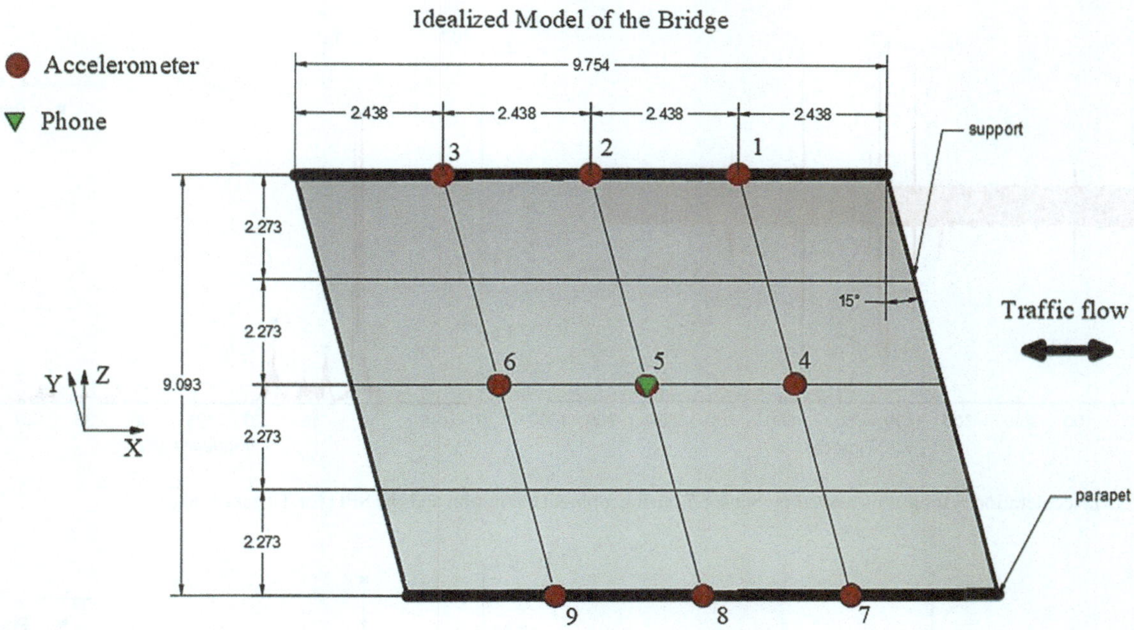

Fig. 46.2 Sensor instrumentation plan for one of the slab bridge span with two setups (all dimensions are in meters)

9.8 m. The thickness of the concrete slab is 0.53 m. The deck slab has voids with a diameter of 0.3 m and spacing of 0.45 m in the direction of traffic. The bridge shown is skewed at an angle of 15 degrees to the main road. The forced and ambient vibration testing of one of the spans are conducted in this study.

The dynamic bridge assessment procedure involves the attachment of nine accelerometers underneath of the span of the bridge at nine measurement points in one set-up as shown in Fig. 46.2. Note one sensor is used as reference in the test setup with Smartphone. The uniaxial accelerometers with a measuring range of ±5 g are used. The response of the bridge to ambient excitations or an impact excitation generated by a modal impact hammer is measured. Ambient vibrations are generated by the passing traffic, wind and walking people and recorded for a total of 15 min. For impact excitation, a large sledge impulse hammer with a force capacity of 22.2 kN is used. Three impact locations are chosen and excited by the impact hammer for five repetitions and for a duration of 15 s. Data is collected with a sampling frequency of 500 Hz.

46.3 Data Collection Using Smartphone

The same data collection technique with built-in accelerometer of a smartphone is used during both bridge testing. Most of current smartphone devices include an accelerometer sensor. Using a smartphone, vibration data can be collected and stored in the device, and can be analyzed in real time or offline. In this study, an iOS-based smartphone that has a sensor capable of measuring accelerations in *x*-, *y*-, and *z*-axis is used. An iOS app (VibSensor) available at the iOS App Store is used for data collection. The app enables quantitative accelerometer and vibration data measurements with easy collection, data storage, and transfer of data. It gives the user following five tools:

- Live Display, which allow you to see vibration data in real time,
- Acquisition, which is timed or vibration activated acquisition, with settable delay, duration, and trigger level. It can collect raw accelerometer data for up to 10 min (600 s) at sampling rates up to 100 Hz.
- Data Storage, where acquired collections are stored on the device with date and time stamp for later retrieval. Collections can be named for easy identification.
- Analysis, that is each collection can be viewed to see the raw accelerometer data, processed vibration and calculated power spectral densities, both in graphical and report format. Units can be selected to be either *g* or m^2/s.
- Email Access, which allows users to transfer raw or processed data in text (csv or tab-delimited) or MATLAB format via e-mail.

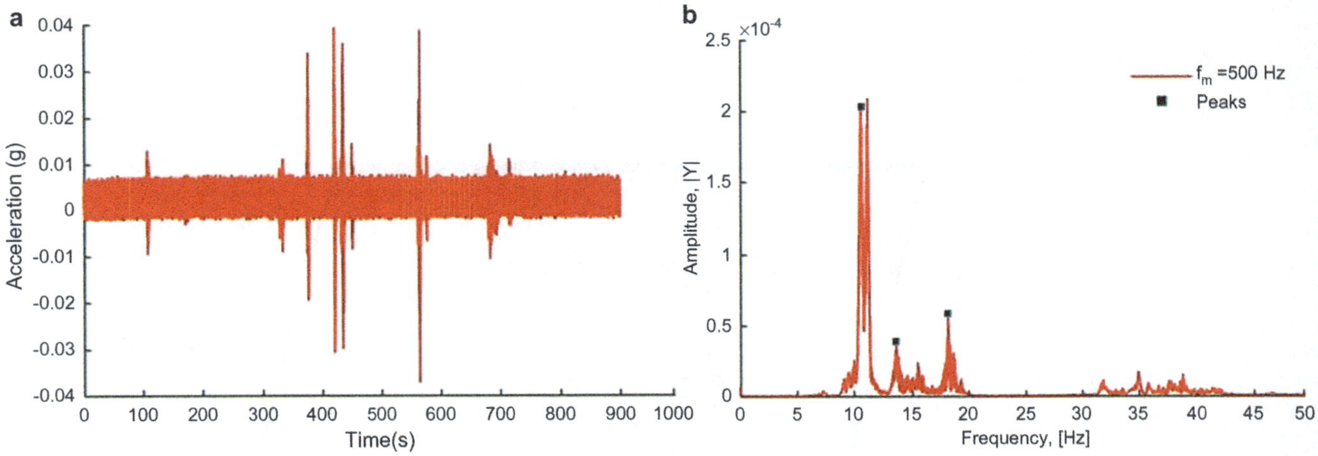

Fig. 46.3 (**a**) Acceleration response measured by Sensor 7 during ambient excitation and (**b**) SPD for T-beam bridge

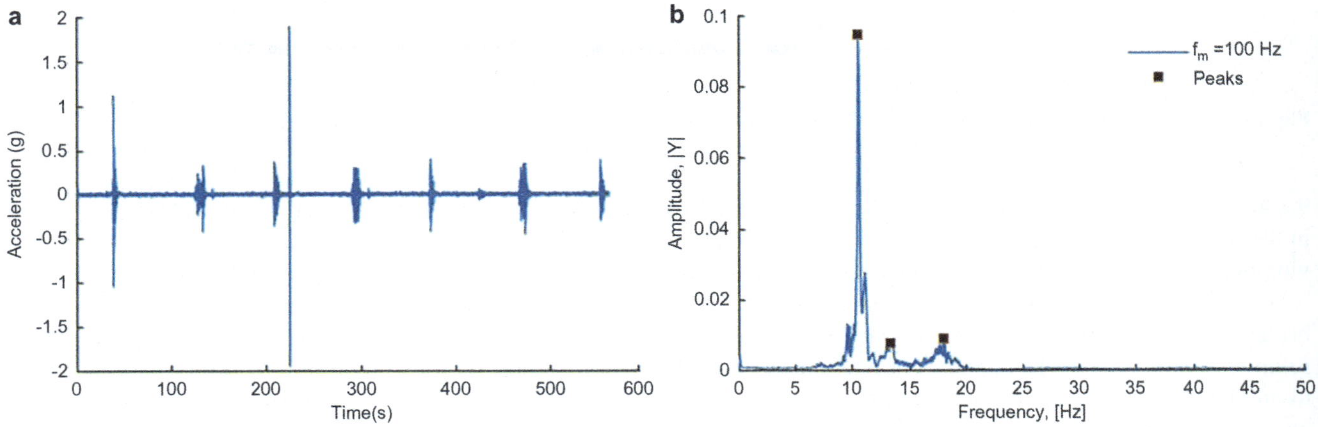

Fig. 46.4 (**a**) Acceleration response measured by smartphone during ambient excitation and (**b**) SPD for T-beam bridge

During field testing, the smartphone is placed on the bridge deck being at the same location of the reference sensor No. 7 that is attached underneath the T-beam bridge and sensor No. 5 for the slab bridge. Data from smartphone sensor is collected at a sampling rate of 100 Hz.

46.4 Modal Identification

46.4.1 T-Beam Bridge

The collected acceleration data is processed to obtain power spectral density of the measured signals and a peak-picking technique is used to identify the first three natural frequencies. Figures 46.3 and 46.4 the raw acceleration data measured by reference accelerometers and smartphone, respectively and corresponding PSD plots for the T-beam bridge during ambient vibrations. The same plots for impact excitations are shown in Figs. 46.5 and 46.6. Table 46.1 summarizes the natural frequencies obtained from ambient vibrations and Impact Hammer excitations, respectively for the T-beam bridge. Furthermore, note that the computed PSDs show that the power present in the smartphone data reaches higher values than those in the reference sensor data. This is due to the fact that traditional accelerometer is placed underneath the bridges whereas the smartphone is placed on top of the deck, receiving more excitation and noise of the traffic.

In order to evaluate the accuracy of the natural frequencies extracted from smartphone data and provided in Table 46.1, the modal properties of the bridge are also computed using the data obtained from 19 sensors, i.e. full instrumentation, and using Enhanced Frequency Domain Decomposition (EFDD) method [6]. The results are provided for both ambient excitation and

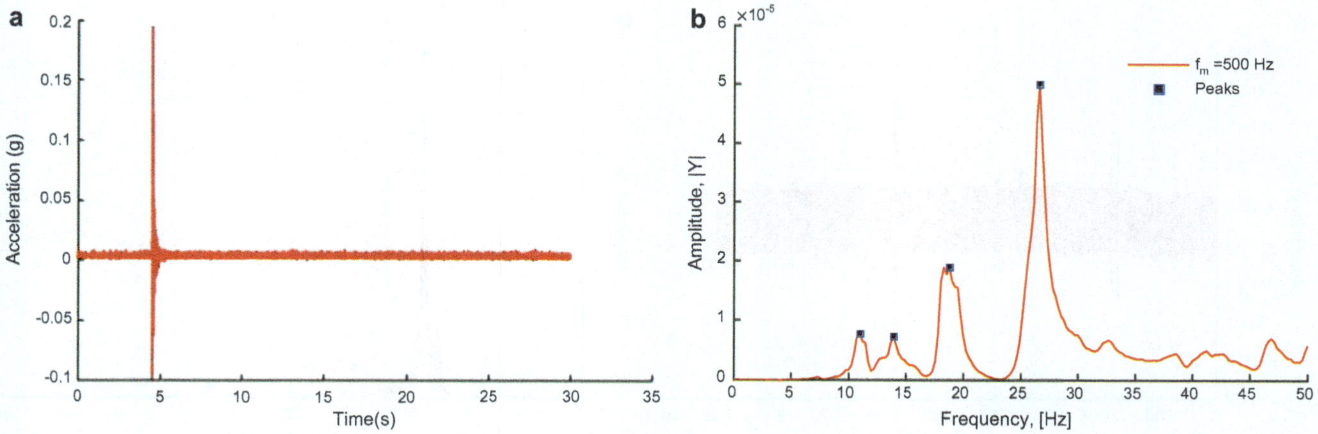

Fig. 46.5 (**a**) Acceleration response measured by Sensor 7 during impact excitation and (**b**) SPD for T-beam bridge

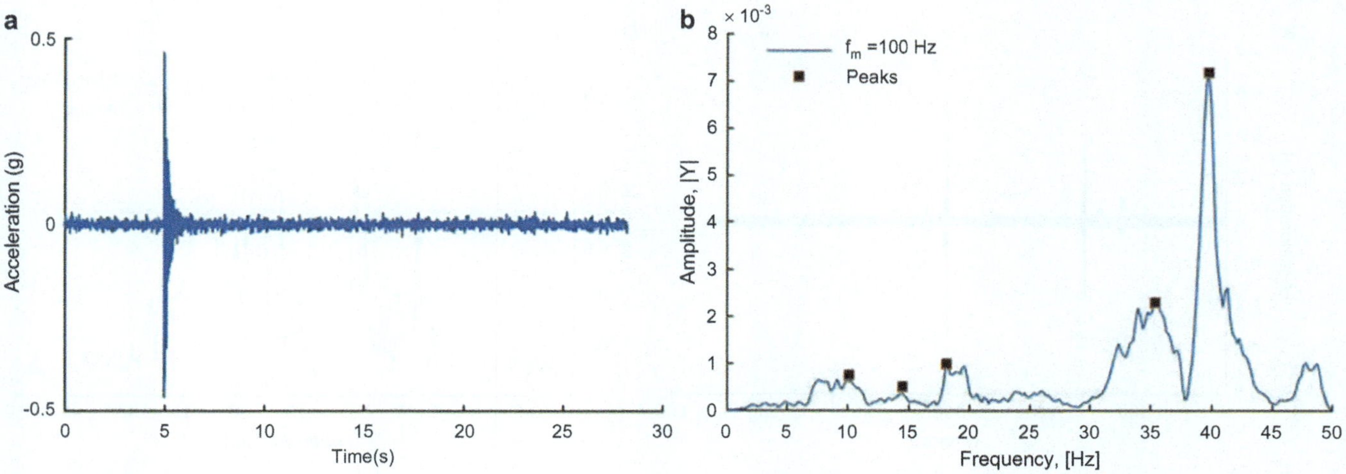

Fig. 46.6 (**a**) Acceleration response measured by smartphone during impact excitation and (**b**) SPD for T-beam bridge

Table 46.1 Natural frequencies obtained from ambient and impact excitations of T-beam bridge

	Modal frequencies (Hz)			
	Ambient test		Impact hammer test	
Mode	Sensor	Smartphone	Sensor	Smartphone
1	10.51	10.53	10.99	10.16
2	13.84	13.40	13.92	14.65
3	18.26	17.91	18.80	18.16

Table 46.2 Natural frequencies obtained from ambient and impact excitations using EFDD

	Modal frequency (Hz)	
Mode	Ambient excitation	Impact hammer excitation
1	10.74 (−2.0%)	10.76 (−5.6%)
2	13.77 (−2.7%)	13.94 (5.1%)
3	18.54 (−3.3%)	18.40 (−1.3%)

impact hammer excitations in Table 46.2. The percent differences between the natural frequencies obtained from smartphone data and from full instrumentation are also provided in Table 46.2. It can be seen that the percent differences between the smartphone and full instrumentation results for the first three natural frequencies range from 2.0% to 3.3% for the ambient excitation tests and range from 1.3% to 5.6% for the impact hammer test.

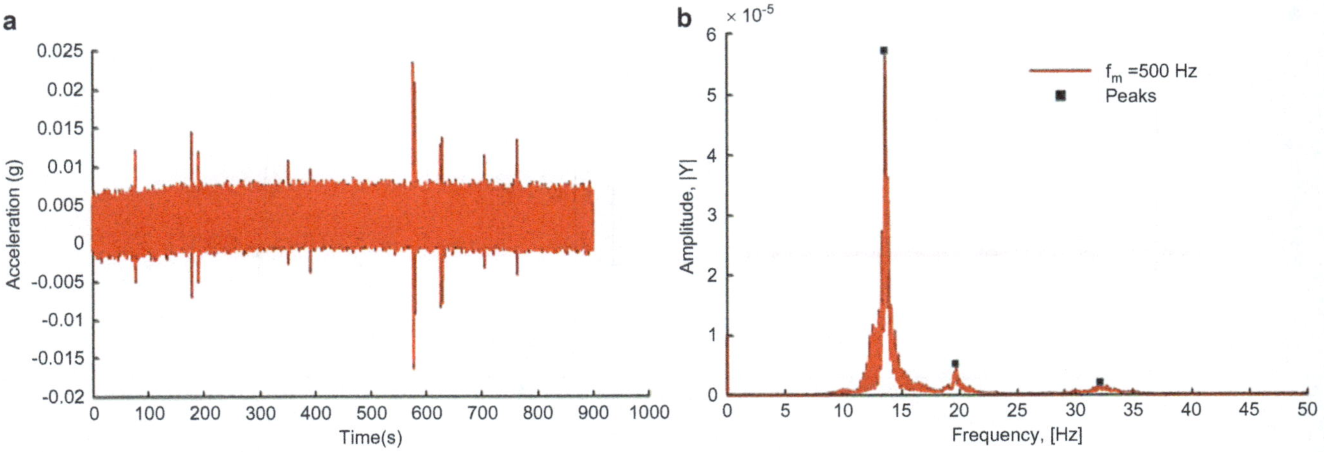

Fig. 46.7 (**a**) Acceleration response measured by Sensor 5 during ambient excitation and (**b**) SPD for slab bridge

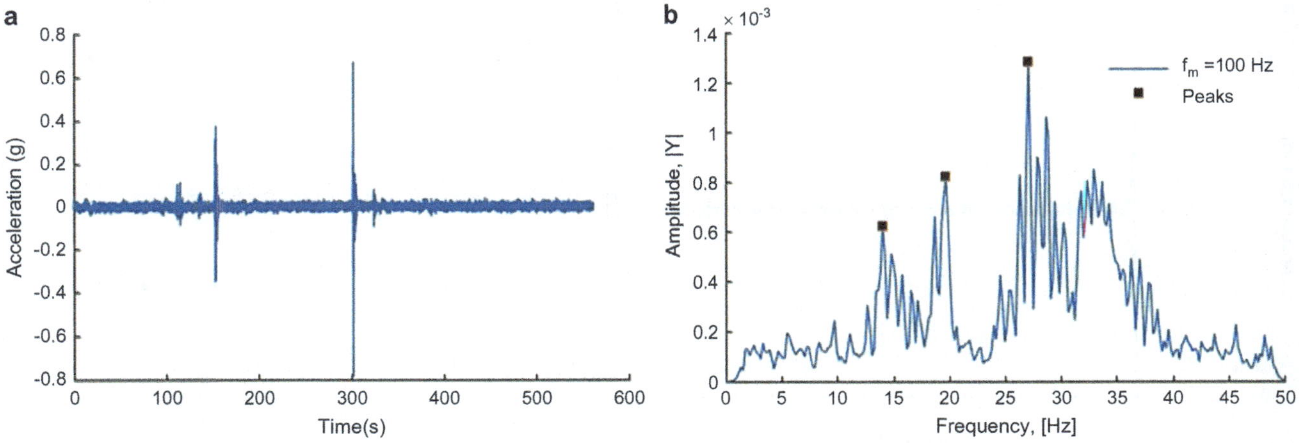

Fig. 46.8 (**a**) Acceleration response measured by smartphone during ambient excitation and (**b**) SPD for slab bridge

46.4.2 Slab Bridge

Similarly, Figs. 46.7 and 46.8 the raw acceleration data measured by reference accelerometers and smartphone, respectively and corresponding PSD plots for the slab bridge during ambient vibrations. The same plots for impact excitations are shown in Figs. 46.9 and 46.10. Table 46.3 summarizes the natural frequencies obtained from ambient vibrations and Impact Hammer excitations, respectively for the slab bridge.

Table 46.4 provides the modal properties of the slab bridge extracted from the data obtained from full instrumentation. Again, both ambient excitation and impact hammer excitations are considered as the source of the vibrations. The percent differences between the natural frequencies obtained from smartphone data and full instrumentation are also provided in Table 46.4. It can be seen that the percent differences between the smartphone and full instrumentation results for the first three natural frequencies range from 0.30% to 15.8% for the ambient vibration tests and 0.35% and 3.85% for the impact hammer tests.

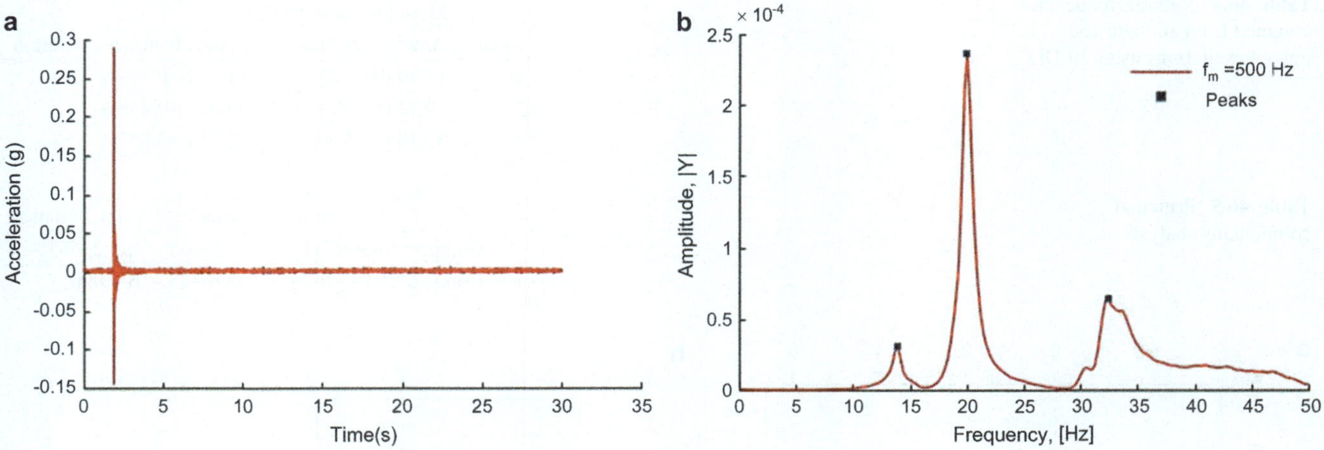

Fig. 46.9 (**a**) Acceleration response measured by Sensor 5 during impact excitation and (**b**) SPD for slab bridge

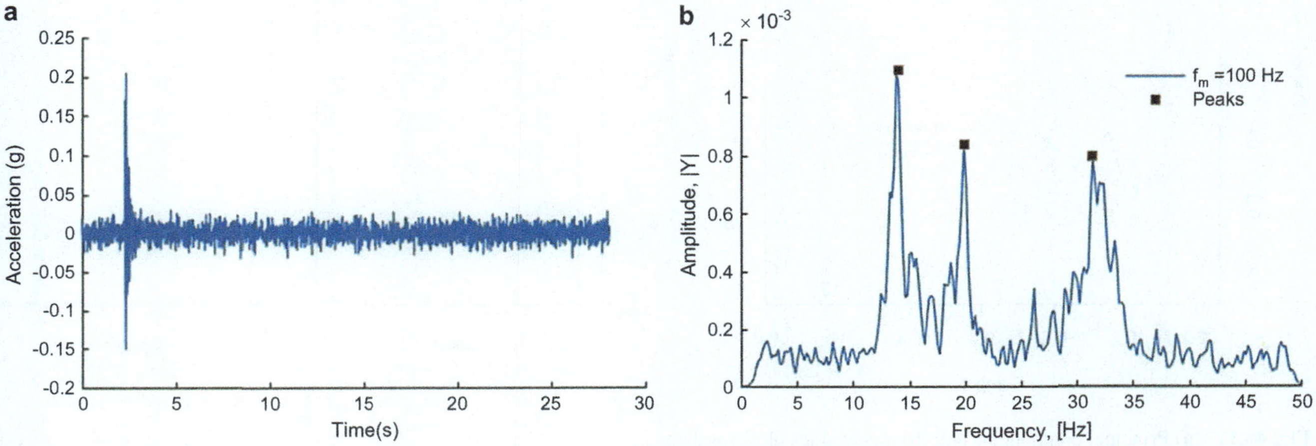

Fig. 46.10 (**a**) Acceleration response measured by smartphone during impact excitation and (**b**) SPD for slab bridge

Table 46.3 Natural frequencies obtained from ambient and impact excitations of slab bridge

	Modal frequencies (Hz)			
	Ambient test		Impact hammer test	
Mode	Sensor	Smartphone	Sensor	Smartphone
1	13.70	13.92	14.16	14.06
2	19.74	19.78	20.02	19.92
3	32.23	27.1	32.23	31.45

46.5 Statistical Analysis of Modal Natural Frequencies

46.5.1 Principal Component Analysis of the Obtained Modal Frequencies

In this section, the Principal Component Analysis (PCA), which is a statistical technique used to emphasize variation and reveal strong patterns in a dataset, is performed. Here, the natural frequencies obtained from the data collected by the smartphone and those collected by the reference sensor are analyzed. Two dimensional sets of data that includes natural frequencies extracted for both bridges using data obtained from smartphone and reference sensor are considered.

In order to evaluate the relation between the selected two variables, i.e. frequencies obtained from smartphone and reference sensor, PCA computes new variables called principal components which are obtained as linear combinations of the original variables. The first principal component is required to have the largest possible variance. The second component is computed under the constraint of being orthogonal to the first component and to have the largest possible inertia. The other principal components are computed likewise. As can be seen from Table 46.5, the eigenvalues of components 1 and

Table 46.4 Natural frequencies obtained from ambient and impact excitations using EFDD

Mode	Modal frequency (Hz)	
	Ambient excitation	Impact hammer excitation
1	13.79 (9.42%)	14.11 (−0.35%)
2	19.72 (0.30%)	20.01 (−0.45%)
3	32.18 (−15.8%)	32.71 (−3.85%)

Table 46.5 Principal components analysis

	Eigenvalues	Variance	Cum. variance
Comp.1	1.984733	99.23665	99.23665
Comp.2	0.015267	0.76335	100.000

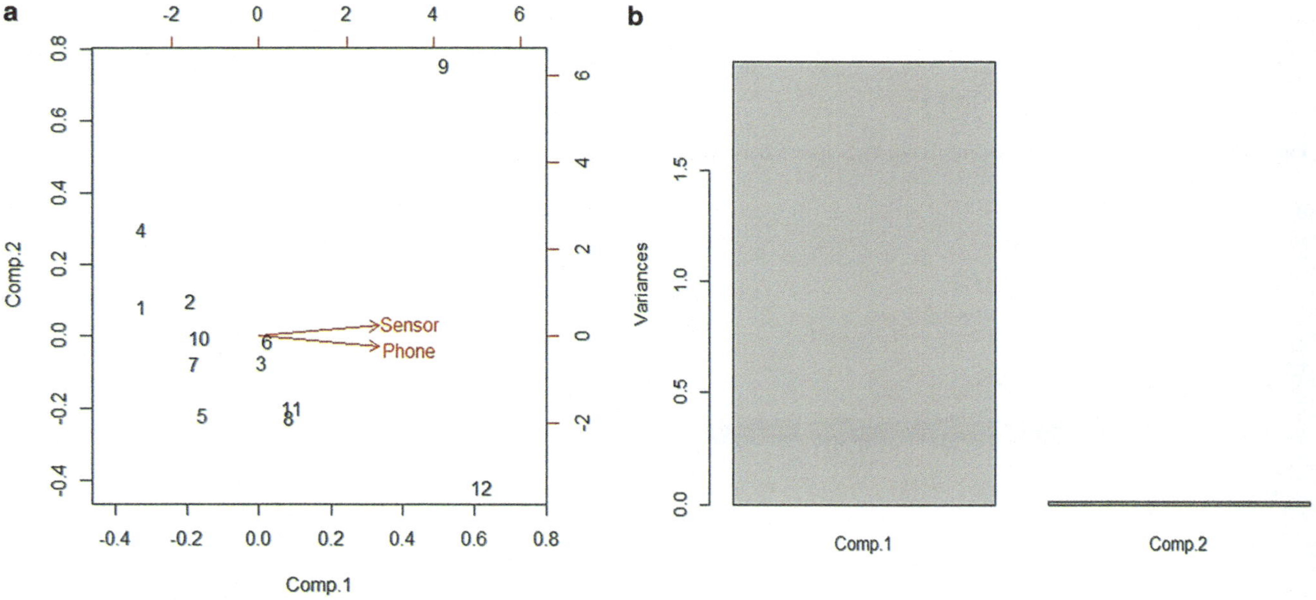

Fig. 46.11 (**a**) Principal components and (**b**) eigenvalues plots for data

2 are 1.985 and 0.015, respectively. This indicates that the eigenvector corresponding to component 1 is the dimension that has almost all the information regarding the natural frequencies, however, the second component which has a very small eigenvalue can be discarded. Therefore, only one component will be considered in the data set.

Figure 46.11 illustrates the principal components and variances obtained for the natural frequencies extracted from ambient vibration and impact hammer tests. As shown in Fig. 46.11a, the directions of the components are not orthogonal indicating that they are highly correlated. The same can also be explained by the variance plot in Fig. 46.11b. An eigenvalue that is greater than 1 indicates that PCs account for more variance than accounted by one of the original variables in standardized data. This is commonly used as a cutoff point for which PCs are retained.

Note that, in general, a good dimension reduction is achieved when the first few PCs account for a large proportion of the variability (80–90%). Figure 46.12a shows the percentage of variances, which is the ratio of the variance accounted for by each component to the total variance in all of the variables. Since the percentage of variances between frequencies extracted from the data using the smartphone and the reference sensor account for more than 90%, there exists a strong correlation between these two variables. The second component (PC2) is completely correlated with the first component, and explains the same variation as the first component (see Fig. 46.12b).

46.5.2 Multivariate Linear Regression Model of the Modal Frequencies

A linear regression model is also developed to explore how the natural frequencies obtained from the two sources, smartphone and reference sensor, are related. Regression analysis is generally used to predict the value of one or more responses from a set of predictors (Table 46.6). It can also be used to estimate the linear association between the predictors and responses. Here,

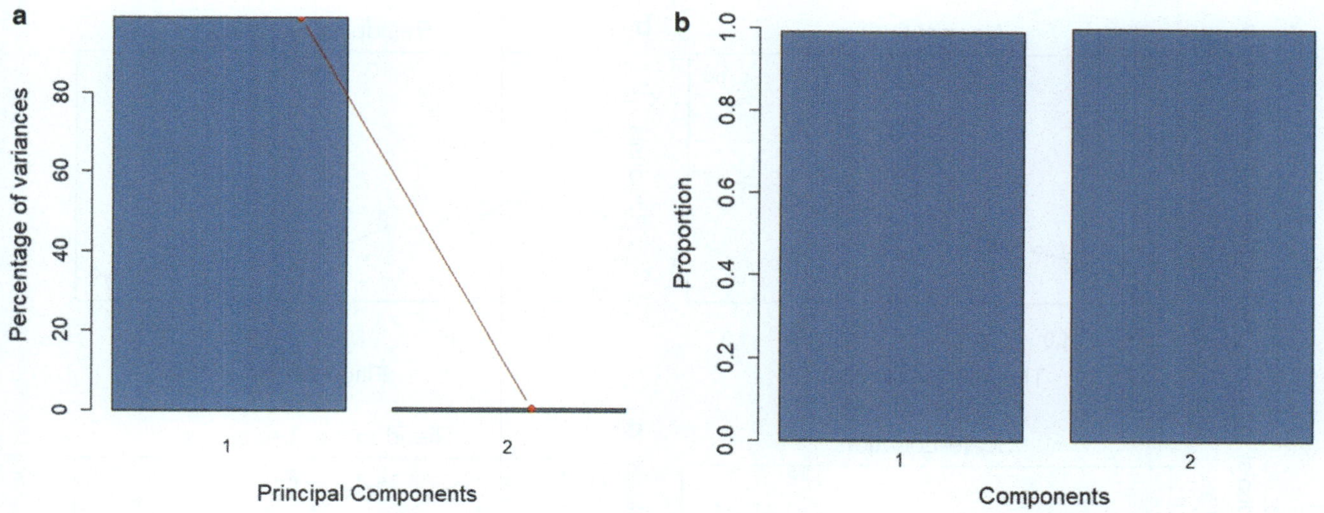

Fig. 46.12 (**a**) Percentage of variance and (**b**) proportion of variance for principal components

Table 46.6 Linear regression analysis-coefficients

Min	Estimate	Std. error	*t* value	Pr(>|*t*|)
Intercept	−1.53341	1.16810	−1.313	0.219
Phone	1.12207	0.06272	17.889	6.37e-09

Residual standard error: 1.331 on 10 degrees of freedom
Multiple R-squared: 0.9697, adjusted R-squared: 0.9667
F-statistic: 320 on 1 and 10 DF, p value: 6.367e-09

the R-squared, which is a measure of the linear relationship between predictor variable (natural frequencies from smartphone) and target variable (frequencies from reference sensor), is computed to be 0.9697. That indicates the frequencies obtained from smartphone and reference sensor are highly correlated, which can also be seen from the Q–Q plot in Fig. 46.13a. The Q–Q plot shows normally distributed residuals except the point 9 falling away from the actual data and being the only point not fitting this data. Figure 46.13b–d illustrates residuals versus fitted, residual versus leverage and scale-location plots of the same data. The results presented in these plots reinforce and prove that point 9 is an outlier. Note that point 9 corresponds the third natural frequency of the slab bridge obtained from the ambient excitation data recorded with smartphone. Nevertheless, the overall analysis reveals the high correlation between two data sets.

46.6 Conclusions

The vibration response of two in-service RC bridges are collected using both traditional accelerometers and smartphone sensors. The bridges are excited by both ambient sources (mainly passing traffic) and an impact hammer. The measured acceleration data and the computed natural frequencies indicate that both bridges are well excited by the traffic on the bridges. The results revealed that the first three natural frequencies of the tested bridges can be quite accurately extracted using the data collected from a smartphone sensor. This can provide considerable advantages in modal identification of such structures due to the ease of installation of a smartphone, which only requires a tape to hold it tight to the pavement. The natural frequencies obtained from the data collected from smartphone can also be used to compute rating factor of these type of bridges using a vibration-based methodology [7, 8]. One of the limitation of using the smartphone in this study is that the available app recorded accelerations with a sampling frequency of 100 Hz, indicating that any modal frequency higher than the Nyquist frequency (50 Hz) cannot be captured. In addition, the statistical models developed in the paper reveal how well correlated the frequencies obtained from traditional sensor and smartphone sensor.

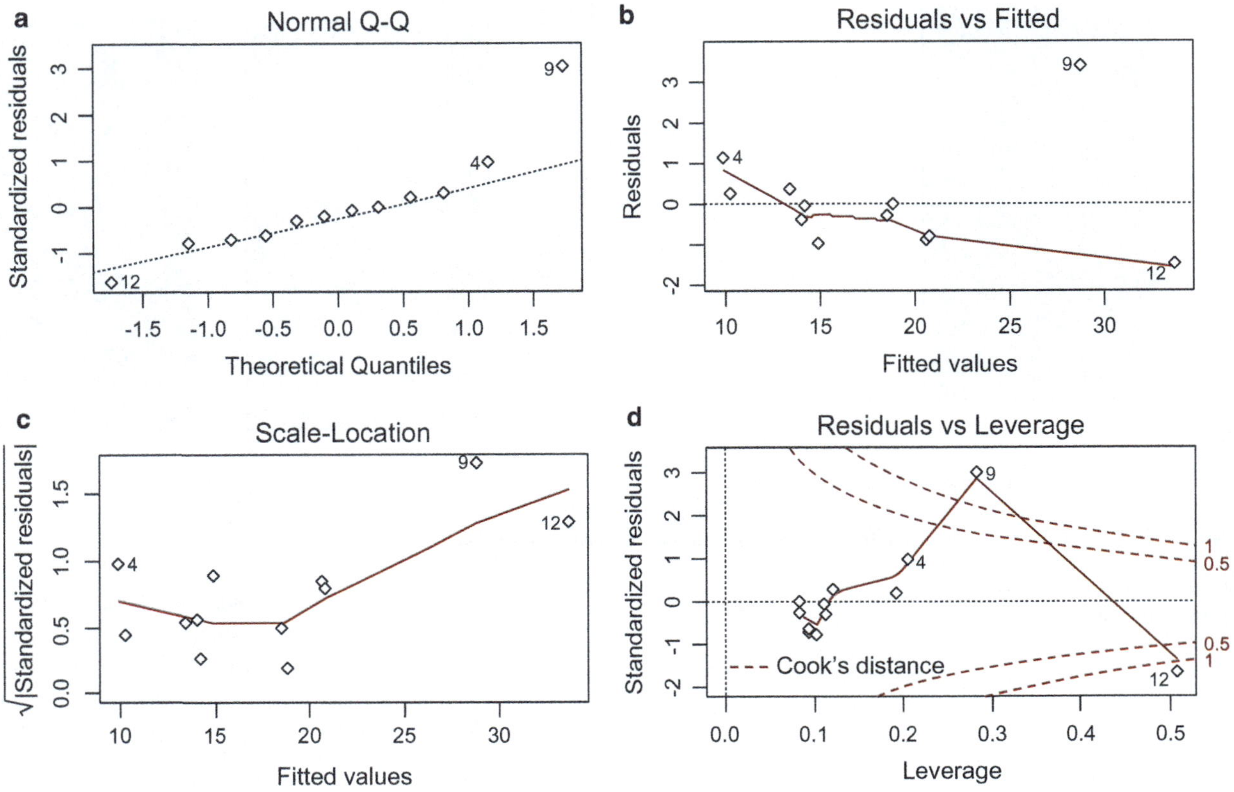

Fig. 46.13 Linear regression model: (**a**) Residuals vs fitted, (**b**) normal Q–Q, (**c**) scale location, (**d**) residuals vs leverage

Acknowledgements This material is based upon the work supported by the Virginia Department of Transportation.

References

1. Vu, V.H., Thomas, M., Lakis, A.A., Marcouiller, L.: Identification of modal parameters by operational modal analysis for the assessment of bridge rehabilitation. In: 2nd, Proceedings of the International Operational Modal Analysis Conference (2007)
2. Zhang, L., Brincker, R., Andersen, P.: An overview of operational modal analysis: major development and issues. In: 1st, International Operational Modal Analysis (2005)
3. Alwash, M., Sparling, B.F., Wegner, L.D.: Influence of excitation on dynamic system identification for a multi-span reinforced concrete bridge. Adv. Civ. Eng. **2009**, 1–19 (2009)
4. Castellanos, S., Marulanda, J., Preciado, M., Cruz, A., Thomson, P.: Identification of the operational frequencies of 300+ bridges using smartphones. In: SPIE Smart Structures and Materials + Nondestructive Evaluation and Health Monitoring, April 8, 2016, pp. 98040C–98040C. International Society for Optics and Photonics. Bellingham, Washington, USA
5. Ozer, E., Feng, M.Q.: Synthesizing spatiotemporally sparse smartphone sensor data for bridge modal identification. Smart Mater. Struct. **25**(8), 085007 (2016)
6. Brincker, R., Ventura, C., Andersen, P.: Damping estimation by frequency domain decomposition. In: 19th International Modal Analysis Conference (2001)
7. Bagheri, A., Alipour, M., Usmani, S., Ozbulut, O.E., Harris, D.K.: Structural stiffness identification of skewed slab bridges with limited information for load rating purpose. In: Dynamics of Civil Structures, Volume 2: Proceedings of the International Modal Analysis Conference (IMAC)-XXXV, pp. 243–249, Garden Grove, CA (January 2017)
8. Alipour, M., Harris, D.K., Ozbulut, O.E.: Vibration testing for bridge load rating. In: Dynamics of Civil Structures, Volume 2: Proceedings of the International Modal Analysis Conference (IMAC)-XXXIV, pp. 175–184, Orlando, FL (January 2016)

Chapter 47
Model Updating and Damage Assessment of a RC Structure Using an Iterative Eigenvalue Problem

Michele Tondi, Seyedsina Yousefianmoghadam, Andreas Stavridis, Babak Moaveni, and Marco Bovo

Abstract This study discusses the model updating performed on a two-story reinforced-concrete masonry-infilled building using the dynamic response obtained during ambient vibration measurements. Five infills were demolished at four stages creating four damage states. The numerical model of the building is developed based on the obtained material properties and geometry. The model updating process minimizes a target function defined as the least-square optimization of an inverse eigenvalue problem. The reference model is then updated using the experimental results in each stage of damage to assess the damage. The dynamic properties obtained from the updated model match well their experimentally-estimated counterparts. Also, the model can reliably locate the damage introduced to the structure.

Keywords Damage Identification · Reinforced concrete building · Differential Evolution algorithm

47.1 Introduction

Structural health monitoring methods aiming at identifying damage as loss of stiffness using ambient- and forced-vibration measurements and linear FE models have been widely studied [1, 2]. However, few studies involve damage identification of a full-scale civil structure at different damage levels [3]. This study describes the results of a damage identification process on a building at several damage levels.

47.2 Test Structure, Damage States, and Dynamic Tests

The two-story structure, constructed in the 1920s in El Centro, CA, had a reinforced-concrete moment-frame structural system infilled with masonry and concrete panels along the exterior frames. The building had sustained damage during three major earthquakes by 1987, mainly in the first story, and had been repaired and retrofitted in the late 1980s. During the 2010 Baja California earthquake, the second story infills and frames were severely damaged and the building was slated for demolition. During the testing sequence, five infills in the second floor were removed at four stages introducing four levels of damage to the building to investigate the effect of damage evolution on its dynamic response. The initial damage state (DS0), reflects the condition of the structure prior to the experiment which includes the damage induced by the earthquakes mentioned above. The first induced damage state (DS1) corresponds to the condition after the removal of an exterior infill panel on the north side. The second damage state (DS2) resulted once an infill on the west side was removed, the third damage state (DS3) resulted from the removal of another infill on the west-side frame, and the fourth damage state (DS4) was introduced after the removal of two infills in the south-west corner. The locations and sequence of the removed walls are shown in Fig. 47.1.

M. Tondi · M. Bovo
University of Bologna, Bologna, Italy

S. Yousefianmoghadam (✉) · A. Stavridis
University at Buffalo, Buffalo, NY, USA
e-mail: seyedsin@buffalo.edu

B. Moaveni
Tufts University, Medford, MA, USA

S. Pakzad (ed.), *Dynamics of Civil Structures, Volume 2*, Conference Proceedings of the Society for Experimental Mechanics Series, https://doi.org/10.1007/978-3-319-74421-6_47

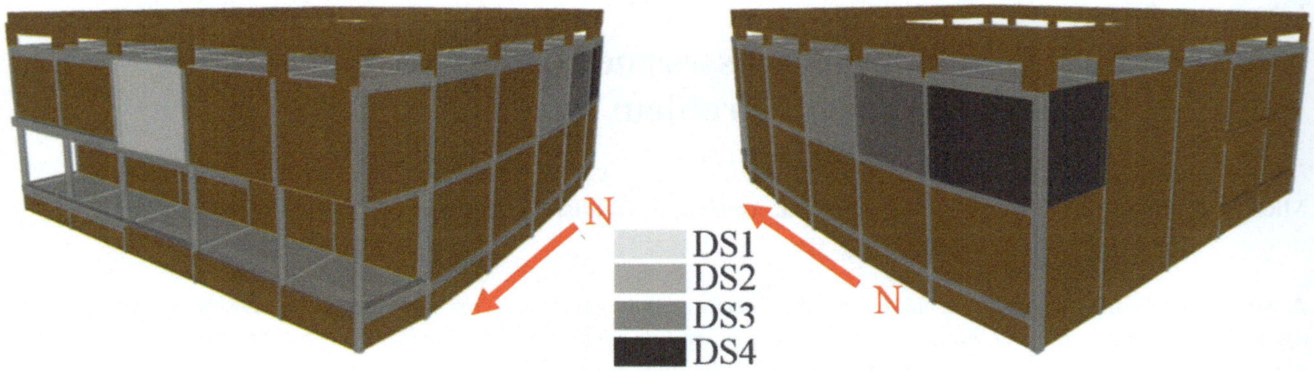

Fig. 47.1 Wall removal sequence and damage states

A total of 97 sensors including accelerometers, string potentiometers (string pots) and LVDTs were installed in the building to measure accelerations and displacements. A series of forced vibration tests were performed on the structure at all damage states to investigate its dynamic properties in the quasilinear and nonlinear ranges using a eccentric mass shaker. Ambient vibrations of the structure were also collected during the testing period. More information about the structure, instrumentation, and dynamic tests can be found in Ref. [4].

47.3 System Identification

The modal parameters of the test structure are estimated at each damage state from the ambient-vibration measurements. The system identification is performed using time-domain and frequency-domain methods, namely the natural excitation technique combined with the eigensystem realization algorithm (NExT-ERA) and the transfer functions between the input and output signals respectively.

47.4 Non-linear Model

The nonlinear model of the structure is developed based on the geometry and material properties obtained from the test structure according the condition prior to the last earthquake. The frames are modeled as beam elements having fiber sections according to the as-built frame detailing while the infills are modeled using truss elements as frame struts. A material law, proposed to simulate the behavior of the infills, consists of a multilinear envelope force-displacement response and a hysteretic rule. The former, referred to as simplified curve, is estimated based on a recently proposed assessment framework [5] for each infill panel. For this study, the model is used to extract the mass and stiffness matrices of the structure.

47.5 Model Updating and Damage Identification Process

The model updating is performed to obtain a model which represents the condition of the structure prior to the dynamic tests (DS1) which is referred to as the reference model in this study. This process is based on minimizing a target function defined as a least-square optimization of an inverse eigenvalue problem. The stiffness matrix of the structure is decomposed into $N + 1$ matrices, in which N is the number of the parameters to be identified. The model updating procedure is discussed in detail in Ref. [6]. The minimum value for the target function is achieved through the application of the Differential Evolution algorithm [7].

The abovementioned procedure has been applied for all of the damage states to obtain the stiffness parameters for the infills. Eight updating parameters, each corresponding to the averaged stiffness of infills on each side of the building in the two stories, have been considered in the model-updating process. The model-updating results in terms of frequencies and

Table 47.1 Comparison of the dynamic properties

Damage state	Frequency (Hz)	Experimental from ambient vibration tests	Numerical	Error (%)	MAC
DS1	Mode 1	2.26	2.26	0.18	0.99
	Mode 2	3.37	3.37	0.10	0.99
DS2	Mode 1	2.14	2.13	−0.22	0.99
	Mode 2	3.07	3.08	0.19	0.99
DS3	Mode 1	2.06	2.06	−0.32	0.98
	Mode 2	2.96	2.97	0.49	0.98
DS4	Mode 1	1.96	1.95	−0.50	0.98
	Mode 2	2.72	2.74	0.76	0.99

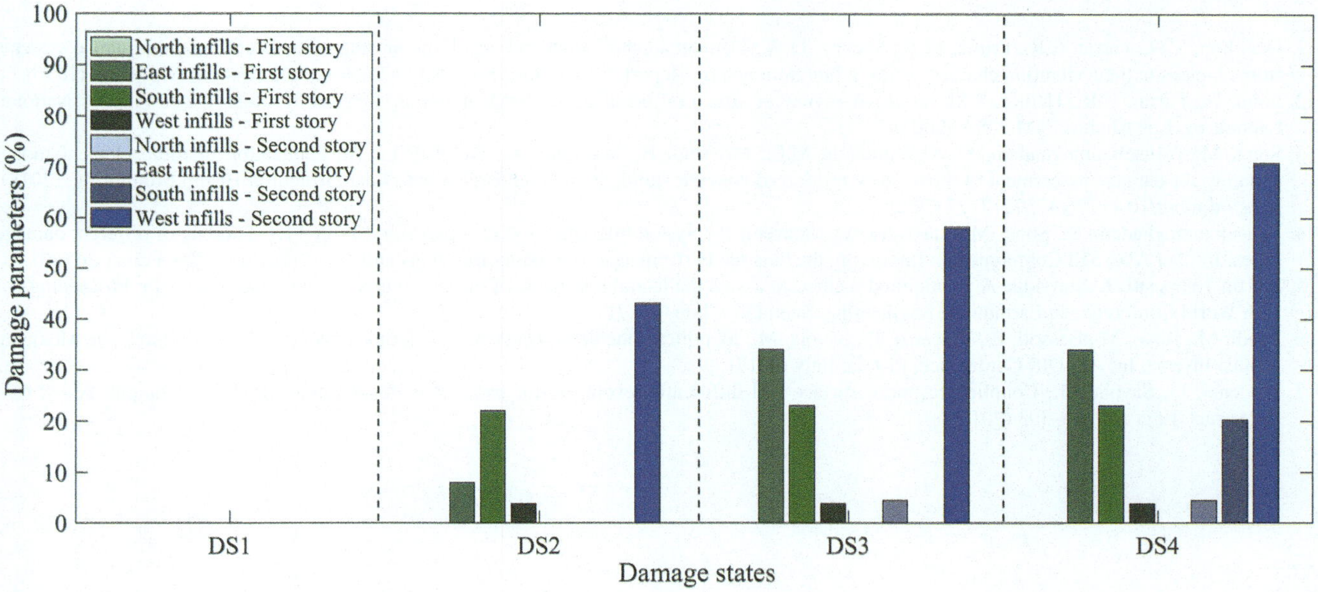

Fig. 47.2 Damage parameters at different damage sates

mode shapes for the two experimentally-identified modes are summarized in Table 47.1 It can be seen that the frequencies and mode shapes of the updated structure are in excellent agreement with their experimentally-identified counterparts with the maximum frequency difference of 0.76% and minimum MAC value of 0.98 at all damage states.

The results of the model updating procedure have been used to quantify the damage in terms of the stiffness reduction in the eight substructures. A damage function has been defined as follows:

$$E_j\left(DS_i\right) = \frac{a_j\left(DS_1\right) - a_j\left(DS_i\right)}{a_j\left(DS_1\right)} \cdot 100\% \tag{47.1}$$

in which $a_j(DS_1)$ is the value for the j-th parameter in the initial damage state, $a_j(DS_i)$ is the value for the same parameter at the i-th damage state, $E_j(DS_i)$ can range between 0% and 100%, in which 0% means undamaged and 100% means completely damaged.

According to Fig. 47.2, which shows the damage identified at each damage state, damage is observed in the infills along the west side in the second story. This is consistent with the wall removal process between DS2 through DS4. Moreover, damage is detected in the infills along the south side in the second story between DS3 and DS4 (resulting in about 20% stiffness reduction) which matches the demolition process. Figure 47.2 also indicates damage in the first story of the structure. This could be contributed to the effort of the model to compensate for the non-updating columns in the second story which might be damaged during the forced-vibration tests.

47.6 Conclusions

The model-updating methodology implemented here results in a model with frequencies and mode shapes in a good agreement with their experimental counterparts as the errors in the estimation of the natural frequencies are in all the cases lower than 1%. The analysis of the damage parameters shows how the procedure can capture the removal of the infill panels in the west and south sides in the second story with good accuracy. The damage-identification process also indicates there may be some damage induced by the shaker to the first-story infills.

References

1. Doebling, S.W., Farrar, C.R., Prime, M.B., Shevitz, D.W.: Damage identification and health monitoring of structural and mechanical systems from changes in their vibration characteristics: a literature review. Report, Los Alamos National Laboratory, Los Alamos, NM, USA (1996)
2. Sohn, H., Farrar, C.R., Hemez, F.M., et al.: A review of structural health monitoring literature: 1996–2001. Report, Los Alamos National Laboratory, Los Alamos, NM, USA (2003)
3. Song, M., Yousefianmoghadam, S., Mohammadi, M.E., Moaveni, B., Stavridis, A., Wood, R.L.: An application of finite element model updating for damage assessment of a two-story reinforced concrete building and comparison with lidar. Structural Health Monitoring. (2018). https://doi.org/10.1177/1475921717737970
4. Yousefianmoghadam, S., Song, M., Stavridis, A., Moaveni, B.: System identification of a two-story infilled RC building at different damage states. In: 2nd ATC-SEI Conference on Improving the Seismic Performance of Existing Buildings and other Structures, San Francisco (2015)
5. Martin Tempestti, J., Stavridis, A.: Simplified method to assess the lateral resistance of infilled reinforced concrete frames. In: Proceedings of 16th World Conference in Earthquake Engineering, Santiago, Chile (2017)
6. Tondi, M., Bovo, M., Bassoli, E., Vincenzi, L., Savoia, M.: Identificazione della Rigidezza dei Tamponamenti in Strutture Intelaiate Mediante Analisi Inversa. In: ANIDIS Conference, Pistoia, Italy (2017)
7. Vincenzi, L., Savoia, M.: Coupling response surface and differential evolution for parameter identification problems. Comput. Aided Civ. Infrastruct. Eng. **30**, 376–393 (2015)

Chapter 48
Modal Properties of a Model of a Chinese Pagoda

Yajie Wu, Xiaobin Song, and Carlos E. Ventura

Abstract China has been building wood frame pagodas since 200 AD during the Han Dynasty. A pagoda is essentially a multi-story tower, and each story is a separate building consisting of all structural components. This paper describes a series of shake table tests conducted on a one-fifth scale model of a seven-story pavilion-style wooden pagoda recently conducted at Tongji University in Shanghai, China. The main purpose of these tests was to investigate the seismic performance of this model pagoda. The model of the pagoda was subjected to different levels of excitation simulating ground shaking generated by earthquakes. White noise excitation was also used to determine the dynamic characteristics of the structure. The results of the modal analyses conducted using the various datasets are presented and discussed in this paper. The parameters studied include modal frequencies, damping and modes shapes of the structure. The variation of these parameters with the increasing levels of ground excitation is investigated, and these are correlated to the nonlinear behaviour and damage observed during each test. A detailed analysis of the mode shapes of the structure indicates that a soft-story behaviour of the first floor of the structure is noticeable with increasing level of shaking. Considering that the pagoda's connections using the Dou-gong style are made of wood only, the behaviour of the structure during extreme levels of excitation was excellent and very minor damage was observed. The mortise-tenon joints of the timber frame dissipated most of the seismic energy.

Keywords Chinse pagoda · Shaking table test · Seismic performance · Modal analyse · Parameter study

48.1 Introduction

The Chinese traditional timber structure is one of the oldest structural forms in Asia. China has been building wood frame pagodas since 200 AD during the Han Dynasty, Japan adopted similar traditional timber structures since 650 BC during Tang Dynasty. Timber pagodas were initially intended as religious monuments in Buddhism, but some of them were also developed as high-rise traditional timber buildings for sight view and other secular public activities. Quite a number of timber pagodas which were built through history have been preserved. In northern China, there is the tallest timber pagoda in the world-Yingxian Wooden Pagoda [1], its history could be traced back to nearly 1000 years ago. And there are more than 20 historical timber pagodas preserved in Japan [2]. These buildings stand as an invaluable legacy to human civilization.

A timber pagoda is essentially a multi-story tower, and each story can be treated as a separate building consisting of all structural components. Such a pagoda is typically formed by use of traditional wood-to-wood connection techniques. Different types of mortise tenon connections are applied to connect beams and column in structures [3–5], and Dou-Gong connections are used to supports upper columns and transferring structural loads from upper floors and overhanging roofs [6]. With a relative high aspect ratio and low moment capacities of connections, timber pagodas are much more flexible than contemporary structures like concrete structures and steel structures.

Health monitoring had been conducted on typical pagodas for obtaining their dynamic characteristics and investigate their dynamic responses [7, 8]. However, the health monitoring is very limited on excitation intensity and earthquake occurrence. A model of a five-story Japanese pagoda was fabricated and was tested on a shaking table using unidirectional excitation, but the model was oversimplified and the property of the pagoda had not been analyzed. This paper describes a series of shake table tests conducted on a one-fifth scale model of a seven-story pavilion-style wooden pagoda recently conducted at Tongji University in Shanghai, China. The main purpose of these tests was to investigate the seismic performance of this

Y. Wu · X. Song
Department of Structural Engineering, Tongji University, Shanghai, China

C. E. Ventura (✉)
Department of Civil Engineering, The University of British Columbia, Vancouver, BC, Canada
e-mail: ventura@civil.ubc.ca

S. Pakzad (ed.), *Dynamics of Civil Structures, Volume 2*, Conference Proceedings of the Society for Experimental Mechanics Series, https://doi.org/10.1007/978-3-319-74421-6_48

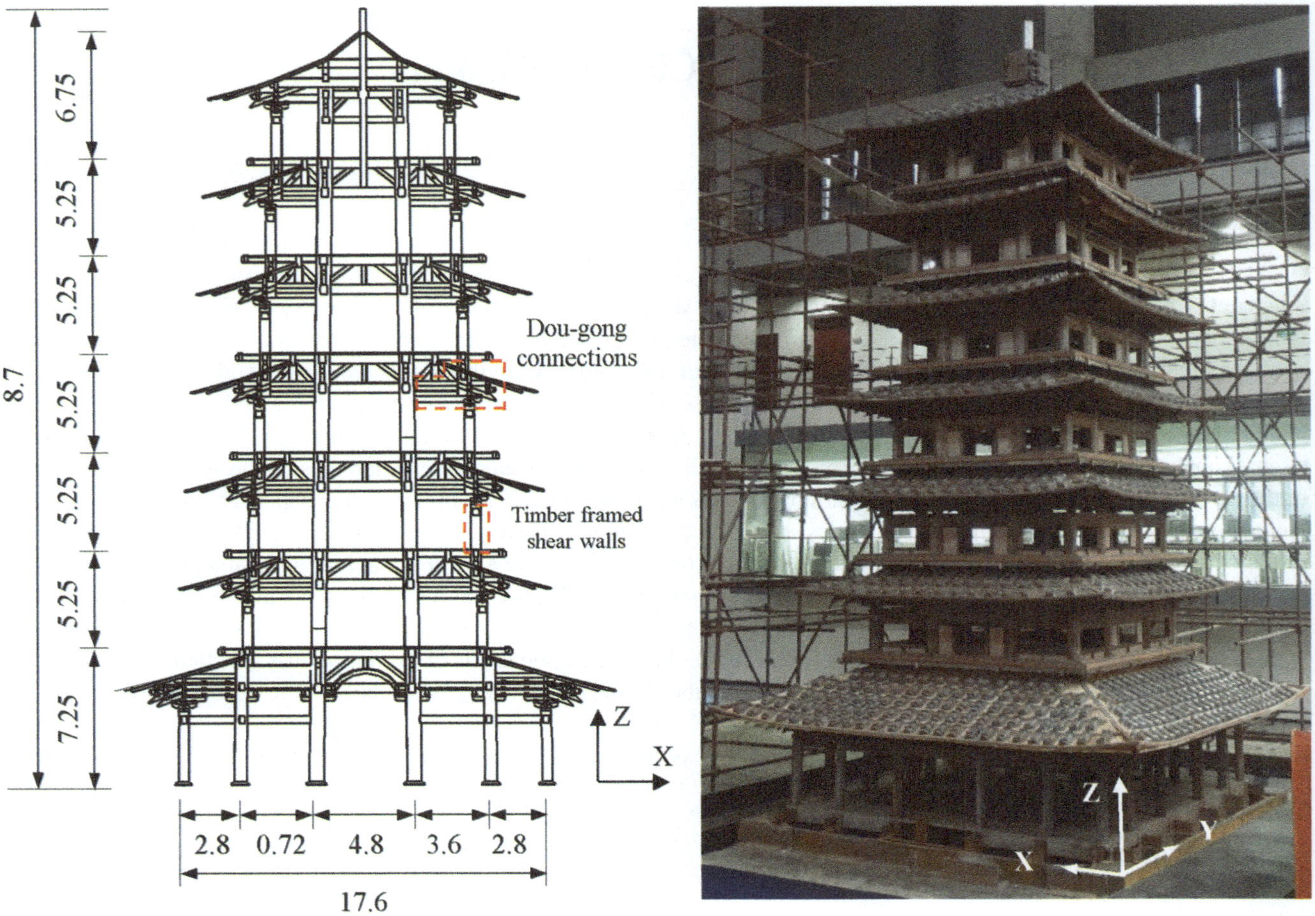

Fig. 48.1 A one-fifth scaled Chinese timber pagoda (unit: m)

model pagoda. The model of the pagoda was subjected to different levels of excitation simulating ground shaking generated by earthquakes. Modal properties under the white noise excitation, including modal frequencies, damping and modes shapes, are identified. The vibration of these modal parameters varying with increasing level of ground motion is investigated.

48.2 Model Description

The pagoda model was scaled down to one-fifth the size of a traditional Chinese seven-story timber pagoda. Figure 48.1 presents the section view and test photograph of the pagoda model. The total height of the model was 8.7 m; the ground story had a height of 1.45 m, and the story height of the second to sixth stories was 1.05 m. The top story had a height of 1.35 m. Each story of the pagoda had three bays in X- and Y-directions, except for the ground story, which had an additional corridor at the periphery. The pagoda basically consisted of a core frame with continuous columns in the central, timber framed shear walls around its periphery as well as floor systems and dou-gong connections between the shear walls and floors. The core frame was assembled with four inclined timber columns connected by main beams. The truss members were placed underneath each floor by mortise-tenon joints. The typical timber framed shear walls were mainly made up of an exterior main beam, exterior columns, ring beams and infill wood panels (Fig. 48.2a). The dou-gong connections were mainly assembled by jointing beams in two principle translational directions, which were set on the top of each column in the timber framed shear walls, supporting the overhanging roofs and upper exterior columns (Fig. 48.2b).

The diameters of the core columns and exterior columns were 130 mm and 100 mm, respectively. Beams connecting with interior columns had a cross section of 70 mm × 100 mm, while other floor beams had a cross section of 45 mm × 65 mm. A total of 17.5-t additional mass was applied onto the overhanging roofs and floor slabs to simulate the tiles and workloads. The metric tonne of 2204.6 lb is the measurement used herein. The self-weight of the pagoda was about 3.1 t. The total weight of the pagoda was 20.6 t.

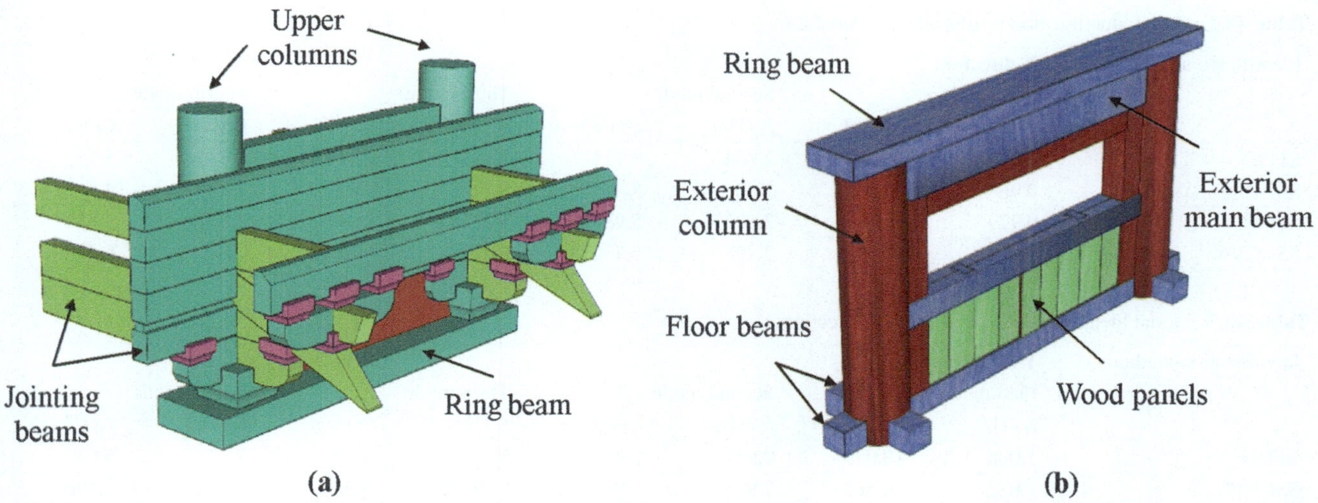

Fig. 48.2 Details of structural components **(a)** Dou-gong connections **(b)** timber framed shear walls

Table 48.1 Main test scenarios

Input order	Excitation level	Input excitation	Design PGA (g)	
			X-direction	Y-direction
1	–	WN1	0.03	0.03
2	Frequently met 7-degree	SHW2	0.07	–
		Kobe, ChiChi	0.07	0.06
3	–	WN2	0.02	0.02
4	Fortification 7-degree	SHW2	0.2	–
		Kobe, ChiChi	0.2	0.17
5	–	WN3	0.02	0.02
6	Rarely met 7-degree	SHW2	0.44	–
7	–	WN4	0.02	0.02
8	Rarely met 7-degree	Kobe	0.44	0.374
9	–	WN5	0.02	0.02
10	Rarely met 7-degree	ChiChi	0.44	0.374
11	–	WN6	0.02	0.02

48.3 Shaking Table Test

White noise (WN) excitations were used to identify the natural frequencies of the structure and detect the stiffness changes after different intensity seismic inputs. In total six white noise excitations were exerted. The acceleration magnitude of the first excitation was initially set as 0.03 g, which caused a substantial vibration of the pagoda. To prevent the pagoda from being damaged in the white noise excitations, the acceleration magnitude of the remaining white noise excitations was adjusted to 0.02 g. One artificial wave record, SHW2, which was established based on the local soil conditions, and two natural earthquake records, Kobe and ChiChi waves, were scaled in time durations and acceleration amplitudes and input to excite the pagoda. Three seismic hazard levels were considered: frequently met 7-degree level, fortification 7-degree level and rarely met 7-degree level specified in Chinese seismic code [9], which corresponded to the input peak ground accelerations (PGAs) of 0.07 g, 0.2 g and 0.44 g, respectively. Because the structure was almost symmetric in plane, the X-direction was chosen as the principal input direction of the two-way inputs. The PGA of Y-direction input was set as 0.85 times of that of X-direction. The main test scenarios are listed in Table 48.1.

Acceleration and displacement sensors were used to obtain the dynamic responses of the pagoda model. The measurement points were placed along the X- and Y-directions at each floor and the top of the pagoda. In each direction, two measuring points were arranged at each side of the structure to monitor the torsion and translational motion along the two translational directions. It was found that the dynamic response of the pagoda was dominant with translational motion. Since the X-direction was the principal input direction, the maximum interstory drift of X-direction reached up to 1/29, and the maximum interstory drift of Y-direction reached up to 1/71. The pagoda did not have obvious main structural damage after the test. Detailed dynamic responses along the height of the structure can be found in Song et al. [10].

Table 48.2 Modal identification results along X-direction

Identification method	X-direction							
	First mode		Second mode		Third mode		Fourth mode	
	f_1 (Hz)	ζ_1 (%)	f_2 (Hz)	ζ_2 (%)	f_3 (Hz)	ζ_3 (%)	f_4 (Hz)	ζ_4 (%)
EFDD	0.953	10.42	2.804	8.714	5.183	6.883	7.736	5.153
SSI-UPC	0.968	2.251	2.805	5.445	5.193	9.508	7.887	7.519
SSI-PC	0.979	4.341	2.822	5.988	5.259	9.659	7.952	7.854
SSI-CVA	1.013	9.931	2.793	6.436	5.517	7.502	7.845	6.256

Table 48.3 Modal identification results along Y-direction

Identification method	Y-direction							
	First mode		Second mode		Third mode		Fourth mode	
	f_1 (Hz)	ζ_1 (%)	f_2 (Hz)	ζ_2 (%)	f_3 (Hz)	ζ_3 (%)	f_4 (Hz)	ζ_4 (%)
EFDD	1.008	10.029	2.945	6.489	5.246	6.685	7.73	5.039
SSI-UPC	1.022	6.582	2.905	5.931	5.373	10.856	7.926	5.496
SSI-PC	1.03	7.642	2.916	6.549	5.408	9.583	8.025	8.193
SSI-CVA	1.024	6.126	2.887	6.449	5.288	7.481	8.044	5.719

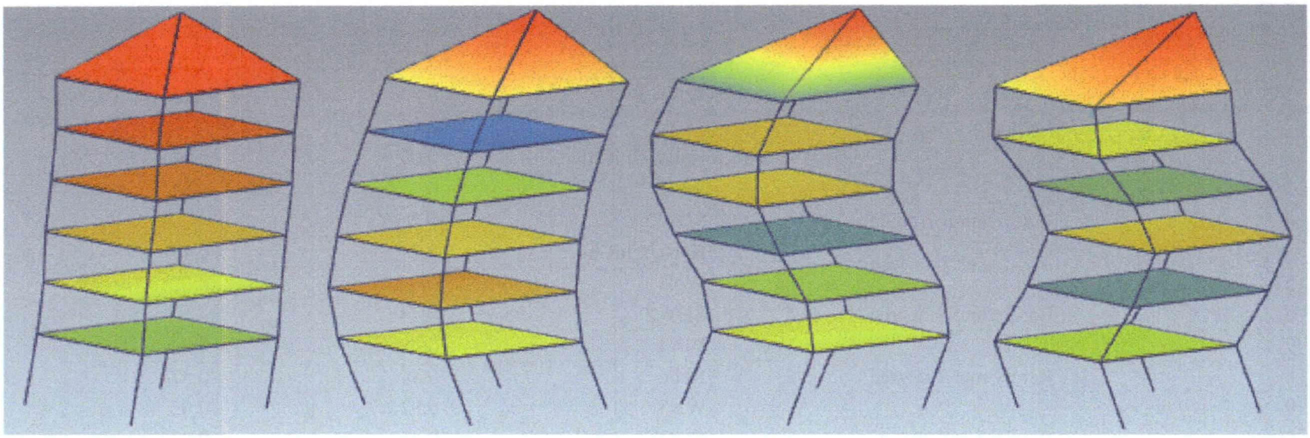

(a) (b) (c) (d)

Fig. 48.3 Mode shapes along X- and Y-directions. (a) First mode shape; (b) second mode shape; (c) third mode shape; (d) fourth mode shape

48.4 Modal Analysis Under White Noise Excitation

The modal analysis of the pagoda model was conducted based on the operational modal analysis software ARTeMIS. Multi modal identification methods, including Enhanced Frequency Domain Decomposition (EFDD), Stochastic Subspace Identification (SSI) with Unweighted Principal Component (UPC) algorithm, Principal Component (PC) algorithm and Canonical Variate Analysis (CVA) algorithm, were applied to identify the dynamic characteristics of the model. The modal analysis was initially carried out along X- and Y-directions with unidirectional output data, respectively. The identified results are shown in Tables 48.2 and 48.3. The mode shapes along X- and Y-directions are illustrated in Fig. 48.3. It can be found from Tables 48.2 and 48.3 that each modal frequency identified by use of different methods were consistent, while the corresponding modal damping ratios changed within a relative wide range. The natural frequencies along X-direction were close to those along Y-direction, indicating that the stiffness along the two translation directions was almost the same.

Modal analysis based on bi-directional output data was also conducted. According to the unidirectional identification results, the pagoda should have four pairs of close modes, either parallel to translational directions or along 45° to translational directions. However, the bi-directional data analysis results indicated that none of the identification methods could identify those close modes. Figure 48.4 illustrates the identification results by use of SSI-UPC method. The stabilization diagram indicated approximately the same four modal frequencies as those in unidirectional modal analysis, but it did not indicate close modes, which means with bi-directional data these methods were not able to obtain the full

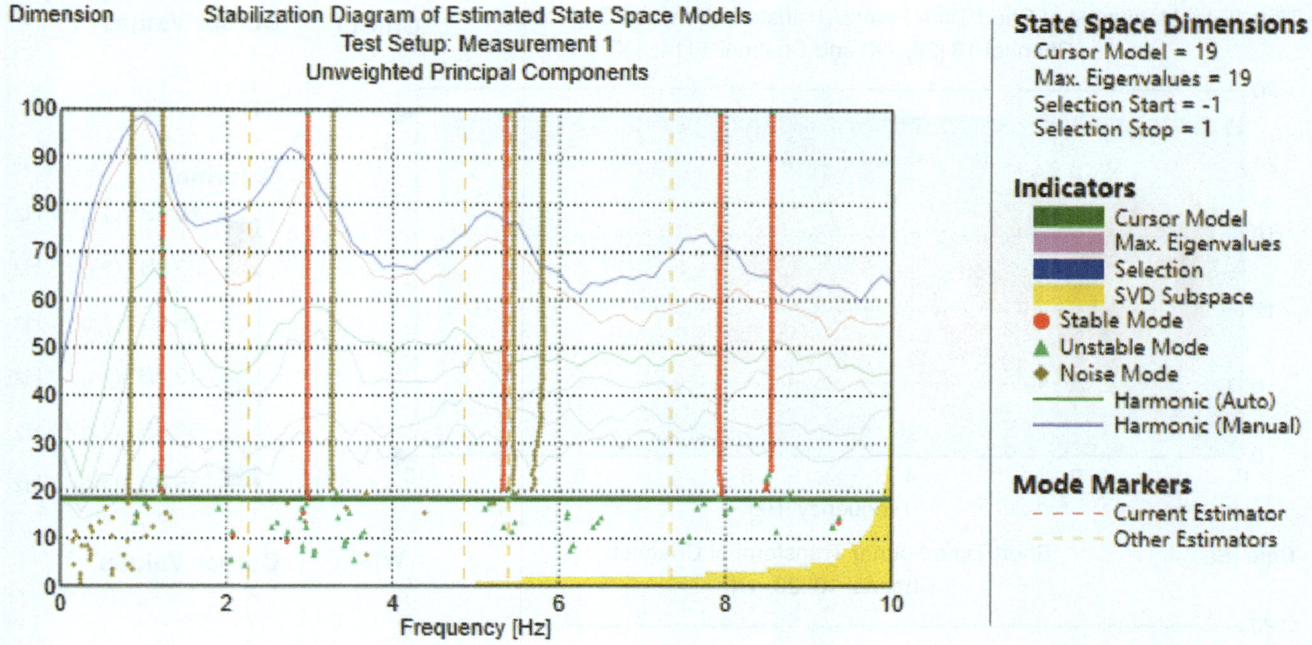

Fig. 48.4 Identification result by use of SSI-UPC method

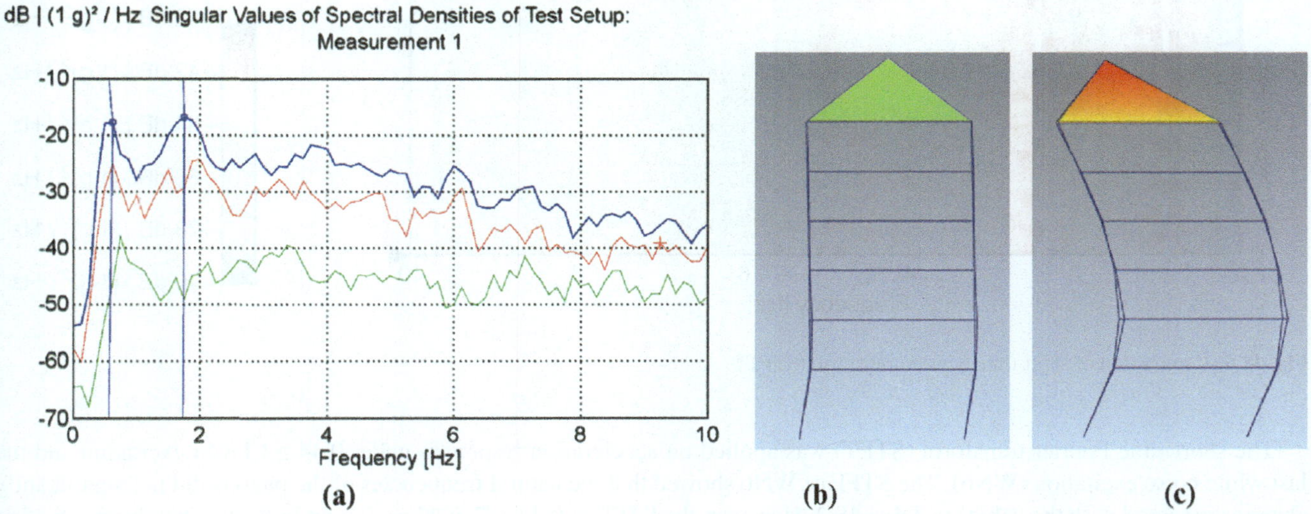

Fig. 48.5 Operational shape analysis. (a) Singular value of spectral densities; (b) first operational shape with soft story; (c) second operational shape

spatial modal shapes of the pagoda. Two possible reasons could be drawn. One is that the white noise exerted in laboratory was the band-limited white noise, which means the input was not stationary. The other is that the excitation duration was less than 120 s, leading to a lack of resolution in identifying very close modes.

48.5 Operational State Analysis Under Earthquakes

Singular values of spectral densities under 0.44 g ChiChi excitation were calculated, as shown in Fig. 48.5a. The mode shapes from the selected two peaks indicate the operational shapes during earthquake responses, as illustrated in Fig. 48.5b, c. Figure 48.5a indicated a significant shift of frequencies compared with the white noise identification results. Different from the first mode shape identified from white noise excitation, the timber pagoda was found with a noticeable soft-story behaviour of the first floor (Fig. 48.5b).

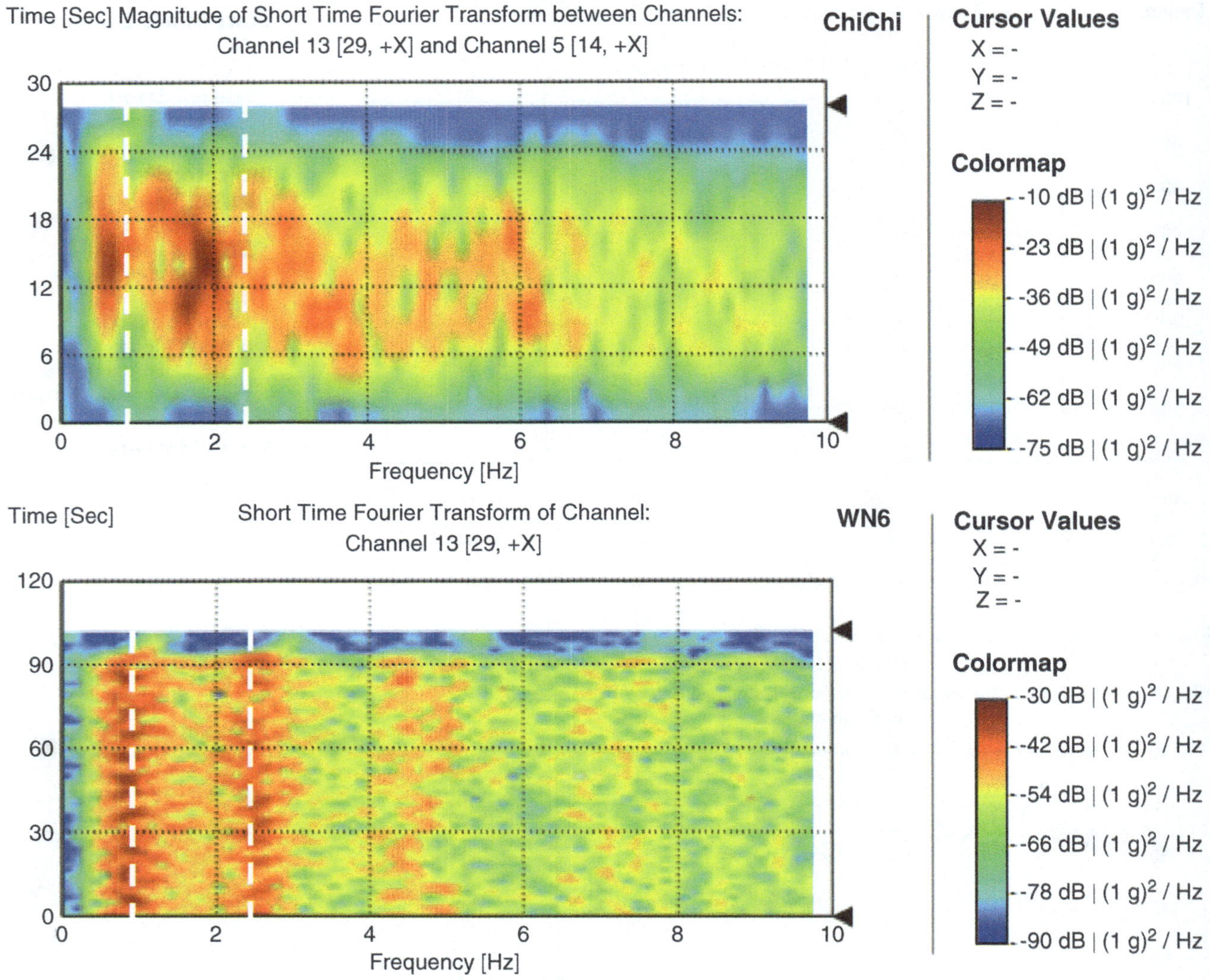

Fig. 48.6 Comparison of short-time Fourier transform (STFT)

The short-time Fourier transform (STFT) was applied on acceleration responses under 0.44 g ChiChi excitation and the last white noise excitation (WN6). The STFF of WN6 showed that the natural frequencies of the pagoda did not significantly change compared with the values in Table 48.2. However, the STFF of 0.44 g ChiChi excitation indicated that during shaking the frequencies of the pagoda obviously decreased, and then the stiffness gradually restored at the last stage of the response (Fig. 48.6).

48.6 Conclusions

A shaking table test was conducted on a one-fifth scaled timber pagoda. White noise excitations and different levels of earthquake excitations were input to investigate the seismic performance of this model pagoda. The modal analysis was conducted based on white noise excitation. With unidirectional output data consistent natural frequencies were identified using different method along each translational direction, but close modes could not be separated with bi-directional data. A noticeable soft story behaviour of the first story of the structure was found when carrying out a detail analysis on the model shapes. He short-time Fourier transform indicated that the frequencies of the structure decreased significantly during shaking but increased at the last stage of the response.

References

1. Lam, F., He, M., Yao, C.: Example of traditional tall timber buildings in China–the Yingxian pagoda. Struct. Eng. Int. **18**(2), 126–129 (2008)
2. Fujita, K., Hanazato, T., Sakamoto, I.: Earthquake response monitoring and seismic performance of five storied timber pagoda. In: 13th World Conference on Earthquake Engineering (2004)
3. Chen, Z., Zhu, E., Pan, J., Wu, G.: Energy-dissipation performance of typical beam-column joints in Yingxian Wood Pagoda: experimental study. J. Perform. Constr. Facil. **30**(3), 04015028 (2015)
4. Chen, C., Qiu, H., Lu, Y.: Flexural behaviour of timber dovetail mortise-tenon joints. Constr. Build. Mater. **2016**(112), 366–377 (2016)
5. Luo, L., Song, X.B., Wu, H.B., Cai, J.N.: Experimental study on mechanical performance of wood pegged semi mortise and tenon connections. In: Proceedings of the 14th World Conference on Timber Engineering, Vienna, Austria (2016)
6. Song, X.B., Wu, Y.J., Gu, X.L., Luo, L: Mechanical behavior of traditional Chinese wood-to-wood connections: a brief review. In: Proceedings of the 14th World Conference on Timber Engineering, Vienna, Austria (2016)
7. Hanazato, T., Minowa, C., Niitsu, Y., Nitto, K., Kawai, N., Maekawa, H., Morii, M.: Seismic and wind performance of five-storied Pagoda of timber heritage structure. Adv. Mater. Res. **13**, 79–95 (2010)
8. Minowa, C., Kwawai, N., Maekwaw, H.: Observation of wind and earthquake responses of national heritage five story wooden pagoda. In: Proceedings of the 11th World Conference on Timber Engineering, Riva del Garda, Italy (2010)
9. GB 50011-2010: Code for Seismic Design of Buildings. China Architecture & Building Press, Beijing, China (2010.) (in Chinese)
10. Song, X.B., Wu, Y.J., Luo, L., Gu, X.L.: Shaking table test on a traditional pavilion-style wooden pagoda. J. Build. Struct. **38**(2), 10–19 (2017.) (in Chinese)

Chapter 49
Advanced Fourier-Based Model of Bouncing Loads

Vitomir Racic, Jun Chen, and Aleksandar Pavic

Abstract Contemporary design guideline pertinent to vibration serviceability of entertaining venues describes bouncing forces as a deterministic and periodic process presentable via Fourier series. However, fitting the Fourier harmonics to a comprehensive database of individual bouncing force records established in this study showed that such a simplification is far too radical, thus leading to a significant loss of information. Building on the conventional Fourier force model, this study makes the harmonics specific to each individual and takes into account imperfections in the bouncing process. The result is a numerical generator of stochastic bouncing force time histories which represent reliably the experimentally recorded bouncing force signals.

Keywords Vibration serviceability · Human-induced vibrations · Human-induced excitation · Stadia

49.1 Introduction

There has been a growing number of vibration serviceability issues of floors, staircases and assembly in entertaining venues when exited by active people. One of the reasons is that crowds has become livelier than ever. In the same time substantial developments in workmanship and structural materials have reduced the mass and stiffness of a structure, yielding a natural frequency often up to 5 Hz. This falls within the range of the body motion rates for common activities, such as bouncing and walking, yielding a large and often resonant vibration response. The final reason is a lack of adequate design guidance relevant to crowd rhythmic excitation. The IStructE/DCLG/DCMS joint Working Group design guidance on crowd dynamic loading of grandstands [1] is a step in the right direction but still not perceived as the final version. The vital refinement should address modelling the actual nature of human activities and the corresponding loads. Although there can be no absolute certainty on how a random group of people would act, the guidance is grounded on a conservative deterministic representation of crowd dynamic loading. More adequate models would portray it as a stochastic process, similar to the existing models of wind, wave or earthquake loading, all of them characterised by considerable uncertainty and randomness. The later feature this study specifically aims to address for bouncing lads.

A reliable load model of bouncing crowds needs a reliable model of individual bouncing forces. Measured individual force time histories are characterised by considerable inter-subject variability and are invariably near-periodic [2, 3], indicating their narrow band nature (Fig. 49.1). However, they are commonly assumed identical, perfectly periodic and presentable via Fourier series $F(t)$:

$$F(t) = G\sum_{i=1}^{m}\alpha_i \cos\left(2\pi i f_{\text{b}} t - \varphi_i\right) \tag{49.1}$$

G is the body weight in the same unit as the force $F(t)$ (typically N). Coefficients α_i and φ_i are the dominant Fourier amplitudes and phase angles corresponding to m integer multiples of the bouncing rate f_b (Fig. 49.1b).

V. Racic (✉)
Department of Civil and Environmental Engineering, Politecnico di Milano, Milan, Italy
e-mail: vitomir.racic@polimi.it

J. Chen
Department of Structural Engineering, Tongji University, Shanghai, People's Republic of China

A. Pavic
College of Engineering, Mathematics and Physical Sciences, University of Exeter, Exeter, UK

S. Pakzad (ed.), *Dynamics of Civil Structures, Volume 2*, Conference Proceedings of the Society for Experimental Mechanics Series, https://doi.org/10.1007/978-3-319-74421-6_49

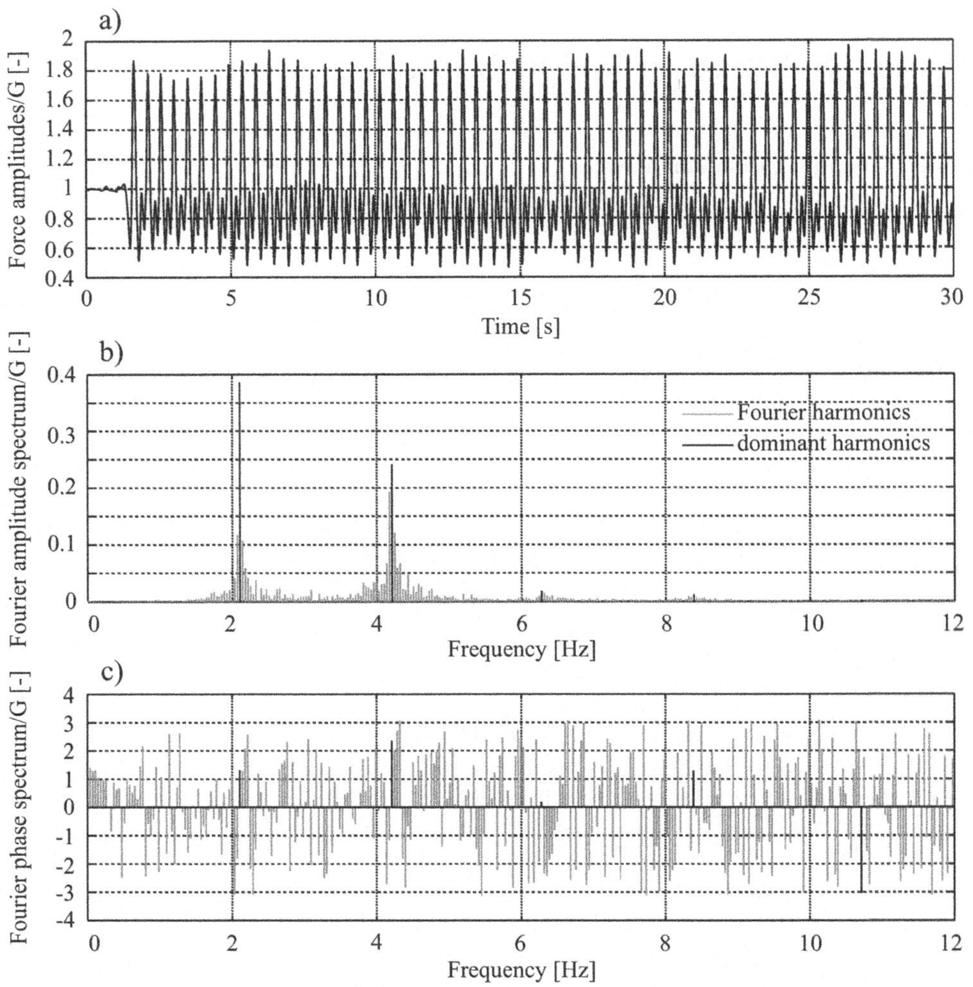

Fig. 49.1 Example of measured bouncing (**a**) force-time history and its Fourier (**b**) amplitude and (**c**) phase spectra

It is widely accepted that the modelling strategy based on Fourier harmonics leads to significant loss of information during the data reduction process [4–9]. For example, Brownjohn et al. [5] demonstrated that neglecting the energy around dominant Fourier harmonics (Fig. 49.1b), which is a result of uneven footfalls, yields up to 50% error in predicted vibration response. More recent study by Van Nimmen [10] showed that precision of simulated resonant vibration response primarily depends on whether variability of timing between successive footfalls is taken into account. A model of variability of successive bouncing intervals and variability of the corresponding force amplitudes are presented later in this study.

49.2 Experimental Data Collection

The data collection was carried out in the Light Structures Laboratory in the University of Sheffield, UK (Fig. 49.2).

Each participant was engaged in 12 bouncing tests, thereby generated 12 force signals. During each test a participant was asked to bounce to a steady metronome beat which was randomly selected from the frequency range 1.2–4.5 Hz with the increment of 0.3 Hz. A test lasted between 25 and 45 s, being shorter for the higher frequencies so the participant would not tire much. Rests were allowed between the tests.

The corresponding force signals were recorded by an AMTI BP-400600 force plate [11] fixed to the laboratory heavy floor (Fig. 49.2). In total, 80 volunteers (51 males and 29 females, body mass 72.7 ± 15.4 kg, height 171.2 ± 9.2 cm, age 26.4 ± 7.1 years) generated 960 vertical bouncing force signals of kind illustrated in Fig. 49.3. Sixty signals were cast aside as some test subjects could not follow very high metronome rates. All signals were sampled at $f_s = 200$ Hz.

Fig. 49.2 Experimental setup

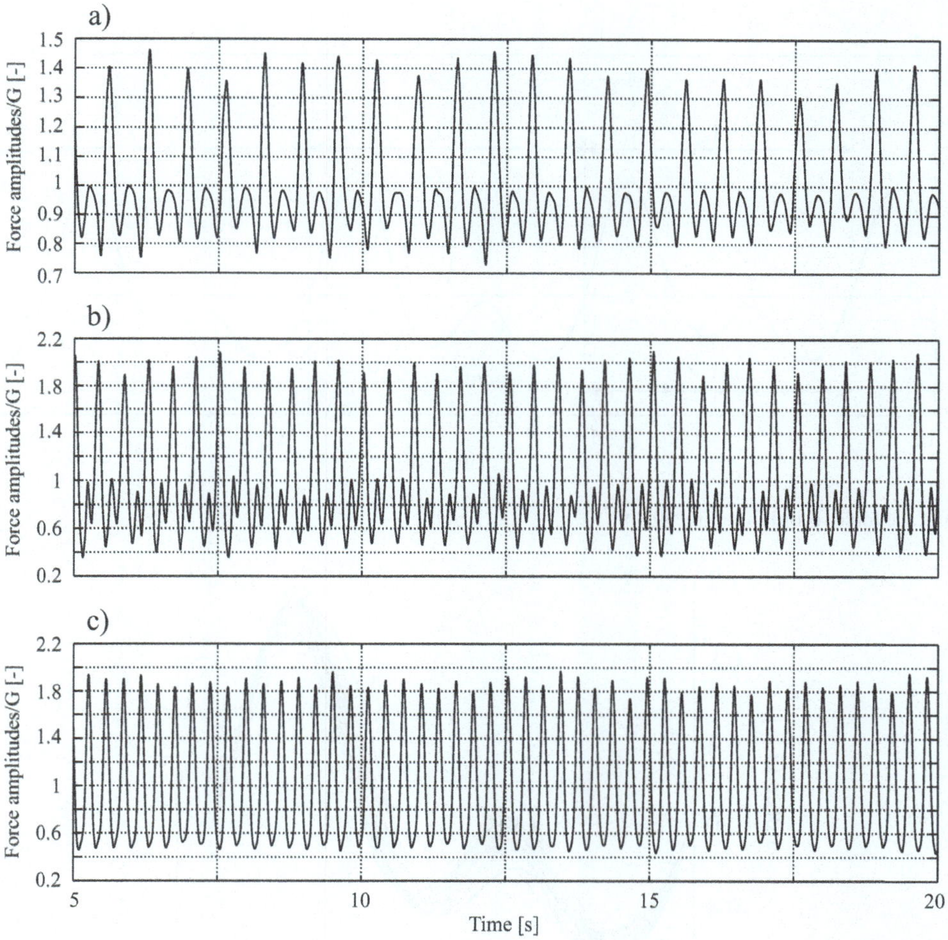

Fig. 49.3 Examples of bouncing force records generated at given metronome beats (**a**) 1.5 Hz, (**b**) 2.4 Hz and (**c**) 3.3 Hz

49.3 Modelling Shape of Force Signals

This section aims to determine the "repetitive" shape of individual force signals between the successive bouncing cycles, so called "template shape". Since the vertical force amplitudes oscillate around body weight G of the test subject, the point where the amplitude is equal to the body weight and has a negative gradient was selected as starting (and completing) event of a bouncing cycle (Fig. 49.4).

From the 30 s long force signal illustrated partly in Figs. 49.3b and 49.4 and yielding about 70 cycles, the central portion comprising 66 successive cycles was extracted for further analysis. The selected cycles have been normalised by the body weight G, lined up at their origins and resampled to f_s/f_b data points. Figure 49.5 confirms that there is a common shape that nonlinearly distorts along time and amplitude axes for successive cycles. Therefore, the template was determined as their numerical average after the common events had been aligned using dynamic time warping (DTW) method [12]. The procedure is elaborated in Racic and Chen [13] and illustrated here in Fig. 49.6.

The shape of the template cycle $Z(t)$ can be modelled as a sum of its first four Fourier harmonics (Eq. 49.2). The results of the curve fitting are summarised in Fig. 49.7 and Table 49.1.

$$Z(t) = \sum_{i=1}^{4} \alpha_i \cos\left(2\pi i f_b t - \varphi_i\right) \tag{49.2}$$

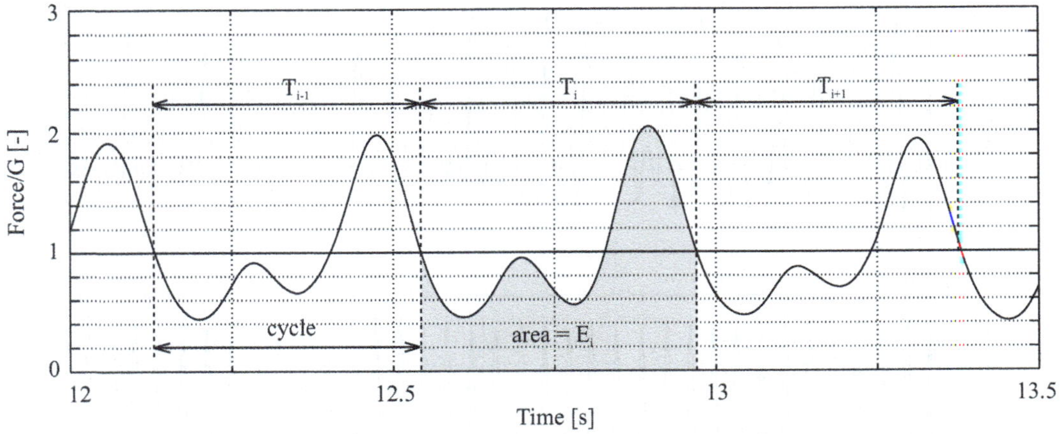

Fig. 49.4 A portion of the force record from Fig. 49.3b

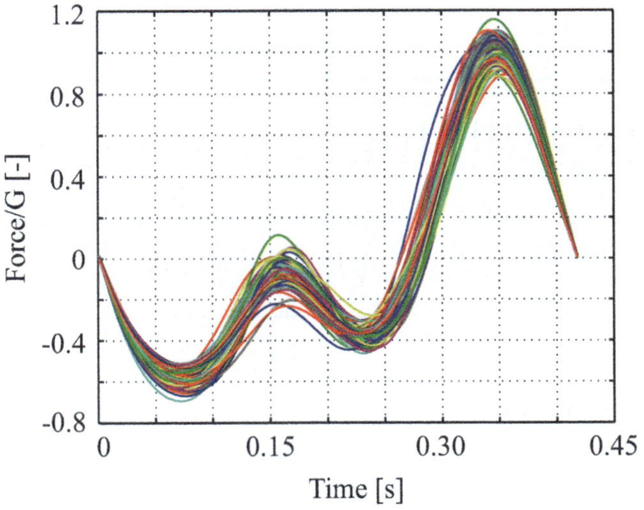

Fig. 49.5 Resampled G-normalised cycles for DTW

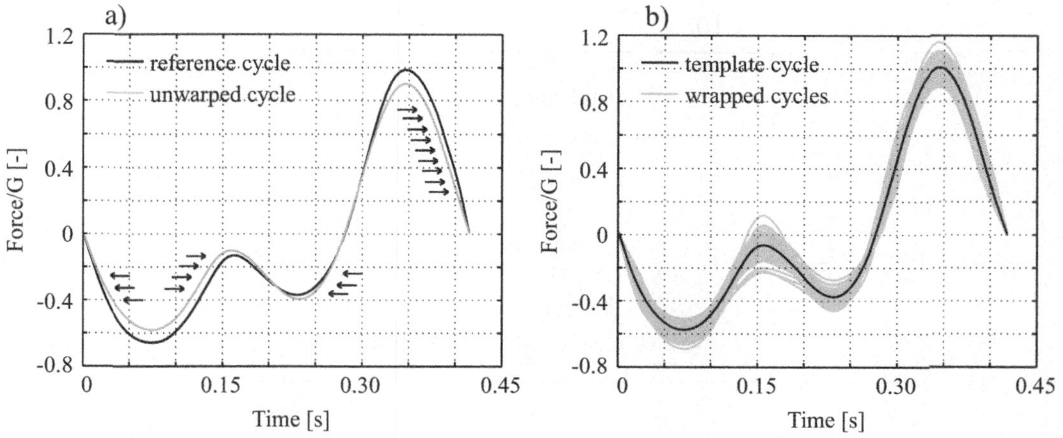

Fig. 49.6 (**a**) Illustration of DTW, and (**b**) template cycle

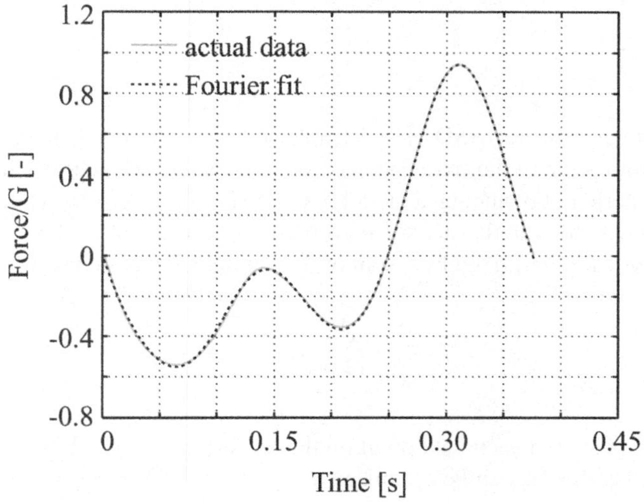

Fig. 49.7 Fitting the template shape

Table 49.1 Results of curve fitting template shape $Z(t)$

Parameters	$f_b = 2.40$ Hz			
	$i = 1$	$i = 2$	$i = 3$	$i = 4$
α_i [−]	0.547	0.420	0.066	0.021
φ_i [rad]	−1.180	−1.847	2.258	2.577

49.4 Variability of Cycle Intervals

Variability of the consecutive cycle intervals T_i ($i = 1, \ldots, 66$) can be represented by a dimensionless series τ_i:

$$\begin{aligned} \tau_i &= \frac{T_i - \mu_T}{\mu_T} \\ \mu_T &= \text{mean}\,(T_i) \end{aligned} \tag{49.3}$$

Mean of τ_i is zero and its auto spectral density (ASD) can be calculated as [14]:

$$S_\tau\,(f_m) = \frac{A_\tau^2\,(f_m)}{2\Delta f},\ f_m = \frac{m}{66},\ m = 0, \ldots, 32 \tag{49.4}$$

where $A_\tau(f_m)$ is a single-sided discrete Fourier amplitude spectra and Δf=1/66 is the spectral line spacing (Fig. 49.8).

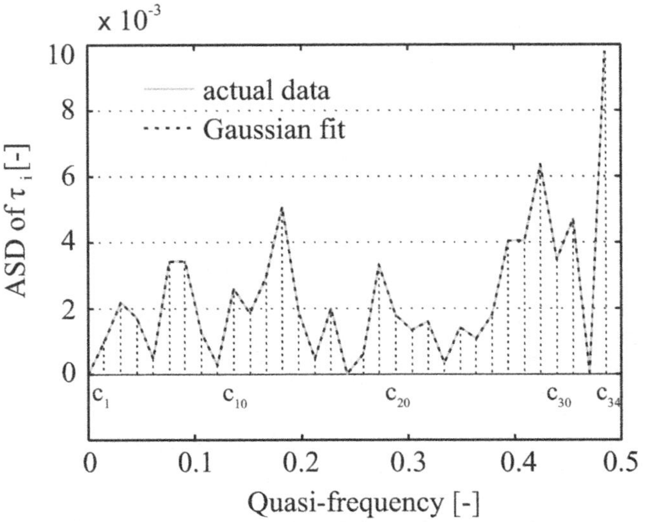

Fig. 49.8 ASD of τ_i series and its curve fit

The variance of τ_i is the integral of the ASDs [14]. Unlike random number generators, such as probability density functions, the ASD contains information of the frequency content of τ_i series. Hence, assuming that a test subject maintains the same bouncing style for any period of bouncing, a model of the ASD can be used to synthesise artificial τ'_k $(k = 1, \ldots, N)$ series of arbitrary length (e.g. $N \gg 66$) with the statistical properties of the actual τ_i series.

The ASD $S_\tau(f_m)$ can be analytically described by a sum of 33 equidistant Gaussians (Fig. 49.8):

$$S'_\tau(f) = \sum_{j=1}^{33} W_j e^{-\frac{(f-c_j)^2}{2b^2}} \tag{49.5}$$

If the Gaussian centres c_j are placed in each data point on the quasi-frequency axis and all Gaussian bells have the same predefined widths $b = \Delta f$, their heights W_j can be optimised using the non-linear least-square method [15] to fit exactly the actual ASD (Fig. 49.8).

To create a series of synthetic cycle intervals T'_k $(k = 1, \ldots, N)$, Eq. (49.5) first calculates $S'_\tau(f_n)$ values at equidistant data points $f_n = n\Delta f'$, where $n = 0, \ldots, N/2 - 1$ and $\Delta f' = 1/N$. These are then used in Eq. (49.4) to compute the corresponding Fourier amplitudes $A'_\tau(f_n) = \sqrt{2\Delta f' S'_\tau(f_n)}$. Assuming the uniform distribution of the phase angles in the range $[-\pi, \pi]$, $A'_\tau(f_n)$ are then fed into the inverse FFT algorithm to generate τ'_k series. Different τ'_k series with the same variance and the frequency content can be created by varying the sets of random phase angles. Finally, multiplying τ'_k by μ_T and adding the offset μ_T, yields a synthetic T'_k series (Fig. 49.9). Working under assumption that a test subject keeps the same bouncing style in a narrow range of bouncing rates, μ_T can take a slightly different value μ'_T to generate cycle intervals corresponding to rate $1/\mu'_T$.

49.5 Variability of Force Amplitudes

Energy of bouncing cycles E_i can be defined as the integral of the weight -normalised force amplitudes over the corresponding cycle intervals T_i (Fig. 49.4).

The relationship between the two parameters (Fig. 49.10) can be described by the following linear regression model:

$$E_i = \rho_1 T_i + \rho_0 + \Delta E_i \tag{49.6}$$

Parameters $\rho_1 = 0.972$ and $\rho_0 = 0.012$ are regression coefficients and ΔE_i is a disturbance term, typically modelled as a random Gaussian noise [16].

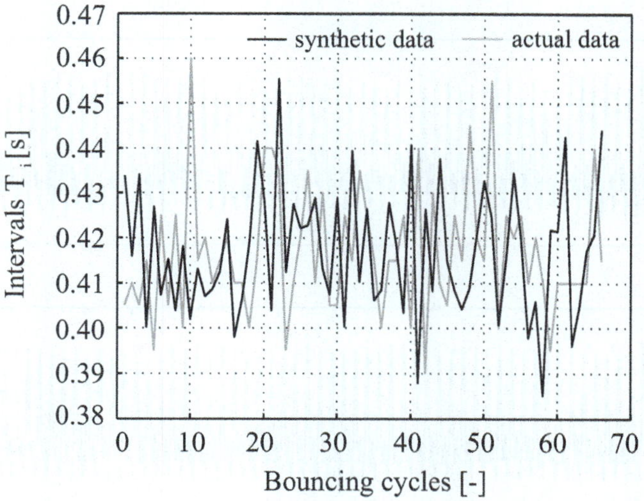

Fig. 49.9 Measured and an example of synthetic cycle intervals

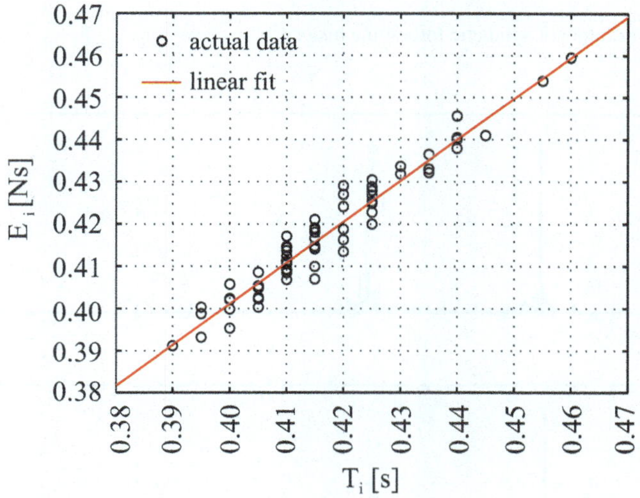

Fig. 49.10 Cycle energy E_i vs. cycle intervals T_i

Hence, having generated synthetic cycle intervals T'_k ($k = 1, \ldots, N$), the corresponding energies E'_k can be calculated using Eq. (49.6). The energies are then assigned to a sequence of N bouncing cycles by scaling the amplitudes of the template shape with factors ξ_k:

$$\xi_k = \frac{E'_k}{E_{\text{tc}}} \tag{49.7}$$

where E_{tc} is the energy of the template shape. The empirical evidence is provided in the next section.

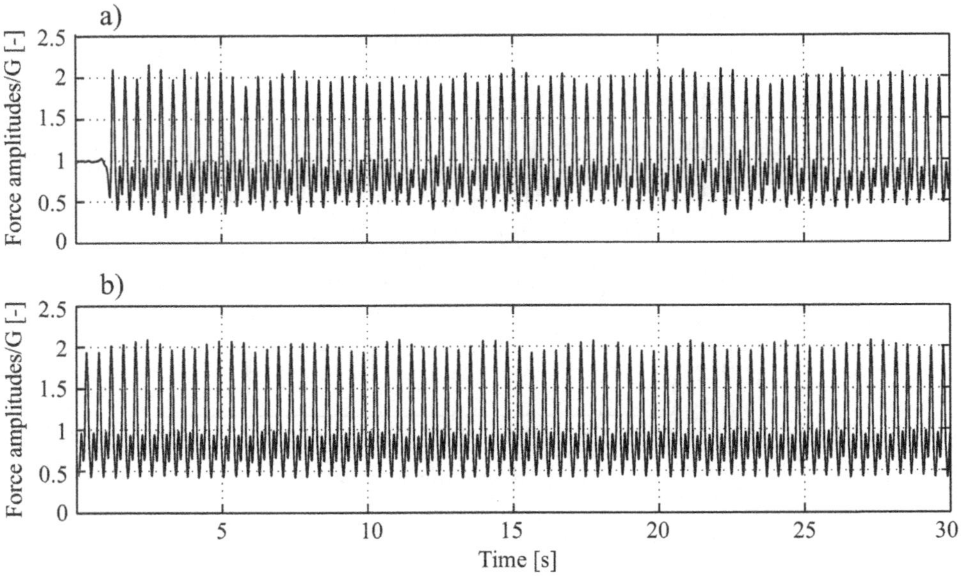

Fig. 49.11 (**a**) Measured and (**b**) an example of synthetic force-time history

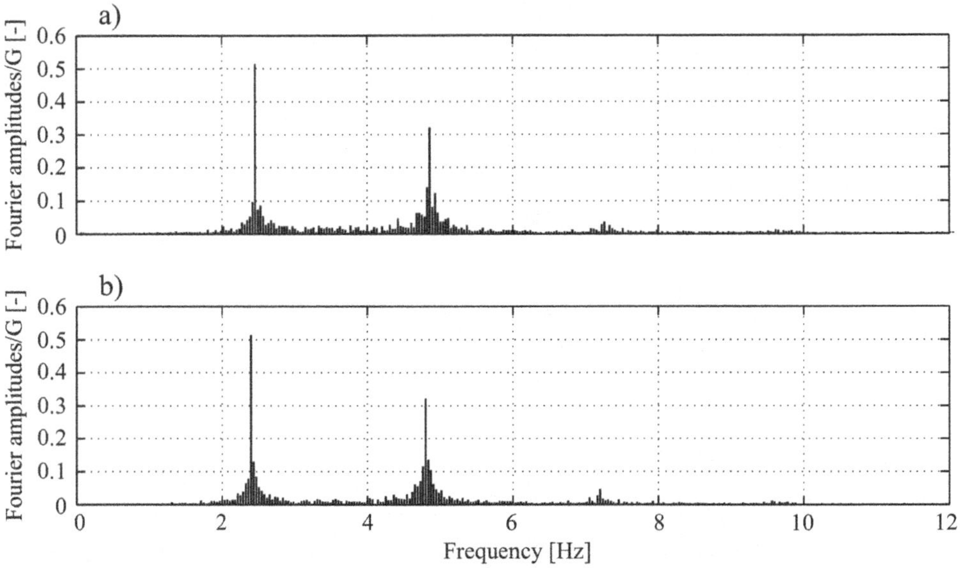

Fig. 49.12 Discrete Fourier amplitude spectra of the time series shown in Fig. 49.11

49.6 Artificial Force Signals

Figures 49.11 and 49.12 illustrate an example of synthetic force signal. Comparison of Figs. 49.11a, b and 49.12a, b highlights the close match with the actual force record in both time and frequency domain.

49.7 Generator of Random Force Signals

Parameters of the template shape $Z(t)$, the ASD $S'_\tau(f)$, disturbance term $\Delta E_i(t)$ and the regression coefficients ρ_1 and ρ_0, were extracted from each of the 900 force signals recorded in Section 2 and stored in in 900 MATLAB structure files [17], here called "mat files". The mat files were classified into 0.3 Hz wide clusters which centres correspond to the 12 bouncing

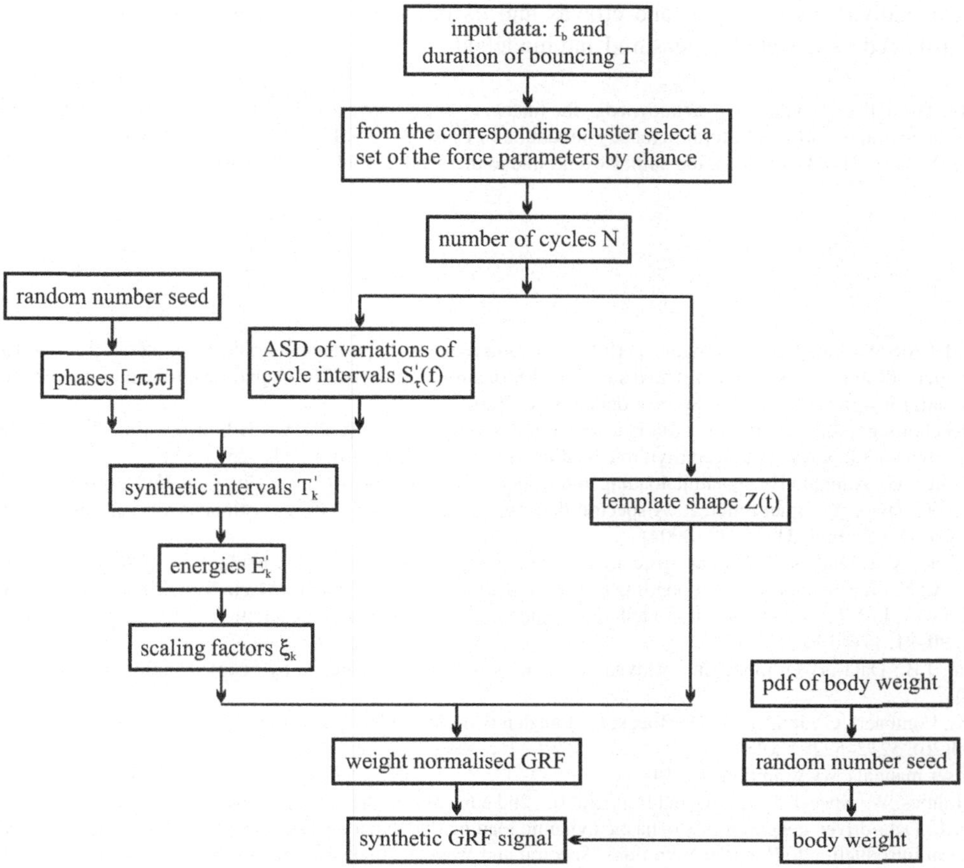

Fig. 49.13 Flow chart

rates in the range 1.2–4.5 Hz given during the data collection. It can be assumed that the modelling parameters stored in any mat file within a cluster can be used to generate synthetic force signals at any bouncing rate within the cluster's frequency range.

The flow chart in Fig. 49.13 presents the algorithm to generate an artificial force signal. A more elaborate explanation behind the process as well as the examples of synthetic forces can be found in Racic and Chen [13].

49.8 Summary and Conclusions

This study combines the traditional Fourier modelling approach with novel models of variability of timing and amplitudes/energy of the successive bouncing cycles, yielding a numerical generator of random near-periodic bouncing force time histories that look like measured. The modelling parameters are extracted from a large database of bouncing force records, classified and stored in narrow clusters with respect to the bouncing rate. Assuming that the forces generated at very close rates have similar shape, amplitudes and level of variability, modelling parameters within a cluster can be used to generate synthetic forces at any rate within the cluster's frequency range. The key modelling parameters are random variables, so two identical forces can be generated only by chance.

However, there is still a room for improvement. The current version of the model runs on the modelling parameters extracted from the force records measured on a rigid laboratory floor. Hence, it can be used to study only cases of incipient dynamic instability, i.e. when the vibration level is not too much perceptible to the occupant and therefore does not affect his/her bouncing motion. An elaborate database of bouncing force records generated by many individuals bouncing at different rates on more or less vibrating surfaces still waits to be established. So is the case with bouncing loads generated by groups of different sizes, with or without the motion of the structure itself. There are many more factors affecting human-induced loads, such as different auditory, visual and tactile stimuli. Moreover, there is still a gap in the knowledge on the interaction between individuals in groups and crowds and its influence on the corresponding net dynamic loads on the structure, which still need to be studied, measured and modelled.

Acknowledgements The authors would like to acknowledge the financial support provided by PRIN 2015-2018 "Identification and monitoring of complex structural systems" and National Natural Science Foundation of China 347 (51478346) and State Key Laboratory for Disaster Reduction of Civil Engineering (SLDRCE14-B-16). Also, the author would like to thank all test subjects for participating in the data collection.

References

1. IStructE/DCLG/DCMS Working Group: Dynamic performance requirements for permanent grandstands subject to crowd action: recommendations for management, design and assessment. The Institution of Structural Engineers, The Department for Communities and Local Government and The Department for Culture Media and Sport, London, UK (2008)
2. Sim, J.H.H., Blakeborough, A., Williams, M.: Statistical model of crowd jumping loads. ASCE J. Struct. Eng. **134**(12), 1852–1861 (2008)
3. Parkhouse, J.G., Ewins, D.J.: Crowd-induced rhythmic loading. Struct. Build. **159**(SB5), 247–259 (2006)
4. Rainer, J.H., Pernica, G., Allen, D.E.: Dynamic loading and response of footbridges. Can. J. Civ. Eng. **15**(1), 66–71 (1988)
5. Brownjohn, J.M.W., Pavic, A., Omenzetter, P.: A spectral density approach for modelling continuous vertical forces on pedestrian structures due to walking. Can. J. Civ. Eng. **31**, 65–77 (2004)
6. Zivanovic, S., Pavic, A., Reynolds, P.: Human-structure dynamic interaction in footbridges. Bridg. Eng. **158**(BE4), 165–177 (2005)
7. Racic, V., Pavic, A.: Stochastic approach to modelling near-periodic jumping force signals. Mech. Syst. Signal Process. **24**, 3037–3059 (2010)
8. Racic, V., Brownjohn, J.M.W.: Mathematical modelling of random narrow band lateral excitation of footbridges due to pedestrians walking. Comput. Struct. **90–91**, 116–130 (2012)
9. Racic, V., Morin, J.B.: Data-driven modelling of dynamic excitation of bridges induced by people running. Mech. Syst. Signal Process. **43**, 153–170 (2014)
10. Van Nimmen, K., Lombaert, G., Jonkers, I., De Roeck, G., Van den Broeck, P.: Characterisation of waking loads by 3D inertial motion tracking. J. Sound Vib. **33**(20), 5212–5226 (2014)
11. AMTI: AMTI user manuals. www.amti.biz (2014)
12. Holmes, J.R., Holmes, W.: Speech Synthesis and Recognition, 2nd edn. Taylor and Francis, London (2001)
13. Racic, V., Chen, J.: Data-driven generator of stochastic dynamic loading due to people bouncing. Comput. Struct. **158**, 240–250 (2015)
14. Newland, D.E.: An Introduction to Random Vibrations, Spectral and Wavelet Analysis, 3rd edn. Pearson Education Limited, Harlow (1993)
15. Bates, D.M., Watts, D.G.: Nonlinear Regression and its Applications. Wiley, New York (1998)
16. Bishop, C.M.: Pattern Recognition and Machine Learning, 4th edn. Springer, New York (2006)
17. MathWorks: Matlab user guides. www.mathworks.com (2014)

Chapter 50
Defining Groupings and Classification of Human Gait Using Correlation of Ground Reaction Force Measurements

Ellis Kessler, Pablo A. Tarazaga, and Robin Queen

Abstract Classification of a person's gait through quantitative methods has wide reaching applications in security, marketing, and healthcare. This study uses ground reaction force (GRF) measurements of healthy subjects and patients with osteoarthritis (OA) to define groupings and classify subjects. In addition to grouping classes into known qualities (gender, diagnosis, etc.), this work introduces classes based on groups coming directly from correlating GRFs from different subjects. Using correlation allows new groupings to be established and more accurate classification results because continuous force time histories are compared as opposed to conventional discrete force data (i.e. peaks). Two new classes are introduced from the data which can be classified with a 92% accuracy, and the physical meaning of these classes is investigated. Comparison of a single healthy person's walking pattern to multiple classes builds a signature that could be used to identify specific individuals. Additionally, patients suffering from OA do not correlate well with healthy groupings and can be distinguished from healthy subjects. This allows for the possibility of using GRFs to track patients' rehabilitation. It is expected that as a patient progresses through a rehabilitation program and begins to recover, their walking patterns will become more consistent and be more highly correlated with healthy groupings.

Keywords Gait · Classification · Correlation · Grouping · GRF

50.1 Introduction

The idea that a person can be identified through biomechanical differences in gait has gained growing interest lately. Previous work has focused on classifying gender, identifying specific individuals, and identifying neurodegenerative diseases using data measured from methods ranging from video surveillance, to force or acceleration measurements using floor mounted sensors [1–4]. After collecting data, methods such as nearest neighbors, hidden Markov models, and machine learning have been applied to classify trials into the chosen groups. All works reviewed herein only work on classification or identification into known qualitative groups such as gender, diagnosis, or individual [1–4].

In this work, quantitative groups are introduced which arise solely from similarities in measurements of gait between individuals. Prior work in the literature breaks gait down into discrete features, while others, which use the entire time histories often use black box methods such as machine learning for classification [1, 4]. Results from work that represent gait as discrete features are easy to visualize since there are few dimensions, however the use of discrete features limits the information that can be obtained from the continuous gait cycle. On the other hand, results from work using time series histories and methods such as machine learning produce results which are difficult to understand and interpret. In this work, time histories of gait measurements are used to retain all possible information and cross correlation of time history vectors was chosen as a simple method to compare whole time histories. Classification based on correlation is still intuitive to understand since it is simply an inner product of two vectors yet able to take into account a time history for richer results.

E. Kessler (✉) · P. A. Tarazaga · R. Queen
Department of Mechanical Engineering, Virginia Tech Smart Infrastructure Lab (VTSIL), Blacksburg, VA, USA
e-mail: ellisk1@vt.edu

S. Pakzad (ed.), *Dynamics of Civil Structures, Volume 2*, Conference Proceedings of the Society for Experimental Mechanics Series, https://doi.org/10.1007/978-3-319-74421-6_50

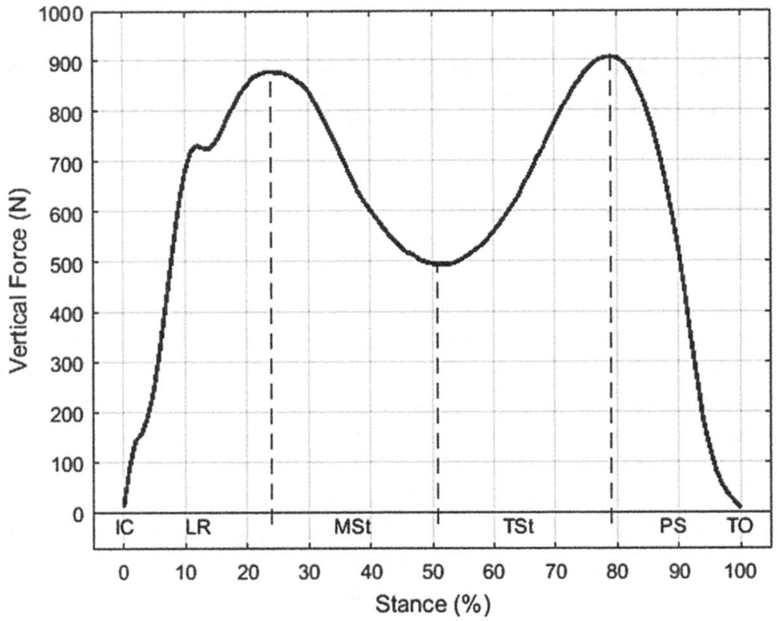

Fig. 50.1 An example of GRF from Subject 1 showing the different discrete points during the stance phase of gait

50.1.1 Ground Reaction Forces

As a person walks, they interact with the environment below them by imparting a force over the course of each stride. Gait is often defined as the specific way a subject walks, and the forces imparted between the floor and a person walking are called ground reaction forces (GRFs). An example of a vertical GRF time history is given in Fig. 50.1. No person can take two identical steps, so thus here is inter-subject variability in each step from a single subject. There is an even larger intra-subject variability between different subjects. The idea of classification into any kind of group relies on the hypothesis that there are similarities in gait and the resulting environmental interactions within a given class.

Forces generated while walking vary in direction and magnitude during the course of a stride. The force can be decomposed into three GRF projections: vertical, medial-lateral, and anterior-posterior. In order to easily extend lessons learned from the study of GRFs to other sensing options, we will focus on the vertical component of the GRF since this is the easiest direction to instrument with accelerometers and other methods. GRFs are generally normalized in time by defining the stance phase of gait (heel strike to toe-off on a single foot), where the stance phase is broken down into equally spaced samples (traditionally 100 samples) between the initial contact with the ground (heel strike) and final contact with the ground (toe-off). There is a point in time where both feet are imparting force on the ground (double support), however these effects are taken out in this work because each step is isolated to a single force plate. A GRF can be broken into six parts: initial contact (IC), loading response (LR), midstance (MSt), terminal stance (TSt), preswing (PS), and toe-off (TO). One method of discretizing GRFs is to take the forces at the three transitions (shown as dashed lines in Fig. 50.1) as representative of the GRF as a whole [5]. This method cannot account for the actual curvature during each section of the stance phase and loses much of the information in how the GRF transitions from one portion of the stance phase to another.

50.2 Experiment and Data Synthesis

Data was first collected on nine healthy subjects. Each subject was asked to walk at a self-selected pace over a floor instrumented with four force plates (AMTI, Watertown, MA) arranged as shown in Fig. 50.2. The force plates are 600 mm by 600 mm, and were sampled at 1920 Hz. Each subject walked down and back for a series of three trials each. During each trial, subjects walked down the middle of a row of three force plates, giving six possible GRFs per trial. Subjects were told to not alter their walking style to accommodate the distance between the plates, so some subjects had a harder time landing

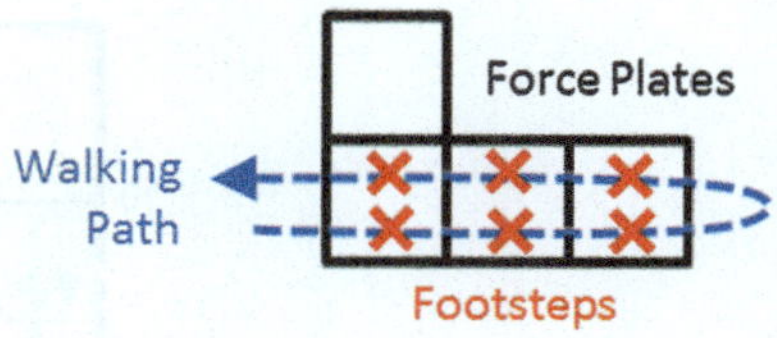

Fig. 50.2 Diagram of force plates showing walking path and ideal step locations

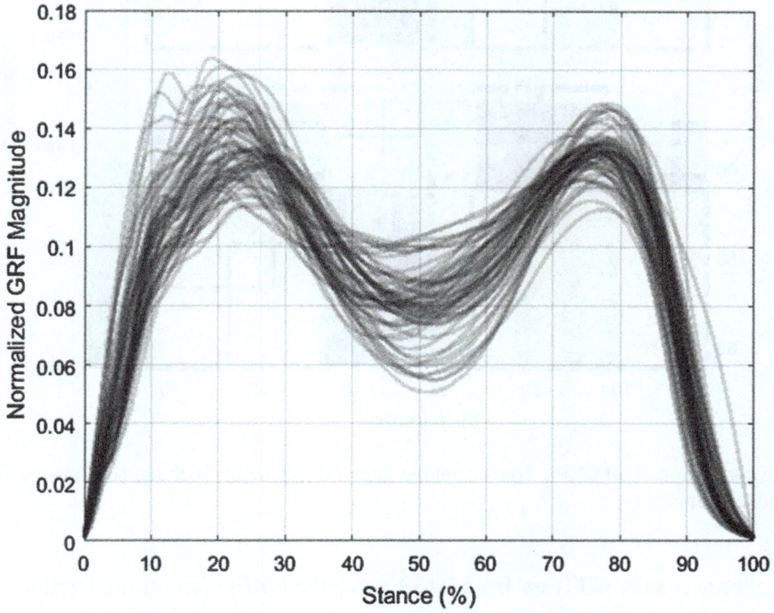

Fig. 50.3 All 62 normalized GRF measurements from all nine subjects

their foot solely on one plate. Adjustments were made by the researchers so that each subject was able to contact the plates without targeting the plate. Only steps which were fully on one plate were used during the analysis, so out of a possible 18 steps per subject only 4–10 per person were kept.

Each trial kept was normalized to the stance phase and interpolated to be represented by a vector $\mathbb{R}^{101}$. All vectors were normalized to have a magnitude of one to remove the effects of subject weight from comparisons. Magnitude normalization was chosen over body-weight normalization so that a correlation of 1 would be perfect agreement between any two trials.

50.3 Group Identification and Classification

There are clear groups that subjects could be broken up into such as gender or diagnosis, however is there a way to use GRF data directly to separate subjects into groups with similar biomechanical information? Further classification beyond qualitative information about subjects could give more insight into how different people walk. Additionally, with enough classes to compare against it, it will become easier to distinguish two people that do not fall into all the same classes.

50.3.1 Correlation Matrix to Find Groups

All 62 healthy GRFs were compared to search for similarities between how subjects walk Fig. 50.3. The GRFs were numbered as x_i where each subject's steps were all together. For example the first 10 are all from subject 1, and so on. A correlation matrix called C was built where each entry was:

$$C_{ij} = x_i^T x_j. \tag{50.1}$$

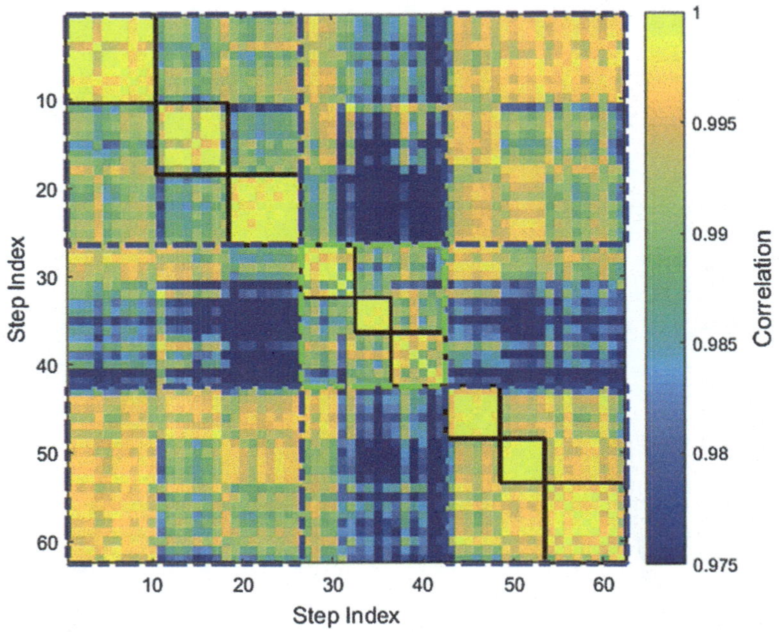

Fig. 50.4 The matrix showing correlation of all GRFs. Trials from the same subject are in order and are shown with black boxes. The two groups are shown as blue and green boxes respectively

The entries of this correlation matrix will vary from 0 to 1 since the GRFs were normalized to a magnitude of 1. In fact, a value of 1 would indicate that the two GRFs being compared are exactly the same (as is the case along the diagonal). Other areas of higher correlation will indicate that steps being compared are similar. Each black square in Fig. 50.4 represent a single subject. The correlation inside each of these squares is high, showing that there is a high correlation between a single subject's different steps.

Figure 50.4 also shows two larger groupings of high correlation. These two groups are shown with blue and green boxes respectively. Subjects 1, 2, 3, 7, 8, and 9 display high correlation between their trials and are therefore put into class 1. Subjects 4, 5, and 6 also display high correlation and are put into class 2. Subject 4 does appear to fit into both groupings more than any other, but is placed in group 2 because of more consistent correlation with the whole group. From the cloud plot in Fig. 50.5 the two groups show clear visual separation in their GRF time histories. Class 1 (blue) appears to be people who have a softer heel strike and a more steady force throughout their stance, while class 2 (green) contains subjects who have a much higher peak from the heel strike and therefore a lower trough between heel strike and toe-off. From visual inspection, we already see that biomechanical differences in walking, have automatically been identified by the correlation groupings.

50.3.2 Classification of Healthy Subjects

To test the hypothesis that the groups found in the previous section can be classified, a correlation classification framework was used to compare each step to both classes. The normalized average of all steps from a given class was defined as $\overline{x}_{\text{class}}$ to represent that class. Each step as a test step, x_{test}, was correlated with each class to find the agreement between the class and test step. The "error", e, was then defined as unity minus the correlation as

$$e = 1 - \overline{x}_{\text{class}}^{T} x_{\text{test}}. \tag{50.2}$$

In order to avoid overfitting, when each step was compared to the both classes, no steps from that subject were used in finding the mean for either class. The error will range from zero if the step is identical to the average step of that class, to an error of one if the step is orthogonal to the average step.

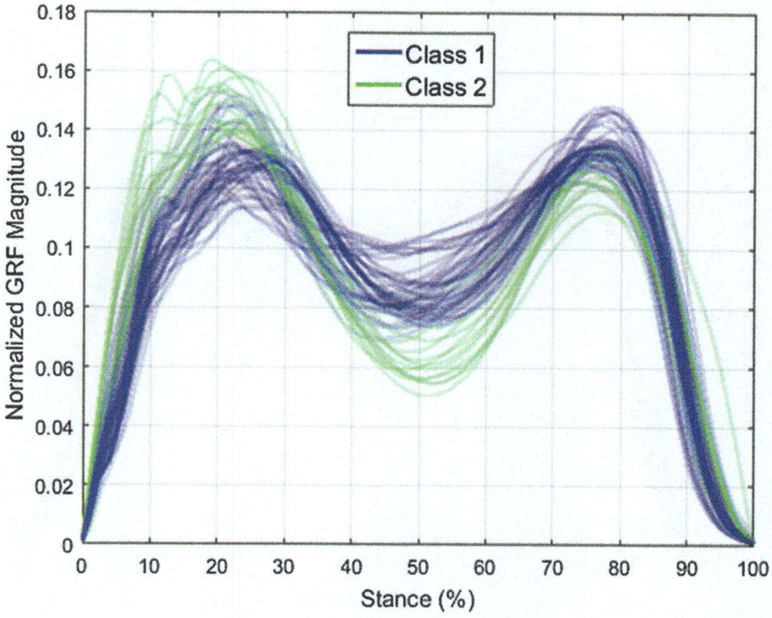

Fig. 50.5 All GRFs plotted and colored by which class they are placed in

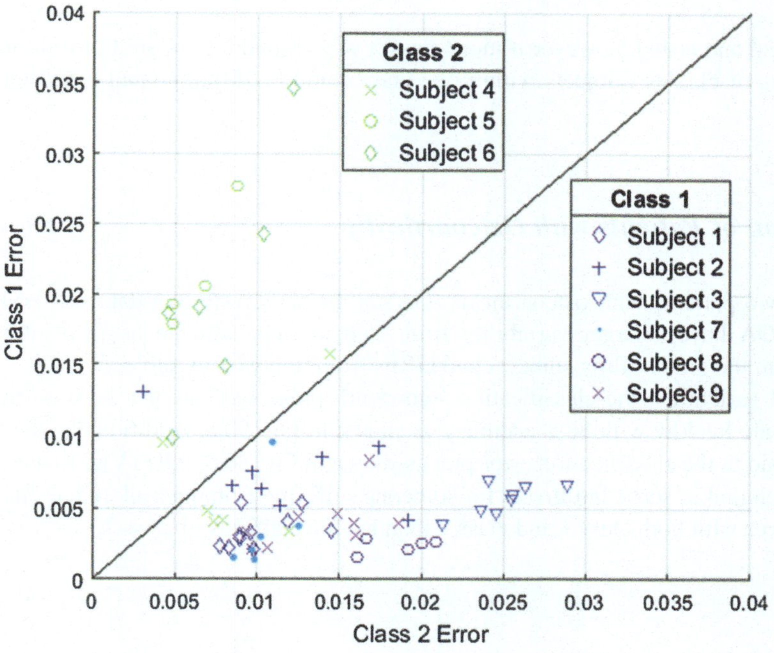

Fig. 50.6 Errors of each trial with each class

Figure 50.6 shows the results of all classification comparisons, and also shows a line plotted along the diagonal where the error with each class is equal. Any markers lying above the line have a lower class 2 error than class 1 error and therefore are classified as class 2, while markers below the line are classified as class 1. This is an extremely simple classification framework, yet it is capable of differentiating between class 1 and 2 with 92% accuracy. Out of the five misclassifications four of them came from subject 4. As noted from correlation matrix, subject 4 appeared to correlate well with both groups so it is not surprising that classification of this subject was the least accurate of all nine subjects. The high classification accuracy proves that the groups developed in the previous section have physical meaning and are useful for classification of subjects.

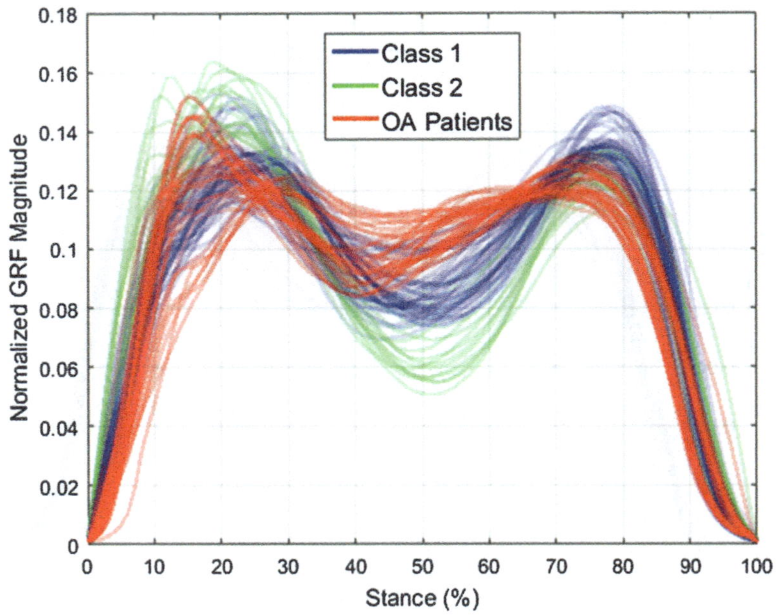

Fig. 50.7 All GRFs plotted, including the OA patients (in red)

Two classes were found and tested, however if more classes were found a clear split in walking styles could be seen. For example with two groups of classes, subjects within one class could be differentiated based on which class they fit from another group.

50.3.3 Classification of Patients with Osteoarthritis

Data was recorded for two patients with osteoarthritis (OA) in the same way that data was collected for the nine healthy subjects. Patients with OA have a larger variability from step to step than the healthy subjects. Figure 50.7 shows a comparison of GRFs from the nine healthy subjects and GRFs from the two OA subjects.

Despite the increased variability, the classification framework presented can also be used to classify between patients with OA and those who are healthy with 88% accuracy as shown in Fig. 50.8. Additionally, the higher variability of GRFs translates to a wider spread in the classification error plot as shown in Fig. 50.9. All OA trials were classified into class 1 (the softer hitting class), which makes sense intuitively for someone suffering from painful joints. The OA patients, however, did have a higher average error with both class 1 and class 2 than healthy subjects in class 1.

50.3.4 Discussion

The classification of healthy subjects in Fig. 50.6 shows that subject's gait is rather consistent and can be separated into the two groups with high accuracy. Although one subject seemed to fit both classes, most stayed in the same class for all steps. From this result, we would expect that most new healthy subjects will also fall into one group or the other. Additionally, another set of groups could be found to account for subjects like subject 4 who did not fit the two classes presented. More samples are needed to fully understand the limits of these preliminary results.

There is a need to track the progress of patients during physical therapy. Figures 50.8 and 50.9 show that patients with OA can be differentiated from healthy subjects, and that OA patients also do not correlate as highly with the groupings defined from healthy subjects. Thus, impaired gait can be diagnosed and monitored. During effective rehabilitation subjects walking patterns should progress to correlate better with healthy classes. As healthy gait is recovered, trials from the patient should group closer in class 1 or 2 depending on the subject's normal healthy walking pattern. This could lead to an easily measurable and analyzed quantitative metric of a person's rehabilitation process.

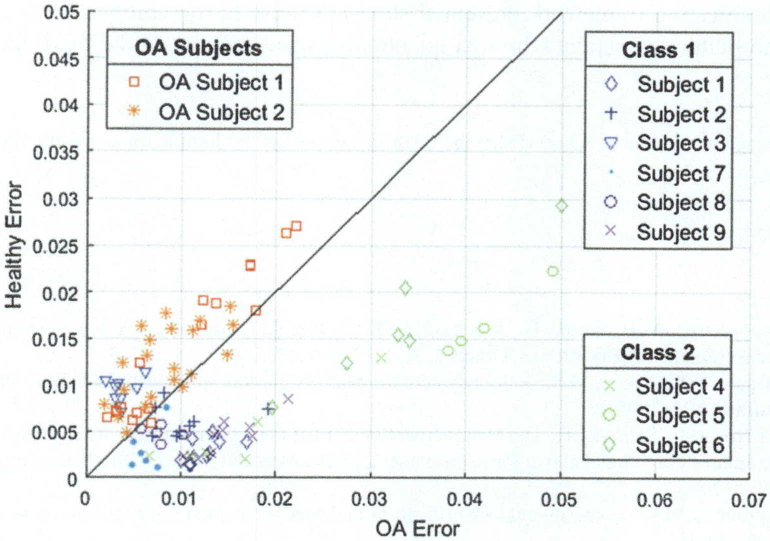

Fig. 50.8 Classification between OA and healthy classes for all trials, including OA patients

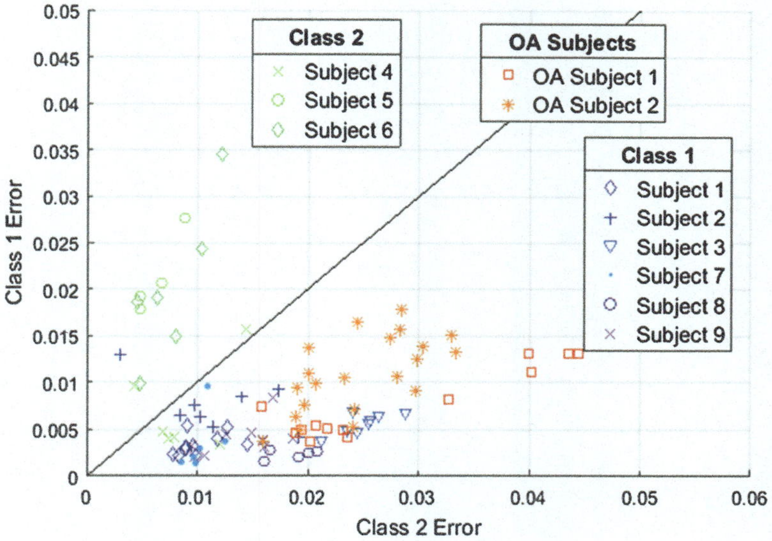

Fig. 50.9 Classification between class 1 and 2 for all trials, including OA patients

50.4 Conclusion

In this paper we have shown that ground reaction force (GRF) measurements of single steps can be accurately classified using normalized correlation. Classification between patients with osteoarthritis (OA) and healthy subjects showed 88% accuracy. Further, this work introduced a new method for determining groupings quantitatively as opposed to qualitative groups such as disease diagnosis. Correlations between all steps from all healthy subjects showed two main groupings emerge where subjects fall into one group or the other. Visually these groups appear to represent people with either soft or hard heel strikes and toe-offs in the course of their step. The correlation classification framework gives 92% accuracy in classifying healthy subjects into these groupings, proving that there is a measurable difference between the two walking styles. Patients with OA did not correlate as strongly as healthy subjects with either of these groupings, suggesting that classification using GRFs could both diagnose and track the rehabilitation of injuries or diseases which alter a patient's gait.

Two mutually exclusive new groupings were introduced in this work, but more groups could give more insight into walking patterns and further ability to distinguish between individuals. Groupings introduced in this work could also be extended to other data measured from steps, such as underfloor accelerometer measurement or motion capture data. In the end, the grouping and classification framework presented shows promise in the ability to classify individuals by walking pattern and track the rehabilitation of patients through quantitative measurement of their gait through measurements from their environment.

Acknowledgements Dr. Tarazaga would like to acknowledge the financial support of the John R. Jones Faculty Fellowship.

References

1. Bales, D.B., Tarazaga, P.A., Kasarda, M.E., Batra, D., Woolard, A.G., Poston, J.D., Malladi, V.V.N.S.: Gender classification of walkers via underfloor accelerometer measurements. IEEE Internet Things J. **3**(6), 1259–1266 (2016)
2. Middleton, L., Buss, A.A., Bazin, A.L., Nixon, M.S.: A floor system for gait recognition. In: Fourth IEEE Workshop on Automatic Identification Advanced Technologies, Buffalo, NY (2005)
3. Addlesee, M.D., Jones, A., Livesey, F., Samaria, F.: The ORL active floor. IEEE Pers. Commun. 35–41 (October 1997)
4. Orr, R.J., Abowd, G.D.: The smart floor: a mechanism for natural user identification and tracking. In: Proceedings of the Conference on Human Factors in Computing Systems (2000)
5. Racic, V., Pavic, A., Brownjohn, J.M.W.: Experimental identification and analytical modelling of human walking forces: literature review. J. Sound Vib. **326**(1–2), 1–49 (2009)

The manufacturer's authorised representative in the EU is Springer Nature Customer Service Centre GmbH, Europaplatz 3, 69115 Heidelberg, Germany. If you have any concerns regarding our products, please contact ProductSafety@springernature.com

Printed and bound by CPI Group (UK) Ltd, Croydon, CR0 4YY
30/09/2026
02225477-0002